# Lecture Notes in Physics

## Volume 954

# The Lecture Notes in Physics

The series Lecture Notes in Physics (LNP), founded in 1969, reports new developments in physics research and teaching - quickly and informally, but with a high quality and the explicit aim to summarize and communicate current knowledge in an accessible way. Books published in this series are conceived as bridging material between advanced graduate textbooks and the forefront of research and to serve three purposes:

- to be a compact and modern up-to-date source of reference on a well-defined topic;
- to serve as an accessible introduction to the field to postgraduate students and nonspecialist researchers from related areas;
- to be a source of advanced teaching material for specialized seminars, courses and schools.

Both monographs and multi-author volumes will be considered for publication. Edited volumes should however consist of a very limited number of contributions only. Proceedings will not be considered for LNP.

Volumes published in LNP are disseminated both in print and in electronic formats, the electronic archive being available at springerlink.com. The series content is indexed, abstracted and referenced by many abstracting and information services, bibliographic networks, subscription agencies, library networks, and consortia.

Proposals should be sent to a member of the Editorial Board, or directly to the responsible editor at Springer:

Dr Lisa Scalone
Springer Nature
Physics
Tiergartenstrasse 17
69121 Heidelberg, Germany
lisa.scalone@springernature.com

More information about this series at http://www.springer.com/series/5304

Kang-Sin Choi • Jihn E. Kim

# Quarks and Leptons From Orbifolded Superstring

**Second Edition**

 Springer

Kang-Sin Choi
Scranton Honors Program
Ewha Womans University
Seoul, Korea (Republic of)

Jihn E. Kim
Department of Physics and Astronomy
Seoul National University
Seoul, Korea (Republic of)

The National Academy of Sciences
Natural Sciences Division, Section 1
Seoul, Korea (Republic of)

ISSN 0075-8450          ISSN 1616-6361    (electronic)
Lecture Notes in Physics
ISBN 978-3-030-54004-3          ISBN 978-3-030-54005-0    (eBook)
https://doi.org/10.1007/978-3-030-54005-0

This Springer imprint is published by the registered company Springer Nature Switzerland AG.
The registered company address is: Gewerbestrasse 11, 6330 Cham, Switzerland

# Preface to the Second Edition

One-and-a-half decades after the first edition, a lot of progress has been made in the application of string compactification. More realistic models have been built, more unified aspects of string theory have been understood, and more mathematical tools have become available. We now understand better the position of the orbifolded string theories in the moduli space and their connections to other constructions. Previously known symmetries are rederived from stringy and geometric effects. The landscape scenario has been heavily used in understanding the smallness of the cosmological constant.

To meet the recent development, we completely rewrote Chaps. 8, 9, 10, 11, 15, and 16. We formally define the string theory on non-prime orbifolds by adding honest calculations in Chaps. 8 and 9. A complete understanding on the superpotential is now available, whose fundamentals are described in Chap. 10. In Chap. 15, we use new mathematical tools to understand the physics in orbifold singularities. We include a detailed explanation of flavor symmetry in Chap. 16. A new development in F-theory compactification is briefly included in Sect. 17.4. We have also tried to make the whole book more coherent and the exposition clearer.

This book was originally designed to be a toolkit on the compactification of the heterotic string. So, we provide more ready-made formulae. We hope that this revision helps everyone to more deeply understand the unified aspects of string theory and practically build more realistic models.

We are grateful to Stefan Groot Nibbelink, Hirotaka Hayashi, Tatsuo Kobayashi, Bumseok Kyae, Seung-Joo Lee, Hans-Peter Nilles, Felix Plöger, Stuart Raby, Sául Ramos-Sánchez, Michael Ratz, Soo-Jong Rey, Patrick K. S. Vaudrevange, Akin Wingerter and Piljin Yi, for helping us learn and clarify ideas.

Seoul, Republic of Korea
Seoul, Republic of Korea
May 2020

Kang-Sin Choi
Jihn E. Kim

# Preface to the First Edition

Using the successful standard model of particle physics but without clear guidance beyond it, it is a difficult task to write a physics book beyond the standard model from a phenomenological point of view. At present, there is no major convincing inner-space-related experimental evidence against the standard model. The neutrino oscillation phenomena can be considered part of it by including a singlet field in the spectrum. Only the outer-space observations on matter asymmetry, dark matter, and dark energy hint at the phenomenological need for an extension, yet the theoretical need has been with us for almost three decades, chiefly because of the gauge hierarchy problem in the standard model.

Thus, it seems that going beyond the standard model hinges on the desirability of resolving the hierarchy problem. At the field theory level, it is fair to say that the hierarchy problem is not as desperate as the nonrenormalizability problem present in the old V–A theory of weak interactions on the road to the standard model. An extension beyond the standard model can easily be ruled out as witnessed in the case of technicolor. However, a consistent framework with supersymmetry for a resolution of the hierarchy problem has been around for a long time. Even its culprit "superstring" has been around for 20 years, and the most remarkable thing about this supersymmetric extension is that it is still alive. So, the time is ripe for phenomenologists to become acquainted with superstring and its contribution toward the minimal supersymmetric standard model in four spacetime dimensions.

This book is a journey toward the minimal supersymmetric standard model (MSSM) down the orbifold road. After some field theoretic orbifold attempts in recent years, there has been renewed interest in the physics of string orbifolds and it is time to revisit them. In this book, we take the viewpoint that the chirality of matter fermions is essential toward revealing the secrets of Nature. Certainly, orbifolds are an easy way to get the chirality from higher dimensions.

Strings and their orbifold compactification are presented for the interests of phenomenologists, sacrificing mathematical rigor. They are presented in such a way that an orbifold model can be constructed by applying the rules included here. At the end of Chap. 10, we construct a $\mathbb{Z}_{12}$ orbifold that contains all imaginable complications. Also, we attempt to correct any incompleteness in the rules presented before in the existing literature. In the final chapter, we tabulate the simplest and most widely used orbifold $\mathbb{Z}_3$ with $\mathcal{N} = 1$ supersymmetry, completely in the phenomenological sense of obtaining three families. These tables encompass all noteworthy models available with two Wilson lines. As three Wilson line $\mathbb{Z}_3$ orbifolds do not automatically give three families, these tables in a practical manner close a chapter on $\mathbb{Z}_3$ orbifolds.

This book is not as introductory as a textbook, nor is it as special as a review article on a superstring topic. Instead, we aim at an interim region, so that a phenomenologist can read and directly commence building an orbifold model.

We thank Kyuwan Hwang for his help in constructing the $\mathbb{Z}_3$ orbifold tables. We are also grateful to Kiwoon Choi, Ki-Young Choi, Luis Ibañez, Gordy Kane, Hyung Do Kim, Jewan Kim, Seok Kim, Tatsuo Kobayashi, Bumseok Kyae, Oleg Lebedev, Andre Lukas, Stefan Groot Nibbelink, Hans-Peter Nilles, Fernando Quevedo, Stuart Raby, Michael Ratz, and Hyun Seok Yang, for providing valuable suggestions in the course of writing this book.

Seoul, Republic of Korea                                                        Kang-Sin Choi
Seoul, Republic of Korea                                                          Jihn E. Kim
November 2005

# Conventions

Mostly, we adopt the conventions of Green, Schwarz, and Witten [2]. We use normalization $\alpha' = \frac{1}{2}$ for closed strings, but sometimes it is made explicit if necessary. We denote worldsheet time and space coordinates as $\tau$ and $\sigma$, respectively. The left movers of worldsheet fields are functions of $\tau + \sigma$ and their oscillators are tilded like $\tilde{\alpha}_n$. The right movers are functions of $\tau - \sigma$ and their corresponding oscillators are untilded.

We denote the spacetime coordinates as follows:

$$x^M = (x^\mu, x^m, x^I), \tag{0.1}$$

where Greek indices $\mu = 0, 1, 2, 3, \ldots$ denote noncompact and lower Latin $m$ denote compact dimensions. If we do not compactify, then the full string lies along $\mu = 0, \ldots, 9$ directions. Upper Latin indices are for the current algebra. A torus is made by modding out the translations:

$$x^m \sim x^m + 2\pi \, Re_i^m,$$

where $x^m$ are orthogonal coordinates and the shape is determined by $e_i^m$.

Sometimes, it is useful to complexify the coordinates

$$z^a = \frac{1}{\sqrt{2}}(x^{2a} + ix^{2a+1}), \quad z^{\bar{a}} \equiv \overline{z^a} = \frac{1}{\sqrt{2}}(x^{2a} - ix^{2a+1}),$$

where $z^a, z^{\bar{a}}, a = 2, 3, 4, \bar{a} = \bar{2}, \bar{3}, \bar{4}$ for holomorphic and antiholomorphic complex indices, respectively. They are complexification of $x^i$ such that $z^2, z^{\bar{2}}$ directions are spanned by coordinates $x^4, x^5$. The gauge space index, the uppercase Latin in $x^I$, $I = 10, \ldots, 25$, remains untouched.

A three-component twist vector $\phi = (\phi_2 \ \phi_3 \ \phi_4)$ parametrizes the point group action of orbifolds. When necessary, we also include the noncompact component as $\phi = (\phi_1; \phi_2 \ \phi_3 \ \phi_4)$.

For the zero point energy, it is renamed $c = -a, \tilde{c} = -\tilde{a}$ to cope with the literature.

# Contents

# List of Symbols

| Symbol | Name, equation | Value |
|---|---|---|
| $g_{\mu\nu}$ | Spacetime metric tensor | $\eta_{\mu\nu} = \mathrm{diag}.(1,-1,-1,\ldots,-1)$ |
| $h_{\alpha\beta}$ | Worldsheet metric tensor | $\eta_{\alpha\beta} = \mathrm{diag}.(1,-1)$ |
| | GeV s | $1.519255 \times 10^{24}$ |
| | GeV cm | $0.5067689 \times 10^{14}$ |
| $c$ | Speed of light $= 1$ | $299{,}792{,}453$ m/s |
| $\hbar$ | Planck constant $= 1$ | $1.064571596(82) \times 10^{-34}$ J s |
| $M_P$ | Planck mass $= (8\pi G_N)^{-1/2}$ | $2.436 \times 10^{18}$ GeV |
| $M_{\mathrm{GUT}}$ | Grand unification scale $\simeq M_U$ | $\sim 2.5 \times 10^{16}$ GeV |
| $v_{\mathrm{ew}}$ | Electroweak symmetry breaking scale | 246 GeV |
| $G_F$ | Fermi coupling constant | $1.16639 \times 10^{-5}$ GeV$^{-2}$ |
| $\sin^2 \theta_W$ | Weak mixing angle | $0.23113(15)$ |
| $M_Z$ | Z boson mass $= \frac{1}{2} g v\, s\theta_W$ | $91.1876(21)$ GeV |
| $\rho$ | Rho parameter $\left( = \frac{M_W^2}{M_Z^2 \cos^2 \theta_W} \right)$ | $1.0012(23)$ |
| $M_S$ | Scale for SUSY breaking source | $\sim 0.5 \times 10^{11}$ GeV in SUGRA |
| $M_{\mathrm{SUSY}}$ | Observable sector SUSY splitting scale | $\sim$TeV |
| $H_0$ | Hubble constant $\left( \sqrt{\rho_{\mathrm{energy}}/3M_P^2} \right)$ | $2.1332 h \times 10^{-42}$ GeV $\left( h = 0.71^{+0.04}_{-0.03} \right)$ |
| $\rho_c$ | Critical density $(0.81 h^2 \times 10^{-46}$ GeV$^4)$ | $1.88 h^2 \times 10^{-29}$ g cm$^{-3}$ |
| $t$ | Cosmic time in RW universe | $\sim (10^{-3}\, \mathrm{GeV}/T)^2$ s |
| $\mathbf{Z}_N$ | Cyclic symmetry of order $N$ | |
| $\tilde{\mathrm{N}}$, N | Oscillator | |
| $N$ | Order | |
| $y$ | Extra dimension coordinate | |
| $\mathcal{N}$ | Number of 4D supersymmetries | |
| $\theta$ | Point group (orbifold) action | |

| Symbol | Name, equation | Value |
|---|---|---|
| $\phi$ | Twist vector | |
| $\alpha'$ | Inverse tension, Regge slope $\frac{1}{2}\ell_s^2$ | |
| $L_n$ | Virasoro operator | |
| $P$ | Momentum lattice vector | |
| $V$ | Shift vector from point group action | |
| $a$ | Shift vector from Wilson line | |
| $f(\eta)$ | Zero point energy per a real degree | $f(0) = -\frac{1}{24},\ f(\frac{1}{2}) = \frac{1}{48}$ |
| $\tilde{c}, \tilde{\alpha}, z$ | For left movers | Eq. (7.15) |
| $c, \alpha, \bar{z}$ | For right movers | Eq. (7.21) |
| $\tau$ | Modulus defining torus | |
| $\mathscr{T}, \mathscr{S}$ | Modular transformations | |
| $\Gamma$ | Gauge group root lattice | |
| $\vartheta$ | Jacobi elliptic (theta) function | |
| $\eta$ | Dedekind (eta) function | |
| $P_m$ | Projection operator | |
| $\tilde{\chi}_{mn}$ | Multiplicity in $(m, n)$ twisted sector | |
| $\alpha$ | Root, made of simple roots $\alpha^i$ | |
| $\Lambda_i$ | Fundamental weight | |
| $\Lambda$ | Orbifold lattice | |
| $k$ | Level of affine Lie algebra | |

# Introduction and Summary

During and since the second half of the twentieth century, enormous progress has been made in understanding our universe in terms of fundamental particles and their interactions, namely in the language of quantum field theory. The advent of the standard model (SM) of particle physics has been the culmination of quantum field theory in all its full glory. The beginning of this successful particle physics era was opened with the unexpected discovery of *parity violation* in weak interaction phenomena [1] and the Brout–Englert–Higgs–Guralnik–Hagen–Kibble particle (the Higgs boson in short) closed the discovery series of the SM particles in 2012 [2, 3].

It had long been known that weak interactions change the electromagnetic charge, i.e. electron ($e$) to electron type neutrino ($\nu_e$), neutron ($n$) to proton ($p$). But, until the mid-1950s it had never occurred to the leading minds [4] that "parity might be violated," chiefly because the atomic and nuclear transitions did not reveal any such possibility before that time. For nuclear transitions, both weak and electromagnetic phenomena contribute but at that time there were not sufficient data to fully conclude on the nature of parity operation in weak interactions [1]. For atomic transitions, the fundamental interaction is of electromagnetic origin and the experimental confirmation of parity conservation in atomic phenomena convinced most physicists that parity is conserved in the universe. In hindsight, parity conservation should have been imposed only on electromagnetic interactions, as the discovery of parity violation in weak interactions started a new era for weak interactions. There is still no experimental evidence that strong and electromagnetic interactions violate parity. Therefore, we know that parity violation in weak interactions is at the heart of making our universe as it is now, because the SM assumes from the outset the existence of massless chiral fields.[1]

Soon afterward, the parity violating weak interactions were neatly summarized as a four fermion (charged current) × (charged current) $(CC \times CC)$ weak interaction

---

[1]Massless compared to the Planck mass $M_P$.

© Springer Nature Switzerland AG 2020
K.-S. Choi, J. E. Kim, *Quarks and Leptons From Orbifolded Superstring*,
Lecture Notes in Physics 954, https://doi.org/10.1007/978-3-030-54005-0_1

where the charged current $J_\mu^{CC}$ is of the "V–A" type [5,6],

$$\mathscr{H}_{\text{weak}} = \frac{G_F}{\sqrt{2}} J^{CC\dagger\mu}(x) J_\mu^{CC}(x), \tag{1.1}$$

where

$$G_F \simeq 1.1664 \times 10^{-5} \, \text{GeV}^{-2}. \tag{1.2}$$

The "V–A" charged current of weak interactions indicates three important aspects: (1) only the left-handed fermions participate in the charge changing weak interactions, (2) the CC weak interaction has only one coupling constant $G_F$ compared to 34 couplings of Fermi's $\beta$ decay interactions [7], and (3) being current, the fundamental interaction at a deeper level may need a vector boson. Here, we note that the chirality nature of weak interactions discovered through the CC weak interaction is still the mystery among all mysteries of particle physics in the search for a fundamental theory at a very high energy scale using the low energy SM. Several years after this effective low-energy (CC) × (CC) four-fermion interaction was proposed, a modest attempt via a more fundamental interaction through a heavy spin-1 charged intermediate vector boson (IVB) $W_\mu^\pm$ was put forth [8]. The charged IVB coupling to the charged current was given by

$$\frac{g}{2\sqrt{2}} J_\mu^{CC} W^\mu + \text{h.c.} \tag{1.3}$$

The mass of $W_\mu$ was supposed to be heavy so that the four-fermion interaction mediated by the IVB is weak compared to the strong interaction scale, i.e. $M_W \gg 1 \, \text{GeV}$. However, this IVB idea had several problems which have since been resolved by the SM of particle physics.

As far as currents are concerned, the first prediction of the SM was the existence of the weak neutral current (NC) [9–11] in addition to the old CC observed in the $\beta$ decay phenomena. The weak NC introduces another parameter known as the weak mixing angle $\sin^2 \theta_W$ in the SM determined around [12],

$$\sin^2 \theta_W \simeq 0.233. \tag{1.4}$$

A spin-1 field coupling to the fermion current had already been known in electromagnetic interactions, i.e. the photon $A_\mu$ coupling to the electromagnetic current through $e \bar{\psi} \gamma^\mu \psi A_\mu$. This electromagnetic interaction can be formulated in terms of U(1) gauge theory [13], where one uses the covariant derivative $\mathscr{D}_\mu$ instead of the ordinary partial derivative $\partial_\mu$,

$$\partial_\mu \longrightarrow \mathscr{D}_\mu \equiv \partial_\mu - ieA_\mu$$

which introduces the minimal gauge coupling of $A_\mu$ to charged fields. In quantum mechanics, the additive conservation of electromagnetic charge implies a global U(1) symmetry and generalizing it to a local U(1) leads to the above covariant derivative. This is our first example of how a bigger symmetry might be discovered from a representation of matter, i.e. starting from the electron in the above example of quantum electrodynamics.

Consider the generalization of this gauge principle to the IVB. Since the IVB changes the electromagnetic charge, we must start from a defining state in the Hilbert space which contains at least two components differing by one unit of the electromagnetic charge. This doublet is a kind of matter which, in the doublet representation, necessarily introduces a non-Abelian gauge group. This is our second example in which matter can indicate a bigger symmetry. In general, one can introduce the covariant derivative using the non-Abelian gauge fields $A^i_\mu (i = 1, 2, \cdots, N_A)$, with the size $N_A$ (e.g. 3 for SU(2)) dependent upon the matter representation. Yang and Mills were the first to show that a consistent construction along this line needs nonlinear couplings between gauge fields.

In the late 1960s the standard model of particle physics was constructed, employing the non-Abelian gauge group. The group structure is SU(2) × U(1) [9–11], and the covariant derivative is

$$\mathscr{D}_\mu = \partial_\mu - igT^i A^i_\mu - ig'YB_\mu, \qquad (1.5)$$

where $T^i (i = 1, 2, 3)$ are the SU(2) generators and $Y$ is the electroweak hypercharge generator. The Gell-Mann–Nishijima type definition of the electromagnetic charge is $Q_{\rm em} = T_3 + Y$. All leptons and quarks are put into left-handed doublets and right-handed singlets, and the charged current IVB mediation violates parity symmetry by construction from the outset. For example, the left-handed electron and its neutrino are put into a doublet $l_L = (\nu_e, \ e)^T_L$, where $L(R)$ represents the left(right)-handed projection $L = \frac{1+\gamma_5}{2}$, or $\psi_L = \frac{1+\gamma_5}{2}\psi$. Since the quarks carry the additional degree called *color* coming in three varieties, the first family $(l_L, e_R, q_L, u_R, d_R)$ contains 15 two-component chiral fields. In addition, these 15 fields repeat three times, making a total of 45 chiral fields, all of which have been observed in high energy accelerators.

The representations of the SM are written in such a way that the intermediate vector boson $W^+$ transforms the lower elements of $l$ and $q$ to their upper elements. For example, $e_L$ to $\nu_{eL}$ and $d_L$ to $u_L$, and hence there exists the coupling

$$\frac{g}{\sqrt{2}} \bar{\nu}_e \gamma^\mu \frac{1 + \gamma_5}{2} e W^+_\mu.$$

For each representation, we can assign the $Y$ quantum number to match the electromagnetic charges of the fields in the representation through the following formula:[2]

$$Q_{em} = T_3 + Y. \tag{1.6}$$

Thus, the standard model is certainly left $(L)$–right $(R)$ asymmetric in that the interchange $L \leftrightarrow R$ does not give the original representation. This is called a chiral theory.[3] In a chiral theory, one cannot write down a mass term for the fermions. Under the SM gauge group $SU(2) \times U(1)_Y$, for example, one cannot write down a gauge-invariant mass term for $e(l = \mathbf{2}_{-\frac{1}{2}}, e_R = \mathbf{1}_{-1}$ where the weak hypercharges are written as subscripts in the usual way). The SM is designed such that chiral fermions can obtain mass after the gauge group $SU(2) \times U(1)_Y$ is spontaneously broken down to $U(1)_{em}$, and then one has to consider only the gauge invariance of the unbroken gauge group $U(1)_{em}$. This makes it possible to write

$$-m_e \bar{e} e = -m_e (\bar{e}_R e_L + \bar{e}_L e_R).$$

This way of rendering mass to SM chiral fields is assumed throughout this book, and the fundamental question is *how such chiral fields arise in the beginning*. For spontaneous symmetry breaking leading to $G \to H$,[4] one needs a singlet member under the Lorentz group and a singlet under the unbroken gauge group $H$, but there should also be a non-singlet under $G$. In the Hilbert space, such a member as a fundamental field is a neutral scalar transforming nontrivially under both $SU(2)$ and $U(1)$. The simplest such representation is a spin-0 Higgs doublet with $Y = \frac{1}{2}$ [10, 11],

$$\phi = \begin{pmatrix} \phi^+ \\ \phi^0 \end{pmatrix}. \tag{1.7}$$

A more complicated mechanism for spontaneous symmetry breaking is the use of a composite field which is a neutral scalar transforming nontrivially under both $SU(2)$ and $U(1)_Y$. The simplest such composite field is one that assumes a new confining force, the so-called *techni-color* confining around the TeV scale, and composites of techni-quarks realize this idea [15, 16]. This neutral scalar component can develop a vacuum expectation value (VEV) which certainly breaks $G$ but leaves $H$ invariant. Breaking the gauge symmetry through the VEV of scalar fields is the *Higgs mechanism* [17–20]. The SM is a chiral theory based on $SU(2) \times U(1)_Y$ with the above Higgs mechanism employed.

---

[2]$l_L$ has $Y = -\frac{1}{2}$, $e_R$ has $Y = -1$, $q_L$ has $Y = \frac{1}{6}$, $u_R$ has $Y = \frac{2}{3}$, and $d_R$ has $Y = -\frac{1}{3}$.
[3]The converse is not necessarily true; the $SU(2)_L \times SU(2)_R \times U(1)$ model is $L - R$ symmetric but chiral [14].
[4]In the above example, $G = SU(2) \times U(1)_Y$ and $H = U(1)_{em}$.

Below the spontaneous symmetry breaking scale, the unbroken gauge symmetry is $H$, and the gauge bosons of $G$ corresponding to $G/H$ obtain mass of order (gauge coupling) × (VEV). This process applied to the SM renders three IVB ($W^{\pm}$, $Z$) masses at the electroweak scale: $M_W \simeq 80\,\mathrm{GeV}$, $M_Z \simeq 91\,\mathrm{GeV}$. Of course, the photon $A_{\mu}$ remains massless. The success of the SM is not only giving $W^{\pm}$ and $Z^0$ mass but also making the photon massless, which amounts to the electromagnetic charge operator $Q_{\mathrm{em}}$ annihilates the vacuum: $Q_{\mathrm{em}}|\mathrm{VEV}\rangle = 0$ [21]. The origin of the fermion masses in the SM is not from the $SU(2) \times U(1)_Y$ invariant mass term, which cannot be written down anyway, but originates from the gauge-invariant Yukawa couplings of the fermions with the spin-0 Higgs doublet [10]. Then, fermion masses are given by (Yukawa couplings) · (VEV). Therefore, a variety of fermion masses is attributed to the variety of the Yukawa couplings.

One can glimpse that the essence of the above description of nature in terms of the SM is that *the theory is chiral* until the SM gauge group is spontaneously broken at the electroweak scale,

$$v_{\mathrm{ew}} \simeq 246\,\mathrm{GeV}. \tag{1.8}$$

Since the fundamental theory may be given at the Planck scale

$$M_P = \frac{1.22 \times 10^{19}\,\mathrm{GeV}}{\sqrt{8\pi}} = 2.44 \times 10^{18}\,\mathrm{GeV}, \tag{1.9}$$

our chief aim in the construction of the SM is to obtain the correct chiral spectrum from a fundamental theory such as from string theory given near the Planck scale. As we move toward a chiral theory from an ultra-violet completed theory, the parity violating weak interaction phenomena guide us to the SM. In this sense, how parity violation is realized in the SM is a key in breaking the chirality. The simplest SM introduces parity violation in the definition of the SM. Another school adopts "spontaneous" breaking of parity symmetry, starting with the so-called left–right symmetry [22]. In terms of simple group, the prototype chiral model is the $SU(5)$ grand unified theory (GUT) by Georgi and Glashow [23].

In the search for a fundamental theory, two approaches can be taken. One can be the accumulation of low energy observed evidence and the building of a theoretically satisfactory gigantic model describing all these phenomena. This is a bottom-up approach in which the model cannot be excluded experimentally and is hence physically sound. The other approach is to find a theoretically satisfactory model given near the Planck scale and compare its low-energy manifestation with experimental data. This is known as the top-down approach. Sometimes, the bottom-up approach is mingled together with the top-down approach because a fundamental theory can never be achieved using the bottom-up approach alone. In any case, one needs guidance for such a theory. From the theoretical point of view, the best guidance is the *symmetry principle*. Before the discovery of the Higgs boson[5] in 2012, the top-down approach has gained momentum. But, after the Higgs discovery, physics related to the Higgs boson is so accurate and ample enough that the first goal is to establish what particles are there just above a few TeV region. Nevertheless, above tens of TeV region the symmetry guided top-down approach might be the available theoretical track.

Looking back at the construction of the SM, it started from matter representation $|\Psi\rangle$ in the Hilbert space where $|\Psi\rangle$ symbolically stands for the L-handed electron doublet $l$ and the R-handed electron singlet $e_R$. If we include quarks also in the matter, $|\Psi\rangle$ will include them as well. In this Hilbert space, operations by the weak charge and the electromagnetic charge are treated in a similar fashion, thus the SM is dubbed with the phrase, "unified theory of weak and electromagnetic interactions." The key point to observe here is the role of matter representation $|\Psi\rangle$. It is the representation on which symmetry charges act. In this book we will generalize this symmetry concept, and adopt the *unification theme*: *unify all the matter representations if it is possible.*

The first top-down approach toward a more fundamental theory beyond the SM was the GUT. In one attempt, among the representations in $|\Psi\rangle$ the lepton doublet $l$ and the charge conjugated field $d_L^c$ of the R-handed $d_R$ quark are unified into a single representation [23]. Other SM representations are grouped together. This attempt succeeded in unifying the SM group into a simple group SU(5). Another early attempt was to combine the quark doublet $q$ and the lepton doublet $l$ together into a single representation [14]. Then, the remaining SM representations are matched together with the attempted extended gauge group. This attempt succeeded with a semi-simple group $SU(4) \times SU(2)_L \times SU(2)_R \times U(1)$. It follows that, in these GUTs the strong and electroweak couplings are necessarily the same when the unification is valid.

Apparently, the strong, weak, and electromagnetic couplings observed at low energy are not the same at the electroweak scale, and at first glance this idea of unification with the identical gauge couplings seems to contradict the observed phenomena. However, the size of the coupling constant looks different at different energy scales of the probing particle. This is due to the fact that a renormalizable theory intrinsically introduces a mass scale $\mu$, and the energy dependence of the coupling is described by the renormalization group equation. Therefore, one can construct a GUT such that the gauge couplings are unified at a scale, say at $M_{GUT}$, which is supposed to be superheavy so that the electroweak coupling and the strong coupling constants are sufficiently separated at the energy scales ($\sim 100\,\text{GeV}$) probed by the current accelerators [24]. For a significant separation through logarithmic dependence, one needs an exponentially large $M_{GUT}$ [24] which should be smaller than the Planck mass so that gravitational corrections might be insignificant. Here, we should not forget that the construction of the simplest SU(5)

---

[5]In this book, we use this simple word instead of the correct Brout–Englert–Higgs–Guralnik–Hagen–Kibble boson (BEHGHK boson).

was possible after realizing that one can collect all the pieces of the 15 Chiral fields with one kind of chirality, i.e. in terms of the L-handed fields.[6] These GUTs render the SM fermions below the GUT symmetry breaking scale so that massless SM fermions can survive down to the electroweak scale. Even though GUTs are basically top-down approaches, they have one notably testable prediction: a proton decays at the experimentally verifiable level with the proton lifetime with the current lower limit of $\tau_p > 10^{33}$ s.

Another attempt can be Kaluza–Klein (KK) [25, 26] type higher dimensional theories. But a naive torus compactification of the internal spaces leads to vectorlike fermions (the heavy KK mode and the vectorlike fermions), implying no massless fermions at low energy.

There are several good reasons to go beyond the SM. Probably, the most important reason for the extension is to understand *the family problem* why the fermion families repeat three times. Of course, an extension to understand the family problem must retain the good chiral property. The family problem has been considered with both global and gauge horizontal (or family) symmetries. However, the horizontal symmetry must be broken since different families obviously have different mass scales. If it were a global symmetry, we would expect Goldstone bosons (familons) after the spontaneous global symmetry breaking. If it were a gauge symmetry, we would expect flavor changing phenomena at some level and one has to be clever enough to forbid gauge anomalies. In general, these horizontal symmetries are very complicated if not impossible. If a horizontal gauge symmetry is considered, it is better to unify it in a GUT-like gauge group so that even the horizontal gauge couplings are also unified with the other gauge couplings. In this respect, the grand unification of families (GUF) has attracted some attention [27]. However, in contrast to GUTs, the grand unification of families in four dimensions (4D) has not produced any conspicuous predictions which can be tested. They are hidden at the unknown super heavy mass scales.

Even though the 4D gauge theories with spontaneous symmetry breaking are very successful, as witnessed by the success of the SM, there exists a fundamental problem in 4D gauge models. The problem lies with the introduction of a specific set of representations for the fermions and Higgs fields in the SM, or in GUTs, or in GUFs. Let us call this *the representation problem*, and it becomes more acute when one tries to understand it together with *the family problem*. The specific choice of chiral representations in the GUTs or in the SM cannot be answered in a 4D gauge theory framework. For example, if we try to embed three families in a spinor representation of a big orthogonal group such as in SO(18), there would remain a question, "Why is there only one spinor in SO(18)?" On the other hand, the group representation for the spin-1 gauge fields is uniquely fixed by the adjoint representation. There is no other choice. If the matter representation is determined

---

[6]About the same time as this theory was put forth, the concept of supersymmetry was born from similar knowledge of Weyl spinors.

as uniquely as the gauge boson is, then the representation problem would be understood.

Even if we go beyond 4D, the representation problem in field theory is still not understood. The only constraint for matter representations is the anomaly cancellation, which does not fix the representation uniquely. Therefore, the theory must be more restrictive than field theory to understand the representation problem.

In this respect, string theory, either bosonic or fermionic, is one alternative to pursue. But to have matter (the spin-$\frac{1}{2}$ fermions), we can only consider the fermionic string or superstring, which is possible in 10D. This superstring theory is far more restrictive than higher dimensional field theories. To apply it to the 4D phenomenology, we must hide the six extra dimensions.

It is known that there are two ways to hide the extra dimensions, one is by the so-called compactification and the other is by introducing a warp factor as in the Randall–Sundrum type-II model [28]. This book is restricted to compactification.

Consider the simplest extra dimensional generalization, i.e. to 5D. If one compactifies the extra dimension, the original 5D fields split into 4D fields. The effective 4D fields might be massive, but massless fields might also exist. Consider, for example, a 5D theory with spin-2 graviton ($g_{MN}(M, N = 0, 1, 2, 3, 4)$) only in 5D. By compactifying the fifth dimension on a circle, one obtains the well-known KK spectrum. At low energy the massless modes are a spin-2 graviton ($g_{\mu\nu}(\mu, \nu = 0, 1, 2, 3)$), a spin-1 gauge boson ($g_{\mu 4}$), and a spin-0 dilaton ($g_{44}$). The spin-1 gauge boson and spin-0 dilaton were originally components of 5D graviton. The compactification radius $R$ determines the masses of the KK modes (integer) $\times R^{-1}$. If one introduces a massless 5D fermion (8 real components in $\psi_L$ or in $\psi_R$) in addition, then by compactifying on a circle $S_1$ one obtains two 4D Weyl fermions at each KK level as shown in Fig. 1.1b. If the representation is vectorlike (in 5D, both 4D $\psi_L$ and 4D $\psi_R$ should be introduced), then it is a general strategy to remove them at a high energy scale due to interactions unless

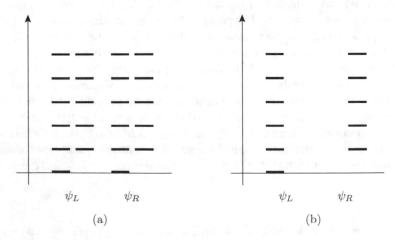

$\psi_L$        $\psi_R$                    $\psi_L$        $\psi_R$

(a)                                          (b)

**Fig. 1.1** KK mode spectrum

some symmetry forbids such a large mass. So, if two Weyl fermions with opposite quantum numbers are matched at each level, they are removed at high energy. In this sense, two lowest level fermions of Fig. 1.1a with zero KK mass can obtain a superheavy mass. Thus, the naive compactification on a circle would not allow a chiral spectrum at low energy. Our hope is to obtain a chiral fermion at low energy through a mechanism in which there remains an unmatched Weyl fermion as shown in Fig. 1.1b. However, it has been proven that if the gauge bosons originate from the gravity multiplet $g_{MN}(M, N = 0, 1, 2, 3, \cdots, D)$ in $D$ dimensions, then a chiral theory such as that shown in Fig. 1.1b does not arise at low energy [29]. Therefore, the KK idea, even though beautiful, is not relevant for a low-energy chiral theory like the standard model.

If 4D SM gauge bosons arise *from gauge bosons in higher dimensions*, then a realization of Fig. 1.1b does not arise from a simple torus compactification on a circle $S_1$. To arrive at Fig. 1.1b, one must impose careful boundary conditions on the 5D fermion wave functions such that some parts of the spectrum are removed. So, we consider the orbifold, which achieves this idea beautifully, by modding out $S_1$ by some discrete group such as $\mathbb{Z}_2$.

We observed that the SM gauge bosons must arise from spin-1 gauge bosons in higher dimensions, not from the higher dimensional gravity multiplet. This leads us to the consideration of higher dimensional gauge theories. These higher dimensional gauge fields are required to contain the SM gauge bosons. In higher dimensions, we also introduce fermions which contain the three SM families. Since we introduce these fermions, one has to worry about the gauge anomaly if $D =$ even. Then the gauge anomalies [30] and gravitational anomalies [31] in even dimensions must be taken into account. As mentioned above, however, the higher dimensional field theories do not explain the representation problem except for the requirement of having no anomaly.

In the mid 1980s a breakthrough was made in the search for higher dimensional theories. To introduce fermions, $\mathcal{N} = 1$ supersymmetry must be considered. Green and Schwarz observed that in 10D no anomaly appears if the gauge group is SO(32) or $E_8 \times E_8$ [32]. Basically, these large gauge groups are required to cancel any anomaly arising from the 10D gravitino. The 10D gravitino carries 496 units of anomaly. If one wants to cancel the gravitino anomaly by gauge fermions, the gauge group must match the dimension of the adjoint representation as 496. SO(32) and $E_8 \times E_8$ fulfil this requirement. In addition, the second rank antisymmetric tensor field $B_{MN}$ is required. Soon, the heterotic string theory was constructed and these large gauge groups are realized as the oscillation modes of the closed strings and winding modes [33, 34].

This string motivated 10D $\mathcal{N} = 1$ gauge field theory introduces fermions as superpartners of the gauge bosons which form the adjoint representation of the gauge group. An adjoint representation is a real representation. If one naively compactifies on a torus, say a six dimensional torus $T_6$, the real representation keeps the reality and one cannot obtain chiral fermions in 4D. To reach at the phenomenologically successful SM, it is of utmost importance in extra dimensional field theory to obtain chiral fermions after compactification. Under torus compact-ification, the 10D $\mathcal{N} = 1$ supersymmetry becomes $\mathcal{N} = 4$ supersymmetry in 4D. Supersymmetries with $\mathcal{N} \geq 2$ introduce vectorlike fermions in 4D which are not needed at the electroweak scale. If one must introduce a supersymmetry to understand the gauge hierarchy problem, the only allowable one in 4D is the $\mathcal{N} = 1$ supersymmetry. Therefore, the first set of criteria is to keep only $\mathcal{N} = 1$ supersymmetry after compactification and to allow chiral fermions. In orbifold compactification, which is the main subject of this book, the requirement for the $\mathcal{N} = 1$ supersymmetry in 4D usually allows chiral fermions.

Phenomenologically, the $E_8 \times E_8$ heterotic string has been discussed much more than the SO(32) heterotic string. The 15 chiral fields of the SM plus one more chiral field are embeddable in a spinor representation of SO(10). The adjoint representation of the SO(32) group cannot contain the spinor representation of SO(10). Thus, making a reasonable spectrum after compactification from the SO(32) heterotic string is not the obvious first step. On the other hand, the adjoint representation of $E_8$ contains the spinor representation of SO(10), and it would seem easy to classify the SM fermions through spinor representation of SO(10) (or for that matter, in terms of $\mathbf{10} \oplus \bar{\mathbf{5}}$ of SU(5)). For this reason, we will restrict our discussion to the $E_8 \times E_8$ heterotic string only.

Suppose that 10D compactifies as $M_{10} \to M_4 \times B_6$, i.e. to a flat 4D Minkowski space $M_4$ times a compact 6D internal space $B_6$. Candelas et al. have shown that the condition for the $\mathcal{N} = 1$ supersymmetry in 4D is to require the SU(3) holonomy of the internal space [35]. This SU(3) symmetry can be embedded in the gauge group also. Spaces with SU($N$) holonomy are called the Calabi–Yau spaces. In the example considered in Ref. [35], one obtains a 4D $\mathcal{N} = 1$ supersymmetric model with the gauge group $E_6 \times E_8'$. The net number of chiral fermions of the $E_6$ sector are 36 copies of $\mathbf{27}$. Each $\mathbf{27}$ of $E_6$ contains one family of the SM. Thus, this Calabi–Yau space gives too many families. Nevertheless, it provides an example of how one obtains chiral fermions and $\mathcal{N} = 1$ supersymmetry in 4D. Actually, Calabi–Yau manifolds are rather complicated spaces.

For reducing the $\mathcal{N} = 4$ supersymmetry down to $\mathcal{N} = 1$, a simpler method known as orbifold was introduced by Dixon, Harvey, Vafa, and Witten [36–38]. The orbifold method uses discrete groups on top of torus compactification. For a given 6D internal torus, all possible discrete group actions have been classified. Generically they do not act freely. The fixed points lead to singularities of the quotient space which is called an orbifold. The orbifold singularities can be eliminated by cutting out the fixed points and gluing them in smooth "disk-like" surfaces such that the resulting smooth space has SU(3) holonomy. Therefore, an orbifold can be considered a singular limit of a good manifold such as the Calabi–Yau. To a low-energy observer, therefore, the orbifold is as good as the Calabi–Yau space. Another merit of the orbifold is that compactification through orbifolding can be systematically found.

The chief merit of the orbifold is that it allows chiral fermions at low energy, even though one starts to compactify with such a simple idea as a 6D torus. The existence of chiral fermions is at the heart of the standard model construction and, hence, the making of our universe as it is now before chemistry and biology were able to play their roles.

Among nine orbifolds cited in [37], $\mathbb{Z}_{12-I}$ can be considered to be the simplest one because there are only three fixed points. The simple looking $\mathbb{Z}_3$ orbifold is not so simple in the sense that it has 27 fixed points. But, the $\mathbb{Z}_3$ can be considered as the center of the holonomy group SU(3) of the Calabi–Yau space, which is needed to have $\mathcal{N} = 1$ supersymmetry in 4D [35]. Most studied non-prime orbifolds $\mathbb{Z}_{6-II}$ [39] and $\mathbb{Z}_{12-I}$ [40] have $\mathbb{Z}_3$ in one two-torus. In a $\mathbb{Z}_3$ orbifold with up to two Wilson lines,[7] the spectrum is always a multiple of 3 which was the main motivation for extensive studies in the early days of orbifold construction of SMs [41, 42].

The models obtained by the orbifold compactification can be considered as 4D string models, but their roots are in 10D superstring. There is another method of obtaining a 4D superstring model, using free fermionic formulation [43]. However, the free fermionic formulation lacks the geometrical interpretation present in the orbifold compactification. Yet there exists another method using self-dual lattices [44].

There are a few important issues in the current SM:

(1) Flavor phenomenology,

(2) Strong CP problem,

(3) Three chiral families, (1.10)

(4) Gauge hierarchy.

Since the representation **248** of the 10D heterotic string group $E_8 \times E_8$ is big enough, (3) can be an easy solution from the compactification, which is one of the main objectives of exploring the orbifold technique here. The issue (1) is the data fitting to the Cabibbo–Kobayashi–Maskawa (CKM) [45, 46] and Pontecorvo–Maki–Nakagawa–Sakata (PMNS) matrices [47], which is not pursued here except introducing the weak CP violation phase from compactification. The issue (2) must be a consequence of compactification. The issue (4) is the Higgs boson related one. Our book is to understand this gauge hierarchy problem from the supersymmetry perspective.

In the next chapter, we start out our journey by reviewing how the successful chiral field theory known as the SM was achieved. In Chap. 3, we present the general concept on orbifold. Then, we introduce field theoretic orbifolds in Chap. 5. The basics on string theory is presented in Chap. 6. In Chap. 7, we start to present orbifolding of the 10D string theory. In Chap. 8, a more formal introduction to string orbifold, with necessity from modular invariance. After introducing the partition function, we explain how the formulae for the massless states and Kaluza–

---

[7]A Wilson line is a closed line integral along the direction tangent to a gauge field, $\oint dx^\mu A_\mu$. A famous example is the Aharonov–Bohm phase. In this book, it is meant as a closed line integral in the compactified space, $\oint dx^i A_i$.

Klein spectra are derived from the partition function. We also briefly review the discussions on the stringy threshold corrections to gauge couplings. Discussion on twisted strings in non-prime orbifold is added, especially on the behavior in the higher twisted sectors in non-prime orbifold. Then, we continue to discuss orbifolds using the Jacobi theta functions, Yukawa couplings, and group theoretic properties of the twisted sector spectra. In Chap. 10, interactions on orbifolds are presented. In Chap. 13, a gross view of orbifold phenomenology is presented. In Chaps. 14–15, we present several orbifold constructions at the level that a devoted reader can construct his own model. They include $\mathbb{Z}_{12-I}$, $\mathbb{Z}_{6-II}$, and $\mathbb{Z}_2 \times \mathbb{Z}_2$. Here, GUTs from orbifolds, especially from $\mathbb{Z}_{12-I}$ are discussed. In Chap. 15, Calabi–Yau spaces are discussed. Since the flavor problem is the last hurdle to overcome in string compactification, we review phenomenology on flavor physics in Chap. 16. In Chap. 17, we present four other methods attempting to obtain the SM from superstring, the fermionic construction, magnetized brane and intersecting brane models. In Appendix A, we present some relations of Jacobi theta functions, used in Chaps. 10, 12, 13, and 14. Here, we list useful tables for model building which include the orbifolds with $\mathcal{N} = 1$ supersymmetry, the degeneracy factor and vacuum energy needed for orbifold construction, and the conditions on Wilson lines. Finally, some useful numbers in group theory are presented in Appendix A.

# References

1. T.D. Lee, C.-N. Yang. Question of parity conservation in weak interactions. Phys. Rev. **104**, 254–258 (1956)
2. G. Aad et al., Observation of a new particle in the search for the Standard Model Higgs boson with the ATLAS detector at the LHC. Phys. Lett. **B716**, 1–29 (2012)
3. S. Chatrchyan et al., Observation of a new boson at a mass of 125 GeV with the CMS experiment at the LHC. Phys. Lett. **B716**, 30–61 (2012)
4. W. Pauli, General remarks on parity non-conservation. Talk at Conf. on Nuclear Structure 416, J. Kvasil, Nuclear Structure and Nuclear Processes.
5. R.P. Feynman, M. Gell-Mann, Theory of Fermi interaction. Phys. Rev. **109**, 193–198 (1958). [,417 (1958)]
6. E.C.G. Sudarshan, R.E. Marshak, Chirality invariance and the universal Fermi interaction. Phys. Rev. **109**, 1860–1860 (1958)
7. E. Fermi, An attempt of a theory of beta radiation. 1. Z. Phys. **88**, 161–177 (1934)
8. T.D. Lee, C.-N. Yang, Implications of the intermediate boson basis of the weak interactions: existence of a quartet of intermediate bosons and their dual isotopic spin transformation properties. Phys. Rev. **119**, 1410–1419 (1960)
9. S.L. Glashow, Partial symmetries of weak interactions. Nucl. Phys. **22**, 579–588 (1961)
10. S. Weinberg, A model of leptons. Phys. Rev. Lett. **19**, 1264–1266 (1967)
11. A. Salam, Weak and electromagnetic interactions. Conf. Proc. **C680519**, 367–377 (1968)
12. J.E. Kim, P. Langacker, M. Levine, H.H. Williams, A theoretical and experimental review of the weak neutral current: a determination of its structure and limits on deviations from the minimal SU(2)-L x U(1) electroweak theory. Rev. Mod. Phys. **53**, 211 (1981)
13. H. Weyl, Gravitation and electricity. Sitzungsber. Preuss. Akad. Wiss. Berlin (Math. Phys.) **1918**, 465 (1918). [,24 (1918)]
14. J.C. Pati, A. Salam, Unified lepton-hadron symmetry and a gauge theory of the basic interactions. Phys. Rev. **D8**, 1240–1251 (1973)

15. L. Susskind, Dynamics of spontaneous symmetry breaking in the Weinberg-Salam theory. Phys. Rev. **D20**, 2619–2625 (1979)
16. S. Weinberg, Implications of dynamical symmetry breaking. Phys. Rev. **D13**, 974–996 (1976). [Addendum: Phys. Rev. D 19, 1277 (1979)]
17. P.W. Higgs, Broken symmetries, massless particles and gauge fields. Phys. Lett. **12**, 132–133 (1964)
18. P.W. Higgs, Broken symmetries and the masses of gauge bosons. Phys. Rev. Lett. **13**, 508–509 (1964)
19. F. Englert, R. Brout, Broken symmetry and the mass of gauge vector mesons. Phys. Rev. Lett. **13**, 321–323 (1964)
20. G.S. Guralnik, C.R. Hagen, T.W.B. Kibble, Global conservation laws and massless particles. Phys. Rev. Lett. **13**, 585–587 (1964)
21. T.W.B. Kibble, Symmetry breaking in non-Abelian gauge theories. Phys. Rev. **155**, 1554–1561 (1967)
22. R.N. Mohapatra, J.C. Pati, A natural left-right symmetry. Phys. Rev. **D11**, 2558 (1975)
23. H. Georgi, S.L. Glashow, Unity of all elementary particle forces. Phys. Rev. Lett. **32**, 438–441 (1974)
24. H. Georgi, H.R. Quinn, S. Weinberg, Hierarchy of interactions in unified gauge theories. Phys. Rev. Lett. **33**, 451–454 (1974)
25. Th. Kaluza, Zum Unitätsproblem der Physik. Sitzungsber. Preuss. Akad. Wiss. Berlin (Math. Phys.) **1921**, 966–972 (1921). [Int. J. Mod. Phys. D 27, no. 14, 1870001 (2018)]
26. O. Klein, Quantum theory and five-dimensional theory of relativity. Z. Phys. **37**, 895–906 (1926). (In German and English)
27. H. Georgi, Towards a grand unified theory of flavor. Nucl. Phys. **B156**, 126–134 (1979)
28. L. Randall, R. Sundrum, An alternative to compactification. Phys. Rev. Lett. **83**, 4690–4693 (1999)
29. E. Witten, Fermion quantum numbers in Kaluza-Klein theory (1983). [Conf. Proc. **C8306011**, 227 (1983)]
30. P.H. Frampton, T.W. Kephart, The analysis of anomalies in higher space-time dimensions. Phys. Rev. **D28**, 1010 (1983)
31. L. Alvarez-Gaume, E. Witten, Gravitational anomalies. Nucl. Phys. **B234**, 269 (1984). [,269 (1983)]
32. M.B. Green, J.H. Schwarz, Anomaly cancellation in supersymmetric D=10 gauge theory and superstring theory. Phys. Lett. **B149**, 117–122 (1984)
33. D.J. Gross, J.A. Harvey, E.J. Martinec, R. Rohm, The heterotic string. Phys. Rev. Lett. **54**, 502–505 (1985)
34. D.J. Gross, J.A. Harvey, E.J. Martinec, R. Rohm, Heterotic string theory. 1. The free heterotic string. Nucl. Phys. **B256**, 253 (1985)
35. P. Candelas, G.T. Horowitz, A. Strominger, E. Witten, Vacuum configurations for superstrings. Nucl. Phys. **B258**, 46–74 (1985)
36. L.J. Dixon, J.A. Harvey, C. Vafa, E. Witten, Strings on orbifolds. Nucl. Phys. **B261**, 678–686 (1985) [,678 (1985)]
37. L.J. Dixon, J.A. Harvey, C. Vafa, E. Witten, Strings on orbifolds. 2. Nucl. Phys. **B274**, 285–314 (1986)
38. L.E. Ibanez, H.P. Nilles, F. Quevedo, Orbifolds and Wilson lines. Phys. Lett. **B187**, 25–32 (1987)
39. T. Kobayashi, S. Raby, R.-J. Zhang, Searching for realistic 4d string models with a Pati-Salam symmetry: orbifold grand unified theories from heterotic string compactification on a Z(6) orbifold. Nucl. Phys. **B704**, 3–55 (2005)
40. J.E. Kim, B. Kyae, Flipped SU(5) from Z(12-I) orbifold with Wilson line. Nucl. Phys. **B770**, 47–82 (2007)
41. L.E. Ibanez, J.E. Kim, H.P. Nilles, F. Quevedo, Orbifold compactifications with three families of SU(3) x SU(2) x U(1)**n. Phys. Lett. **B191**, 282–286 (1987)

42. J.A. Casas, C. Munoz, Three generation SU(3) x SU(2) x U(1)-Y models from orbifolds. Phys. Lett. **B214**, 63–69 (1988)
43. I. Antoniadis, C.P. Bachas, C. Kounnas, Four-dimensional superstrings. Nucl. Phys. **B289**, 87 (1987)
44. W. Lerche, D. Lust, A.N. Schellekens, Chiral four-dimensional heterotic strings from self-dual lattices. Nucl. Phys. **B287**, 477 (1987)
45. N. Cabibbo, Unitary symmetry and leptonic decays. Phys. Rev. Lett. **10**, 531–533 (1963). [,648 (1963)]
46. M. Kobayashi, T. Maskawa, CP violation in the renormalizable theory of weak interaction. Prog. Theor. Phys. **49**, 652–657 (1973)
47. Z. Maki, M. Nakagawa, S. Sakata, Remarks on the unified model of elementary particles. Prog. Theor. Phys. **28**, 870–880 (1962)

# Standard Model and Beyond

**2**

## 2.1    The Standard Model

The standard model (SM) consists of the confining color gauge theory SU(3) for strong interactions and the spontaneously broken electroweak gauge theory $SU(2)_L \times U(1)_Y$. In this subsection, we introduce the SM, concentrating on the issues relevant for our string orbifold construction and phenomenological issues after the discovery of the Higgs boson [1, 2].

The symmetry principle is the heart of particle physics. It has its origins in Heisenberg's SU(2) isospin, and flourished in the 1960s under the name of $SU(3)_{\text{flavor}}$. The triple tensor product of the fundamental representation **3** of $SU(3)_{\text{flavor}}$ gives

$$\mathbf{3} \otimes \mathbf{3} \otimes \mathbf{3} = \mathbf{1} \oplus \mathbf{8} \oplus \mathbf{8} \oplus \mathbf{10}.$$

The low lying octet baryons, $p, n, \Lambda$, and $\Sigma$, are assigned to one **8** of the above representation. This classification works also for spin-0 and spin-1 mesons, $\mathbf{3} \otimes \bar{\mathbf{3}} = \mathbf{1} \otimes \mathbf{8}$. This old $SU(3)_{\text{flavor}}$ classification was very successful in classifying low lying hadrons, and the basic ingredients **3** for all these classifications are called quarks; up, down, and strange quarks, $\mathbf{3} = (u, d, s)^T$ [3].[1] The quarks were assumed to carry spin-$\frac{1}{2}$ because three quarks make up spin-$\frac{1}{2}$ baryons and quark–antiquark pairs make up integer spin mesons. This development in probing the subnuclear structure was probably the cornerstone of the discovery of the standard model, but a few more new ideas needed to be added. First, a new degree of freedom, color, was introduced. Second, local groups were considered. Third, the flavor group became bigger with discoveries of more quarks, $c, b, t$. And fourth, the $SU(3)_{\text{flavor}}$ was understood to be nothing but 1 GeV is the strong interaction scale compared to the relatively small

---

[1]G. Zweig also introduced SU(3) triplets, see [4].

© Springer Nature Switzerland AG 2020
K.-S. Choi, J. E. Kim, *Quarks and Leptons From Orbifolded Superstring*,
Lecture Notes in Physics 954, https://doi.org/10.1007/978-3-030-54005-0_2

**Fig. 2.1** The Young tableaux
for **56** of SU(6)

current quark mass scale of three flavors of quarks, $3 = (u, d, s)^T$. As for the electroweak interaction, only a part of SU(3)$_{\text{flavor}}$ is useful for localized (gauged) symmetry.

The introduction of color degrees of freedom was motivated from the old SU(6) classification of hadrons. This old SU(6) is a generalization of the above Gell–Mann's flavor SU(3)$_{\text{flavor}}$ together with the rotation group SU(2). For the S-wave composites, the representations of the rotation group are just spins. Thus, the fundamental representation **6** of SU(6) is composed of the direct product representation of the flavor triplet, i.e. $u, d, s$ quarks, and the spin doublets, $| \uparrow \rangle$ and $| \downarrow \rangle$, $\mathbf{6} = (u^\uparrow, u^\downarrow, d^\uparrow, d^\downarrow, s^\uparrow, s^\downarrow)^T$. Following Gell–Mann, one may try to classify low lying hadrons in terms of tensor products of **6** and $\bar{\mathbf{6}}$, but in the case of SU(6) we note that different spin representations can come in one SU(6) representation. For mesons, we consider $\mathbf{6} \otimes \bar{\mathbf{6}} = \mathbf{1} \oplus \mathbf{35} = \mathbf{1}_{s=0} \oplus (\mathbf{8}_{s=0} + \mathbf{3}_{s=1} + \mathbf{24}_{s=1})$, where we indicated the spin $s$ as subscripts. Thus, the representation **35** of SU(6) consists of the pseudoscalar meson octet ($\pi, K^\pm, K^0, \overline{K}^0, \eta$), the vector meson singlet ($\sim \phi$),[2] and the vector meson octet ($\rho, K^{*\pm}, K^{*0}, \overline{K}^{*0}, \sim \omega$). This appears to work rather well.

However, as for baryons, we expect

$$\mathbf{6} \otimes \mathbf{6} \otimes \mathbf{6} = \mathbf{20} \oplus \mathbf{56} \oplus \mathbf{70} \oplus \mathbf{70'}. \tag{2.1}$$

We may try to embed the low lying baryons, the baryon octet ($p, n, \Lambda, \Sigma$) and the baryon decuplet ($\Delta_{(I=\frac{3}{2})}, Y^*_{(I=1)}, \Xi^*_{(I=\frac{1}{2})}, \Omega_{(I=0)}$), where isospin $I$ is shown as subscript, in one representation of (2.1). Indeed, the representation **56** can house all these low lying baryons since the baryon octet is spin-$\frac{1}{2}$ and the baryon decuplet is spin-$\frac{3}{2}$. This appears to work as well. But the **56** of SU(6), namely the Young tableaux given in Fig. 2.1, is completely symmetric under any exchange of its composite **6**.

Since we interpreted **6** as a spin-$\frac{1}{2}$ quark, the notion that a pair of spin-$\frac{1}{2}$ objects must anticommute under their exchange is grossly violated by **56**. This symmetry problem was found earlier in the prediction of $\Omega^-$ particle [5]. It is basically the completely symmetric wave function of $\Omega^- \sim s^\uparrow s^\uparrow s^\uparrow$ under any exchange of its constituents $s^\uparrow$. One cannot resolve this dilemma without the introduction of a new degree of freedom [6][3] which is named *color* in modern theory.

Thus, one assumes that baryons made of three quarks must be completely antisymmetric under the exchange of the color index. The complete symmetric

---

[2]The symbol $\sim$ implies that the singlet–octet mixing has to be considered.

[3]The violation of the spin–statistics relation was noted earlier [5, 7].

property of **56** of SU(6) then becomes a fortune instead of a disaster when combined with this completely antisymmetric property of color exchange, and the spin–statistics theorem is satisfied. In other words, the baryon wave function must behave as $\Psi_{\text{baryon}} \sim \epsilon_{\alpha\beta\gamma}q^{\alpha}q^{\beta}q^{\gamma}$, where $\alpha\beta\gamma$ are the color indices, 1, 2, 3, or red, green, yellow, respectively. Since a baryon is supposed to be made of three quarks, we need just three indices for color, and $\epsilon_{\alpha\beta\gamma}q^{\alpha}q^{\beta}q^{\gamma}$ must be a singlet under transformation in this new color space. Then we conclude that the new color space is SU(3).[4] The gauge theory of color SU(3) is called quantum chromodynamics, or simply QCD. The gauge bosons of this SU(3)-color are called gluons $G_{\mu}^{a}$ ($a = 1, 2, \cdots, 8$). The reason why only the color singlet states appear as low lying hadrons is the key issue among QCD problems, and is known as the confinement problem. Over the last four decades, strong interactions have successfully been described by QCD. After the initial introduction of SU(3) as the color interaction with integer-charged quarks [6], however, the present day QCD is based on the fractionally charged quarks with unbroken SU(3)$_{\text{color}}$ [8–11] as proven by the $\pi^{0} \to 2\gamma$ decay rate.[5]

The flavor group we considered for SU(3)$_{\text{flavor}}$ is now generalized as SU(6)$_{\text{flavor}}$ since there exist at least 6 quarks, $u, d, c, s$, and $t, b$. The weak interaction gauges only a part of this SU(6)$_{\text{flavor}}$ since it is known that different families exhibit exactly the same kind of tree level charged current (CC) weak interactions. Concerned with the CC weak interactions, the leptons are not different from quarks. Therefore, for gauge interactions it will be enough to consider just the first family doublets, $(\nu_{e}, e)^{T}$ and $(u^{\alpha}, d^{\alpha})^{T}$. Now the electroweak gauge group is taken as SU(2)×U(1) to unify the electromagnetic interaction with the weak interaction [12–14], resulting in the unified theory aptly named the *electroweak theory*. In the electroweak theory, the interactions are known to be chiral; only the left-handed fields participate in the CC interactions, which was the famous old "V-A" theory introduced in the previous chapter. Thus, the electroweak representation of leptons and quarks is asymmetrical under the chirality exchange, $L \leftrightarrow R$,

$$l_{L} \equiv \begin{pmatrix} \nu_{e} \\ e \end{pmatrix}_{L}, \quad e_{R}, \quad q_{L} \equiv \begin{pmatrix} u^{\alpha} \\ d^{\alpha} \end{pmatrix}_{L}, \quad u_{R}^{\alpha}, \quad d_{R}^{\alpha}, \tag{2.2}$$

where $\alpha$ runs over the three color indices and the weak hypercharges $Y$ for the representations are

$$Y(l_{L}) = -\tfrac{1}{2}, \; Y(e_{R}) = -1, \; Y(q_{L}) = \tfrac{1}{6}, \; Y(u_{R}^{\alpha}) = \tfrac{2}{3}, \; Y(d_{R}^{\alpha}) = -\tfrac{1}{3}. \tag{2.3}$$

---

[4]Note that it cannot be SO(3) since SO(3) does not allow complex representations the needed property for the existence of both quarks and antiquarks.

[5]With a judicious symmetry breaking of SU(3)$_{\text{color}}$, one can obtain integer-charged quarks. For pure strong interaction phenomena, e.g. for QCD $\beta$ functions, only the electromagnetic charges of quarks do not matter.

The electromagnetic charge is given by $Q_{em} = T_3 + Y$. The chirality of the left-handed doublet and the hypercharge $Y$ are customarily written explicitly as subscripts in the electroweak group, $SU(2)_L \times U(1)_Y$. The gauge sector of this theory has two gauge couplings, the $SU(2)_L$ coupling $g$ and the $U(1)_Y$ coupling $g'$ as shown in Eq. (1.5). The ratio of these couplings defines the weak mixing angle $\tan \theta_W = g'/g$. The group $SU(2)_L \times U(1)_Y$ is spontaneously broken down to $U(1)_{em}$, and three gauge bosons obtain mass. These massive spin-1 gauge bosons $W^\pm$ and $Z^0$ are considered as the mediators of the weak force. If the spontaneous symmetry breaking occurs only through the vacuum expectation value (VEV) of Higgs doublet(s), the tree level ratio of these masses is related to the weak mixing angle by

$$\sin^2 \theta_W = 1 - \frac{M_W^2}{M_Z^2} \, . \tag{2.4}$$

The same weak mixing angle appears in neutral current (NC) neutrino scattering experiments,

$$\frac{G_F}{\sqrt{2}} \bar{v}_\mu \gamma^\alpha (1 + \gamma_5) v_\mu \sum_i (\bar{q}_i Q_Z (1 + \gamma_5) q_i + \bar{q}_i Q_Z (1 - \gamma_5) q_i) \, , \tag{2.5}$$

where $Q_Z = T_3 - Q_{em} \sin^2 \theta_W$. Phenomenologically, $\sin^2 \theta_W$ can be determined by several independent processes. It turned out that experimentally determined $\sin^2 \theta_W$ is close to the relation (2.4). Thus, any deviation from the tree level relation obtained for the doublet breaking is important and for this purpose we introduce another phenomenologically important parameter $\rho$ defined by

$$\rho = \frac{M_W^2}{M_Z^2 \cos^2 \theta_W} \, . \tag{2.6}$$

$\rho$ becomes 1 at tree level if the spontaneous symmetry breaking occurs only through the vacuum expectation value (VEV) of Higgs doublet(s). The experimentally determined weak mixing angles from the gauge boson mass ratio and the weak NC experiments coincide. The coincidence of these mixing angles given in (2.4) and (2.5) and the $\rho$ parameter being close to 1, implies that the electroweak symmetry breaking occurs through VEVs of a Higgs doublet(s). Large electron–positron (LEP) collider experiments at CERN decisively confirmed this Higgs doublet condition. The dominant loop contribution to $\rho$ parameter is from the heavy top quark [15]. The global fit, including the radiative corrections, gives $\rho = 1.00039 \pm 0.00019$ from a global fit [16] and $\sin^2 \theta_W (M_Z)^{lept} = 0.23101 \pm 0.00052$ from a compilation of leptonic data shown in Fig. 2.2 [17].

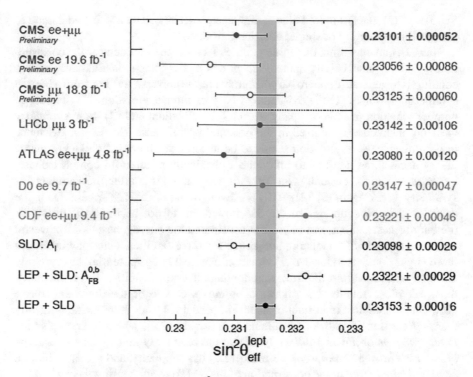

**Fig. 2.2** Comparison of the measured $\sin^2 \theta_W$ in the muon and electron channels [17]

The electroweak representation given in Eq. (2.2) repeats three times, thus we say that there are three families of fermions plus a doublet of Higgs bosons:

$$\begin{pmatrix} v_e \\ e \end{pmatrix}_L, \quad e_R, \quad \begin{pmatrix} u^\alpha \\ d^\alpha \end{pmatrix}_L, \quad u^\alpha_R, \ d^\alpha_R$$

$$\begin{pmatrix} v_\mu \\ \mu \end{pmatrix}_L, \quad \mu_R, \quad \begin{pmatrix} c^\alpha \\ s^\alpha \end{pmatrix}_L, \quad c^\alpha_R, \ s^\alpha_R$$

$$\begin{pmatrix} v_\tau \\ \tau \end{pmatrix}_L, \quad \tau_R, \quad \begin{pmatrix} t^\alpha \\ b^\alpha \end{pmatrix}_L, \quad t^\alpha_R, \ b^\alpha_R$$

$$\mathbf{H}_1 = \begin{pmatrix} H^0 \\ H^- \end{pmatrix} \tag{2.7}$$

which make up the needed matter spectrum of the SM at low energy. The standard model is a gauge theory SU(3)×SU(2)×U(1) with the representation (2.7). For nonzero neutrino masses observed by underground detectors, one usually introduces

$SU(3) \times SU(2) \times U(1)$ singlet fields at high energy scale(s). The SM is understood to include this possibility of singlet neutrino introduction.

Naively, one may think that three families seem to be too much for the construction of our universe, chiefly made of $e$, $p$, and $n$. But there exist Sakharov's three conditions to generate nonzero baryon number in the universe starting from a baryon symmetric universe: (1) the existence of baryon number violation, (2) the baryon number violation must accompany $\mathscr{C}$ and $\mathscr{C}\mathscr{P}$ violation, and (3) these symmetry violating interactions occurred in the nonequilibrium phase. So for our existence in the universe, $\mathscr{C}\mathscr{P}$ violation seems to be necessary. The $SU(2)_L \times U(1)_Y$ gauge theory needs three families to introduce a physically meaningful complex phase, i.e. $\mathscr{C}\mathscr{P}$ violation, basically via Yukawa couplings [18], after the spontaneous symmetry breaking of $SU(2)_L \times U(1)_Y$. Even though the $\mathscr{C}\mathscr{P}$ phase from the complex Yukawa couplings of the SM turned out to not be the $\mathscr{C}\mathscr{P}$ phase for the baryogenesis, still many physicists considered the baryogenesis as a powerful argument for the $\mathscr{C}\mathscr{P}$ violation and hence for three families. One may introduce more Higgs doublets to have $\mathscr{C}\mathscr{P}$ violation from the Higgs potential, but certainly the $\mathscr{C}\mathscr{P}$ violation from the Yukawa couplings is one possibility. Experimentally, the weak $\mathscr{C}\mathscr{P}$ violation by Yukawa couplings seems to be the dominant source [16]. One of the most important problems in particle physics is understanding why there appear three families of fermions as shown in (2.7), which is known as the *family problem* or *flavor problem*. Note that the seemingly innocuous *introduction of color $\mathbf{3}$'s (also $\bar{\mathbf{3}}$'s) and weak $\mathbf{2}$'s* in Eq. (2.7) has simplicity, and probably hides a *profound implication* in the construction of the SM from superstring theory.

The SM gauge group is $SU(3) \times SU(2) \times U(1)$ where the color group $SU(3)$ is unbroken and the electroweak group $SU(2)_L \times U(1)_Y$ is spontaneously broken. Non-Abelian gauge theories are known to be asymptotically free if the matter content is not too large [9, 10]. QCD is asymptotically free if the number of quarks is less than 16.5 in the one loop estimation. So, at high energy, the QCD coupling $\alpha_c$ becomes small and QCD is perturbatively calculable. On the other hand, at low energy of $E \leq$ 1 GeV, QCD becomes strong and all complications due to its strong nature occur, such as the chiral symmetry breaking and confinement. A complete understanding of non-perturbative nature is needed to know the effects of QCD at low energy. One such non-perturbative phenomenon is given by the instanton solution of non-Abelian gauge theories [19]. The instanton solution recommends the use of the so-called $\theta$-vacuum [20, 21]. In the $\theta$-vacuum, one must consider the $\mathscr{C}\mathscr{P}$ violating interaction, $(g_c^2 \bar{\theta}/16\pi^2)\mathrm{Tr}G\tilde{G}$, where $G$ is the gluon field strength and $\tilde{G}$ is its dual. However, we know that $\mathscr{C}\mathscr{P}$ violations have been observed only in weak interaction phenomena, not in strong interactions. This restricts $\bar{\theta}$ phenomenologically, from the observed upper bound of the neutron electric dipole moment, $\bar{\theta} < 10^{-10}$. So, "Why is $\bar{\theta}$ so small?" becomes a good theoretical problem in the SM: *the strong $\mathscr{C}\mathscr{P}$ problem* [22, 23]. At the level of the SM with (2.7), this strong $\mathscr{C}\mathscr{P}$ problem is not understood.

In this preceding concise review of the SM, we have already exposed three fundamental problems of the SM,

- Coupling unification problem: Why are there three different gauge couplings, or three factors of SU(3)×SU(2)×U(1) in the SM?
- Family problem: Why are there three families?
- Strong $\mathscr{C}\mathscr{P}$ problem: Why is $\bar{\theta}$ so small?

This book is devoted to some theoretical developments regarding the coupling unification problem and the family problem from the currently popular fundamental theory, superstring. The strong $\mathscr{C}\mathscr{P}$ problem can probably be understood from the introduction of a very light axion, also from superstring, however this is outside the focus of this book.

The number of fields in one family appearing in Eq. (2.2) is 15, and since we have represented them as L- and R-handed fields, these are 15 chiral fields. In Lagrangian formulation, one can use a field $\psi$ or its charge conjugated field $\psi^c$. For L- and R-handed quantum fields, $\psi_L = \frac{1+\gamma_5}{2}\psi$ and $\psi_R = \frac{1-\gamma_5}{2}\psi$, the charge conjugated fields have the opposite chiralities, for example, $(\psi_R)^c = (\psi^c)_L$, which can easily be checked from the definition of the charge conjugation operation on the Weyl fields $\psi_{L,R}$. So rather than using Eq. (2.2), let us use only one chirality of the L-handed fields,

$$l_L \equiv \begin{pmatrix} v_e \\ e \end{pmatrix}_L, \; e_L^c, \; q_L \equiv \begin{pmatrix} u \\ d \end{pmatrix}_L, \; u_L^c, \; d_L^c, \qquad (2.8)$$

where color multiplicity must be understood for the quarks, and $c$ refers to the charge conjugated field, i.e. $Y(e^c) = 1, Y(u^c) = -\frac{2}{3}, Y(d^c) = \frac{1}{3}$. This is a simple rewriting of the 15 chiral fields of Eq. (2.2). Note that it is simpler to view all fermions on the same footing, e.g. as left-handed (L), as we have done here. For spin-0 scalars, we do not have such a distinction in terms of chirality.

Chirality seems to be the essential property constructing our universe where the SM fields live long in a vast space. Understanding the chirality of the SM fermions is the overriding theme of this book.

In the SM spectra, Eq. (2.7), there is a spin-0 boson doublet $\mathbf{H}_1$. Spontaneous symmetry breaking by $\langle \mathbf{H}_1 \rangle \neq 0$ renders all the particles of the SM masses, and the final comer in the serial search of new particles was the Higgs boson $h$ at 125 GeV appeared at the LHC of CERN in 2012. Parametrizing the Higgs potential in terms of $\lambda$ and $\mu^2$ as

$$V(\mathbf{H}_1) = \frac{\lambda}{4}(\mathbf{H}_1^\dagger \mathbf{H}_1)^2 + \frac{\mu^2}{2}\mathbf{H}_1^\dagger \mathbf{H}_1 + \text{constant}, \qquad (2.9)$$

$$v_{\text{ew}} = \sqrt{\frac{2|\mu^2|}{\lambda}}, \qquad (2.10)$$

where the electroweak scale VEV, $\langle \mathbf{H}_1 \rangle = (v_{\text{ew}}/\sqrt{2}, 0)^T$, is given for $\mu^2 < 0$. From the strength of the Fermi constant $G_F$, $v_{\text{ew}}$ is numerically 246 GeV. Since the mass of the Higgs boson is $\sqrt{2}|\mu| = \sqrt{\lambda}\, v_{\text{ew}}$ in our parametrization, 125 GeV mass of the Higgs boson implies $\lambda \simeq \frac{1}{4}$ at the electroweak scale. $\lambda = \frac{1}{4}$ is large, which helps toward a positive $\beta$ function. Due to the large top Yukawa coupling, $\lambda$ decreases as $E$ increases. So, if $\lambda$ is smaller than a certain value, the positive contribution to the $\beta$ function from $\lambda$ is not enough and $\lambda$ turns to negative above certain energy scale. Above this energy scale, vacuum decays, which led to the problem on vacuum stability. So, the SM gives a lower limit for the Higgs boson mass from the vacuum stability. In Fig. 2.3, we show the evolution of $\lambda$ for a few parameters of the SM [24].

For the flavor phenomenology in the SM, we may try to understand the CKM and the PMNS matrices from string compactification. For the parametrization of these matrices, it is advantageous if the determinants of these unitary matrices are real [25] in which case the CP violation barometer Jarlskog determinant [26] is

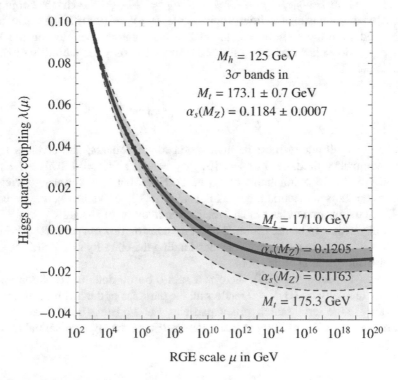

**Fig. 2.3** Evolution of $\lambda$ [24]

$J = |\text{Im } V_{13} V_{22} V_{31}|$. For a real determinant, $V_{\text{CKM}}$ is close to

$$V = \begin{pmatrix} 0.974395 & 0.22481 & 1.41 \times 10^{-3} \\ +8.6794 \times 10^{-5}i\text{ '} & +5.66 \times 10^{-6}i\text{ '} & -3.33 \times 10^{-3}i\text{ '} \\ -0.224672 & 0.97352 & 4.23 \times 10^{-2} \\ -1.416 \times 10^{-4}i\text{ '} & -7.46 \times 10^{-5}i\text{ '} & +5.32 \times 10^{-6}i \\ 8.132 \times 10^{-3} & -4.151 \times 10^{-2} & 0.99910 \\ -3.24 \times 10^{-3}i\text{ '} & -7.42 \times 10^{-4}i\text{ '} & -4.502 \times 10^{-5}i \end{pmatrix} \tag{2.11}$$

from which we obtain $J = 3.16 \times 10^{-5}$ which can be compared to the value $J = 3.18 \times 10^{-5}$ given in [16]. For the PMNS matrix, data are not accurate enough to present with a sufficient accuracy. But we note that the CP phase in the PMNS matrix got a lot of attention. It will be a reasonable success if the matrix (2.11) is obtained from string compactification.

In Fig. 2.4, we present all the particles of the SM at the electroweak scale.

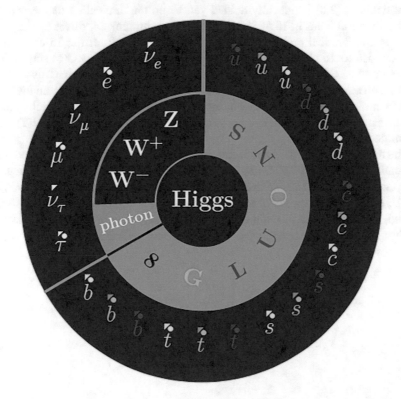

**Fig. 2.4** Particles in the SM where those in the gray disk are massless and all the other particles obtain mass by the Higgs mechanism. Spins are 0, 1, and $\frac{1}{2}$, respectively, in the annuli from the center. Helicities are denoted by triangle for L-handed and bullet for R-handed. Neutrinos have only the L-handed helicity. There are 45 chiral fermions

## 2.2     Grand Unified Theories

The history of particle physics, quantum field theory, and quantum mechanics has been the extension of the realm of symmetry operation in Hilbert space. In the previous section, we have explored this extension of symmetry. Considering only the first quark family members $u$ and $d$, Heisenberg's isospin symmetry is the unification of individual $u$ and $d$ (or proton and neutron) in one doublet representation of the rotation operation in an extended (internal) symmetry space, i.e. the isospin space. In the previous section, we further extended the isospin space to (isospin)×(color). This extension was applied to matter fields which can be the bases of the Hilbert space. Thus, we put forward the following in this discussion,

**The Theme of Unification** "Put all the matter fields on equal footing in an extended space."

Under this unification theme, the standard model SU(3)×SU(2)×U(1) with 15 chiral fields in (2.8) is not fully unified yet. From the unification theme, we may put the 15 chiral fields in **15** of SU(15). However with rank 14, this is too large and also has a theoretical problem of having a gauge anomaly. If the 15 chiral fields are extended to 16 chiral fields, by adding one SU(3)×SU(2)×U(1) singlet field, one can find an SO(10) gauge theory with a spinor representation.[6] This perfectly agrees with our unification theme, and also does not have the gauge anomaly. In this SO(10) unified theory, there is only one gauge coupling since all the chiral fields are put in a single representation. In other words, the gauge groups SU(3)×SU(2)×U(1) are unified in SO(10).

The unified theories of electromagnetic, weak, and strong forces are called grand unification theories (GUTs). Georgi and Glashow looked for the minimal rank gauge group which unifies SU(3)×SU(2)×U(1) in one simple group, unifying SU(3) and SU(2), but allows the possibility that there may be more than one representation. Indeed, they succeeded in finding the minimal unification group SU(5) where the 15 chiral fields of (2.8) are grouped together but still split into two [28],

$$
\overline{\mathbf{5}}_F = \begin{pmatrix} d^c \\ d^c \\ d^c \\ \nu_e \\ -e \end{pmatrix}_L, \quad \mathbf{10}_F = \begin{pmatrix} 0 & u^c & -u^c & u & d \\ -u^c & 0 & u^c & u & d \\ u^c & -u^c & 0 & u & d \\ -u & -u & -u & 0 & e^c \\ -d & -d & -d & -e^c & 0 \end{pmatrix}_L. \tag{2.12}
$$

[6]H. Georgi found this SO(10) model several hours before the SU(5) [27].

The Higgs doublet $\mathbf{H}_1$ in (2.7), which is actually $\mathbf{H}_2 \sim i\sigma_2 \mathbf{H}_1^*$, is unified into $\mathbf{5}_H$ with a new color triplet of scalars, $\mathbf{h}$,

$$\mathbf{5}_H = \begin{pmatrix} \mathbf{h}^1 \\ \mathbf{h}^2 \\ \mathbf{h}^3 \\ \mathbf{H}^+ \\ \mathbf{H}^0 \end{pmatrix}. \tag{2.13}$$

The rank of SU(5) is 4 which is the same as that of SU(3)×SU(2)×U(1). Thus, to break SU(5) down to SU(3)×SU(2)×U(1), one needs a Higgs field which can have a VEV in the center of SU(5), in order not to reduce the rank. The simplest such choice is an adjoint Higgs field $\mathbf{24}_H$. The next simple choice is $\mathbf{75}_H$. Let us introduce $\mathbf{24}_H$ only for the GUT symmetry breaking, and let its VEV be $V_U$. Then, below the scale $V_U$, the effective fields are only those in the SM and the colored scalar $h$. The heavy fields we introduce in this minimal GUT at the scale $V_U$ are the so-called $X$ and $Y$ gauge bosons each with mass $M_X = M_Y = \frac{1}{2} g_5 V_U$, where $g_5$ is the unification coupling constant, and twelve real Higgs fields originally introduced in $\mathbf{24}_H$. From (2.12) one can see that $X$ and $Y$ gauge bosons couple to a lepton and a quark, and hence are called "lepto–quark" gauge bosons, and also to two quarks $uu$ or $ud$, rendering them "di-quark" gauge bosons. Therefore, the lepto–quark gauge bosons mediate proton decay, and their masses should be extremely heavy, $M_{X,Y} \geq 10^{15}$ GeV.

The prototype SU(5) GUT model classifies the 15 chiral fields neatly in $\mathbf{10}_F \oplus \bar{\mathbf{5}}_F$. But a more important implication is that the electromagnetic charge is quantized, $3 Q_{\text{em}}(u) + 3 Q_{\text{em}}(d) + Q_{\text{em}}(e) = 0$, i.e. $Q_{\text{em}}(p) = -Q_{\text{em}}(e)$. This is possible because quarks and leptons are put in the same representation.

Another equally important implication is that the gauge coupling is unified at the GUT scale $M_{X,Y}$. The running of gauge coupling constants below the GUT scale is [29],

$$\frac{8\pi^2}{g_i^2(M_Z)} = \frac{8\pi^2}{g_U^2(M_U)} + b_i \ln \frac{M_U}{M_Z}, \tag{2.14}$$

where we can assume $M_U = M_X$ and $g_U$ is the SU(5) coupling at the unification point $M_U$, and $b_i$ are the coefficients of $\beta$ functions of the gauge groups, SU(3), SU(2)$_L$ and U(1)$_Y$,

$$\begin{aligned} b_1 &= +\tfrac{3}{5} \operatorname{Tr} Y^2 \delta_{FB}, \\ b_2 &= -\tfrac{11}{3} \cdot 2 + \tfrac{2}{3} \sum_j l(R_j(SU(2)_L)) \delta_{FB}, \\ b_3 &= -\tfrac{11}{3} \cdot 3 + \tfrac{2}{3} \sum_i l(R_i(SU(3))) \delta_{FB}, \end{aligned} \tag{2.15}$$

where

$$\mathrm{Tr}\, T_{\mathbf{R}}^a T_{\mathbf{R}}^b = l(\mathbf{R})\delta_{ab}, \tag{2.16}$$

is the index of representation $R$. For example, $l$(fund. representation) $=$ $\frac{1}{2}$, $l$(adj. representation) $= N$ for SU($N$) groups. In these formulae, $l$ is normalized for a chiral fermion; thus $\delta_{FB} = 1$ for L and R fermions and summed over L and R. For complex bosons $\delta_{FB} = \frac{1}{2}$, and for real bosons $\delta_{FB} = \frac{1}{4}$. Also, every generator, including the U(1) generator, is normalized for (2.16) to be $\frac{1}{2}$ for the fundamental representation. If the electroweak hypercharge is embedded in a GUT group, the hypercharge is properly normalized. The proportionality of the electroweak hypercharge generator $Y$ to the normalized U(1) generator $T_{24}$ in SU(5) is parametrized by the normalization constant $C$,

$$Y = CT_{24}, \quad \text{with} \ C^2 = \frac{5}{3}, \tag{2.17}$$

where

$$T_{24} = \sqrt{\frac{3}{5}} \begin{pmatrix} -\frac{1}{3} & 0 & 0 & 0 & 0 \\ 0 & -\frac{1}{3} & 0 & 0 & 0 \\ 0 & 0 & -\frac{1}{3} & 0 & 0 \\ 0 & 0 & 0 & \frac{1}{2} & 0 \\ 0 & 0 & 0 & 0 & \frac{1}{2} \end{pmatrix} = \sqrt{\frac{3}{5}} Y. \tag{2.18}$$

Since the gauge interaction gives $g'Y = g_1 T_{24}$, we note that $g_1 = Cg'$. We can calculate the weak mixing angle (2.4)

$$\sin^2 \theta_W^0 = \frac{g'^2}{g_2^2 + g'^2} = \frac{1}{1 + C^2} \tag{2.19}$$

with $g_1 = g_2 = g_3 = g_5 = g_U$, by running the bare values down to the electroweak scale, as we will see in the following section. Since $Q_{\mathrm{em}} = T_3 + Y$, and using the fact that the index (2.16) is independent of $a$, we arrive at the handy result [29],

$$\sin^2 \theta_W^0 = \frac{\mathrm{Tr}\, T_3^2}{\mathrm{Tr}\, Q_{\mathrm{em}}^2}. \tag{2.20}$$

Note that we have assumed that all the SM fields belong to *complete* multiplets over which the trace is taken. If some fields fail to form a multiplet, the relation is no longer valid. As an exercise, we note the following for $N_g$ families of fermions of (2.8),

$$b_3 = -11 + N_g \frac{2}{3} \cdot \frac{1}{2}(1_{u_L} + 1_{d_L} + 1_{u_L^c} + 1_{d_L^c}) = -11 + \frac{4}{3}N_g$$

$$b_2 = -\frac{22}{3} + N_g \frac{2}{3} \cdot \frac{1}{2}(3_{(u,d)_L} + 1_{(v_e,e)_L}) = -\frac{22}{3} + \frac{4}{3}N_g$$

$$b_1 = N_g \frac{3}{5} \cdot \frac{2}{3}\left\{2\left(-\frac{1}{2}\right)^2_{(v_e,e)_L} + 1_{e_L^c} + 2 \cdot 3\left(\frac{1}{6}\right)^2_{(u,d)_L} + 3\left(-\frac{2}{3}\right)^2_{u_L^c}\right.$$

$$\left. +3\left(\frac{1}{3}\right)^2_{d_L^c}\right\} = \frac{4}{3}N_g,$$

where in the subscript of each number we show the representation contributing to that number. Except for the gauge bosons, a complete multiplet (2.8) gives the same contribution to $b_i$, i.e. $\frac{4}{3}N_g$. Thus, the necessary differentiation of the weak SU(2) coupling and the QCD coupling is implementable in the GUT from the gauge boson contributions [29]. Which GUT actually succeeds in fitting the electroweak data is the key issue in building the GUT model.

In Fig. 2.5, we present a schematic behavior of the running of gauge couplings from the GUT scale $M_U$ down to the electroweak scale $M_Z$. For a sufficient separation of $\alpha_3$ and $\alpha_2$ at the electroweak scale, we need a large logarithm, or

$$\frac{M_U}{M_Z} \geq 10^{12}. \tag{2.21}$$

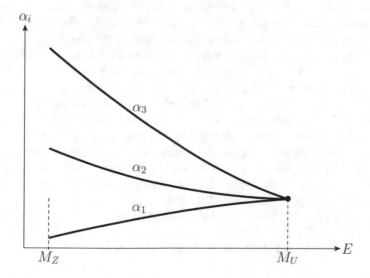

**Fig. 2.5** Running of gauge couplings

**Fig. 2.6** Dimensional
transmutation

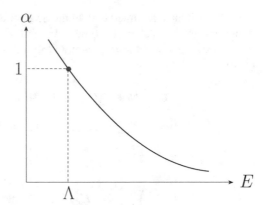

A good theory must explain this huge number, which is the *gauge hierarchy problem* [30].

The *gauge hierarchy problem* consists of two parts: (1) "How are two different scales implied by (2.21) introduced?" and (2) "Given those two different scales, are quantum corrections safe to guarantee the hierarchy of (2.21)?" This book is an exploration of bigger theories originally motivated to answer the gauge hierarchy problem, and hence we will come back to this gauge hierarchy question again and again. Supersymmetry was used to answer the second problem, forbidding the quadratic divergence of the Higgs mass [31]. For the first problem, recently a solution has been found with a confining $SU(5)'$ in the hidden sector [32].

An exponential function of a reasonably large number can introduce a huge number. So, the best way to understand the gauge hierarchy problem is to introduce an exponential function. Dimensional transmutation of dimensionless gauge couplings $\alpha$ can introduce an exponentially small mass. When a dimensionless coupling $\alpha$ is given, the renormalization group evolution of $\alpha$ finds a mass scale, where $\alpha$ becomes order 1 as shown in Fig. 2.6, whose scale is defined as the coupling $\Lambda$, having the mass dimension. It is used in asymptotically free theories, i.e. for non-Abelian gauge theories. In QCD, $\Lambda_{QCD}$ is determined around 300 MeV. In this way, the techni-color idea was introduced to have the electroweak scale around 3 TeV, exponentially smaller than the GUT scale [11, 33]. The techni-color idea, extended to extended-techni-color [34] to have a flavor hierarchy, failed miserably and the SUSY idea has been tried several decades since 1981 [31] for the hierarchy solution. But, the SUSY solution so far has been satisfied merely with the stable hierarchy, i.e. once the hierarchical parameters are introduced then the solution is stable. Recently, a model for dynamical breaking of SUSY has been found which can lead to a gigantic step toward the gauge hierarchy solution [32].

Note that there is one merit to having the large hierarchy (2.21). Since there exist the super heavy $X$ and $Y$ gauge bosons, the proton decay rate is proportional to $M_X^{-4}$, giving the proton lifetime for the dominant mode as [35, 36]

$$\tau(p \rightarrow e^+ + \pi^0) = O(10^3 - 10^4) \left( \frac{M_X}{m_p} \right)^4 \frac{1}{m_p}, \qquad (2.22)$$

where $m_p$ is the proton mass.[7] Thus in fact, a large hierarchy is needed from the requirements of proton longevity and differentiation of strong and electroweak couplings. The current experimental bound on the above partial lifetime is $1.6 \times 10^{33}$ years [37].

We note that there also exists the colored scalar **h** in $5_H$, in Eq. (2.13). This colored scalar **h** triggers proton decay through Yukawa couplings $10_F \overline{5}_F \overline{5}_H$ and $10_F 10_F 5_H$. Even though the Yukawa couplings of the first family is $O(10^{-6})$, mixing with heavier families introduces larger Yukawa couplings for proton decay through **h** exchanges. Thus, **h** in $5_H$ and $\overline{h}$ in $\overline{5}_H$ must be removed at the GUT scale. Below the GUT scale, there must survive three families of fermions and a Higgs doublet ($\mathbf{H}_1$ and/or $\mathbf{H}_2$).[8]

The unification theme of our discussion does not stop at the minimal GUT, SU(5). As noted before, SO(10) is much nicer than SU(5) from the standpoint of the unification theme. However, SU(5) is favorable for two reasons. First of all, SO(10) requires several intermediate steps in the process of breaking it down to $SU(2)_L \times U(1)_Y$, and hence it is more complicated. These intermediate steps are

$$
SO(10) \longrightarrow
\begin{cases}
SU(5)_{GG} \\
SU(5)' \times U(1) \ [\text{flipped SU(5), or anti} - \text{SU(5)}] \\
SU(4) \times SU(2)_L \times SU(2)_R \ [\text{Pati} - \text{Salam}] \\
SU(4) \times SU(2) \times U(1) \\
SU(3) \times SU(2) \times SU(2) \times U(1)
\end{cases}
\tag{2.23}
$$

Secondly, even though SO(10) is superior to SU(5) in terms of the unification theme, it still cannot include three families in one spinor representation **16**. SU(5) is favorable because it is minimal. We need to unify families in the GUT scheme. This family unified GUT is an obvious generalization of GUT from the unification theme, "Find a bigger space where all matter fields can be put into a single representation of a GUT group." Georgi formulated the family unification with the hypothesis [38],

- *Survival hypothesis*: If a GUT group $G$ breaks down to a subgroup $G_{\text{sub}}$ at a scale $M_G$, all real representations of $G_{\text{sub}}$ are removed at the scale $M_G$. The surviving fermions are complex representations of $G_{\text{sub}}$.

Therefore, the search of chiral theories is reduced to the problem of finding groups which allow complex representations. Of course, the full complex fermionic representation(s) should not lead to gauge anomalies. In string theory, every physical parameter is calculable in principle and we may not need the survival hypothesis.

---

[7]Note that $\text{GeV}^{-1} \simeq 0.658 \times 10^{-24}$ in natural units.

[8]The bold faced notation for the color triplet and weak doublet implies that they are not representing one particle. Separating these is the doublet–triplet splitting problem.

So, because of the anomaly problem, an obvious possibility in 4D is to look for complex representations in groups other than $SU(N)$. There are only two classes which achieve this objective,

$$SO(4n+2) : \text{ with spinor representation } \mathbf{4}^n$$

$$E_6 \qquad : \text{ with the fundamental representation } \mathbf{27}. \qquad (2.24)$$

Again, the $E_6$ model [39] houses only one family. As noted before the **16** of SO(10) houses one family. Therefore, for the family unification we are left with SO(14), SO(18), etc., in the above category of (2.24). If SO(14) breaks down trivially to SO(10) at a scale $M_{14}$, then below $M_{14}$ the SO(14) spinor **64** reduces to $2(\mathbf{16} \oplus \overline{\mathbf{16}})$ of SO(10) which is real under SO(10) and hence, removed at the scale $M_{14}$ by the survival hypothesis. There remains no family. However, if SO(14) directly breaks down to $SU(3) \times SU(2) \times U(1)$ in a skewed way, there can survive light fermions [40], but they are not of the form given in (2.7). This is the reason why the study of $SU(N)$ groups is necessary. But for the anomaly cancellation, here we cannot build a model with one fermion representation. One interesting GUT model with three SU(5) families in which SU(N) representation appears only once is the SU(11) model [38]

$$SU(11): \ \psi^{\alpha\beta\gamma\delta} \oplus \psi_{\alpha\beta\gamma} \oplus \psi_{\alpha\beta} \oplus \psi_{\alpha},$$

where the SU(11) indices $\alpha$, $\beta$, etc., are anti-symmetrized. Generalizing the single appearance of irreducible representations to relative primes for the multiplicities of the representations, there can be numerous GUT models with three SU(5) families after the application of survival hypothesis [41, 42].

So far, we have only discussed simple gauge groups for GUT. But one may consider semi-simple groups for grand unified models. In fact, the Pati–Salam model $SU(4) \times SU(2)_L \times SU(2)_R$ [43] was considered before SU(5). But there is no rationale that gauge couplings of SU(4) and SU(2) are unified above some scale. An intermediate group can be between a real GUT at $M_U$ and the SM at $M_Z$. Even though this model does not achieve gauge coupling unification, it puts quarks and leptons in the same representation and therefore can be called a GUT. Namely in **4** of SU(4), a quark triplet and a lepton are put together and hence the lepton is called the fourth color. Therefore, the possibility of proton decay exists in principle, but in one version of the Pati–Salam model where $SU(4) \times SU(2)_L \times SU(2)_R$ is unified to $SU(4) \times SU(4)$, the rate is much more suppressed since the nontrivial operator for proton decay appears only at dimension nine operators, instead of dimension six operators shown above in SU(5) [44]. So the breaking scale of SU(4) can be much lower than that of the SU(5) GUT. Another interesting factor group for GUT is $SU(5) \times U(1)$ the so-called flipped SU(5) or anti-SU(5). This factor group is not semi-simple. Here also, the SU(5) coupling and U(1) coupling are not unified, but can be called a GUT since quarks and leptons are put in the same representation. It can be a subgroup of SO(10) [45].

If a semi-simple group is considered as a GUT, then a discrete symmetry must exist for the exchange of factor groups, thus guaranteeing the equality of gauge couplings. The most interesting GUT in this category is the so-called *trinification*, SU(3)×SU(3)×SU(3), with the representation

$$\mathbf{27}_{\text{tri}} = (\bar{\mathbf{3}}, \mathbf{3}, \mathbf{1}) \oplus (\mathbf{1}, \bar{\mathbf{3}}, \mathbf{3}) \oplus (\bar{\mathbf{3}}, \mathbf{1}, \mathbf{3}). \tag{2.25}$$

This model cannot separate out any SU(3) and hence there exists an exchange symmetry of gauge groups, and the three SU(3) gauge couplings are unified above the unification scale $M_U$. In the trinification, quarks and leptons are not put in the same representation, but still it is called a GUT because gauge couplings are unified. In many aspects, the trinification model is very similar to the $E_6$ model.[9]

In this book, we will discuss mainly the SU(5) model, the SO(10) model, the Pati–Salam model, the flipped SU(5) model, and the trinification model. We attempt to resolve family unification with a new method appearing in the compactification process.

## 2.3 Supersymmetry

### 2.3.1 Global Supersymmetry

Supersymmetry is probably the most spectacular symmetry ever since the discovery of isospin. It can be introduced in many different ways. In our search of an enlarged symmetry of matter spectrum, we note that the SM spectrum (2.7) includes Higgs scalars also. Surprisingly, the chiral (L-handed) lepton doublet $l$ and $\mathbf{H}_1$ have the same quantum numbers except for the spin. Even though it is not necessary, we may try to put a chiral fermion and a complex scalar in a bigger Hilbert space so that they belong to the same representation, in our pursuit of unification.[10] Due to the fact that the spectrum (2.7) contains spin-0 particles, we open up the possibility of unifying fermions and bosons. The new space is called *superspace* [47] and the supercharge $Q$ is the generator for the transformation of fermions to bosons and vice versa,

$$Q|\text{F}\rangle = |\text{B}\rangle, \quad Q|\text{B}\rangle = |\text{F}\rangle. \tag{2.26}$$

From the definition of supercharge in (2.26), we note that $Q$ must be spinorial because both $Q|\text{B}\rangle$ and $|\text{F}\rangle$ must transform like a spinor under rotation. The spinorial charge $Q$ has two components since our motivation for unification started so that L-handed leptons in (2.7) could be obtained from complex scalars by operating $Q$. So, $Q$ transforms like a L-handed Weyl fermion. Here follows the chirality which we

---

[9]For trinification references, see [46].

[10]Phenomenologically however, the Higgs fields are known to not be in the same representation as the chiral lepton doublet in the supersymmetrized version.

need desperately for the scalars, and if we succeed in this F and B unification, that alone can be a merit. The Lorentz algebra $SO(3,1)$ is the same as $SU(2) \times SU(2)$; thus the Lorentz group $SO(3,1)$ can be locally viewed as $SU(2) \times SU(2)$. The Weyl fermion transforms as $(\mathbf{2}, \mathbf{1})$ or as $(\mathbf{1}, \mathbf{2})$ under $SU(2) \times SU(2)$. Let $(\mathbf{2}, \mathbf{1})$ be the L-handed field and $(\mathbf{1}, \mathbf{2})$ be the R-handed field. The indices for $SU(2)_L$ are denoted as undotted $\alpha = \{1, 2\}$ and the indices for $SU(2)_R$ are named as dotted $\dot{\alpha} = \{1, 2\}$.[11] Since $Q$ transforms like a L-handed Weyl spinor, its commutation relation with the angular momentum generator is

$$[J^{\mu\nu}, Q_\alpha] = -i(\sigma^{\mu\nu})_\alpha^\beta Q_\beta. \tag{2.27}$$

For the R-handed Weyl fermions and their accompanying complex scalars, the supercharge must transform nontrivially under $SU(2)_R$. Let us call this supercharge $\overline{Q}$ which changes the dotted indices,

$$[J^{\mu\nu}, \overline{Q}_{\dot{\alpha}}] = -i(\overline{\sigma}^{\mu\nu})_{\dot{\alpha}}^{\dot{\beta}} \overline{Q}_{\dot{\beta}}. \tag{2.28}$$

In fact, $\overline{Q}$ is the charge conjugated of $Q$, since as we have seen in the previous section of this chapter, L changes to R and vice versa under charge conjugation. The total number of complex components in $Q$ and $\overline{Q}$ is four. Thus, we can form the following representation for the supercharge,

$$\mathscr{Q} = \begin{pmatrix} Q \\ \overline{Q} \end{pmatrix}. \tag{2.29}$$

There exists the famous no-go theorem by Coleman and Mandula [49] which says that the fermion–boson symmetry and the spacetime symmetry cannot be unified in a naïve way, just by making the internal space bigger, i.e. making the Lie group bigger. Supersymmetry escapes this Coleman–Mandula theorem by not making just the Lie group bigger, but by introducing an algebra outside the Lie algebra. It is called the graded Lie algebra where one adds anti-commutators for some generators. For supercharges $Q$ and $\overline{Q}$, one introduces the following anti-commutators:

$$\{Q_\alpha, \overline{Q}_{\dot{\beta}}\} = 2(\sigma^\mu)_{\alpha\dot{\beta}} P_\mu \tag{2.30}$$

$$\{Q_\alpha, Q_\beta\} = 0, \quad \{\overline{Q}_{\dot{\alpha}}, \overline{Q}_{\dot{\beta}}\} = 0. \tag{2.31}$$

With the space translation generators $P_\mu$, one introduces commutators as in Eq. (2.28),

$$[P_\mu, Q_\alpha] = 0, \quad [P_\mu, \overline{Q}_{\dot{\alpha}}] = 0, \quad [P_\mu, P_\nu] = 0. \tag{2.32}$$

---

[11] See, for example [48].

Indeed, this expansion of symmetry by introducing anti-commuting supercharges is known to be possible and the fermion–boson symmetry introduced in this way is *supersymmetry*. The parameter for the supersymmetry transformation, $\epsilon$, must transform like a spinor under rotation so that $\overline{Q}\epsilon$ transforms like a scalar. Supersymmetry (SUSY) was introduced in 1971 [50], but the linear realization (2.26) of supersymmetry, what we use today, is due to the work of Wess and Zumino [51].

The time component of the algebra (2.30) gives the Hamiltonian in terms of supercharges

$$H = P^0 = \frac{1}{4}(Q_1\overline{Q}_1 + \overline{Q}_1 Q_1 + Q_2\overline{Q}_2 + \overline{Q}_2 Q_2) \tag{2.33}$$

which implies that the energy eigenvalues are nonnegative. If supersymmetry is unbroken, the supercharges annihilate the vacuum $|0\rangle$, and the vacuum energy is zero,

$$\text{Global SUSY}: \quad E_{\text{vac}} \geq 0, \quad \text{equality for unbroken SUSY}. \tag{2.34}$$

The supersymmetry we have discussed above is $\mathcal{N} = 1$ supersymmetry. If we introduce more supercharges, they define extended supersymmetry, $\mathcal{N} = 1, 2, \cdots$. For example, for $\mathcal{N}=2$ the total $S_z$ interval between $(S_z)_{\text{max}}$ and $(S_z)_{\text{min}}$ is $2 \cdot \frac{1}{2} = 1$ because one application of a supercharge changes spin by a half unit. Not to include spin greater than 2, thus we obtain $\mathcal{N}=8$ as the maximum extended supersymmetry. In the remainder of this section, *only $\mathcal{N} = 1$ supersymmetry is considered, where the introduction of chirality is possible.*[12]

Although we do not use superfield formalism [47] here, we list its powerful constraints on the form of Lagrangian. Introducing an anti-commuting coordinate $\theta^\alpha$ ($\alpha = 1, 2$) and $\bar{\theta}_{\dot{\alpha}}$ ($\dot{\alpha} = 1, 2$), one can also introduce a quantum field as a function of $x, \theta$, and $\bar{\theta}$. A polynomial of $\theta$ includes only three terms, $1, \theta$, and $\theta^2$. A similar polynomial results from $\bar{\theta}$. In view of the $SU(2)_L \times SU(2)_R$ property of the Lorentz group (viz. Eqs. (2.27), (2.28)), $\theta(\bar{\theta})$ is a doublet under $SU(2)_L(SU(2)_R)$. For fermions, we introduce an L-handed chiral field $\psi$, which is a singlet under $SU(2)_R$. This $\psi$ can make a singlet of $SU(2)_L$ (viz. $\mathbf{2} \times \mathbf{2} = \mathbf{1} + \mathbf{3}$), by taking the antisymmetric combination with $\theta$, $\theta\psi \equiv \epsilon_{\alpha\beta}\theta^\alpha \psi^\beta$ where $\epsilon_{\alpha\beta}$ is the Levi-Civita tensor of $SU(2)_L$. Thus, $SU(2)_L$ singlets can be a scalar function of $\phi(x)$ and spin$\frac{1}{2} \cdot$ spin $-\frac{1}{2}$, which is denoted as $\Phi$, a function of the forms $\theta\psi(x)$ and $F(x)\theta\theta$.

---

[12]For example, for $\mathcal{N}=2$ matter fermions are located at $S_z = \frac{1}{2}$ and $S_z = -\frac{1}{2}$ with the same gauge quantum number. These $S_z = \pm\frac{1}{2}$ representations form a vector-like representation under the gauge group.

A (L-handed) chiral superfield $\Phi(x, \theta)$ is defined as a function of $\theta$ only; thus a *chiral superfield* has expansion[13]

$$\Phi(x, \theta) = \phi(x) + \theta^\alpha \psi_\alpha + \theta^\alpha \theta^\beta \epsilon_{\alpha\beta} F^2. \tag{2.35}$$

The chiral superfield $\Phi$ contains a spin-0 boson and a spin-$\frac{1}{2}$ fermion. A superfield containing both $\theta$ and $\bar{\theta}$ has more degrees. Here, it is possible to introduce a spin-1 boson in the superfield $V$, with the reality condition $V = V^\dagger$. $V(x, \theta, \bar{\theta})$ is called a vector superfield. This superfield formalism is quite useful in analyzing the form of allowed actions. As a result, first the globally supersymmetric Lagrangian is known to depend only on three functions $K$, $W$, and $f$:

1. Kähler potential $K(\Phi, \Phi^*)$, a Hermitian function, determines mainly the kinetic terms.
2. Superpotential $W(\Phi)$, a holomorphic function, determines the potential $V(\Phi)$.
3. Gauge kinetic function $f_{ab}(\Phi)$, a holomorphic function, is the coefficient of gauge kinetic terms.

In particular, the scalar potential is always positive definite

$$V(\phi, \phi^*) = \sum_i \left| \frac{\partial W(\phi)}{\partial \phi_i} \right|^2 + \frac{1}{2} f_{ab} D^a D^b, \tag{2.36}$$

where we replaced the superfield $\Phi$ with its scalar field $\phi$ and $D^a = G^i T_i^{aj} \phi_j$ is the D-term. This is another expression of Eq. (2.34). Second, there is the following *non-renormalization theorem*. Because of the holomorphicity, the superpotential $W(\Phi)$ does not receive loop corrections other than those of wave function renormalization, at all orders of perturbation theory. Also, the gauge kinetic function $f_{ab}(\Phi)$ does not receive higher order correction beyond one loop order.[14] These are quite restrictive compared to non-supersymmetric models.

### 2.3.2  Local Supersymmetry, or Supergravity

In this subsection, we introduce some formulae which are needed in later chapters. Thus far, we have considered the *global supersymmetry* where the parameter $\epsilon$ for supersymmetry transformation is independent of $x_\mu$. If supersymmetry is the symmetry of the action, it must be severely broken since the superpartner of the electron has not been discovered up to 1 TeV [16]. If the global supersymmetry is broken at the scale $M_S$, the vacuum energy must be of the order $M_S^4$, in view of

---

[13]Similarly, from the consideration of SU(2)$_R$ an R-handed chiral field $\Phi^\dagger(\bar{z})$ can be given.

[14]The gauge kinetic function includes the anomaly term which is complete by one-loop order [52].

(2.34), and the cosmological constant problem phenomenologically excludes global supersymmetry from square one. There is no escape from this problem in the global SUSY case.

This leads us to the necessary introduction of gravity through the localization of the supersymmetry parameter, $\epsilon(x)$. The resulting theory is *supergravity*. If the vacuum energy problem is not resolved in supergravity as well, then this would be just an academic exercise. Fortunately, supergravity allows the possibility of introducing a zero cosmological constant after supersymmetry breaking [53].

In the supergravity Lagrangian, $K$ and $W$ appear with a single function $G$

$$G(\phi, \phi^*) = -3 \log(-K/3) + \log |W|^2 \tag{2.37}$$

where we set the Planck mass $M_P = 1$. The oddly looking coefficients are for convenient calculations. The function (2.37) has a symmetry

$$3 \log(-K/3) \rightarrow 3 \log(-K/3) + h(\phi) + h^*(\phi^*),$$
$$W \rightarrow e^{-h} W. \tag{2.38}$$

Covariant and contravariant indices are used for holomorphic and antiholomorphic scalars, respectively. The Kähler metric of the (sigma model) target space is defined as

$$G^i = \frac{\partial G}{\partial \phi_i}, \quad G_i = \frac{\partial G}{\partial \phi^{*i}}, \quad G^i_j = \frac{\partial^2 G}{\partial \phi_i \partial \phi^{*j}}. \tag{2.39}$$

Then, the bosonic Lagrangian is given by

$$e^{-1}\mathscr{L} = \frac{\mathscr{L}}{\sqrt{g}} = -\frac{1}{2}R + G^i_{\ j}D_\mu \phi_i D^\mu \phi^{*j} + \frac{1}{4}\text{Re}(f_{ab})F^a_{\mu\nu}F^{b\mu\nu}$$
$$+ \frac{1}{8}\text{Im}(f_{ab})\epsilon^{\mu\nu\rho\sigma}F^a_{\mu\nu}F^b_{\rho\sigma} + V(\phi, \phi^*) \tag{2.40}$$

with the following scalar potential [54,55]

$$V(\phi, \phi^*) = e^K \left[ (G^{-1})_j^{\ i} D_i W^* D^j W - 3|W|^2 \right] + \frac{1}{2}f_{ab}D^a D^b, \tag{2.41}$$

where

$$D^i W = \frac{\partial W}{\partial \phi_i} + \frac{\partial K}{\partial \phi_i}W, \quad D_i W^* = \frac{\partial W^*}{\partial \phi^{*i}} + \frac{\partial K}{\partial \phi^{*i}}W^*, \tag{2.42}$$

and on-shell D-term

$$D^a = G^i T_i^{aj} \phi_j. \tag{2.43}$$

In contrast to the global supersymmetry case (2.36), here the superpotential is not positive definite.

The supersymmetry breaking condition can be read from the transformation laws of fermionic fields. For supersymmetry breaking, only their scalar component(s) can assume VEV that does not violate the Lorentz symmetry,

$$\delta_\epsilon \Psi \sim -e^{G/2}\frac{1}{W^*}D_i W^*\epsilon - \frac{1}{8}\frac{\partial f_{ab}}{\partial \phi^{*j}}\lambda_a\lambda_b\epsilon \tag{2.44}$$

$$\delta_\epsilon \lambda \sim \frac{i}{2}g \, \mathrm{Re} \, f_{ab}^{-1} G^i (T_b)_i^j \phi_j \epsilon. \tag{2.45}$$

The right-hand side of (2.44) is called the F-term of $\Psi$. Supersymmetry can be broken if the F-term is nonvanishing, either by (1) the first term assuming VEV or (2) the second term through the gaugino ($\lambda_a$) condensation by some strong force. The supersymmetry breaking scale for Case (1) is

$$M_S^2 = e^{G/2}\frac{1}{W^*}D_i W^*, \tag{2.46}$$

and a similar expression holds for Case (2).

Supergravity formulated with an AdS curvature $\overline{\Lambda} < 0$ has a negative vacuum energy $\overline{\Lambda}$. With a broken SUSY, a positive constant, (viz. ((2.34), (2.41))), is added to $\overline{\Lambda}$, making it possible for the vacuum energy in the broken phase to be made zero, $V_0 = 0$. For example, if the SUSY is broken by the nonzero $F$ term only, then the flat space condition requires, $G_{0k}G_0^k = 3$. In this flat limit, the gravitino mass is

$$m_{3/2} = M_P e^{G_0/2}, \tag{2.47}$$

where $M_P$ is the Planck mass, $2.44 \times 10^{18}$ GeV. Thus, supergravity saves us from the disaster of a huge cosmological constant. But, it must be remembered that *we achieved the flat space by a fine-tuning*, since the initial curvature $\overline{\Lambda}$, with which we formulated the theory, is an arbitrary number.

Now let us introduce the $\mathcal{N} = 1$ supergravity for the matter content (2.7). The R-handed fields are understood to be the charge conjugated L-handed fields as in (2.8). The Yukawa couplings are contained in the cubic terms of the superpotential $W$. Since all fields in (2.7) have superpartners, we do not bother to distinguish between the complex scalars and L-handed fermions. In the superpotential $W$, the fields denote the first component, i.e. scalar component. The scalar partners of fermions in (2.7) are called sfermions, e.g. squark, selectron, and sneutrino. On the other hand, the fermionic partners of (2.7) scalars carry the suffix "ino," e.g. Higgsinos. Supersymmetrization of Higgs bosons require the Higgsinos to be L-handed so that they can couple to other L-handed fields. We said before that scalars do not have a left- or right-handed distinction. But, here we change this statement. Scalars in supersymmetric theory have chiralities which are determined by the chirality of their fermionic partner. If we consider gauge bosons, their fermionic partners

$(s = \frac{1}{2})$ are called gauginos, e.g. gluino, wino, zino, photino, and bino. Here, bino means the partner of the U(1)$_Y$ gauge boson $B_\mu$. To give mass to an electron, we can consider a superpotential, $l_L e_L^c \mathbf{H}_1$. Down-type quarks can obtain mass by the term, $q_L d_L^c \mathbf{H}_1$, however, up-type quarks cannot obtain mass by $\mathbf{H}_1$. They need the $Y = +\frac{1}{2}$ Higgsino doublet $\mathbf{H}_2$. Unlike the case without supersymmetry, $i\sigma_2 \mathbf{H}_1^*$ cannot serve for the up-type quark mass since the charge conjugated field is R-handed and $W$ does not allow couplings of both chiralities. Therefore, we need another L-handed Higgs doublet $\mathbf{H}_2$ for the up-type quark masses, $q_L u_L^c \mathbf{H}_2$. It is also needed to cancel the gauge anomaly since we considered the complex fermion through supersymmetrization of $\mathbf{H}_1$. The addition of $\mathbf{H}_2$ makes the representation $\mathbf{H}_1 \oplus \mathbf{H}_2$ real under SU(3)×SU(2)×U(1). The $\mathcal{N} = 1$ supersymmetric gauge model with the spectrum given in (2.7) and with the addition of $\mathbf{H}_2$ is the minimal supersymmetric standard model (MSSM). Namely, the Higgs sector becomes a bit bigger

$$\text{MSSM} \ni \mathbf{H}_1 \oplus \mathbf{H}_2. \tag{2.48}$$

So far, we have taken the strategy of extending the symmetry and ultimately introducing supersymmetry. Supersymmetry helps in solving some important problems as well, such as the second gauge hierarchy problem. One way to phrase the gauge hierarchy problem is, "Why is the mass of the electroweak Higgs boson $\mathbf{H}_1$ so much smaller than the mass of the GUT Higgs boson $\mathbf{24}_H$?" If we start from a small ratio for these masses for $\mathbf{H}_1$ and $\mathbf{24}_H$, the radiative corrections should not destroy this smallness. With supersymmetry, indeed this stability problem is understood because the relevant Yukawa couplings and quartic couplings are related. Both kinds of couplings are related by the superpotential $W$. As depicted in Fig. 2.7, boson loops positively contribute and fermion loops negatively contribute to the Higgs boson mass so that the quadratic divergence is cancelled. Thus, the second gauge hierarchy problem is understood with a SUSY extension. This leads to the MSSM being a serious contender for interactions around the TeV scale. If the MSSM were to solve the gauge hierarchy problem, the divergence problem of the Higgs boson mass should only be forced a few TeV above the upper region of the electroweak scale. If the SUSY breaking scale $M_S$ is raised any more than a few TeV, there will be a wide range of energy scales, from $M_Z$ to $M_S$, where SUSY is not helpful for the stabilization of the Higgs boson mass. With $M_S > 100$ TeV, it is called the *little hierarchy problem*.

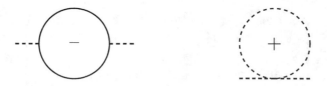

**Fig. 2.7** Fermion and boson loops cancel the quadratic divergences

In the MSSM, the superpartners of the three families must be raised to the SUSY scale $M_S$ since they have not yet been discovered. This is realized by making the supersymmetry spontaneously broken. If it is broken softly, then only soft SUSY breaking masses appear below the scale $M_S$. Soft masses are the non-supersymmetric scalar masses and the gaugino masses. These soft SUSY breaking parameters can be given if the SUSY breaking mechanism is known. One popular scenario is the gravity-mediated SUSY breaking where soft masses at the electroweak scale are given as functions of the gravitino mass, $m_{3/2}$. For a review, see [55].

### 2.3.3    SUSY GUT

This leads us to the obvious unification, the supersymmetrized GUT or SUSY GUT. Again, the simplest SUSY GUT is SUSY SU(5) [56]. The matter content $\mathbf{10}_F$ and $\overline{\mathbf{5}}_F$ are supersymmetrized. Also, two Higgs representations are needed, one housing $\mathbf{H}_1$ and the other housing $\mathbf{H}_2$,

$$
\mathbf{5}_H = \begin{pmatrix} \mathbf{h}^1 \\ \mathbf{h}^2 \\ \mathbf{h}^3 \\ \mathbf{H}_1^+ \\ \mathbf{H}_1^0 \end{pmatrix}, \quad \overline{\mathbf{5}}_H = \begin{pmatrix} \overline{\mathbf{h}}_1 \\ \overline{\mathbf{h}}_2 \\ \overline{\mathbf{h}}_3 \\ \mathbf{H}_2^0 \\ \mathbf{H}_2^- \end{pmatrix} \tag{2.49}
$$

which are all L-handed fields. Here, $\mathbf{h}^4$ and $\mathbf{h}^5$ form the Higgs doublet $\mathbf{H}_1$, and $\overline{\mathbf{h}}_4$ and $\overline{\mathbf{h}}_5$ form the Higgs doublet $\mathbf{H}_2$. To break the GUT group, we also need a L-handed adjoint Higgs field $\mathbf{24}_H$. Then, we can consider a renormalizable superpotential

$$
W = m\overline{\mathbf{5}}_H^T \mathbf{5}_H + \overline{\mathbf{5}}_H^T \mathbf{24}_H \mathbf{5}_H + \mathrm{Tr}\mathbf{24}_H^3 + M\mathrm{Tr}\mathbf{24}_H^2, \tag{2.50}
$$

where all fields are L-handed, $m$ and $M$ are GUT scale masses, and the couplings have been suppressed. After assigning a huge VEV to $\mathbf{24}_H$, we need to make $\mathbf{H}_1$ and $\mathbf{H}_2$ survive to the electroweak scale and $h$ and $\overline{h}$ escape at the GUT scale. This can be done by a fine-tuning of the couplings in the above superpotential. For example, the above superpotential ends up requiring the following form after assigning a VEV $m_1$ to $\mathbf{24}_H$,

$$
\overline{\mathbf{5}}_H^T \begin{pmatrix} m+2m_1 & 0 & 0 & 0 & 0 \\ 0 & m+2m_1 & 0 & 0 & 0 \\ 0 & 0 & m+2m_1 & 0 & 0 \\ 0 & 0 & 0 & m-3m_1 & 0 \\ 0 & 0 & 0 & 0 & m-3m_1 \end{pmatrix} \mathbf{5}_H .
$$

One must fine-tune $m$ and the VEV of $\mathbf{24}_H$ $m_1$ so that $m = 3m_1$, in order to obtain the electroweak Higgs doublets $\mathbf{H}_1$ and $\mathbf{H}_2$. This is the problem of separating $\mathbf{H}_1$ and $\mathbf{H}_2$ from the triplets $\boldsymbol{h}$ and $\overline{\boldsymbol{h}}$, which is known as the *doublet–triplet splitting problem*. It is one of the most difficult hierarchy problems in SUSY models.

Assuming the following extremely simplified symmetry breaking pattern,

$$SU(5) \longrightarrow [SU(3) \times SU(2) \times U(1)]_{\text{SUSY}} \text{ at } M_U,$$

$$\longrightarrow [SU(3) \times SU(2) \times U(1)]_{\text{non–SUSY}} \text{ at } M_{\text{SUSY}},$$

$$\longrightarrow SU(3) \times U(1)_{\text{em}} \text{ at } M_Z, \tag{2.51}$$

we can estimate how much the gauge couplings are differentiated at the electroweak scale $M_Z$. Conversely, given the experimentally measured couplings at the electroweak scale, we can check whether or not they meet at the unification point $M_U$. The current situation is shown in Fig. 2.8 with the one-loop evolution of gauge couplings. Originally, with crude data on $\alpha_c$ [57] they do not seem to meet [58]. However, with the LEP data on $\alpha_c$ [16] they meet within the experimental error bounds [59–61]. Thus, SUSY GUT models with the MSSM spectrum below $M_U$ seem to have some truth to them.

So far in the last few decades, supersymmetry has been the most popular scenario for the solution of the gauge hierarchy problem, and is the reason why the phenomenological aspects of MSSM have been so vigorously studied. But the MSSM has some serious theoretical problems related to the following issues;

- $15(+1)$ *chiral fermions* in one family,
- number of fermion families $\geq 3$, probably exactly 3,
- $\mathcal{N} = 1$ supersymmetry,

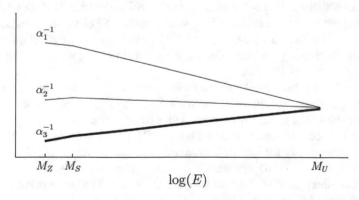

**Fig. 2.8** Running of gauge couplings of Fig. 2.5 with one-loop beta function and experimental inputs. We used $M_Z = 91.187\,\text{GeV}$, $M_S = 10^3\,\text{GeV}$, $M_U = 1.4 \times 10^{16}\,\text{GeV}$. $\alpha_1^{-1}(M_Z) = 1/0.01681$, $\alpha_2^{-1}(M_Z) = 1/0.03358$, $\alpha_3^{-1}(M_Z) = 1/(0.1200 \pm 0.0028)$, and $\sin^2\theta_W(M_Z) = 0.231$ [16]

- gauge hierarchy,
- doublet–triplet splitting,
- one pair of Higgsinos, $\mathbf{H}_1$ and $\mathbf{H}_2$,
- the hypercharge quantization, $\sin^2 \theta_W \simeq \frac{3}{8}$ at GUT scale,
- absence of strong CP violation,

as well as more detailed questions such as that pertaining to the hierarchies of Yukawa couplings related to Eq. (2.11). The Yukawa hierarchy problem is probably the most important issue at the TeV scale, which will likely be verified by future high energy experiments. In this book, however, we focus on searches of theoretical structures with which all or some of the above problems can be understood.

## 2.4    Extra Dimensions

### 2.4.1    Field Theory

In the 4D quantum field theory framework, we exhausted most possibilities of unification which are not obviously excluded by experimental data. This leads us to extending spacetime itself to higher dimensions, by adding extra dimensions to the familiar 4D model. In fact, the addition of extra dimension(s) beyond 4D is an older idea than any of the extensions we have discussed so far, and was first addressed by Kaluza and Klein in the 1920s [62,63]. Trying to interpret a photon as a gauge field in the metric, Kaluza said that, "$\cdots$ such an interpretation of $F_{\mu\nu}$ is hardly supported unless otherwise one makes an extremely odd decision of a new fifth dimension of the world." If we are to introduce *extra dimensions*, they must be cleverly hidden from us. The Kaluza–Klein (KK) idea is to compactify extra dimensions to such a small scale that the resolution power of 4D observers cannot see the structures of the extra dimensions. This is the famous *compactification* idea. Kaluza's excuse of introducing the extra dimensions was to view gravity and electromagnetism on the same footing. But Kaluza's view of $g_{\mu\nu}$ and $A_\mu^a$ on the same footing has failed in fact because of the chirality problem. Our excuse for introducing six extra dimensions, is to the unification of families.

Starting in the late 1990s, extra dimensions were studied in the quantum field theory framework. The chief motivation for the field theoretic study was to understand the gauge hierarchy problem with extra dimensions.

Arkani-Hamed, Dimopoulos, and Dvali (ADD) [64] introduced a TeV scale fundamental mass $M_5$ for gravity with extra dimensions, where the full extended space is called *bulk*. ADD assumed a flat bulk. The electroweak scale quantum fields are confined to a 4D boundary called the *brane*. Our perception of a huge 4D Planck mass $M_P$ is blamed for the large size of the extra dimensions compared to the Planck scale. This idea works even for extra dimensions as large as $100\,\mu\text{m}$ because of poor gravity experiments at small scales. This idea is designed to answer the gauge hierarchy problem by introducing only TeV scale masses $M_5$ and $M_Z$ in the Lagrangian, i.e. the hierarchy problem is not there from the outset. On the other hand, Randall and Sundrum (RS) considered the AdS bulk, i.e. with a negative bulk cosmological constant. In the RS-I model with two brane boundaries [65], the SM fields are put on the SM brane, B2, where the brane tension is negative while there exists another brane, B1, where the brane tension is positive. If the distance between these two branes is $d$ in the 5D example, by two fine-tunings the mass parameters at the SM brane are $\sim M_5 e^{-kd}$ where $k$ is the mass determined by $M_5$ and the 5D cosmological constant $\Lambda_5$. Thus, the relevant mass parameters at the SM brane can be exponentially small compared to the fundamental mass $M_5 \sim M_P$, and $d$ of $O(100 M_5^{-1})$ can give the electroweak scale. This is an attractive proposal for an exponential hierarchy, but serious cosmological problems arise from putting the SM fields in the negative tension brane [66]. Furthermore, from our theme of unification, this proposal can be at best just one side view of the full structure.

The ADD and RS-I proposals use the old compactification idea. Randall and Sundrum proposed another idea of cleverly hiding extra dimensions even though they are not compactified. It is the RS-II model [67] where only one positive tension brane is placed at $y = 0$ which is the brane for the SM fields. Again, the effect of gravity falls off exponentially $e^{-k|y|}$ in the bulk and the deep bulk ($|y| > k^{-1}$) is effectively hidden from us. Since all the fields (the SM fields and GUT fields) are put at the same brane, the RS-II model does not give a rationale for the gauge hierarchy solution. Most likely, this RS-II may be a way to understand the more serious hierarchy problem, the cosmological constant problem, through self-tuning solutions [68].

## 2.4.2 String Theory

Quantum field theory in four spacetime dimensions (4D) has a limited ability to unify all forces in Nature even though its favorite baby, the SM, is known to be very successful phenomenologically below the electroweak scale $\sim 100$ GeV. The SM has 19 free parameters (gauge couplings and flavor parameters such as given in Eq. (2.11)) which are tuned such that all the observed electroweak data are explained; at present this tuning is known to be possible without a major discrepancy with the data. But the SM does not explain why the 19 parameters take those phenomenologically required values. This is the uniqueness problem of why the SM takes the specific set having those required values out of numerous other possible sets. One can envision the existence of a truly unifying fundamental theory which is sometimes called the theory of everything (TOE). Nature may not allow such a theory. However, if such a theory exists, one should be able to calculate those parameters of the SM. We have observed that such a unification is possible in GUTs for the case of gauge couplings. In addition, 4D quantum field theory at present cannot incorporate gravity in the scheme because of its inability to treat the divergences appearing in quantum gravity. At the very least, these two problems, the uniqueness problem and the quantum gravity problem, hint to another theory beyond the 4D quantum field theory for TOE if it indeed exists.

String theory can be a candidate for TOE, at least for answering the above problems. As for the uniqueness problem, it has only one parameter, the string tension $\alpha'^{-1}$. If everything works out fine, then all the SM parameters are calculable in principle. As for the gravity problem, all string theories include closed strings which contain an excitation responsible for gravity. Removal of infinities in string theory can be explained after exploiting the one loop amplitude. Intuitively, one can imagine that string is better behaved than a point particle theory because string is an extended object. String theory with $\alpha'^{-1} \sim 1\,\text{GeV}^2$ was initially considered for the theory for strong interactions[15] before the development of QCD. But we know that strong interaction is due to the confining $SU(3)_c$ gauge theory. If string is useful for physics at all, it must be for gravity with $\alpha'^{-1} \sim M_P^2$ [70,71].

In quantum mechanics (the first quantization), identical particles are treated by the permutation symmetry which is put in by hand. This is similar to declaring that all $0.511\,\text{MeV}$ mass fermionic particles in the universe are identical. They are the identical particle, electron. Quantum field theory (the second quantization) gives a cute interpretation for the existence of identical particles in Nature, i.e. why an electron on Earth is identical to an electron in Andromeda Galaxy. Quantum field theory assumes that in the universe there is only one electron quantum field $\psi_e$. Both the electron on Earth and the electron in Andromeda are created by the same quantum field $\psi_e$ and hence they cannot be different kinds of particles. But why are there so many quantum fields in the SM? String theory answers this question by saying that there is only one string field $X$ in the universe (in bosonic string). Different modes of excitations correspond to different quantum fields and string theory can explain the existence of different kinds of particles in Nature. In this sense, string theory contains the basic logic for TOE.

One predictive power of string theory is that it fixes the number of spacetime dimensions where it lives. Bosonic string theory, which is the simplest string theory of all, is possible only in 26 spacetime dimensions (26D). But, quarks and leptons, which are fermions, are not present in the 26D bosonic string theory. To incorporate fermions, one considers superstring theory which is possible only in ten spacetime dimensions (10D). In the above determination of the number of spacetime dimensions, the quantum idea for string at the level of the first quantization is used for a consistent string theory. Still the second quantization of string is not yet fully developed if it is even needed at all. String theory used in particle physics is the first quantized version. At present, it is not known what kind of bonus on the theme of unification will reveal if one succeeds in further generalizing the first quantized string theory. Superstring theories accompany supersymmetry. Because quarks and leptons are present, it is better for the fundamental dimension of spacetime to be 10D rather than 26D if string is the fundamental object in the universe, and with this logic supersymmetry is essential at the fundamental level in 10D. But the spacetime around us seems to be 4D and we must find a way out from 10D to 4D. A useful way to obtain 4D is by compactifying six extra dimensions so that its smallness is not perceived by us. The main aim of this book is to discuss this process of

---

[15] See, for example [69].

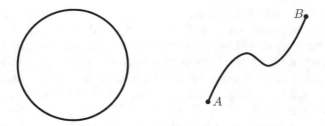

**Fig. 2.9**  Closed and open strings

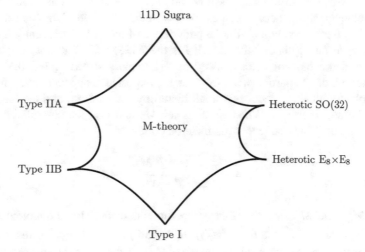

**Fig. 2.10**  M-theory vacua

compactification, using *orbifolds*. Here, the question why we live in 4D is not answered theoretically. Simply, we use the fact that our universe is 4D. But there exists a cosmological argument that 4D extended to become large while the extra six dimensions did not [72].

There are several superstring theories in 10D. Strings come in two varieties, closed strings and open strings as shown in Fig. 2.9.

One cannot construct an open string only theory since $A$ and $B$ of the open string in Fig. 2.9 can be joined together to give a closed string. Thus, any superstring theory must contain closed strings and string theory with closed strings can be a theory for gravity.

As for the uniqueness problem, superstring does not satisfy this criterion as was hoped in the beginning. Namely, the number of possible superstring theories is not one. It is known that there exist five 10D superstring theories which are distinguished by what kind of string(s) is used: Type-I, Type-IIA, Type-IIB [73], SO(32) heterotic, and most importantly $E_8 \times E_8'$ heterotic string [74–76]. However, in the mid-1990s string duality was found, in which different superstrings are related by duality. In fact, six theories (five 10D superstrings and one 11D supergravity) are related by duality, which is pictorially shown in Fig. 2.10. The six theories are considered to be located at specific corners of their mother theory called the M-theory. At present, we do not know in detail how the M-theory looks.

Among these, the $E_8 \times E'_8$ heterotic string has attracted the most attention [77–80] since the group is sufficiently large and the symmetry breaking chain $E_8 \rightarrow E_6 \rightarrow$ SO(10)$\rightarrow$ SU(5) is the desirable one, at least at the level of classification of the 15 chiral fields. As noted in Sect. 2.2, the spinor **16** of SO(10) correctly houses the 15 chiral fields. Regarding the doublet–triplet splitting problem of Sect. 2.3, it was found that a $\mathbf{Z}_3$ orbifold compactification does not introduce the extra color triplets $h$ and $\overline{h}$, thus solving the doublet–triplet splitting problem [81]. In addition, string theory has very rich structures, possibly solving all the problems listed in Sect. 2.3.

Even though string theory escapes the two fundamental problems we mentioned earlier, it still must explain all the 19 parameters of the SM if it is really the TOE. Therefore, in string theory the search for the MSSM is of utmost importance. But, this dream has not been realized experimentally so far at 1–2 TeV region. Superpartners of SM particles, if they exist, are expected to appear above a few TeV region ($\approx M_S$) which introduces a small hierarchy between $v_{\text{ew}}$ and the mass scale $M_S$ of SM superpartners. When we consider $M_S$ and $v_{\text{ew}}$, the naturalness criteria suggest to satisfy the tree level minimization condition,

$$\frac{M_Z^2}{2} = \frac{M_{H_d}^2 - \tan^2 \beta \, M_{H_u}^2}{\tan^2 \beta - 1} - \mu^2 \tag{2.52}$$

where $M_{H_d}^2$ and $M_{H_u}^2$ are the (mass)$^2$ coefficients of the Higgs doublets $H_d$ and $H_u$ in the potential $V$, $\tan \beta = \langle H_u^0 \rangle / \langle H_d^0 \rangle$, and $\mu$ is the Higgsino mass term in the superpotential. Starting with a positive $M_{H_u}^2$ at the ultimate unification scale, it becomes negative at the electroweak scale, triggering the electroweak symmetry breaking. The degree of fine-tuning is to look at how severely the above relation is violated in terms of the naturalness parameter $\Delta_{\text{ew}}$ [82, 83]. In the community, $\Delta_{\text{ew}} \gtrsim 100$ are looked for at present.

String theory has to be tested by experiments for it to be a good physical theory. But the size of string for gravitation is considered to be extremely small ($\sim 10^{-31}$ cm) if another large parameter is not introduced.[16] In this case, it is impossible to see 'string' by exciting it, and hints of string can come only indirectly. Such hints are the SM forbidden processes, e.g. proton decay, some level of flavor changing neutral current processes, etc. But these SM forbidden processes at low energy are discussed basically by effective field theory and hence they cannot pinpoint their origin, even if discovered, to string or GUTs for example. In this sense, we are in dilemma of proving the existence of "string." Maybe, the best we can hope is that string theory gives a *consistent framework* for the appearance of the SM. On this road of consistency check supplied with some prejudice on the gauge

---

[16]Ref. [64] considers the possibility of a large extra dimension.

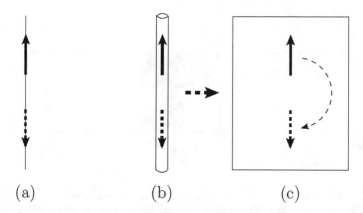

(a) (b) (c)

**Fig. 2.11** The chirality of a massless particle in even dimension $2n$ embedded in a higher odd dimension $2n + 1$: (a) D = 4, (b) D = 5, and (c) opening the cylinder of (b)

hierarchy problem, we anticipate the first experimental window to string at a little hierarchy scale as verification of a supersymmetric standard model.

### 2.4.3 Compactification

We conclude this lengthy chapter by showing qualitatively that if we start with extra dimensions, then the chiral fields necessary at the electroweak scale need some kind of orbifold-like compactification.[17]

The chirality is given by $(1 \pm \Gamma_{2n+1})/2$ in even spacetime dimensions, $2n$. Consider the chirality in 4D. In Fig. 2.11a, we consider an arrow in 2D (i.e. one space dimension), which can be considered a $S_z = \frac{1}{2}$.[18] In one space dimension, the arrow can be put in either of two ways, as shown with the solid arrow and the dashed arrow in Fig. 2.11a. They are different. But if the arrow came from a torus compactification from 3D (i.e. two space dimensions), it is in fact Fig. 2.11b. It appears that the 2D directions are different in the compactified case 2.11b, but in fact by opening up the compactified dimension as in Fig. 2.11c, we note that the solid arrow can be transformed to the dashed arrow by a 3D rotation. Stated in terms of Weyl spinors in 4D, $\psi_L$(solid) and $\psi_R$(dashed), the Weyl spinor $\psi_L$ cannot be transformed to $\psi_R$. But if these are embedded in 5D, a 5D rotation transforms one to the other, which means that they can belong to the same representation in 5D. Thus, a 5D spinor contains two 4D Weyl spinors, $\psi_L$ and $\psi_R$. The rigorous mathematical statement on this situation can be found in any textbook on superstring.[19] This

---

[17]Some manifolds such as the Calabi–Yau have singular limits which become orbifolds.

[18]The angular momentum can be given from 4D, and the directions in 2D become helicities in 4D.

[19]See [84].

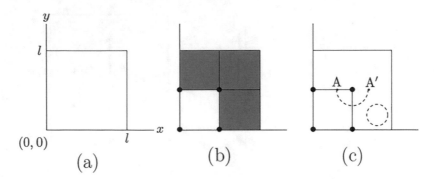

**Fig. 2.12** A 2D torus moded out by $\mathbb{Z}_2 \times \mathbb{Z}'_2$: (**a**) 2D torus, (**b**) the moded torus with fixed points and fixed lines, and (**c**) untwisted (dashed closed loop) and twisted (dashed arc A-A′) strings

shows that even if we start from a chiral theory in higher dimensions, a *naïve torus compactification washes out the chirality in 4D*, since we must pass through 5D in the process.

To obtain chirality, one must twist tori. Because of the simplicity in drawing figures in 2D, consider two internal dimensions compactified on a 2D torus as shown in Fig. 2.12a. For the torus compactification Fig. 2.12a, $x \equiv x + l$ and $y \equiv y + l$, there is no chiral fermions left in 4D as discussed above. So, an attempt should be made to mode out the torus by a discrete group. In Fig. 2.12b, we show the $\mathbb{Z}_2 \times \mathbb{Z}'_2$ moding: $x \equiv -x$ and $y \equiv -y$. Then, there are four fixed points under the shifts and reflections, shown as bullets in Fig. 2.12b. Using the shift and reflection symmetries, it is sufficient to consider the *fundamental domain* which is shown as the white square in Fig. 2.12b. The fixed points are at the corners of the fundamental domain. This geometry most probably with fixed points is called an *orbifold*.[20] Closed strings moving in this geometry can be untwisted or twisted as shown in Fig. 2.12c.

The untwisted string is obviously closed. The twisted string in Fig. 2.12c does not appear closed at first glance, but it is in fact closed around the fixed point in the orbifold geometry since A and A′ are identical, which can be shown by the allowed shifts and reflections. Thus a string appears just sitting around that fixed point. The spacetime dimension of fixed points are 4D. We have seen that 4D *does not necessarily require* that a 4D Weyl fermion accompany the opposite chirality partner. Since a 4D chiral field (the effective field for a string) sitting at a fixed point can be present without its chiral partner at the same fixed point, it is possible to have a chiral theory with orbifold geometry after compactification. But the final effective 4D theory must be free of gauge anomalies, which means that the sum of untwisted and twisted sector fields together give an anomaly-free theory. Therefore, as for the untwisted fields also, they can carry an anomaly, i.e. they can be chiral. This happens because the moding group $\mathbb{Z}_2 \times \mathbb{Z}'_2$ keeps only a half of two massless Weyl

---

[20]Some orbifolds do not have fixed points.

fields in the compactified 4D, or in other words, a half of the massless spectrum is projected out from the Hilbert space. *In string orbifolds, it is unambiguously determined which chiral fields should be at the fixed points and which should be in the bulk* [78–80]. Unlike string orbifolds, there is no guiding principle in field theoretic orbifolds, "how to put the chiral fields at the brane and in the bulk," except for the condition of total anomaly cancellation. For field theoretic orbifolds, the argument is the same as the above until we try to put localized fields at the fixed points and some in the bulk. One can consider the dashed curves of Fig. 2.12c as the same altitude points of the wave function sitting at the fixed point. In field theoretic orbifolds, one extra dimension (5D) has been considered extensively. On the other hand, in closed string theory with one compactification scale, it is difficult to achieve an effective 5D theory.

In this book, we aim to explore *the most probable orbifold compactification toward the MSSM* from the $E_8 \times E_8'$ heterotic string. To obtain three families, the moding discrete group $\mathbb{Z}_3$ has been considered in detail in Vol. I. In Vol. II, the $\mathbb{Z}_3$ orbifold is discussed as an easy example and we discuss $\mathbb{Z}_{12-I}$ in some detail. We focus to present all model-building toolkits for everyone's use.

In Chap. 3, we give the definition of orbifolds. Then, before presenting the full-fledged string theory orbifolds, we will introduce the orbifold in field theory in Chap. 5 in order to first present a general understanding of orbifolds. Recently, field theoretic orbifolds have been studied extensively, after showing that the doublet–triplet splitting is also possible in the field theoretic orbifold compactification.

## References

1. G. Aad et al., Observation of a new particle in the search for the Standard Model Higgs boson with the ATLAS detector at the LHC. Phys. Lett. **B716**, 1–29 (2012)
2. S. Chatrchyan et al., Observation of a new boson at a mass of 125 GeV with the CMS experiment at the LHC. Phys. Lett. **B716**, 30–61 (2012)
3. M. Gell-Mann, A Schematic Model of Baryons and Mesons. Phys. Lett. **8**, 214–215 (1964)
4. G. Zweig, CERN TH-401, 412 (1964)
5. S. Okubo, Note on unitary symmetry in strong interactions. Prog. Theor. Phys. **27**, 949–966 (1962)
6. M.Y. Han, Y. Nambu, Three triplet model with double SU(3) symmetry. Phys. Rev. **139**, B1006–B1010 (1965)
7. O.W. Greenberg, Spin and unitary spin independence in a paraquark model of baryons and mesons. Phys. Rev. Lett. **13**, 598–602 (1964)
8. W.A. Bardeen, H. Fritzsch, M. Gell-Mann, Light cone current algebra, $\pi^0$ decay, and $e^+e^-$ annihilation, in *Topical Meeting on the Outlook for Broken Conformal Symmetry in Elementary Particle Physics Frascati, Italy, May 4–5* (1972)
9. H.D. Politzer, Reliable perturbative results for strong interactions? Phys. Rev. Lett. **30**, 1346–1349 (1973)
10. D.J. Gross, F. Wilczek, Ultraviolet behavior of nonAbelian gauge theories. Phys. Rev. Lett. **30**, 1343–1346 (1973)
11. S. Weinberg, Implications of dynamical symmetry breaking. Phys. Rev. **D13**, 974–996 (1976). [Addendum: Phys. Rev. D19, 1277 (1979)]
12. S.L. Glashow, Partial symmetries of weak interactions. Nucl. Phys. **22**, 579–588 (1961)

13. S. Weinberg, A model of leptons. Phys. Rev. Lett. **19**, 1264–1266 (1967)
14. A. Salam, Weak and electromagnetic interactions. Conf. Proc. **C680519**, 367–377 (1968)
15. M.J.G. Veltman, Limit on mass differences in the Weinberg model. Nucl. Phys. **B123**, 89–99 (1977)
16. M. Tanabashi et al., Review of particle physics. Phys. Rev. **D98**(3), 030001 (2018)
17. A. Bodek, Electroweak precision measurements with the CMS detector. PoS **EPS-HEP2017**, 429 (2017)
18. M. Kobayashi, T. Maskawa, CP violation in the renormalizable theory of weak interaction. Prog. Theor. Phys. **49**, 652–657 (1973)
19. A.A. Belavin, A.M. Polyakov, A.S. Schwartz, Yu.S. Tyupkin, Pseudoparticle solutions of the Yang-Mills equations. Phys. Lett. **B59**, 85–87 (1975). [,350 (1975)]
20. C.G. Callan Jr., R.F. Dashen, D.J. Gross, The structure of the gauge theory vacuum. Phys. Lett. **B63**, 334–340 (1976). [,357 (1976)]
21. R. Jackiw, C. Rebbi, Vacuum periodicity in a Yang-Mills quantum theory. Phys. Rev. Lett. **37**, 172–175 (1976). [,353 (1976)]
22. J.E. Kim, Light pseudoscalars, particle physics and cosmology. Phys. Rep. **150**, 1–177 (1987)
23. J.E. Kim, G. Carosi, Axions and the strong CP problem. Rev. Mod. Phys. **82**, 557–602 (2010)
24. G. Degrassi, S. Di Vita, J. Elias-Miro, J.R. Espinosa, G.F. Giudice, G. Isidori, A. Strumia, Higgs mass and vacuum stability in the Standard Model at NNLO. JHEP **08**, 098 (2012)
25. J.E. Kim, D.Y. Mo, S. Nam, Final state interaction phases obtained by data from CP asymmetries. J. Korean Phys. Soc. **66**(6), 894–899 (2015)
26. C. Jarlskog, Commutator of the quark mass matrices in the standard electroweak model and a measure of maximal CP violation. Phys. Rev. Lett. **55**, 1039 (1985)
27. H. Georgi, The state of the art—gauge theories. AIP Conf. Proc. **23**, 575–582 (1975)
28. H. Georgi, S.L. Glashow, Unity of all elementary particle forces. Phys. Rev. Lett. **32**, 438–441 (1974)
29. H. Georgi, H.R. Quinn, S. Weinberg, Hierarchy of interactions in unified gauge theories. Phys. Rev. Lett. **33**, 451–454 (1974)
30. S. Weinberg, Gauge symmetry breaking. Conf. Proc. **C750926**, 1–26 (1975)
31. E. Witten, Dynamical breaking of supersymmetry. Nucl. Phys. **B188**, 513 (1981)
32. J.E. Kim, B. Kyae, A Model of Dynamical SUSY Breaking. Phys. Lett. **B797**, 134807 (2019)
33. L. Susskind, Dynamics of spontaneous symmetry breaking in the Weinberg-Salam theory. Phys. Rev. **D20**, 2619–2625 (1979)
34. S. Dimopoulos, L. Susskind, Mass without scalars. Nucl. Phys. **B155**, 237–252 (1979). [2, 930 (1979)]
35. A.J. Buras, J.R. Ellis, M.K. Gaillard, D.V. Nanopoulos, Aspects of the grand unification of strong, weak and electromagnetic interactions. Nucl. Phys. **B135**, 66–92 (1978)
36. P. Langacker, Grand unified theories and proton decay. Phys. Rep. **72**, 185 (1981)
37. M. Shiozawa et al., Search for proton decay via p $\longrightarrow e + \pi^0$ in a large water Cherenkov detector. Phys. Rev. Lett. **81**, 3319–3323 (1998)
38. H. Georgi, Towards a grand unified theory of flavor. Nucl. Phys. **B156**, 126–134 (1979)
39. F. Gursey, P. Ramond, P. Sikivie, A universal gauge theory model based on E6. Phys. Lett. **60B**, 177–180 (1976)
40. J.E. Kim, A model of flavor unity. Phys. Rev. Lett. **45**, 1916 (1980)
41. P. Frampton, S. Nandi, SU(9) Grand unification of flavor with three generations. Phys. Rev. Lett. **43**, 1460 (1979)
42. P.H. Frampton, Unification of flavor. Phys. Lett. **B89**, 352–354 (1980)
43. J.C. Pati, A. Salam, Unified lepton-hadron symmetry and a gauge theory of the basic interactions. Phys. Rev. **D8**, 1240–1251 (1973)
44. J.C. Pati, A. Salam, Is baryon number conserved? Phys. Rev. Lett. **31**, 661–664 (1973)
45. S.M. Barr, A new symmetry breaking pattern for SO(10) and proton decay. Phys. Lett. **B112**, 219–222 (1982)
46. J.E. Kim, SU(3) trits of orbifolded $E_8 \times E_8'$ heterotic string and supersymmetric standard model. JHEP **08**, 010 (2003)

47. A. Salam, J.A. Strathdee, Supergauge transformations. Nucl. Phys. **B76**, 477–482 (1974)
48. J. Wess, J. Bagger, *Supersymmetry and Supergravity*. Princeton Series in Physics, 2nd edn. (Princeton, New Jersey, 1992)
49. S. Coleman and J. Mandula, All Possible Symmetries of the S Matrix. Phys. Rev. **159**, 1251 (1967)
50. Yu.A. Golfand, E.P. Likhtman, Extension of the algebra of Poincare group generators and violation of p invariance. JETP Lett. **13**, 323–326 (1971). [Pisma Zh. Eksp. Teor. Fiz. **13**, 452 (1971)]
51. J. Wess, B. Zumino, A Lagrangian model invariant under supergauge transformations. Phys. Lett. **B49**, 52 (1974)
52. S.L. Adler, W.A. Bardeen, Absence of higher order corrections in the anomalous axial vector divergence equation. Phys. Rev. **182**, 1517–1536 (1969). [,268(1969)]
53. S. Deser, B. Zumino, Broken supersymmetry and supergravity. Phys. Rev. Lett. **38**, 1433–1436 (1977)
54. E. Cremmer, S. Ferrara, L. Girardello, A. Van Proeyen, Yang-Mills theories with local supersymmetry: Lagrangian, transformation laws and super-Higgs effect. Nucl. Phys. **B212**, 413 (1983)
55. H.P. Nilles, Supersymmetry, supergravity and particle physics. Phys. Rep. **110**, 1–162 (1984)
56. S. Dimopoulos, H. Georgi, Softly broken supersymmetry and SU(5). Nucl. Phys. **B193**, 150–162 (1981)
57. J.E. Kim, P. Langacker, M. Levine, H.H. Williams, A theoretical and experimental review of the weak neutral current: a determination of its structure and limits on deviations from the minimal $SU(2)_L$ x U(1) electroweak theory. Rev. Mod. Phys. **53**, 211 (1981)
58. S. Dimopoulos, S. Raby, F. Wilczek, Supersymmetry and the scale of unification. Phys. Rev. **D24**, 1681–1683 (1981)
59. U. Amaldi, W. de Boer, H. Furstenau, Comparison of grand unified theories with electroweak and strong coupling constants measured at LEP. Phys. Lett. **B260**, 447–455 (1991)
60. P. Langacker, M.-x. Luo, Implications of precision electroweak experiments for $M_t$, $\rho_0$, $\sin^2 \theta_W$ and grand unification. Phys. Rev. **D44**, 817–822 (1991)
61. C. Giunti, C.W. Kim, U.W. Lee, Running coupling constants and grand unification models. Mod. Phys. Lett. **A6**, 1745–1755 (1991)
62. Th. Kaluza, Zum Unitätsproblem der Physik. Sitzungsber. Preuss. Akad. Wiss. Berlin (Math. Phys.) **1921**, 966–972 (1921). [Int. J. Mod. Phys. **D27**, no. 14, 1870001 (2018)]
63. O. Klein, Quantum theory and five-dimensional theory of relativity. Z. Phys. **37**, 895–906 (1926). (In German and English)
64. N. Arkani-Hamed, S. Dimopoulos, G.R. Dvali, The hierarchy problem and new dimensions at a millimeter. Phys. Lett. **B429**, 263–272 (1998)
65. L. Randall, R. Sundrum, A large mass hierarchy from a small extra dimension. Phys. Rev. Lett. **83**, 3370–3373 (1999)
66. J.M. Cline, C. Grojean, G. Servant, Cosmological expansion in the presence of extra dimensions. Phys. Rev. Lett. **83**, 4245 (1999)
67. L. Randall, R. Sundrum, An alternative to compactification. Phys. Rev. Lett. **83**, 4690–4693 (1999)
68. J.E. Kim, B. Kyae, H.M. Lee, Randall-Sundrum model for selftuning the cosmological constant. Phys. Rev. Lett. **86**, 4223–4226 (2001)
69. P. Frampton, *Dual Resonance Models* (W. A. Benjamin, Reading, 1974)
70. J. Scherk, J.H. Schwarz, Dual models and the geometry of space-time. Phys. Lett. **B52**, 347–350 (1974)
71. J. Scherk, J.H. Schwarz, Dual models for nonhadrons. Nucl. Phys. **B81**, 118–144 (1974)
72. R.H. Brandenberger, C. Vafa, Superstrings in the early universe. Nucl. Phys. **B316**, 391–410 (1989)
73. J.H. Schwarz, Superstring theory. Phys. Rep. **89**, 223–322 (1982)
74. D.J. Gross, J.A. Harvey, E.J. Martinec, R. Rohm, The heterotic string. Phys. Rev. Lett. **54**, 502–505 (1985)

75. D.J. Gross, J.A. Harvey, E.J. Martinec, R. Rohm, Heterotic string theory. 1. The free heterotic string. Nucl. Phys. **B256**, 253 (1985)
76. D.J. Gross, J.A. Harvey, E.J. Martinec, R. Rohm, Heterotic string theory. 2. The interacting heterotic string. Nucl. Phys. **B267**, 75–124 (1986)
77. P. Candelas, G.T. Horowitz, A. Strominger, E. Witten, Vacuum configurations for superstrings. Nucl. Phys. **B258**, 46–74 (1985)
78. L.J. Dixon, J.A. Harvey, C. Vafa, E. Witten, Strings on orbifolds. Nucl. Phys. **B261**, 678–686 (1985). [,678(1985)]
79. L.J. Dixon, J.A. Harvey, C. Vafa, E. Witten, Strings on orbifolds. 2. Nucl. Phys. **B274**, 285–314 (1986)
80. L.E. Ibanez, H.P. Nilles, F. Quevedo, Orbifolds and Wilson lines. Phys. Lett. **B187**, 25–32 (1987)
81. L.E. Ibanez, J.E. Kim, H.P. Nilles, F. Quevedo, Orbifold compactifications with three families of SU(3) x SU(2) x U(1)$^n$. Phys. Lett. **B191**, 282–286 (1987)
82. R. Barbieri, G.F. Giudice, Upper bounds on supersymmetric particle masses. Nucl. Phys. **B306**, 63–76 (1988)
83. H. Baer, V. Barger, P. Huang, A. Mustafayev, X. Tata, Radiative natural SUSY with a 125 GeV Higgs boson. Phys. Rev. Lett. **109**, 161802 (2012)
84. J. Polchinski, *String Theory*, vol. II (Cambridge University Press, Cambridge, 1988), p. 430

# Orbifold

**3**

We introduce orbifold geometry. An orbifold is obtained from a manifold by identifying points under a discrete symmetry group. Roughly speaking, this identification
is done by cutting the points related by symmetry and gluing the resulting edges.
We are familiar with this cutting and gluing when we make a cone from a disc.

Later, we will define field and string theories on orbifold. By associating the
orbifold symmetry, we can introduce a projection restricting the wavefunction.
Thus, we may break large symmetry of the UV complete theory to obtain a realistic
model. One important application is selecting exclusive chirality, whose importance
is emphasized in Chap. 1. We may also break supersymmetry and gauge symmetry,
so we may design a model close to the Standard Model. We will also meet another
orbifold of the moduli space of the modular group.

Orbifold allows fixed points, which are invariant under the symmetry action.
They are singular, but string theory is well-defined on the fixed points. Moreover,
states can be localized on such fixed points. They may explain the family structure
and give rise to interesting interaction.

Most of the times, we are interested in toroidal orbifolds, that is, a torus modded
out by rotation by $2\pi/N$ compatible to the torus. We will construct it from Euclidean
space. After defining orbifolds in an abstract form, we will consider some examples
and try to understand the geometrical meaning of them. The references [1–3] contain
extensive treatments on this topic.

## 3.1 Orbifold Geometry

An *orbifold* is obtained by modding out a manifold $\mathcal{M}$ by its discrete symmetry
subgroup $\mathsf{G}$

$$\mathcal{M}/\mathsf{G}. \tag{3.1}$$

© Springer Nature Switzerland AG 2020
K.-S. Choi, J. E. Kim, *Quarks and Leptons From Orbifolded Superstring*,
Lecture Notes in Physics 954, https://doi.org/10.1007/978-3-030-54005-0_3

By modding, we regard two points $x \in \mathcal{M}$ and $gx$ ($g \in \mathsf{G}$) as the same point. This is done by assigning an equivalence relation between two elements related by actions of $\mathsf{G}$. Roughly speaking, an orbifold is formed from a manifold by identifying points following the symmetry. We first look into one of the simplest orbifolds of torus. Then, we obtain toroidal orbifold by further orbifolding by discrete rotation compatible to the torus.

### 3.1.1 Torus

The most familiar example of orbifold is torus. The one dimensional torus is a circle $T^1 = S^1$, obtained by identifying the coordinate in the real space $\mathbb{R}$

$$x \sim x + 2\pi R, \tag{3.2}$$

where $R$ is the radius. Formally, it may be rephrased as in (3.1) that the torus is obtained by modding out a real line by integer

$$T^1 = \mathbb{R}/\mathbb{Z}, \tag{3.3}$$

because any coordinate differs by an integer multiple $2\pi R$ is regarded same.

We may also form a two dimensional torus $T^2$ by similar identification as (3.2). However, a two-torus is more than a direct product of two circles $S^1 \times S^1$, because in general the two directions $x^1$ and $x^2$ may neither be orthogonal nor are of the same length. A convenient parameterization is to let the coordinate dependence fixed

$$x^i \sim x^i + 2\pi R, \quad i = 1, 2$$

and parameterize the geometry in the basis vectors. To reflect this, we introduce real, orthogonal coordinates $x^m, m = 1, 2$ in $\mathbb{R}^2$ and basis vectors having components $e_i = (e_i^m), m = 1, 2$ and its inverse $e_m^i$ so that

$$x^i \equiv x^m e_m^i, \quad m = 1, 2.$$

This means that we define $T^2 = \mathbb{R}^2/\Lambda_2$ using the lattice

$$\Lambda_2 = \{ m_1 e^1 + m_2 e^2 \,|\, m_1, m_2 \in \mathbb{Z} \}. \tag{3.4}$$

The actual radii $R_i$ of the circles are related by the lengths of the vectors $e^i$

$$R_1 = R|e^1|, \quad R_2 = R|e^2|,$$

and the shape of torus is described by a *complex structure*

$$iU \equiv \frac{|e^2|}{|e^1|}e^{i\phi}. \tag{3.5}$$

with the angle $\phi$ between the two basis vectors. In the complex coordinates, the two basis numbers are related as $e_2 = iUe_1$. We say the tori are equivalent if the defining lattices are equivalent. Two equivalent but different bases can be made identical by an appropriate linear combination of the bases. We shall come back to this issue later.

We can generalize this to a $d$ dimensional torus by

$$T^d = \mathbb{R}^d/\Lambda, \tag{3.6}$$

where the lattice $\Lambda$ is now $d$ dimensional

$$\Lambda = \left\{ \sum_{i=1}^{d} m_i e^i \,\middle|\, m_i \in \mathbb{Z} \right\}. \tag{3.7}$$

In coordinates, the torus is made by identification

$$x^i \sim x^i + 2\pi R. \tag{3.8}$$

Here, again the coordinate periodicities are fixed as in (3.4), the actual geometric information is contained in the basis. Started with a natural orthonormal coordinate system $\mathbb{R}^d$, the lattice is generated by basis vectors $e^i, i = 1, 2, \ldots, d$ having components $e_i = (e_i^m), m = 1, 2, \ldots, d$. Inverting this, we have $e_i^m$ and its inner product makes the metric

$$G_{ij} \equiv e_i^m e_j^n \delta_{mn} \equiv e_i \cdot e_j. \tag{3.9}$$

Note that this has the same structure of the metric tensor constructed from vielbeins. We can switch between the vectors in the lattice and orthogonal space by multiplying $e_i^m$ or its inverse $e_m^i$. Thus, the definition (3.8) can be rewritten as

$$x^m \sim x^m + 2\pi R e_i^m, \quad i = 1, 2, \ldots, d, \tag{3.10}$$

Therefore, the volume of torus is

$$\text{Vol } T^d = (2\pi R)^d \sqrt{\det G}. \tag{3.11}$$

The torus is flat in the sense that we have a constant metric and the curvature scalar vanishes.

## 3.1.2 Toroidal Orbifold

We are interested in general action transforming a point $x$ in $\mathbb{R}^d$ as

$$g : x^m \to \theta^m{}_n x^n + v^m \tag{3.12}$$

with summation convention. We may compactly write it as

$$gx = (\theta, v)x = \theta x + v. \tag{3.13}$$

The set of such actions forms a *space group* S having the following properties [2]:

$$(\theta, v)(\omega, u) = (\theta\omega, v + \theta u),$$

$$(\theta, v)^{-1} = (\theta^{-1}, -\theta^{-1}v). \tag{3.14}$$

$$[(\theta, v)(\omega, u)](\rho, w) = (\theta, v)[(\omega, u)(\rho, w)]$$

with the identity element $(1, 0)$.

It looks the same as the Euclidean or the Poincaré group. However, we will only consider a discrete rotation. We define *point group* P as the subgroup of SO($d$) of rotation generated by $\theta$. An order $N$ of the rotation is the minimum natural number satisfying

$$\theta^N = 1. \tag{3.15}$$

This is isomorphic to the cyclic group $\mathbb{Z}_N$. The point group is therefore

$$\mathsf{P} = \{1, \theta, \ldots, \theta^{N-1}\}. \tag{3.16}$$

We may have more than one generators if the rotation is done on a part of the tori of $T^d$.

We define a toroidal orbifold by modding out $\mathbb{R}^d$ by the space group action S

$$\mathcal{O} = \mathbb{R}^d / \mathsf{S}. \tag{3.17}$$

Restricting the space group S to translational $\{(1, m_i e_i), i = 1, 2, \ldots, d\}$, we obtain the lattice $\Lambda$, and the resulting orbifold is torus $T^d / \Lambda$. Throughout this book, we will make an orbifold by modding out torus by the above rotation $\theta$. Then, we need a modified point group that is compatible with the torus. For instance, a rotation by $\theta$ at the origin should be the same as another rotation by $\theta$ at $e_1$. For this, we observe that for every $\theta$ there is a unique vector $v$ making $(\theta, v)$ an element in S, up to translation $\Lambda$ defining the torus. To see it, compare two elements with the same $\theta$

$$(\theta, v)(\theta, u)^{-1} = (\theta, v)(\theta^{-1}, -\theta^{-1}u) = (1, v - u). \tag{3.18}$$

Then, $v - u$ should belong to the lattice $\Lambda$; otherwise, the product is not compatible with torus. Thus, we may label the point group action by $\theta$ only. This generalized point group is defined as

$$\overline{\mathsf{P}} = \mathsf{S}/\Lambda \tag{3.19}$$

and commute with $\Lambda$. Thus, the toroidal orbifold is

$$\mathscr{O} = \mathbb{R}^d/(\Lambda \times \overline{\mathsf{P}}) = T^d/\overline{\mathsf{P}}. \tag{3.20}$$

**Fundamental Region**
A set of images due to the action like $\mathsf{S}$ is called orbit. On the above orbifold, all the orbits of $\mathsf{S}$ on the covering space $\mathbb{R}^d$ are identified as the same point. Hence, the name orbifold is given. This (co)set of points is called *fundamental region* or fundamental domain. So, the fundamental region contains exactly one point from each of the orbit. There is no unique way to take a fundamental region; however, every choice should form the same orbifold.

The volume of an orbifold is defined by that of the fundamental region. We can show

$$V(T^d/\mathbb{Z}_N) = V(T^d)/N. \tag{3.21}$$

We can think of the geometry of orbifold by gluing the boundaries of the fundamental region.

### 3.1.3  Fixed Points and Conjugacy Class

An action is *free* if there is a unique inverse. For instance, a rotation is not free because rotation by different angle leaves the origin invariant and its inverse is not well-defined. Formally, if an action $g$ is free, $gx = x$ means $g = 1$. Also, a general space group action in $\mathsf{S}$ is not free because it is a rotation possibly accompanied by a lattice translation identifying different points.

As a result, an orbifold has *fixed points*. A fixed point $f$ is an invariant point under the space group action $h \equiv (\theta^k, v)$

$$f = hf = \theta^k f + v. \tag{3.22}$$

In other words, this fixed point is invariant under the point group action $\theta^k$ up to a lattice translation $v$.

Since $\mathsf{S}$ is a discrete group, the orbifold $\mathbb{R}^d/\mathsf{S}$ inherits the flatness of $\mathbb{R}^d$ as long as the action is free. On the fixed point, it is not flat anymore and we have deficit angle. Because of this, an orbifold cannot be a manifold.

This supplementary translation $v$ is unique and has one-to-one correspondence to the fixed point $f$ through (3.22)

$$v = (1 - \theta^k)f. \tag{3.23}$$

If it can be inverted, that is, when

$$\det(1 - \theta^k) \neq 0,$$

we may represent a fixed point by the element of space group $(\theta^k, v)$. Otherwise, there is a fixed torus (in two or more real dimensions) rather than fixed points. For example, in the $T^2/\mathbb{Z}_2$ case, we may complexify the coordinate and $\theta$ is represented by a number $-1$. We have $\theta^2 = 1$, and the resulting action is trivial, so we have invariant torus.

Now, the question is how many different fixed points are there on the orbifold. We have redundancy. For a given fixed point,

$$f' = \theta^l f + u \equiv gf \tag{3.24}$$

is the equivalent fixed point, because

$$g = (\theta^l, u) \in \mathsf{S}$$

in (3.24) is also an element of the space group $\mathsf{S}$ used for identification. We may rewrite the above relation as

$$f' = gf = ghf = ghg^{-1}gf = ghg^{-1}f'. \tag{3.25}$$

So, the same fixed point is specified by another space group element $ghg^{-1}$. We say $h$ is *conjugated* by $g \in \mathsf{S}$. We define the *conjugacy class* of $h$ by assigning equivalence relation under conjugation

$$[h] = \{ghg^{-1} | g \in \mathsf{S}\}. \tag{3.26}$$

The conjugacy class is a more fundamental concept than the fixed point.

We take examples. Consider an action of translation by $(1, v_0)$. Using (3.14), one can easily check that

$$(\omega, u)(1, v_0)(\omega, u)^{-1} = (1, \omega v_0). \tag{3.27}$$

Therefore, the conjugacy class of $(1 - v_0)$ is

$$[(1, v_0)] = \{(1, \omega v_0) | \omega \in \mathsf{P}\}, \tag{3.28}$$

which provides the set of basis of the lattice compatible with the point group.

Next, consider a general element $(\theta^k, (1 - \theta^k)f)$ and its conjugation by an arbitrary element $(\theta^l, u)$

$$(\theta^l, u)(\theta^k, (1 - \theta^k)f)(\theta^l, u)^{-1}$$
$$= (\theta^l \theta^k \theta^{-l}, -\theta^l \theta^k \theta^{-l} u + \theta^l (1 - \theta^k)f + u) \qquad (3.29)$$
$$= (\theta^k, (1 - \theta^k)(\theta^l f + u)).$$

Note that the rotational element $\theta^k$ is not affected by the conjugation (3.29). It means that there is *one-to-one correspondence between the fixed point $f$ and the conjugacy class*

$$[(\theta^k, (1 - \theta^k)f)] = \{(\theta^k, (1 - \theta^k)(\theta^l f + u)) \mid \theta^l \in \mathsf{P}, u \in \Lambda\}. \qquad (3.30)$$

This set is sometimes loosely denoted as

$$(\theta^k, (1 - \theta)(\theta^l f + \Lambda)). \qquad (3.31)$$

The expression (3.31) will be extensively used in calculating quantities related with fixed points. Consequently, the fixed points are represented by these conjugacy classes, called cosets. In other words, we have unique set of distinctive fixed points if we mod out the fixed points in the ambient space by the conjugacy classes.

## 3.2 One Dimensional Orbifolds

One dimensional orbifold has been extensively studied in the context of field theoretical orbifold, which is the main theme of Chap. 5.

### 3.2.1 $S^1/\mathbb{Z}_2$ Orbifold

The simplest example of nontrivial orbifold is $S^1/\mathbb{Z}_2$. Let us choose $\mathbb{Z}_2$ action as the reflection with respect to the origin

$$y \to -y. \qquad (3.32)$$

It follows that $\pi R + y$ and $\pi R - y$ are also identified because

$$\pi R + y \to -\pi R - y \to \pi R - y, \qquad (3.33)$$

with the last relation coming from the definition of $S^1$. In other words, we make the orbifold $S^1/\mathbb{Z}_2$ by identifying the opposite points of $S^1$ with respect to vertical axis of $S_1$ drawn in 2D as shown in Fig. 3.1. Here, we have two fixed points $y = 0$ and

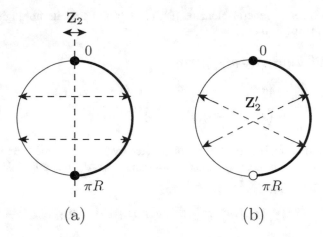

**Fig. 3.1** Two possible $S^1/\mathbb{Z}_2$ orbifolds. (a) $\mathbb{Z}_2$ action is defined by reflection $y \to -y$. Fixed points at $y = 0, \pi R$ are denoted by bullets. (b) $\mathbb{Z}_2'$ action is defined by translation $y \to y + \pi R$. Two points $0, \pi R$ are again identified, and there is no fixed point

$y = \pi R$. Equivalently saying, we have a finite interval $[0, \pi R]$. This interval is the fundamental region; fields can move within this interval $[0, \pi R]$.

### 3.2.2 Another Modding

There is another $\mathbb{Z}_2'$ symmetry action on $S^1$, i.e. the rotation by $\pi$,

$$y \to y + \pi R. \tag{3.34}$$

It is also viewed as a translation by a half of the original interval. It is of order two because twice the action is an identity.

The $\mathbb{Z}_2'$ action is freely acting on the circle $S^1$ and the resulting orbifold has no fixed points. The topology of this orbifold is again the circle with circumference reduced to the half $\pi R$.

Sometimes, this orbifold is called $\mathbb{RP}^1$, meaning the real, projected interval in one dimensional plane. Although it is one dimensional, we can introduce two real numbers $(x, y)$ and identify coordinates up to a scale

$$(x, y) \sim (\lambda x, \lambda y). \tag{3.35}$$

Since coordinates are the same up to a scale factor $\lambda$, we may fix the scale as $R$. Also, because $\lambda = R$ and $\lambda = -R$ lead to the same point, we can equivalently mod it out by $\mathbb{Z}_2$ action (3.34).

**Fig. 3.2** $S^1/(\mathbb{Z}_2 \times \mathbb{Z}_2')$ orbifold. It has two fixed points at $y = 0, \frac{\pi R}{2}$. The thick arc is the fundamental region

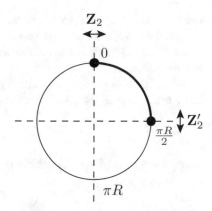

### 3.2.3 $S^1/(\mathbb{Z}_2 \times \mathbb{Z}_2')$ Orbifold

We construct an $S^1/(\mathbb{Z}_2 \times \mathbb{Z}_2')$ orbifold by modding out $S^1$ by one rotation $\theta$ and one translation $e$ as

$$\theta : \quad y \to -y, \quad \theta^2 = 1 \tag{3.36}$$

$$e : \quad y \to y + \pi R, \quad e^2 = 1. \tag{3.37}$$

The resulting geometry is drawn in Fig. 3.2. We may think of virtual axes in this figure. The action *theta* is a reflection around the vertical axis passing thru the origin, and the second action $e$ generates translation by $\pi R$. Combining them, the action $e\theta$ (acting on the right) gives rise to the reflection around the horizontal axis

$$y \xrightarrow{\;\theta\;} -y \xrightarrow{\;e\;} \pi R - y. \tag{3.38}$$

Therefore, we have a new fixed point at $y = \pi R/2$ under the combined action. We do not count the identical points $y = \pi R, y = -\pi R/2$ because the fundamental region is now $[0, \pi R/2]$.

We can understand this orbifold using two $\mathbb{Z}_2$ reflections. For this, we introduce a new coordinate $y' = y + \pi R/2$. Then, the action $h \equiv e\theta$ becomes a reflection around the horizontal axis in Fig. 3.2. Thus, we can form an equivalent $S^1/(\mathbb{Z}_2 \times \mathbb{Z}_2')$ orbifold generated by two reflections

$$g : \quad y \to -y, \quad g^2 = 1 \tag{3.39}$$

$$h : \quad y' \to -y', \quad h^2 = 1. \tag{3.40}$$

This is the most complicated example in one dimension. Whatever (or more complex) discrete group $\mathbb{Z}_N$ we use in 1D, the effect is essentially the same. Everything reduces to two cases we have considered above. At best, the length of interval reduces to $2\pi R/N$. Nevertheless, when we consider a field or string living on this orbifold, we can associate a more complicated $\mathbb{Z}_N$ action for the field or string to satisfy. This is a profitable aspect of orbifold when we try to break many symmetries.

Also, combining more than one $\mathbb{Z}_2$ actions $P$ and $\mathbb{Z}_2'$ actions $T$, one obtains the same orbifold as in the previous example because they commute

$$\theta e = e\theta. \tag{3.41}$$

Thus, the $\mathbb{Z}_2$ and $\mathbb{Z}_2 \times \mathbb{Z}_2'$ orbifolds discussed above are all the orbifolds in one dimension.

## 3.3 Two Dimensional Orbifolds

In two dimensions, we start to see nontrivial geometry of orbifold. They are the best complicated orbifolds that we are able to visualize. We can see also hints on the crystallographic classification.

### 3.3.1 $T^2/\mathbb{Z}_2$ Orbifold

It will be convenient to complexify the coordinate

$$z = x^1 + ix^2.$$

Two basis vectors define two dimensional torus $T^2$

$$z \sim z + e_m, \quad i = 1, 2, \tag{3.42}$$

where unit vectors $e_m$ are not necessarily orthogonal nor of unit length.

We may consider, for example, the orthogonal basis vectors

$$e_1 = 2\pi R_1, \quad e_2 = 2\pi R_2 i. \tag{3.43}$$

Orthogonal because the ratio of the two basis vectors, as complex numbers, has the phase $i$. Eventually, we make a $T^2/\mathbb{Z}_2$ orbifold by the identification

$$z \sim -z, \tag{3.44}$$

meaning that $\theta = -1$. Note that this action does not relate the lengths of the basis vectors. In terms of coordinates $a_1$, $a_2$, this action can be expressed as

$$a_1 e_1 + a_2 e_2 \rightarrow -a_1 e_1 - a_2 e_2. \tag{3.45}$$

**Fig. 3.3** The $T^2/\mathbb{Z}_2$ orbifold in the $z = x_1 + ix_2$ plane. The orbifold fixed points are denoted by bullets. The shaded is one choice of the fundamental region. The physical space may be taken as the two-sided rectangle formed by folding the boxed region along the line between $\frac{1}{2}e_2$ and $\frac{1}{2}e_1 + \frac{1}{2}e_2$ and then gluing together the touching edges

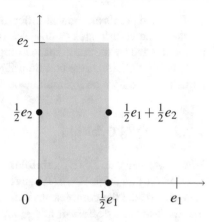

This orbifold is depicted in Fig. 3.3. We have four fixed points

$$0, \quad \frac{1}{2}e_1, \quad \frac{1}{2}e_2, \quad \frac{1}{2}e_1 + \frac{1}{2}e_2, \tag{3.46}$$

which are marked with bullets in Fig. 3.3. In view of (3.31), we can label fixed points by the following space group elements:

$$(\theta, 0), \quad (\theta, e_1), \quad (\theta, e_2), \quad (\theta, e_1 + e_2), \tag{3.47}$$

with which we come back to the original fixed points.

We can take the fundamental region as the box and inside surrounding $0, \frac{1}{2}e_1$, $\frac{1}{2}e_1 + e_2, e_2$. This choice is not unique, and we have many equivalent ones (see Exercise). For example, we can take another box surrounding $0, e_1, e_1 + \frac{1}{2}e_2, \frac{1}{2}e_2$. In any case, the resulting geometry is the same.

**Complex Structure**

Recall that we had the complex structure $iU = iR_2/R_1$ in the sense that we had $e_2 = ie_1$. In the case of $T^2/\mathbb{Z}_2$ orbifold, any complex structure is allowed. If the complex structure is transformed by PSL$(2, \mathbb{Z})$, we have the same orbifold. It is a symmetry group, and we expect target space modular invariance.

If we take a different complex structure, we have a different torus and a different orbifold. For example, we may take $iU = iR_2/R_1 + \frac{1}{2}$ we have $e_2 = 2\pi iR_2 + \pi R_1$. The $\mathbb{Z}_2$ action $z \rightarrow -z$ is still compatible to the lattice. Then, we have different orbifolds: they have different fixed points having different complex coordinates. Unlike the above case, the intuitive "folding picture" does not work for a general $iU$.

The fixed points are now at different locations. However, they are still expressed by the same coordinates (3.46). See the reflection structure (3.45). They are also parameterized by the same space group elements (3.47).

Note that although we can do a $\mathbb{Z}_2$ reflection along any directions $x^1 \to -x^1$ and $x^2 \to -x^2$, we do it simultaneously in the two directions.

### 3.3.2  $T^2/\mathbb{Z}_3$ Orbifold

Now, consider the $T^2/\mathbb{Z}_3$ orbifold. The point group is $\mathbb{Z}_3$ and is generated by rotation $\theta$ by an angle $2\pi/3$. It can represented as $\theta = e^{2\pi i/3}$ in the complex plane.

We also take the complex structure same as the point group element $iU = \theta$, so that the two basis vectors $e_1$ and $e_2$ are related as

$$e_2 = e^{2\pi i/3} e_1 \tag{3.48}$$

forcing equal lengths $|e_1| = |e_2|$. This is a requirement for the compatibility of orbifolding. The unit lattice is depicted in Fig. 3.4a. The basis vectors are the root vectors of SU(3), and hence it is called the "SU(3) or A$_2$ lattice". This amounts to fixing two of three parameters of $G^{ij}$ in (3.9) except the overall size of torus $R_1 = R_2$.

It follows the relations

$$\theta e_1 = e_2, \quad \theta e_2 = -e_1 - e_2. \tag{3.49}$$

So, we obtain the matrix form of the twist in the $e_1, e_2$ basis

$$\theta = \begin{pmatrix} 0 & -1 \\ 1 & -1 \end{pmatrix}. \tag{3.50}$$

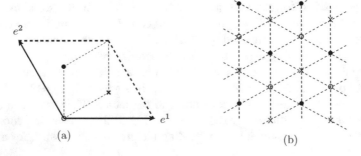

(a)                                                                  (b)

**Fig. 3.4** (a) $\mathbb{Z}_3$ orbifold is formed by identification of two edges of SU(3) torus. Three fixed point are marked by $\circ$, $\times$, $\bullet$. The region inside the lightly dashed parallelogram is the fundamental region. (b) The covering space of $\mathbb{Z}_3$ orbifold. Any two adjacent equilateral triangles form a fundamental region. The unit lattice is a hexagon and respects the $\mathbb{Z}_3$ symmetry

The fixed points are marked by $\circ, \times, \bullet$ as shown in Fig. 3.4. They have coordinates

$$f_\circ = 0, \quad f_\times = \frac{1}{3}(2e_1 + e_2), \quad f_\bullet = \frac{1}{3}(e_1 + 2e_2).$$

They are invariant points under the space group action. Also, they are invariant under the point group action, up to lattice translation $\Lambda$. For example,

$$\theta f_\times + e_1 = \frac{1}{3}(-e_1 + e_2) + e_1 = \frac{1}{3}(2e_1 + e_2) = f_\times.$$

We can always convert the fixed point to the unique space group element using (3.23). For example,

$$e_1 = (1 - \theta)\frac{1}{3}(2e_1 + e_2).$$

Using space group elements, we denote the fixed points as

$$\circ : (\theta, 0), \quad \times : (\theta, e_1), \quad \bullet : (\theta, e_1 + e_2). \tag{3.51}$$

From (3.30), fixed points belonging to the same conjugacy class (3.31) are equivalent. That is, they may be different in the covering space $\mathbb{R}^d$ but are identified as the same point on $\mathbb{R}^d/\mathsf{S}$. One may lie outside the fundamental region but can be moved into the fundamental region by conjugation by $\mathsf{S}$. For instance, a lattice translation by $(1-\theta)\Lambda$ in the translation part of (3.51) gives always the equivalent fixed points.

We have another twisted sector. The second twisted sector is generated by $\theta^2 = e^{4\pi i/3} = e^{-2\pi i/3}$, which is the same amount of rotation but in the opposite direction. We can see that the lattice is also compatible to this action, in the sense that the basis vectors are mapped to another linear combination of basis vectors. Thus, we expect the same properties as discussed above for this $\theta^2$ twisted sector also.

### 3.3.3 Geometry of Orbifold

The orbifold is formed by identifying points under the point group $\overline{\mathsf{P}}$. This is achieved by folding the fundamental region of Fig. 3.4 and gluing the edges. This leads to a "ravioli" (or "pillow")-type manifold that has a similar topology as $S^2$. However, we have singular points, so the topology is not the same. The fixed points are clearly singular in the sense that the curvature diverges on them. Therefore, an orbifold is not a manifold in general. However, in later chapters we will see that fields and strings have well-defined behaviors at the fixed points. Also, we will see that there is well-defined resolution procedure that replaces the fixed points with smooth geometry without losing "good" properties such as unbroken supersymmetry.

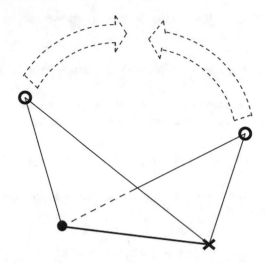

**Fig. 3.5** The orbifold is formed by identifying points under the point group $\overline{P}$. This is achieved by folding the fundamental region of Fig. 3.4 and gluing the edges to form a ravioli

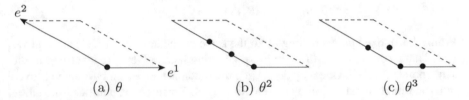

**Fig. 3.6** $T^2/\mathbb{Z}_6$ orbifold is defined on a $G_2$ lattice by $\theta = e^{2\pi i/6}$. Each twisted sector has a different number of fixed points

However, the folding, rather than identification, is somewhat misleading. As discussed in the torus case, there is no change of curvature and it is everywhere flat with constant $G^{ij}$. This situation is depicted in Fig. 3.5.

### 3.3.4 $T^2/\mathbb{Z}_6$ Orbifold

So far, we have dealt with the examples in which the basis vectors of covering torus obey the point group, for example, $\theta e_1 = e_2$. However, this is not mandatory, since the generalized point group $\overline{P}$ action in (3.17) is defined up to lattice translation. In other words, it is only necessary that *the lattice* generated by basis vectors should be compatible with the space group action. A good illustration is provided by $T^2/\mathbb{Z}_6$ orbifold. We make $T^2$ using the $G_2$ lattice of Fig. 3.6, for which the complex structure $iU = \sqrt{3}e^{5\pi i/6}$. The basis vectors are related as

$$e_2 = \sqrt{3}e^{5\pi i/6}e_1. \tag{3.52}$$

Note that the basis vectors are not of equal lengths $|e_2| = \sqrt{3}|e_1|$ and angle between them is $\phi_{12} = 5\pi/6$. From this torus, we make an orbifold under the identification given in (3.52)

$$z \sim e^{2\pi i/6}z. \tag{3.53}$$

We can readily check that

$$\theta e_1 = 2e_1 + e_2 = e_2 + (e_1 + e_2), \tag{3.54}$$

so that the space group action is compatible.

The fixed points are shown in Fig. 3.6 for three twisted sectors. For the first twisted sector, $\theta$, there is only one fixed point, the origin. In the second twisted sector, $\theta^2$, the action is equivalent to $\mathbb{Z}_3$ generated by $\theta^2 = e^{2\pi i/3}$, and we have three fixed points which can be read from Fig. 3.4b. In the third twisted sector, the action is equivalent to $\mathbb{Z}_2$ generated by $\theta^3 = e^{\pi i}$ and there are four fixed points as in the case of Fig. 3.3.

**Equivalent Lattice**

Let us see what Eq. (3.54) means. If we define

$$e^{2\prime} = e_1 + e_2, \tag{3.55}$$

we see that, by $2\pi/3$ rotation, $\theta^2 e_1 = e^{2\prime}$. This is the defining condition for the SU(3) lattice as we saw in the previous subsection. *The $G_2$ lattice generated by $e_1$ and $e_2$ and SU(3) lattice generated by $e_1$ and $e^{2\prime}$ are the same.* Therefore, the resulting orbifold is the same, as is clear from Fig. 3.7.

This can be done by target space modular transform

$$iU' = iU + 1 = \sqrt{3}e^{5\pi i/6} + 1 = e^{2\pi i/3},$$

$$e^{2\prime} = iU'e_1 = (\sqrt{3}e^{5\pi i/6} + 1)e_1 = e_2 + e_1.$$

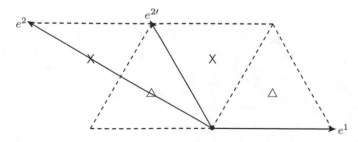

**Fig. 3.7** The $G_2$ lattice is the same as the SU(3) lattice. Fixed points on the $e_2$ axis of the $G_2$ lattice are the same, up to shifts, as the fixed points of the SU(3) lattice

We have $\mathrm{PSL}(2, \mathbb{Z})$ invariance under the target space complex structure. They are compatible to the above space group actions with the point group $\mathbb{Z}_3$ and $\mathbb{Z}_6$. The reason for taking a specific orbifold is that the name of lattice contains the information of the order. We will see that the Coxeter elements of $\mathrm{SU}(3)$, $G_2$ are, respectively, $e^{2\pi i/3}$, $e^{2\pi i/6}$ having the orders 3 and 6.

## 3.4    Classifying the Space Group

We classify the space group. It is done by classifying lattices compatible to the twists. As seen in two dimensions in Sect. 3.3.3, there are only four possible twists $\phi$ compatible with lattices forming crystals. In general, there are only limited number of lattices and twists in a given dimension. The systematic analysis is done in [4–6].

### 3.4.1    Crystallography

Now, the question is to find out how many lattices are allowed in two dimensions. We have already considered the $\mathrm{SU}(3)$ and $G_2$ lattices. This question can be restated in another form: using only one kind of tile, how many polygon types are allowed for tiling the two dimensional space. The lattice point is defined by vertices of tiles, where the adjacent tiles meet. The tile is well-filled only if at the vertex the sum of the angles of a polygon equals exactly to $2\pi$. In other words, every vertex should have $n$-fold rotational symmetry $D_n$. A $n$-gon have a side angle $\pi - 2\pi/n$. Decomposing such polygon as $n$ isosceles triangles and using the fact that the angles of a triangle sum up to $\pi$, one can easily show that the filling condition becomes

$$N\left(\pi - \frac{2\pi}{n}\right) = 2\pi. \tag{3.56}$$

Equation (3.56) admits only four solutions $N = 2, 3, 4, 6$ for which $n = \infty, 6, 4, 3$, respectively. This situation is depicted in Fig. 3.8. Therefore, by discrete rotations we can have only four kinds of orbifold lattices in the two dimensional torus.

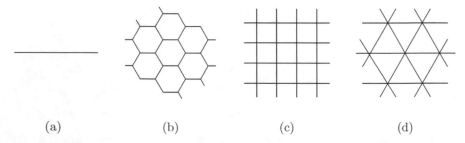

(a)                        (b)                        (c)                        (d)

**Fig. 3.8** There are only four possible crystals (tilings) in two dimensions; thus, there exist four two dimensional orbifolds. Each figure is invariant under two-, three-, four-, and sixfold rotations

### 3.4.2 Finding Lattices for a Given Twist

We want to classify order $N$ irreducible lattices in various dimensions. They provide building blocks of higher dimensional orbifolds. We can generate the basis vectors of the lattice by iteratively rotating a vector $\alpha_1$. We define other basis vectors as

$$\alpha_{i+1} \equiv \theta \alpha_i, \quad i = 1, \ldots, N-2. \tag{3.57}$$

To guarantee order $N$, or $\theta^N = \mathbf{1}_N$, we require $\det \theta = \pm 1$ and the final vector $e_N$ should be rotated to the linear combination of the other vectors

$$\theta \alpha_N \equiv \alpha_0 = v_1 \alpha_1 + v_2 \alpha_2 + \cdots + v_N \alpha_N, \tag{3.58}$$

where $v_i$ are integers and

$$\theta \alpha_0 = \alpha_1. \tag{3.59}$$

In other words, we can represent the twist in the square matrix [5, 7, 8]

$$\theta = \begin{pmatrix} & & & & v_1 \\ 1 & & & & v_2 \\ & 1 & & & v_3 \\ & & \ddots & & \vdots \\ & & & 1 & v_N \end{pmatrix}, \tag{3.60}$$

where the empty elements represent zero. We can solve these constraints and classify the irreducible lattices. We list the results $(v_i)$ in Table 3.1. Most of them are the Coxeter lattices, but we also find exceptions. The order 3 dimension 4 lattice $[SU(3)^2]$ and the order 8 dimension 6 lattice $[SU(4)^2]$ are further identified by permutation of the lattices, so irreducible [5].

### 3.4.3 Coxeter Group

Since we are modding out the torus by generalized point group $\overline{\mathsf{P}}$, the unit lattice should form regular polytopes and the orbits of the basis vector form the vertices of them. Such symmetry group is called *Coxeter group* [9]. We will discuss the Coxeter group in more detail in Chap. 12.

It is a reflection group of finite order, generated by a special element, *Coxeter element $w$* in (12.71). The vertices are generated from one basis vector $e$ as

$$\{e, we, w^2 e, \ldots w^{g-1} e\}. \tag{3.61}$$

**Table 3.1** Representation of the irreducible twists for various orders up to twelve and dimensions up to six

| Dim. | Order | Lattice | The last column ($v_i$) of $\theta$ |
|---|---|---|---|
| 1 | 2 | SU(2) | $(-1)$ |
| 2 | 3 | SU(3) | $(-1, -1)$ |
|  | 4 | SO(5) | $(-1, 0)$ |
|  | 6 | $G_2$ | $(-1, 1)$ |
| 3 | 4 | SU(4) | $(-1, -1, -1)$ |
|  | 6 | SO(7) | $(-1, 0, 0)$ |
| 4 | 3 | $[SU(3)^2]$ | $(-1, 0, -1, 0)$ |
|  | 8 | SO(9) | $(-1, 0, 0, 0)$ |
|  | 12 | $F_4$ | $(-1, 0, 1, 0)$ |
| 5 | 6 | SU(5) | $(-1, -1, -1, -1, -1)$ |
|  | 8 | SO(10) | $(-1, -1, 0, 0, -1)$ |
| 6 | 7 | SU(6) | $(-1, -1, -1, -1, -1, -1)$ |
|  | 8 | $[SU(4)^2]$ | $(-1, 0, -1, 0, -1, 0)$ |
|  | 12 | $E_6$ | $(-1, -1, 0, 1, 0, -1)$ |

The last vectors provides the last column vector in the representation matrix (3.60). The corresponding twist vector can be found in Table 3.2. The lattices $[SU(3)^2]$, $[SU(4)^2]$ are irreducible: see the main text

(We should make a good enough choice of vector $e$ such that $w^c \neq 1, c < g$, and the above elements spans $g$ dimension.) Its order is called the Coxeter number $g$ (12.12)

$$w^g = 1. \tag{3.62}$$

Most of them are identical to the Weyl groups of Lie groups and are denoted by the same symbols.

The twist (3.67) is naturally made of Coxeter elements. Diagonalizing $w$ in the complex basis (3.64), we obtain a twist vector of order $g$. The number of real dimensions the twist acts is nothing but the rank of the Lie group. For example, the SU(3) lattice is generated by the Coxeter element (3.50), which is diagonalized as $e^{2\pi/3}$ and provides the twist $\frac{1}{3}(1)$ in this complex dimension. It turns out that the entries of the twists are $\frac{1}{g}(c_a - 1), a = 1, \ldots, g/2$ for even $g$, where $c_a$ are the dimensions of the Casimir invariant, shown in Table 12.2. If $g$ is odd, always the last entry is $\frac{1'}{2}$. For instance, the Casimir invariants for $SU(n)$ and $E_6$ are, respectively, $2, 3, 4, \ldots n + 1$ and $2, 5, 6, 8, 9, 2$. We can check that they make the twists in Table 3.2.

Each Coxeter group provides a building block of the four or six dimensional twist vector. They are tabulated in Table 3.2, in which the primed element (which is always $\frac{1'}{2}$) needs another primed element of another Coxeter group to form a complete entry $\frac{1}{2}$. For instance, SU(4) × SU(4) Coxeter group can make the twist vector $\frac{1}{4}(2\ 1\ 1)$. If we completely classify the lattices, they should give us all the

**Table 3.2** Coxeter groups, labelled by the name of the Lie group having the equivalent Weyl reflections

| Cox. no. | Groups and twists |
|---|---|
| 2 | $A_1 : \frac{1}{2}(1')$ |
| 3 | $A_2 : \frac{1}{3}(1)$ |
| 4 | $A_3 : \frac{1}{4}(2'\ 1),\ B_2 : \frac{1}{4}(1)$ |
| 5 | $A_4 : \frac{1}{5}(2\ 1),\ H_2 : \frac{1}{5}(1)$ |
| 6 | $A_5 : \frac{1}{6}(3'\ 2\ 1),\ B_3 : \frac{1}{6}(3'\ 1),\ D_4 : \frac{1}{6}(3\ 1),\ G_2 : \frac{1}{6}(1)$ |
| 7 | $A_6 : \frac{1}{7}(3\ 2\ 1)$ |
| 8 | $A_7 : \frac{1}{8}(4'\ 3\ 2\ 1),\ B_4 : \frac{1}{8}(3\ 1),\ D_5 : \frac{1}{8}(4'\ 3\ 1)$ |
| 9 | $A_8 : \frac{1}{9}(4\ 3\ 2\ 1)$ |
| 10 | $A_9 : \frac{1}{10}(5'\ 4\ 3\ 2\ 1),\ B_5 : \frac{1}{10}(5'\ 3\ 1),\ D_6 : \frac{1}{10}(5\ 3\ 1),\ H_3 : \frac{1}{10}(5'\ 1)$ |
| 11 | $A_{10} : \frac{1}{11}(5\ 4\ 3\ 2\ 1)$ |
| 12 | $A_{11} : \frac{1}{12}(6'\ 5\ 4\ 3\ 2\ 1),\ B_6 : \frac{1}{12}(5\ 3\ 1),\ D_7 : \frac{1}{12}(6'\ 5\ 3\ 1),$ |
| | $E_6 : \frac{1}{12}(5\ 4\ 1),\ F_4 : \frac{1}{12}(5\ 1)$ |

We do not have $C_n$s because they are identical to $B_n$s. The rank of the group is the dimension of the lattice. Their actions provide irreducible building blocks of the twists, which are also displayed. The Coxeter numbers determine the order of twists. A primed element $\frac{1'}{2}$ should be combined with another primed element to form a twist $\frac{1}{2}$ in the complex space

possible combinations of twists (see Exercise). However, $H_2$ and $H_3$ cannot for a lattice; see below.

Then, we apply the supersymmetry condition (3.73) to form twist vectors in four and six dimensions. For instance, we could make SU(4) × SU(2) Coxeter group with twist vector $\frac{1}{4}(2\ 1)$, but this is not allowed by the supersymmetry condition (3.73).

### 3.4.4 Coxeter Lattice

Although we distinguish the Coxeter *groups* as the above sense, once we form *lattices*, different groups may form identical lattices. It is because a linear combinations of one basis set can make another. We have seen this in the $T^2/\mathbb{Z}_6$ example in Fig. 3.7. It turns out what remain are $ADE$-type lattices

$$\Lambda_{A_n},\ n \geq 1, \quad \Lambda_{D_n},\ n \geq 4, \quad \Lambda_{E_6}, \quad \Lambda_{E_7}, \quad \Lambda_{E_8}. \tag{3.63}$$

We can show the following redundancies:

$$\Lambda_{C_3} = \Lambda_{A_3} = \Lambda_{D_3},$$

$$\Lambda_{C_n} = \Lambda_{B_n} = \Lambda_{D_n}, \quad n \geq 4,$$

$$\Lambda_{G_2} = \Lambda_{SU(3)},$$

$$\Lambda_{F_4} = \Lambda_{SO(8)}.$$

Note the same dimensions and different orders among groups in each identity. In denoting the lattice, as in Table 3.4, we still distinguish the Coxeter groups to show the order and dimension. Note that although $H_2$, $H_3$ makes regular polytopes of pentagon, icosahedron, respectively, they can make neither periodic lattices nor lattices with a single tile.[1]

## 3.5   Six Dimensional Orbifold $T^6/\mathbb{Z}_N$

Our main interest will be to obtain four dimensions from ten dimensional heterotic string. It is done by compactifying and hiding six internal dimensions $d = 6$, which is the case we mainly discuss here. However, we may also have an intermediate orbifold that spans a part of the internal dimension.

First, we complexify the orthogonal coordinates

$$z^a \equiv x^{2a+2} + ix^{2a+3}. \tag{3.64}$$

In this convention, $z^1 = x^4 + ix^5$, $\bar{z}^{\bar{1}} = x^4 - ix^5$, so that the index $a = 1, 2, 3$ covers $m = 4, 5, 6, 7, 8, 9$ directions. Here, $iU_a$ are of complex structure (3.5)

$$\Lambda_d = \{e^a, iU_a e^a, i = 1, 2, 3\}. \tag{3.65}$$

We will specify later the directions of $e^a$ in concrete cases. That is, we do not necessarily take one of $e^a$ as a unit vector in the $x^i$ direction. We do not let $e^a$ to be orthogonal. For the moment, the only constraint is that they span six dimensional real dimensions. We define a six-torus $T^6$ by modding out $\mathbb{R}^6/\Lambda_d$. That is, we identify

$$z^a \sim z^a + e^a, \quad z^a \sim z^a + iU_a e^a, \quad a = 1, 2, 3. \tag{3.66}$$

The point group action $\theta \in \mathsf{P}$ has a representation in complex numbers as done in the above examples. In six dimensions, it is diagonalized as

$$\theta = \mathrm{diag}(e^{2\pi i\phi_1}, e^{2\pi i\phi_2}, e^{2\pi i\phi_3}), \tag{3.67}$$

along the direction of the lattice defined in (3.65). We form $T^6/\mathbb{Z}_N$ orbifold from the above $T^6$ by modding out further by actions of $\overline{\mathsf{P}}$

$$z^a \sim e^{2\pi i\phi_a} z^a, \quad a = 1, 2, 3, \quad \text{no summation}, \tag{3.68}$$

up to lattice translation. The most convenient choice of the complex structures is to identify the eigenvalues (3.68) as $iU_a = e^{2\pi i\phi_a}$. However, it is not obligatory as

---

[1]They can form aperiodic quasi-crystals, by Penrose tiling.

long as they are compatible

$$\theta e^i = \sum_{j=1}^{6} n_j e^j, \quad n_j \text{ integers}. \tag{3.69}$$

That is, as long as the basis vectors are integrally closed under the space group action. We will see shortly that there are in general more than one bases compatible with the same point group action.

The point group action (3.67) defines the *twist vector*

$$\phi = (\phi^1, \phi^2, \phi^3). \tag{3.70}$$

For six dimensional orbifold, we define three component vectors parameterizing the rotation. For the point group action $\theta$ with order $N$, we require $\theta^N = 1$ restricting each $N\phi_a$ an integer. Sometimes, we apply this in the eight dimensional light-cone coordinate. In that case, we add the "zeroth" component $\phi^0 = 0$ to have

$$\phi = (\phi^0; \phi^1, \phi^2, \phi^3).$$

Then, we consider irreducible building blocks of twist vector [5]. The point group action $\overline{P}$ should take a basis vector into linear combination of the basis vector as (3.69). A twist is said to be *reducible* if it is decomposed into block diagonal form. In Table 3.2, irreducible twists are shown. By combining the twists, we can make six dimensional twists. To make order three twist in six dimensions, we need to employ two dimensional twists $\frac{1}{3}(1)$. An order seven twist can only exist in six dimensions.

### 3.5.1   Supersymmetry Constraint

For a complete discussion, we borrow one physical concept although we are interested in the geometry in this chapter. For the fields living in the orbifold, we impose the condition for supersymmery for many phenomenological reasons.

When we compactify six dimensions, ten dimensional supersymmetry generators can be decomposed into

$$Q_{(10)} = Q_{(4)} \otimes Q_{(6)}. \tag{3.71}$$

Choosing one chirality, the six dimensional part $Q_{(6)}$ transforms as **4** of SO(6) = SU(4). Because the remaining parts $Q_{(4)}$ become the four dimensional generators, the dimension **4** counts the number of supersymmetries.[2] Its spinorial representation in three two-tori is represented by $|s\rangle = |s_1 \, s_2 \, s_3\rangle = |\pm \frac{1}{2}, \pm \frac{1}{2}, \pm \frac{1}{2}\rangle$ with even

---

[2]Later, we will use breaking **4** of SU(4) to **3** of SU(3) to have $\mathcal{N} = 1$.

number of minus signs (see the next chapter). Under point group, it transforms as

$$Q_{(6)} \to \exp(2\pi i s \cdot \phi) Q_{(6)}.$$

The invariant component corresponds to the unbroken supersymmetry generator. For $\mathcal{N} \geq 1$ supersymmetry, we need at least one solution. With the SU(3) rotational freedom, we choose 4D generator and call it $R$-vector

$$s = \left( +\tfrac{1}{2}, -\tfrac{1}{2}, -\tfrac{1}{2} \right) \equiv r. \tag{3.72}$$

We require the invariance

$$\phi_1 - \phi_2 - \phi_3 = 0. \tag{3.73}$$

The number of solutions, $\mathcal{N}$, counts the number of unbroken supersymmetry generators from orbifold compactification.

On the spacetime spinor $\psi$, the twist acts as

$$\psi \to e^{2\pi i \sum J_a \phi_a} \psi, \tag{3.74}$$

where $J_a$ are the rotation generators on the spinor that we will see in the next chapter. A spinor acquires a minus sign when it is rotated by $2\pi$. The order $N$ condition is extended, so that

$$N \sum_a \phi_a = 0 \mod 2. \tag{3.75}$$

We relaxed the condition to even integers because the superstring has fermionic coordinates on which we perform the same action. Then, the order of twist is the same as that of the fermions.

The results are tabulated in Table 3.3 up to $d = 4$ and Table 3.4 up to $d = 6$. The resulting orbifolds preserve, respectively, $\mathcal{N} = 2$ and $\mathcal{N} = 1$ supersymmetries in terms of four dimensional supercharges. The upper limits of the orders are, respectively, 6 and 12. For one twist, it is possible to have more than one lattices.

Another way of making a point group is to combining two point groups of $\mathcal{N} = 2$ supersymmetry. In other words, we can take two twists in Table 3.3. The resulting

**Table 3.3** Possible four dimensional orbifolds with $\mathcal{N} = 2$ supersymmetry

| Order | Coxeter lattice | $\phi$ | $\chi$ |
|---|---|---|---|
| 2 | SU(2)$^4$ | $\frac{1}{2}(1\ 1)$ | 16 |
| 3 | SU(3)$^2$ | $\frac{1}{3}(1\ 1)$ | 9 |
| 4 | SO(5)$^2$ | $\frac{1}{4}(1\ 1)$ | 1 |
| 6 | G$_2^2$ | $\frac{1}{6}(1\ 1)$ | 1 |

They also provide a building block of $T^6/(\mathbb{Z}_M \times \mathbb{Z}_N)$ orbifold

**Table 3.4** Possible six dimensional orbifolds allowing for $\mathcal{N} = 1$ supersymmetry

| P | Coxeter lattice | Twist(s) | $\chi$ |
|---|---|---|---|
| $\mathbb{Z}_3$ | $SU(3)^3$ | $\frac{1}{3}(2\ 1\ 1)$ | 27 |
| $\mathbb{Z}_4$ | $SU(2)^2 \times SO(5)^2$ | $\frac{1}{4}(2\ 1\ 1)$ | 16 |
| | $SU(2) \times SU(4) \times SO(5)$ | | |
| | $SU(4)^2$ | | |
| $\mathbb{Z}_6$-I | $SU(3) \times G_2^2$ | $\frac{1}{6}(2\ 1\ 1)$ | 3 |
| | $[SU(3)^2] \times G_2$ | | |
| $\mathbb{Z}_6$-II | $SU(2) \times SU(6)$ | $\frac{1}{6}(3\ 2\ 1)$ | 12 |
| | $SU(3) \times SO(8)$ | | |
| | $SU(2) \times SU(3) \times SO(7)$ | | |
| | $SU(2)^2 \times SU(3) \times G_2$ | | |
| | $SU(2)^2 \times [SU(3)^2]$ | | |
| $\mathbb{Z}_7$ | $SU(7)$ | $\frac{1}{7}(3\ 2\ 1)$ | 7 |
| $\mathbb{Z}_8$-I | $SO(9) \times SO(5)^*$ | $\frac{1}{8}(3\ 2\ 1)$ | 4 |
| | $[SU(4)^2]$ | | |
| $\mathbb{Z}_8$-II | $SU(2)^2 \times SO(9)$ | $\frac{1}{8}(4\ 3\ 1)$ | 8 |
| | $SU(2) \times SO(10)^*$ | | |
| $\mathbb{Z}_{12}$-I | $E_6$ | $\frac{1}{12}(5\ 4\ 1)$ | 3 |
| | $SU(3) \times F_4$ | | |
| $\mathbb{Z}_{12}$-II | $SU(2)^2 \times F_4$ | $\frac{1}{12}(6\ 5\ 1)$ | 4 |
| $\mathbb{Z}_2 \times \mathbb{Z}_2$ | $SU(2)^6$ | $\frac{1}{2}(1\ 1\ 0), \frac{1}{2}(1\ 0\ 1)$ | 32 |
| $\mathbb{Z}_2 \times \mathbb{Z}_4$ | $SO(5) \times SU(2)^2 \times SO(5)$ | $\frac{1}{2}(1\ 1\ 0), \frac{1}{4}(1\ 0\ 1)$ | 16 |
| $\mathbb{Z}_3 \times \mathbb{Z}_3$ | $SU(3)^3$ | $\frac{1}{3}(1\ 1\ 0), \frac{1}{3}(1\ 0\ 1)$ | 27 |
| $\mathbb{Z}_2 \times \mathbb{Z}_6$-I | $G_2 \times SU(2)^2 \times G_2$ | $\frac{1}{2}(1\ 1\ 0), \frac{1}{6}(1\ 0\ 1)$ | 4 |
| $\mathbb{Z}_2 \times \mathbb{Z}_6$-II | $G_2 \times SU(3) \times G_2$ | $\frac{1}{2}(1\ 1\ 0), \frac{1}{6}(2\ 1\ 1)$ | 3 |
| $\mathbb{Z}_4 \times \mathbb{Z}_4$ | $SO(5)^3$ | $\frac{1}{4}(1\ 1\ 0), \frac{1}{4}(1\ 0\ 1)$ | 8 |
| $\mathbb{Z}_3 \times \mathbb{Z}_6$ | $G_2 \times SU(3) \times G_2$ | $\frac{1}{3}(1\ 1\ 0), \frac{1}{6}(1\ 0\ 1)$ | 3 |
| $\mathbb{Z}_6 \times \mathbb{Z}_6$ | $G_2^3$ | $\frac{1}{6}(1\ 1\ 0), \frac{1}{6}(1\ 0\ 1)$ | 1 |

The lattice within the [ ] bracket involves further modding by outer automorphisms and thus is irreducible. On each line, the lattice has the same order of the entries of the twist vectors, except ones with asterisk (*). The point group is P, and the number of fixed points is $\chi$. $\mathbb{Z}_2 \times \mathbb{Z}_3$ is missing because it is identical to $\mathbb{Z}_6$-II

orbifold is $T^6/(\mathbb{Z}_N \times \mathbb{Z}_M)$. Since each twist preserves supersymmetry and commute, various combinations of them make good twists.

### 3.5.2   $T^6/\mathbb{Z}_3$ Orbifold

The $T^6/\mathbb{Z}_3$ orbifold is specified by a twist vector $\phi$

$$\phi = \left(\frac{2}{3}\ \frac{1}{3}\ \frac{1}{3}\right) \equiv \frac{1}{3}(2\ 1\ 1). \tag{3.76}$$

This is the only possible order $N = 3$ twist in six dimensions. Also, the only possible lattice choice is SU(3) × SU(3) × SU(3) as in Table 3.2. Note that although we made the six dimensional lattice by direct product of two dimensional ones, the orbifold is not. It is because the $\mathbb{Z}_3$ action (3.76) cannot be reduced to the product of two dimensional ones $\mathbb{Z}_3 \times \mathbb{Z}_3 \times \mathbb{Z}_3$.

The geometry is described by the metric $G_{ij}$. The shape of the lattice is fixed $R_1 = |e_1|R = |e_2|R = R_2, iU = e^{2\pi i/3}$, thus $\phi_{12} = 2\pi/3$. It fixes some parameters of two-torus, as in the two dimensional case, but not inter-torus parameters. Now, with the symmetric matrix $G$ it takes the form

$$
R^2 G = \begin{pmatrix}
R_1^2 & * & * & * & * & * \\
-\frac{1}{2}R_1^2 & R_1^2 & * & * & * & * \\
R_1 R_3 \cos\phi_{13} & R_1 R_3 \cos\phi_{23} & R_3^2 & * & * & * \\
R_1 R_3 \cos\phi_{14} & R_1 R_3 \cos\phi_{24} & -\frac{1}{2}R_3^2 & R_3^2 & * & * \\
R_1 R_5 \cos\phi_{15} & R_1 R_5 \cos\phi_{25} & R_3 R_5 \cos\phi_{35} & R_3 R_3 \cos\phi_{45} & R_5^2 & * \\
R_1 R_5 \cos\phi_{16} & R_1 R_5 \cos\phi_{26} & R_3 R_5 \cos\phi_{36} & R_5 R_5 \cos\phi_{46} & -\frac{1}{2}R_5^2 & R_5^2
\end{pmatrix}
$$

$$(3.77)$$

with the redundancy as (3.78). Some angles are related as follows:

$$\cos\phi_{23} = -\cos\phi_{15} - \cos\phi_{14}, \tag{3.78}$$

$$\cos\phi_{24} = \cos\phi_{13}, \tag{3.79}$$

and we may find similar relations involving $\phi_{25}, \phi_{26}, \phi_{45}, \phi_{46}$. Thus, we are left with nine independent parameters

$$R_1, R_3, R_5, \phi_{13}, \phi_{14}, \phi_{15}, \phi_{16}, \phi_{35}, \phi_{36}. \tag{3.80}$$

The fixed points are

$$\left(\theta, \sum_{i=1}^{6} m_i e_m\right),$$

with

$$(m_i, m_{i+1}) = (0,0), (1,0), (1,1), \quad i = 1, 3, 5.$$

### 3.5.3  Holonomy

The orbifold naturally leads to the notion of holonomy group. From a given point of orbifold, we can transport a tangent vector along a closed loop, taking back to the original point. Then, because of geometry, the vector rotates by some amount.

**Fig. 3.9** The $\mathbb{Z}_N$ orbifold
has a holonomy group $\mathbb{Z}_N$.
Transporting a vector, we
have an $N$-fold rotation as the
effect of identification (bold
arrows)

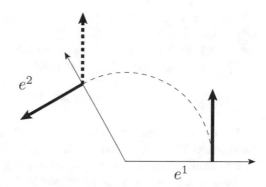

A successive transformation rotates such vector further, and these rotations form
a group. The *holonomy* group of a manifold or orbifold is defined as a group
containing all the possible rotations under the transport.[3]

We see that, for the orbifold formed by the group P, this P itself is a holonomy
group. Consider the $\mathbb{Z}_N$ example. When we parallel transport vector once around a
fixed point, it rotates by an angle $2\pi/N$ because of the identification. See Fig. 3.9.
All the possible rotations form a discrete group $\mathbb{Z}_N$ generated by $2\pi/N$ rotation,
which is nothing but the defining group $\mathbb{Z}_N$. The converse is also true: a toroidal
orbifold can also be defined in terms of a discrete holonomy group P.

The condition on the $\mathbb{Z}_N$ action from the supersymmetry (3.73) can be translated
into that the manifold should have a $\mathbb{Z}_N$ holonomy. In fact, we will see that as long
as the holonomy group P belongs to SU(3), we preserve $\mathcal{N} = 1$ supersymmetry.
All $\mathbb{Z}_N$ groups of Table 3.4 are subgroups of SU(3). They will be discussed along
with the Calabi–Yau manifold in Chap. 15.

### 3.5.4 Homology, Number of Fixed Points

We briefly discuss homological cycle, which is further explained in Sect. 15.1. Think
of a circle on the torus along around one, say $e_1$, direction. This path is wound so
cannot be contractible to a point. The winding path is a topological notion, since
deforming the path does not change the winding number. We regard this equivalent
set of closed paths as homological cycle. Since it is one dimensional, we refer it to
one-cycle. Its winding number is an integer. For a $d$ dimensional torus $T^d$, we have
$d$ independent 1-cycles which are parameterized by $d$ integer numbers. We have

$$H_1(T^d, \mathbb{Z}) = \mathbb{Z}^d.$$

---

[3]The Levi-Civita transportation does not change the length, the "norm" of tangent vector. The
name "holo-" means that the whole "holo-" group preserves the norm.

We have seen that there is a non-contractible loop in the torus along the direction of basis vectors. We can consider higher $n$ dimensional cycles, or $p$-cycles. In the torus, they are combinations of one-cycles. We have $\binom{d}{p}$ such hyperplanes and the winding is also generalized to covering number. We have

$$H_p(T^d, \mathbb{Z}) = \mathbb{Z}^{\binom{d}{p}}, \quad p = 0, \ldots, d.$$

In the orbifold case, we have another kind of non-contractible cycle around the fixed points. We have another kind of closed path, up to a point group element used in the identification. When we contract this loop, it is wound around the fixed point but cannot be shrunk into a point outside the fixed point. Later, we deal with this by smoothening the fixed points.

A point group action $\theta \in \overline{P}$ shuffles among $p$-cycles. Let $\theta_p$ be a $\binom{d}{p} \times \binom{d}{p}$ matrix representation for this $\theta$. For instance, $\theta_1 = \theta$ shuffles the basis vectors. The Lefschetz fixed point theorem states that the number of fixed points is counted by the following alternating sum [1, 2, 10]:

$$\chi = \sum_{p=0}^{d}(-1)^p \operatorname{Tr}\theta_p = \det(1 - \theta). \tag{3.81}$$

The action of $\theta \in \overline{P}$ takes a basis into another integral linear combination of the basis, and $\theta$ has a matrix representation with integral entries. Thus, (3.81) is an integer. We may express $\theta$ in the diagonal, complex basis (3.67) then

$$\chi = \det(1 - \theta) = \prod_{a=1}^{d/2}(1 - e^{\pi\phi_a i})^2 = \prod_{a=1}^{d/2} 4\sin^2(\pi\phi_a), \tag{3.82}$$

with the index $a$ running over the complexified compact dimensions. This relies on the twist vector $\phi$ only, regardless of the form of torus, i.e. the lattice.

We may obtain the number of fixed points in the higher $\theta^j$, $j > 1$ twisted sectors

$$\chi^{(j)} = \det(1 - \theta^j) = \prod_a 4\sin^2(j\pi\phi_a),$$

as long as it does not vanish. If we include noncompact dimensions, or fixed tori than points, then the eigenvalue in this direction is $e^{j\phi_a \pi i} = -1$ making $\chi = 0$. In Tables 3.3 and 3.4, we list examples of orbifolds and their $\chi$.

## Exercises

▶ **Exercise 3.1**

(a) Show that tori having the same complex structure are similar.
(b) Show that, up to overall size, equivalent tori are related to each other by the modular transformation PSL$(2, \mathbb{Z})$

$$iU \rightarrow \frac{aiU + b}{ciU + d}, \quad ad - bc = 1, \qquad (3.83)$$

with all the entries as integers and we neglect overall sign of $iU$.

▶ **Exercise 3.2** A half-cylinder having the boundary as this orbifold is known as cross-cap. What is the topology of the complete cylinder with both boundaries being cross-caps?

▶ **Exercise 3.3** Can the following be the fundamental region of $T^2/\mathbb{Z}_2$ discussed above?

▶ **Exercise 3.4** Show that combination of twists in Table 3.2 leads to twists with order $N = 2, 3, 4, 5, 6, 7, 8, 9, 10, 12, 14, 15, 18, 20, 24, 30$ up to six dimensions.

▶ **Exercise 3.5** We shall show that complete twists of order 4 in four dimensions

$$\frac{1}{4}(j_1 \ j_2), \quad j_1, j_2 = 1, 2, 3$$

can be generated by those in Table 3.2.

(a) Show that a certain twist is a multiple of another.
(b) Using the fact that each entry of the twist is defined modulo one, show that some twists are related.
(c) Show that the independent twists in the sense of (a) and (b) can be made of three twists from $A_1$, $A_3$, $B_2$.

▶ **Exercise 3.6** Construct twists and lattices of $T^8/\mathbb{Z}_N$ preserving one eights of the supersymmetry.

▶ **Exercise 3.7** Show (3.79) and (3.78). Find similar redundancies involving $\phi_{25}, \phi_{26}, \phi_{45}, \phi_{46}$.

# References

1. L.J. Dixon, J.A. Harvey, C. Vafa, E. Witten, Strings on orbifolds. Nucl. Phys. **B261**, 678–686 (1985)
2. L.J. Dixon, J.A. Harvey, C. Vafa, E. Witten, Strings on orbifolds. 2. Nucl. Phys. **B274**, 285–314 (1986)
3. L.J. Dixon, D. Friedan, E.J. Martinec, S.H. Shenker, The conformal field theory of orbifolds. Nucl. Phys. **B282**, 13–73 (1987)
4. D.G. Markushevich, M.A. Olshanetsky, A.M. Perelomov, Description of a class of superstring compactifications related to semisimple lie algebras. Commun. Math. Phys. **111**, 247 (1987)
5. J. Erler, A. Klemm, Comment on the generation number in orbifold compactifications. Commun. Math. Phys. **153**, 579–604 (1993)
6. R.L.E. Schwarzenberger, *N-dimensional Crystallography*, vol. 41 (Pitman Publishing, London, 1980)
7. A. Klemm, R. Schimmrigk, Landau-Ginzburg string vacua. Nucl. Phys. B **411**, 559–583 (1994)
8. M. Kreuzer, H. Skarke, No mirror symmetry in Landau-Ginzburg spectra! Nucl. Phys. B **388**, 113–130 (1992)
9. H.S.M. Coxeter, *Regular polytopes*, 3rd eds. (Dover Publications, New York, 1973)
10. S. Lefschetz, Intersections and transformations of complexes and manifolds. Trans. Am. Math. Soc. **28**(1), 1–49 (1926)

# Spinors

**4**

In this chapter, we introduce spinor properties in higher dimensions. We need the spinor properties in higher dimensions extensively toward obtaining 4D chiral fermions. Therefore, let us define spinors in arbitrary dimensions and study their key properties related to our phenomenological needs. Also we summarize massless supersymmetry multiplets in various dimensions.

## 4.1 Spinors in General Dimensions

We define spinors in various dimensions by naturally extending Dirac spinors. They have different properties depending on the number of dimensions.

### 4.1.1 Rotation and Vector

The group SO(4) is defined as four dimensional isotropic rotation. Its generator $J_{ab}$ has the following components:

$$[J_{ab}]_{jk} = -i(\delta_{aj}\delta_{bk} - \delta_{bj}\delta_{ak}). \tag{4.1}$$

It has $(ab)$ element $-i$ and $(ba)$ element $+i$ and generates rotation around $a$-$b$ plane. Exponentiating it we have the familiar form for the rotation matrix.

It satisfies the following commutation relation:

$$[J_{ab}, J_{cd}] = i(\delta_{ac}J_{bd} + \delta_{bd}J_{ac} - \delta_{bc}J_{ad} - \delta_{ad}J_{bc}). \tag{4.2}$$

It defines a *vector* $v^j$, which transforms as

$$v^j \rightarrow v'^{\,j} = [e^{J_{ab}\theta_{ab}}]^j{}_k v^k. \tag{4.3}$$

© Springer Nature Switzerland AG 2020
K.-S. Choi, J. E. Kim, *Quarks and Leptons From Orbifolded Superstring*,
Lecture Notes in Physics 954, https://doi.org/10.1007/978-3-030-54005-0_4

It is straightforward to extend the above 4D rotational symmetry to an arbitrary $d$ dimensional rotation. Simply, the $a, b$ running is extended to $1, \ldots, d$ and the resulting group is called SO($d$).

There is another important group similar to SO($d$): the Lorentz group denoted as SO($1, d-1$). In the Lorentz group, the definition of inner product is given by the metric

$$\eta_{ab} = \mathrm{diag}(+1, -1, -1, \ldots, -1). \tag{4.4}$$

Now, the angular momentum commutation relation becomes

$$[J_{ab}, J_{cd}] = i(\eta_{ac}J_{bd} + \eta_{bd}J_{ac} - \eta_{bc}J_{ad} - \eta_{ad}J_{bc}). \tag{4.5}$$

In what follows, mostly we will deal with the Lorentz group since we are interested in the spacetime spinors, and we can switch to the rotation group SO($d$) if needed simply by replacing $\eta$s of (4.5) to $\delta$s of (4.2).

Note that the index $a$ of (4.5) runs the number of spacetime dimensions. The representation $v$ in (4.3) is called the "vector representation" which has the dimension $d$.

### 4.1.2  Spinors in General Dimensions

**Another Set of Generators**
There exists another kind of "rotational" generators satisfying the same commutation relation, Eq. (4.5). Consider the familiar case in four dimensions. Although the entire argument will be basis independent, we work in the *chiral basis* of gamma matrices of the standard textbooks [1, 2]

$$\gamma^0 = \begin{pmatrix} 0 & 1_2 \\ 1_2 & 0 \end{pmatrix}, \quad \gamma^i = \begin{pmatrix} 0 & \sigma^i \\ -\sigma^i & 0 \end{pmatrix}, \quad i = 1, 2, 3, \tag{4.6}$$

expressed in terms of the conventional Pauli matrices $\sigma^i$ and the $2 \times 2$ identity matrix $1_2$. If we define

$$S^{\mu\nu} = \frac{i}{4}[\gamma^\mu, \gamma^\nu], \tag{4.7}$$

$S_{\mu\nu}$ satisfies the commutation relation (4.5), replacing $J_{\mu\nu}$ by $S_{\mu\nu}$. Sometimes this group is referred to as Spin(1,3) and it is a double covering group of SO(1,3). Later we will extend it to an arbitrary dimensional Lorentz group SO($1, d-1$).

A *Dirac spinor* is defined to be the object transforming under the SO($1, d-1$) group as

$$\psi_\alpha \to \psi'_\alpha = [e^{S_{ab}\theta_{ab}}]_\alpha{}^\beta \psi_\beta. \tag{4.8}$$

The Dirac spinor is reducible under the transformation (4.8). Let us define the $\gamma^5$ matrix in 4D as

$$\gamma^5 = -i\gamma^0\gamma^1\gamma^2\gamma^3 = \begin{pmatrix} -\mathbf{1}_2 & 0 \\ 0 & \mathbf{1}_2 \end{pmatrix}, \tag{4.9}$$

where only the last equality is dependent on the choice of basis. The projections $\frac{1\pm\gamma^5}{2}$ choose only the upper or lower two-entries,

$$\psi_L = \frac{1}{2}(\mathbf{1}_4 - \gamma^5)\psi, \quad \psi_R = \frac{1}{2}(\mathbf{1}_4 + \gamma^5)\psi \tag{4.10}$$

which are named as "left-handed" (L-handed) and "right-handed" (R-handed) spinors, respectively. One can check that $\gamma^5$ commutes with all the Lorentz group generators, $[\gamma^5, S^{\mu\nu}] = 0$. Therefore, one can decompose the Dirac spinors into two *Weyl spinors* according the eigenvalues of $\gamma^5$. In other words, each $\psi_L$ and $\psi_R(\equiv (\psi^c)_L)$ transform independently. These are the 4D Weyl spinors.

How can one generalize the 4D spinor to five dimensions? First, note that we need one more gamma matrix because one dimension is increased. The essential property of gamma matrices in $d$ dimensions is the following anti-commutation relation known as the Dirac–Clifford algebra:

$$\{\Gamma^M, \Gamma^N\} = 2\eta^{MN}, \quad M \text{ and } N = 0, 1, 2, \ldots, d - 1. \tag{4.11}$$

Fortunately if we define five dimensional gamma matrices as

$$\Gamma^\mu = \gamma^\mu, \quad \Gamma^4 = -i\gamma^5 \tag{4.12}$$

they satisfy the relation (4.11). We adopt the convention that the upper case roman character $M = 0, \ldots, 4$ is used for a higher dimensional index. In effect, we introduced five independent gamma matrices, each with a "vector" index, which satisfy the Dirac–Clifford algebra. These five gamma matrices define five dimensional spinors with which a 5D Dirac equation can be written.

The same extension cannot be applied to six dimensions because we have no independent sixth $4 \times 4$ gamma matrix. We should double the size of matrices. For an even $d$, by direct products of gamma and Pauli matrices, we obtain $d$ dimensional gamma matrices out of $d - 2$ dimensional ones

$$\Gamma^\mu = \gamma^\mu \otimes \sigma^3, \quad \mu = 0, \ldots, d - 3,$$
$$\Gamma^{d-2} = iI \otimes \sigma^1, \tag{4.13}$$
$$\Gamma^{d-1} = iI \otimes \sigma^2,$$

where $\gamma^\mu$ are the gamma matrices in $d - 2$ dimensions and $I$ is the unit matrix of the same size. They form generators acting on spinors as done in (4.7), we extend the group to Spin$(1, d - 1)$.

By a similarity transformation, we can choose the chiral basis in $d$ dimensions. In the chiral basis, the chirality operator is defined as

$$\Gamma = -i^{(d-2)/2}\Gamma^0\Gamma^1 \ldots \Gamma^{d-1} = \begin{pmatrix} \mathbf{1} & 0 \\ 0 & -\mathbf{1} \end{pmatrix}. \tag{4.14}$$

The Weyl spinors are the eigenstates of (4.14) with eigenvalues $\pm 1$, and the L- and R-handed Weyl spinors are given as in Eq. (4.10).

Finally, the gamma matrix (4.14) provides the $(d + 1)$-th component of the $(d + 1)$-dimensional Lorentz group.

### Spinorial Basis

It is most useful to work in the *spinorial basis* [3, 4]. We can regroup gamma matrices into ladder operators,

$$\Gamma^{0\pm} = \frac{1}{2}(\Gamma^0 \pm \Gamma^1),$$

$$\Gamma^{a\pm} = \frac{1}{2}(i\Gamma^{2a} \pm \Gamma^{2a+1}), \quad a = 1, \ldots, (d - 2)/2. \tag{4.15}$$

Then the anti-commutation relation becomes

$$\{\Gamma^{\tilde{a}+}, \Gamma^{\tilde{b}-}\} = \delta^{\tilde{a}\tilde{b}}, \quad \tilde{a} = \{0, a\} \tag{4.16}$$

while the other anti-commutators vanish. So, we can represent it as a direct product of number-$\frac{d}{2}$ $2 \times 2$ matrices. Now, we have the direct product of number-$\frac{d}{2}$ spin-$\frac{1}{2}$ systems whose eigenstates are represented by a set of respective $S_z$ eigenvalues

$$|s_0, s_1, \ldots, s_{d/2-1}\rangle. \tag{4.17}$$

This is the standard method for the construction of spinor states. We can define the "lowest state" by the state annihilated by all the annihilation operators

$$\Gamma^{\tilde{a}-}|--\cdots-\rangle = 0, \tag{4.18}$$

where $+$ and $-$ denote $+\frac{1}{2}$ and $-\frac{1}{2}$, respectively. Now we can construct a tower of states by acting the creation operators on the "lowest state"

$$|s_0, s_1, \ldots, s_{d/2-1}\rangle = \prod_{\tilde{a}=0}^{d/2-1} (\Gamma^{\tilde{a}+})^{s_{\tilde{a}}+\frac{1}{2}}|--\cdots-\rangle. \tag{4.19}$$

Note that the eigenvalues of the chirality operator $\Gamma$ in (4.14) are

$$2^{d/2} s_0 s_1 \ldots s_{d/2-1} . \tag{4.20}$$

Therefore, the positive (negative) chirality representation has even (odd) number of $-\frac{1}{2}$ eigenvalues. The Lorentz transformation can be written simply as

$$|s_i\rangle \rightarrow e^{2\pi i \sum s_j \theta_j} |s_i\rangle, \tag{4.21}$$

where $\theta_j = \theta_{2j,2j+1}$, with $j = 0, \ldots, \frac{d}{2} - 1$. For example, the spinor **16** of SO(10) has irreducible representation[1] with even (or odd depending on conventions) number of $-\frac{1}{2}$s,

$$| + + + + +\rangle$$
$$| - - + + +\rangle, \quad \text{(and 9 more permutations)} \tag{4.22}$$
$$| - - - - +\rangle, \quad \text{(and 4 more permutations)}.$$

## Majorana Spinors

In four dimensions, the charge conjugation matrix $C$ satisfies

$$C^{-1} \gamma^\mu C = -\gamma^{\mu\top}, \tag{4.23}$$

and the charge conjugated spinor is

$$\psi^c \equiv C \overline{\psi}^\top = C \psi^*. \tag{4.24}$$

The charge conjugated spinor $\psi^c$ satisfies the Dirac equation with the opposite charge from that of $\psi$. Note that the 4D rotation matrix transforms as $C^{-1} S^{\mu\nu} C = -S^{\mu\nu\top}$. Now let us extend the charge conjugation matrix to arbitrary dimensions. First we note that the eigenvalue $s_i$ is real. Therefore, the eigenvalues of the following operators are either real or pure imaginary:

$$C_1 \Gamma^0 = \Gamma^3 \Gamma^5 \ldots \Gamma^{d-1},$$
$$C_2 \Gamma^0 = \Gamma C_1, \tag{4.25}$$

where $C_1$ or $C_2$ is taken as the charge conjugation matrix. We check that

$$C^{-1} S^{MN} C = -S^{MN\top}, \tag{4.26}$$

---

[1] The dimension of the representation is the same as that of SO(1,9).

with $C$ being either $C_1$ or $C_2$. From this, we see that spinors $\psi$ and $C\overline{\psi}^\top$ transform the same under the Lorentz group, therefore the Dirac spinor is self-conjugate. On the other hand, the chirality matrix transforms as

$$C^{-1}\Gamma C = (-1)^{(d-2)/2}\Gamma^*. \tag{4.27}$$

For $d = 2$ (mod 4), the chirality of each Weyl representation is not changed under the charge conjugation, i.e. each Weyl field is its own conjugated field. For example, in these dimensions a left-handed Weyl spinor transforms again into a left-handed one under the charge conjugation. In detail, for a left-handed Weyl spinor we have $\Gamma\psi = \psi$, i.e. $\Gamma = 1$. Multiplying $C$ on the complex conjugated relation, we obtain

$$C\Gamma^*\psi^* = C\psi^* \rightarrow C\Gamma^*C^{-1}C\psi^* = \Gamma C\psi^* = \Gamma\psi^c = \psi^c, \tag{4.28}$$

where we used the relation (4.27) and the last equality of (4.28) shows that $\Gamma = 1$ for $\psi^c$ also in $d = 2$ (mod 4). In these $d = 2$ (mod 4) dimensions, the chirality remains so strongly under charge conjugation and hence it appears that the anomaly cancelation in $d = 2$ (mod 4) dimensions needs more care. Especially, there is a potential danger of *gravitational anomalies*. On the other hand, for $d = 0$ (mod 4) each Weyl spinor is the charge conjugated one of the other Weyl spinor.

The Majorana condition defines a Majorana spinor

$$\psi = \psi^c \rightarrow \psi = C\psi^* = C(C\psi^*)^* = CC^*\psi \tag{4.29}$$

whence we obtain the condition $CC^* = 1$ to have a Majorana spinor. Using the properties of gamma matrices, we obtain the following conditions in even dimensions:

$$C_1 C_1^* = (-1)^{(d-2)d/8}, \quad C_2 C_2^* = (-1)^{(d-2)(d-4)/8}. \tag{4.30}$$

In odd dimensions, we have $C^{-1}\Gamma^\mu C = (-1)^{(d-1)/2}\Gamma^{\mu\top}$ and they satisfy (4.26). Therefore, the Majorana condition can be consistently imposed in $d = 2, 3, 4, 8, 9$ dimensions mod 8. The above results are summarized in Table 4.1.

Sometimes we call the spinor pseudo-Majorana if the Majorana condition is satisfied by $C_2$.

## Symplectic Majorana Spinors

In five and six dimensions, we cannot have Majorana spinors because of the conditions (4.30). In five dimensions, the minimal spinor is Dirac, while in six dimensions it is Weyl. Both of these are four-component complex spinors which is identical to the four dimensional Dirac spinor and can be formally decomposed into four dimensional Weyl spinors

$$\begin{pmatrix} \psi_1 \\ \overline{\psi}_2 \end{pmatrix}.$$

**Table 4.1** Spinors in various dimensions

| Dim. $d$ | Majorana condition | Weyl | Majorana–Weyl | Min. rep. |
|---|---|---|---|---|
| 2 | Yes | Self | Yes | 1 |
| 3 | Yes | | | 2 |
| 4 | Yes | Complex | | 4 |
| 5 | No | | | 8 |
| 6 | No | Self | | 8 |
| 7 | No | | | 16 |
| 8 | Yes | Complex | | 16 |
| 9 | Yes | | | 16 |
| 10 | Yes | Self | Yes | 16 |
| 11 | Yes | | | 32 |
| 12 | Yes | Complex | | 64 |

Same patterns are repeated by the period 8. A spinor in $d$ dimensions has real components $2^{\lfloor d/2 \rfloor}$. Further Majorana and/or Weyl condition reduces the number of components by a half. This table is valid for $SO(1, d - 1)$

Here $\psi_1$, $\bar{\psi}_2$ are respectively left and right-handed and the bar in $\bar{\psi}_2$ indicates the chirality. From 4D viewpoint, they are related by the $SU(2)$ R-symmetry in 4D supersymmetry.

We introduce the symplectic Majorana condition between a pair of a component spinors

$$\psi_1^c = -\bar{\psi}_2, \quad \psi_2^c = \psi_1, \tag{4.31}$$

using the charge conjugation (4.24). This is generalized to symplectic (pseudo) Majorana by introducing antisymmetric matrix

$$\psi^i = \Omega^{ij} (\psi^j)^c \tag{4.32}$$

with

$$\Omega^{ij} = -\Omega^{ji}, \quad (\Omega^{ik})^* \Omega^{jk} = -\delta_j^i. \tag{4.33}$$

We may verify that the five dimensional case (4.31) satisfies $\Omega^{ij} = \epsilon^{ij}$. They are useful in constructing supersymmetry in $d = 5, 6, 7$ dimensions [5,6].

**Invariants**

By sandwiching gamma matrices between spinors, we form $SO(d)$ or $SO(1, d - 1)$ tensor representations

$$\zeta^\top C^{-1} \Gamma^{\mu_1} \Gamma^{\mu_2} \dots \Gamma^{\mu_m} \chi, \tag{4.34}$$

where we note that $\zeta^\top C^{-1}$ transforms like $\bar{\zeta}$. It is reducible and decomposed into antisymmetric tensors. We will encounter these tensors when we consider the SO(10) GUT Lagrangian and supersymmetry algebra related with D-branes.

It is easily understood that a gamma matrix converts two spinorial indices to one vectorial one

$$\Gamma^\mu_{\alpha\dot\alpha} \tag{4.35}$$

and the square of gamma matrix is proportional to the unit matrix

$$(\Gamma^\mu)^2 \propto 1 \tag{4.36}$$

in the spinor basis. Therefore, sandwiching gamma matrices between a pair of spinors with the same (opposite) chirality, only those with even (odd) numbers of vector indices survive.

As an exercise, note that **16 · 16** coupling in SO(10) contains $\zeta^\top C^{-1}\Gamma_{\mu_1\mu_2\mu_3}\chi$ transforming like a tensor of 120 dimensions and $\zeta^\top C^{-1}\Gamma_{\mu_1\mu_2\mu_3\mu_4\mu_5}\chi$ transforming like a tensor of 126 dimensions.

The number of independent $\Gamma_{\mu_1\mu_2\mu_3}$ for the vector indices $\mu_i$ is $\binom{10}{3} = 120$. For spinors, we consider $2^{d/2} \times 2^{d/2}$ real matrices. Then, we observe that multiples of gamma matrices

$$\Gamma^{\mu_1\mu_2\cdots\mu_d} \equiv \Gamma^{[\mu_1}\Gamma^{\mu_2}\ldots\Gamma^{\mu_d]}, \tag{4.37}$$

where the indices inside the square bracket are antisymmetrized, provide a complete set of bases for $2^{d/2} \times 2^{d/2}$ real matrices. We can verify it by counting the number of independent matrices of the form (4.37), $1+\binom{d}{1}+\binom{d}{2}+\binom{d}{3}+\cdots+\binom{d}{d} = 2^{d/2} \times 2^{d/2}$. A general $2^{d/2} \times 2^{d/2}$ matrix has the same number of independent elements and hence matrices (4.37) form a complete set. Since $\binom{d}{k}$ matches $\binom{d}{d-k}$, for the state $\binom{d}{k} = \binom{d}{d-k}$, the dimension of representation $\Gamma^{\mu_1\mu_2\cdots\mu_d}$ is $\frac{1}{2}\binom{10}{5} = 126$. These are useful to form a seesaw neutrino mass matrix in SO(10).

Matrices with the number of indices greater than $d/2$ are related to those with less indices by the relation

$$\Gamma^{\mu_1\cdots\mu_s}\Gamma = -\frac{i^{-k+s(s+1)}}{(d-s)!}\epsilon^{\mu_1\cdots\mu_d}\Gamma_{\mu_{s+1}\cdots\mu_d}, \tag{4.38}$$

whose four dimensional counterpart is familiar. Taking Dirac spinors on both sides, we have decomposition

$$2^{\mathbf{d}/2} \times 2^{\mathbf{d}/2} = [0] + [1] + \cdots + [d], \tag{4.39}$$

where $[m]$ denotes antisymmetric tensor with $m$ indices. In even dimensions, for Weyl spinors with the same chiralities, we follow a similar step to arrive at

$$\begin{aligned}
\mathbf{2^{d/2-1}} \times \mathbf{2^{d/2-1}} &= [1] + [3] + \cdots + [d/2], \quad d/2 \text{ odd}, \\
\mathbf{2^{d/2-1}} \times \mathbf{2^{d/2-1}} &= [0] + [2] + \cdots + [d/2], \quad d/2 \text{ even}.
\end{aligned} \tag{4.40}$$

For those with different chiralities,

$$\begin{aligned}
\mathbf{2^{d/2-1}} \times \mathbf{2^{d/2-1}} &= [1] + [3] + \cdots + [d/2], \quad d/2 \text{ even}, \\
\mathbf{2^{d/2-1'}} \times \mathbf{2^{d/2-1}} &= [0] + [2] + \cdots + [d/2], \quad d/2 \text{ odd}.
\end{aligned} \tag{4.41}$$

Adding the numbers in Eqs. (4.40) and (4.41), we obtain (4.39).

## 4.2 Supersymmetry Multiplets

We review massless multiplets of supersymmetry in five and six dimensions, focusing on the symmetry associated with orbifold actions. The massive multiplets can be obtained by merging massless ones by Higgs mechanism [6].

### 4.2.1 Five Dimensions

The minimal spinor thus the minimal SUSY generator in five dimension has eight real components. The supermultiplets can be understood in terms of $\mathcal{N} = 2$ supersymmetry multiplets in four dimensions. There are three important multiplets:

1. Vector multiplet. Denoting the multiplet using the SO(3) little group and Sp(1) $\simeq$ SU(2) R-symmetry group

$$V = \{A_M(\mathbf{3}; \mathbf{1}), \Sigma(\mathbf{1}; \mathbf{1}), \lambda(\mathbf{2}; \mathbf{2})\},$$

where we use the symplectic Majorana notation (4.31) $\lambda \equiv (\lambda^1, \epsilon^{ij}\bar{\lambda}_2)^\top$. It is composed of one vector multiplet and one chiral multiplet in four dimensions

$$V = \{v_v = (A_\mu, \lambda_2), \ v_c = (A_5 + i\Sigma, \lambda_1)\}. \tag{4.42}$$

We observed earlier that $A_\mu$ and $A_5$ are decoupled in 4D. The latter forms a complex field with $\Sigma$. The 5D SU(2) R-symmetry becomes the $SU(2)$ $R$-symmetry in four dimensional $\mathcal{N} = 2$ supersymmetry. Thus still there are symplectic Majorana relation $\lambda_L$ and $-\lambda_R$.

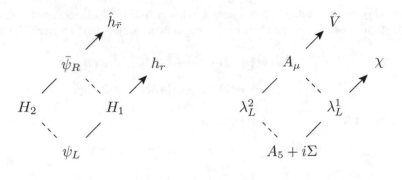

hyper-multiplet $H$         vector-multiplet $V$

**Fig. 4.1** Decomposition of $\mathcal{N}=2$ multiplets into $\mathcal{N}=1$ multiplets

2. Hypermultiplet

$$H = \{h_{\mathbf{R}} = (\phi_1, \psi_L)_{\mathbf{R}}, \ \hat{h}_{\bar{\mathbf{R}}} = (\phi_2, \bar{\psi}_R)_{\bar{\mathbf{R}}}\}. \tag{4.43}$$

It is composed of two $\mathcal{N} = 1$ chiral multiplets, for each being charge conjugate to the other. Since the two submultiplets have opposite chiralities and complex conjugation representations, dimensional reduction on a circle gives vectorlike multiplets. They also form a doublet under the SU(2) R-symmetry, satisfying the symplectic relation $\psi_L$ and $-\psi_R$.

3. Gravity multiplet

$$\{G_{\mu\nu}(\mathbf{5}; \mathbf{1}) + B_\mu(\mathbf{3}; \mathbf{1}) + \psi_\mu(\mathbf{4}; \mathbf{2})\}.$$

These are shown in Fig. 4.1. There are two directions for supersymmetry transformation which are denoted as real lines and dashed lines, respectively.

Since the supersymmetry generators commute with all the other symmetries of Lagrangian, we expect that the transformation properties under the space group are the same for the components of a supermultiplet. But we note from (5.42) that the transformation is different for the fields with the opposite chirality. Since an $\mathcal{N} = 1$ multiplet has a definite chirality, we conclude that *in terms of $\mathcal{N} = 1$, all the multiplets have the same transformation properties, including the project action P for the orbifold.*

### 4.2.2 Six Dimensions

We briefly comment on some properties of six dimensional supersymmetries for later use. Mostly the unique properties rely on the nature of spinors in six dimensions, i.e. the result of the above section.

The minimal spinor, a six dimensional Weyl spinor, is the same as the four dimensional Dirac. Thus we have the same minimal supersymmetry as in five dimensional case. We can similarly define chirality operator as in (4.14)

$$\Gamma = \Gamma^0 \Gamma^1 \Gamma^2 \Gamma^3 \Gamma^4 \Gamma^5.$$

Under charge conjugation, the chirality does not change. The supersymmetry is also chiral. According to classification of Nahm we also have $(\mathcal{N}_L, \mathcal{N}_R) = (1, 0), (2, 0), (1, 1), (2, 2), (4, 0)$ supersymmetry theories, up to chirality redefinition [7]. We have $\mathrm{Sp}(\mathcal{N}_L) \times \mathrm{Sp}(\mathcal{N}_R)$ R-symmetry.

The minimal supersymmetry is $(1, 0)$ supersymmetry. We have four kind of multiplets. They have massless multiplets represented by $SO(3,1)$ little group which is isomorphic to $SU(2) \times SU(2)$. Since it has $R$-symmetry $\mathrm{Sp}(1) \simeq SU(2)$, it would be also convenient to denote the multiplets by its quantum number $SU(2) \times SU(2) \times \mathrm{Sp}(1)$:

1. The gravity multiplet is decomposed to gravity and half-hypermultiplet.

$$g_{MN}(\mathbf{3}, \mathbf{3}; \mathbf{1}) + \psi_M(\mathbf{2}, \mathbf{3}; \mathbf{2}) + B_{MN}^+(\mathbf{1}, \mathbf{3}; \mathbf{1}).$$

Here and in what follows $M, N = 0, 1, 2, 3, 4, 5, \mu, \nu = 0, 1, 2, 3, 4$. They are decomposed into 5D $\mathcal{N} = 1$ multiplets:

$$\{G_{\mu\nu}(\mathbf{5}; \mathbf{1}) + B_\mu(\mathbf{3}; \mathbf{1}) + \psi_\mu(\mathbf{4}; \mathbf{2})\} + \{\psi(\mathbf{2}; \mathbf{1}) + \phi_5(\mathbf{1}; \mathbf{2})\}.$$

We defined $\phi \equiv g_{55}$ and the vector field $B_\mu \equiv B_{\mu5}^+$ is called graviphoton. Note the last line of the above decomposition, which is just a half of the hypermultiplet (4.43). In six dimensions, a real representation like $\mathbf{2}$ of $SU(2)$ or $\mathbf{56}$ of $E_7$ may form a half-hypermultiplet.

2. Tensor multiplet becomes the vector multiplet:

$$B_{MN}^-(\mathbf{3}, \mathbf{1}; \mathbf{1}) + \Psi(\mathbf{2}, \mathbf{1}; \mathbf{2}) + \Phi(\mathbf{1}, \mathbf{1}; \mathbf{1}).$$

They are decomposed into a vector multiplet

$$\{A_\mu(\mathbf{3}; \mathbf{1}) + \phi(\mathbf{1}; \mathbf{1}) + \psi(\mathbf{2}; \mathbf{2})\},$$

where we defined $A_\mu \equiv B_{\mu5}^-$. Since the 5D vector multiplet can have non-Abelian structure, we also find that the tensor multiplet may also have non-Abelian structure. However the rank-two field $B_{M}^- N$ can only have U(1) symmetry so non-Abelian structure cannot be easily described by Lagrangian. Indeed, it admits brany description. This multiplet contains anti-self-dual rank-two tensor $B_{\mu\nu}^-$. Its source is string, sometimes called M-string. It is a string stretched between two M5-branes. The inter-brane distance is parametrized by the scalar $\Phi$ in the same multiplet [8, 9].

3. Vector multiplet

$$A_M(\mathbf{2}, \mathbf{2}; \mathbf{1}) + \lambda(\mathbf{1}, \mathbf{2}; \mathbf{2})$$

also becomes the vector multiplet (4.42)

4. Hypermultiplet

$$\psi(\mathbf{2}, \mathbf{1}; \mathbf{2}) + \phi(\mathbf{1}, \mathbf{1}; \mathbf{4})$$

becomes the complete hypermultiplet in 5D (4.43).

The gaugino and the gravitino are chiral and its charge conjugation cannot transform into the ones with the opposite chirality, thus the theory has a potential chiral and gravitational anomalies. Especially, gauginos contribute to chiral anomaly.

In 6D (1, 1) supersymmetry, we have two parity conserving copies of the above. We have $\mathrm{Sp}(1) \times \mathrm{Sp}(1)$ R-symmetry:

1. Gravity multiplet

$$G_{\mu\nu}(\mathbf{3}, \mathbf{3}; \mathbf{1}, \mathbf{1}) + \psi_\mu(\mathbf{2}, \mathbf{3}; \mathbf{2}, \mathbf{1}) + B_{\mu\nu}^+(\mathbf{1}, \mathbf{3}; \mathbf{1}, \mathbf{1})$$

$$+ B_{\mu\nu}^-(\mathbf{3}, \mathbf{1}; \mathbf{1}, \mathbf{1}) + \bar{\psi}_\mu(\mathbf{3}, \mathbf{2}; \mathbf{1}, \mathbf{2}) + \phi(\mathbf{1}, \mathbf{1}; \mathbf{1}, \mathbf{1})$$

$$+ A_\mu(\mathbf{2}, \mathbf{2}; \mathbf{2}, \mathbf{2}) + \lambda(\mathbf{1}, \mathbf{2}; \mathbf{1}, \mathbf{2}) + \bar{\lambda}(\mathbf{2}, \mathbf{1}; \mathbf{2}, \mathbf{1}).$$

2. Vector multiplet

$$A_\mu(\mathbf{2}, \mathbf{2}; \mathbf{1}, \mathbf{1}) + \lambda(\mathbf{1}, \mathbf{2}; \mathbf{2}, \mathbf{1}) + \phi(\mathbf{1}, \mathbf{1}; \mathbf{2}, \mathbf{2}) + \bar{\lambda}(\mathbf{2}, \mathbf{1}; \mathbf{1}, \mathbf{2}).$$

Noting that the R-symmetry is isomorphic to the SO(4) local rotational symmetry of $T^4$, this theory is obtained by compactifying $\mathcal{N}_{10D} = 1$ supersymmetry in ten dimensions on $T^4$. Ten dimensional gravity and vector multiplets become respectively (1, 1) gravity and vector multiplets.

In 6D (2, 0) supersymmetry, we have $\mathrm{Sp}(2) \simeq \mathrm{SO}(5)$ R-symmetry

1. Gravity multiplet: $G_{\mu\nu}(\mathbf{3}, \mathbf{3}; \mathbf{1}) + \psi_\mu(\mathbf{2}, \mathbf{3}; \mathbf{4}) + B_{\mu\nu}^+(\mathbf{1}, \mathbf{3}; \mathbf{5})$
2. Tensor multiplet: $B_{\mu\nu}^-(\mathbf{3}, \mathbf{1}; \mathbf{1}) + \psi(\mathbf{2}, \mathbf{1}; \mathbf{4}) + \phi(\mathbf{1}, \mathbf{1}; \mathbf{5})$

There is no vector multiplet. This SO(5) R-symmetry can be understood as internal symmetry of M-theory compactified on $T^5$. It is one of the mysterious and interesting supersymmetry.

We have (2, 2) and (4, 0) supersymmetries which contain gravitons only [6].

# References

1. J. D. Bjorken and S. D. Drell, *Relativistic Quantum Fields*. (McGraw-Hill, New York, 1965)
2. M. E. Peskin, D. V. Schroeder, *An Introduction to Quantum Field Theory*. (CRC Press, Boca Raton, Florida, 2018)
3. J. Polchinski, *String Theory. Vol. 2: Superstring Theory and Beyond*. Cambridge Monographs on Mathematical Physics, vol. 12 (Cambridge University Press, Cambridge, 2007)
4. M. Gell-Mann, P. Ramond, R. Slansky, Complex Spinors and Unified Theories. Conf. Proc. **C790927**, 315–321 (1979)
5. Y. Tanii, *Introduction to Supergravity*. Springer Briefs in Mathematical Physics, vol. 1 (Springer, Tokyo, 2014)
6. J.A. Strathdee, Extended Poincare supersymmetry. Int. J. Mod. Phys. A **2**, 273 (1987)
7. W. Nahm, Supersymmetries and their representations. Nucl. Phys. **B135**, 149 (1978)
8. E. Witten, Some Comments on String Dynamics. in *String 95: Future Perspectives in String Theory*, I. Bars et al. eds. (World Scientific, Singapore, 1996)
9. P. S. Aspinwall, D. R. Morrison, U Duality and Integral Structures. Phys. Lett. **B355**, 141–149 (1995)

# Field Theoretic Orbifolds

**5**

If one starts from higher dimensional spacetime than four, it is necessary to obtain an effective four dimensional (4D) theory by hiding the extra dimensions. The idea is that, if these extra dimensions are small enough, the experiments performed so far in four dimensions could not have probed the such space.

We perform dimensional reduction of scalars, gauge bosons, and fermions, first on torus and then on orbifolds. The study of orbifold is started from the dilemma that the simple torus compactification does not allow chiral fermions in 4D. Even though the orbifold started its appearance in physics in string theory, its physics can be studied at field theory level also and in fact it is easier. So, before presenting the orbifold study in string theory [1], we introduce field theory on orbifolds first. It also provides low-energy description in the point particle limit.

We show that, with the orbifolded internal space, it is possible to have chiral fermions in an effective 4D low-energy theory. However, we also consider other possibilities for obtaining 4D chiral fermions using background scalar and magnetic flux. Also we can embed the nontrivial boundary conditions of the orbifolds into the space of unification group and break symmetries. Thus, after the work of Kawamura, the investigation of the field theory on orbifolds got a great deal of interest to obtain 4D chiral fermions toward a realistic construction of the standard model (SM) [2–13]. The anomaly structure of these theories on singular manifold has been also analyzed [14–18] and all the models considered above are anomaly free.

## 5.1 Fields on Orbifolds

We review the behavior of various fields on orbifolds. First we study the Kaluza–Klein (KK) reduction on a circle. Then we see how some zero modes are projected out by associating the global symmetry with the space group actions of orbifolds.

© Springer Nature Switzerland AG 2020
K.-S. Choi, J. E. Kim, *Quarks and Leptons From Orbifolded Superstring*,
Lecture Notes in Physics 954, https://doi.org/10.1007/978-3-030-54005-0_5

### 5.1.1   Scalar Fields

We begin by studying a complex scalar field living in flat five spacetime dimension. It is described by action

$$S = \int d^5 x \eta^{MN} \partial_M \Phi^* \partial_N \Phi. \qquad (5.1)$$

We employ the five dimensional metric $\eta_{MN} = \text{diag}(+1, -1, -1, -1, -1)$, where we index the entire five dimensions (5D) by the uppercase Latin characters and our four spacetime dimensions by Greek characters: $x^M = (x^\mu, x^4)$, $M = 0, 1, 2, 3, 4$. Sometimes we parametrize the extra dimension by $y \equiv x^4$.

**Toroidal Compactification**

Let us form a circle $S^1$ by identification

$$y \sim y + 2\pi R. \qquad (5.2)$$

This breaks the Lorentz group SO(1,4) down to SO(1,3); thus the $y$ dependence decouples and we may write the coordinate $x^M = (x^\mu, y)$. The action becomes

$$S = \int d^4 x dy (\partial_\mu \Phi^* \partial^\mu \Phi - \partial_y \Phi^* \partial_y \Phi). \qquad (5.3)$$

In the last line we evaluated the metric $\eta^{yy} = -1$ instead of raising the indices.

We first find the eigenstates $f^{(n)}(y)$ of the operator $(\partial_y)^2$

$$\left( \frac{d^2}{dy^2} + M_n^2 \right) f^{(n)}(y) = 0. \qquad (5.4)$$

They form the complete basis

$$\frac{1}{2\pi R} \int_0^{2\pi R} dy f^{(n)*}(y) f^{(m)}(y) = \delta_{nm}. \qquad (5.5)$$

Here $L$ is the length of the fundamental domain, which in this case is the circumference $2\pi R$. The first factor in the integrand in (5.5) has complex conjugation, so the eigenfunctions admit phases. If take them all real, we can make mode expansion for the a scalar in the same way. We assume for the moment that the field $\Phi$ obeys the periodicity (5.3). A complete set of eigenfunctions satisfying these is

$$\left\{ f_+^{(0)} = 1, \ f_+^{(n)}(y) = \sqrt{2} \cos \frac{ny}{R}, \ f_-^{(n)}(y) = \sqrt{2} \sin \frac{ny}{R}, \ n = 1, 2, \dots \right\}. \qquad (5.6)$$

All of them have the eigenvalues

$$M_n = \frac{n}{R}, \quad n = 0, 1, 2, \ldots. \tag{5.7}$$

We have an additional relation to make (5.6) complete

$$\frac{1}{2\pi R} \int_0^{2\pi R} dy \sin \frac{my}{R} \cos \frac{ny}{R} = 0. \tag{5.8}$$

Then we make mode expansion

$$\Phi(x, y) = \Phi_+^{(0)}(x^\mu) + \sqrt{2} \sum_{n=1}^\infty \left( \Phi_+^{(n)}(x^\mu) \cos \frac{ny}{R} + \Phi_-^{(n)}(x^\mu) \sin \frac{ny}{R} \right). \tag{5.9}$$

Plugging this back to the original action (5.3), and using the completeness relation (5.5), we obtain

$$S = \int d^4x \left[ \partial_\mu \phi_+^{(0)*} \partial^\mu \phi_+^{(0)} + \sum_{n=1}^\infty \left( \partial_\mu \phi_+^{(n)*} \partial^\mu \phi_+^{(n)} - \frac{n^2}{R^2} \phi_+^{(n)*} \phi_+^{(n)} \right) \right.$$
$$\left. + \sum_{n=1}^\infty \left( \partial_\mu \phi_-^{(n)*} \partial^\mu \phi_-^{(n)} - \frac{n^2}{R^2} \phi_-^{(n)*} \phi_-^{(n)} \right) \right]. \tag{5.10}$$

To make the kinetic function canonical, we have redefined $\sqrt{2\pi R}\Phi_\pm^{(n)} \equiv \phi_\pm^{(n)}$. The resulting field has the correct dimension as well.

As a result, we have a tower of scalar fields with the nonnegative masses $M_n = |n|/R$ given in (5.6). Notably, we have the unique massless complex scalar $\phi_+^{(0)}$, called *zero mode*. We have a pair of complex scalars $\phi_+^{(n)}$, $\phi_-^{(n)}$ for each nonzero $n$. We may visualize the spectrum as in Fig. 1.1a, where one of $\psi_L$ or $\psi_R$ is replaced by $\Phi$. In the small radius limit $R \to 0$, these massive modes become heavy and decouple. In the low-energy limit, some lowest modes $n$ below the interested energy scale $\Lambda$ ($\gg n/R$) are relevant.

## Up to Global Symmetry

The invariance of the action under the translation (5.2) does not imply that the field itself is invariant under (5.2). It is sufficient to be invariant up to a global symmetry of the action

$$\Phi(x, y + 2\pi R) \to e^{2\pi i Rc} \Phi(x, y). \tag{5.11}$$

Note that $Rc$ is defined modulo 1 and can always be made $0 \leq c < 1/R$. Then, the mode expansion is changed as

$$\Phi(x^\mu, y) = \frac{1}{\sqrt{2\pi R}} \sum_{n=0}^\infty \left( \Phi_+^{(n)}(x^\mu) \cos\left(\frac{ny}{R} + cy\right) + \Phi_-^{(n)}(x^\mu) i \sin\left(\frac{ny}{R} + cy\right) \right),$$

(5.12)

where we put a phase $i$ in front of the sine function for convenience. Thus the mass of $\phi(n)$ is shifted to

$$M_n^2 = \left(\frac{n}{R} + c\right)^2, \quad n = 0, 1, 2, \ldots$$

(5.13)

If we can assign different phase $c$ to different fields, for example, to bosons and fermions, the noninvariant field is projected out so that we can break supersymmetry [19]. We will see that we have more degrees of freedom such as global or gauged symmetry, and we have much freedom to choose the phase.

## Fields on Orbiolds

Now we compactify the fifth dimension on the $S^1/\mathbb{Z}_2$ orbifold as discussed in Chap. 3 and shown in Fig. 5.1. The point group is $\mathbb{Z}_2$ generated by the action

$$\theta : y \rightarrow -y.$$

(5.14)

This impose an additional boundary condition $\theta$

$$\Phi(x, -y) = \eta \Phi(x, y),$$

(5.15)

**Fig. 5.1** $S^1/\mathbb{Z}_2$ orbifold. The fundamental region $[0, \pi R]$ is surrounded by fixed points (bullets)

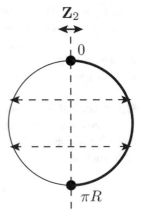

where $\eta$ is an eigenvalue. It becomes a parity $\eta = \pm 1$ from the consistency of $\mathbb{Z}_2$, $\eta^2 = 1$. This will affect the choice of the basis functions, which are solution to (5.4).

For the even case $\eta = 1$, the modes are expanded by orthonormal basis

$$\left\{ f^{(0)}(y) = 1, \ f^{(n)}(y) = \sqrt{2} \cos \frac{ny}{R}, n = 1, 2, \ldots \right\}, \qquad (5.16)$$

satisfying (5.4) and (5.5). This time, the size of the fundamental region is $L = \pi R$. We have

$$\Phi(x^\mu, y) = \sqrt{2} \sum_{n=0}^{\infty} \Phi_+^{(n)}(x^\mu) \cos \frac{ny}{R}. \qquad (5.17)$$

As a result, we projected out half of the tower. The spectrum is visualized in Fig. 1.1b, as the $\psi_L$ column.

We can alternatively take the parity odd $\eta = -1$ giving the expansion

$$\Phi(x^\mu, y) = \sqrt{2} \sum_{n=1}^{\infty} \Phi_-^{(n)}(x^\mu) \sin \frac{ny}{R}. \qquad (5.18)$$

We draw the spectrum as the $\psi_R$ column in Fig. 1.1b.

We obtain the four dimensional action similar to (5.10). The KK tower become half. Note that the zero mode exists only for the even eigenfunction $\Phi_+^{(0)}(x)$. Because we see only the lowest mode in the low-energy limit, we can explain the absence of some fields, by assigning an appropriate parity $\eta$ in (5.15).

## Bulk and Branes

We call the five dimensional space as *bulk*. We sometimes assume a hypersurface at the fixed points of $y$ which are called *branes*. We use the word *p*-brane for the $p$ spatial-dimensional objects. In the example of Fig. 5.1, they are 3-branes because the spatial dimensions of $y = 0$ and $y = \pi R$ are three.

The word "brane" originates from Dirichlet brane (D-brane) [20], which we will shortly review in Chap. 17. It has the following features

- *Charge*. An extended supersymmetry naturally possesses spatially extended objects with central charge.
- *Tension*. It has tension, positive or negative, thus acts as a source of gravity.
- *Field localization*. In string theory, an open string can end at the brane thus its low- energy state looks as localized field.

A priori, orbifold fixed points do not have these properties but are made to mimic them by hand. There are some known mechanisms for localizing fields.[1]

## 5.1.2  Gauge Fields

Similar features obtained above are shared by gauge fields. Consider an Abelian gauge field $A_M$ in the flat spacetime with the action

$$
\begin{aligned}
S &= -\frac{1}{4g_5^2} \int d^5x \eta^{MN} \eta^{PQ} F_{MP} F_{NQ} \\
&= -\frac{1}{4g_5^2} \int d^4x dy \left( F_{\mu\nu} F^{\mu\nu} - 2(\partial_\mu A_y - \partial_y A_\mu)(\partial^\mu A_y - \partial_y A^\mu) \right).
\end{aligned}
\tag{5.19}
$$

As before, the minus sign in the last term comes from the metric $\eta^{yy} = -1$. We have omitted a gauge fixing term. The gauge field appearing here enters in the covariant derivative as

$$
\mathscr{D}_M = \partial_M - i A_M,
\tag{5.20}
$$

which is different from the one in (1.5): here we absorbed the gauge coupling. The resulting action (5.19) is convenient to track the gauge coupling dependence, since the five dimensional gauge coupling $g_5$ only appears as overall normalization. It has a dimension of $(\text{length})^{1/2}$ thus nonrenormalizable. It should be completed by a theory like string theory.

We may couple this gauge field to the above scalar theory as

$$
S = \int d^4x dy (\mathscr{D}_M \Phi)^* \mathscr{D}^M \Phi,
\tag{5.21}
$$

The total action inherits the invariance under the gauge transformation

$$
\Phi(x^\mu y) \to e^{i\alpha(x^\mu, y)} \Phi(x^\mu y),
\tag{5.22}
$$

$$
A_\mu(x^\mu, y) \to A_\mu(x^\mu, y) - \partial_\mu \alpha(x^\mu, y),
\tag{5.23}
$$

$$
A_y(x^\mu, y) \to A_y(x^\mu, y) - \partial_y \alpha(x^\mu, y),
\tag{5.24}
$$

if the phase is also periodic $\alpha(x^\mu, y + 2\pi R) = \alpha(x^\mu, y)$.

---

[1]Localization of gravity in the warped background is discussed in Ref. [21] and of gauge fields in Ref. [22].

## Compactification on Circle $S^1$

We compactify the $y$-direction on a circle as before by identification $y \sim y + 2\pi R$. As a consequence of the broken 5D Lorentz symmetry, here $A_y$ is decoupled to become a scalar in the 4D effective theory. We can expand as

$$A_\mu(x, y) = A_{+\mu}^{(0)}(x^\mu) + \sqrt{2} \sum_{n=0}^{\infty} \left( A_{+\mu}^{(n)}(x^\mu) \cos \frac{ny}{R} + A_{-\mu}^{(n)}(x^\mu) \sin \frac{ny}{R} \right),$$

$$A_5(x, y) = \sqrt{2} \sum_{n=0}^{\infty} \left( A_{+y}^{(n)}(x^\mu) \cos \frac{ny}{R} + A_{-y}^{(n)}(x^\mu) \sin \frac{ny}{R} \right).$$

$$(5.25)$$

This time, the gauge field is real, so we impose the orthogonality condition similar to (5.5). However, the fundamental region is the whole circle, so we have different normalization. Again, for each nonzero $n$, we have a pair of vectors $A_{+\mu}^{(n)}$, $A_{-\mu}^{(n)}$ and a pair of real scalars $A_{+y}^{(n)}$, $A_{-y}^{(n)}$.

Integrating out $y$ gives

$$S = -\frac{2\pi R}{4g_5^2} \int d^4x \left[ \sum_{n=0}^{\infty} \left( F_{+\mu\nu}^{(n)} F_{+}^{(n)\mu\nu} + F_{-\mu\nu}^{(n)} F_{-}^{(n)\mu\nu} \right) \right.$$

$$- 2 \sum_{n=0}^{\infty} \left( \partial_\mu A_{+y}^{(n)} - \frac{n}{R} A_{+\mu}^{(n)} \right) \left( \partial^\mu A_{+y}^{(n)} - \frac{n}{R} A_{+}^{(n)\mu} \right) \tag{5.26}$$

$$\left. - 2 \sum_{n=0}^{\infty} \left( \partial_\mu A_{-y}^{(n)} - \frac{n}{R} A_{-\mu}^{(n)} \right) \left( \partial^\mu A_{-y}^{(n)} - \frac{n}{R} A_{-}^{(n)\mu} \right) \right].$$

The normalization determines four dimensional coupling

$$g_4^2 = \frac{g_5^2}{2\pi R},$$

making it dimensionless. We observe that there is no zero mode scalar. The nonzero modes $n \neq 0$ gauge fields acquired the masses $n/R$ by absorbing scalars $A_y^{(n)}$ via the Higgs mechanism. For these, we may define massive gauge fields by fixing $A_\mu^{(n)} + (R/n)\partial_\mu A_y^{(n)} \to A_\mu^{(n)}$, removing the degrees of freedom $A_y^{(n)}$. It is possible because the gauge transformation parameter, a five dimensional real scalar, obeys the same periodicity and can be also reduced in the same way and we have as many gauge symmetry as the number of KK towers. The spectrum of the tower $A_\mu^{(n)}$ is visualized as in Fig. 1.1a, where we take $\psi_L$, $\psi_R$ as two polarizations of $A_\mu^{(n)}$.

If we observe infinite tower of gauge fields with the same quantum number except the regular masses (5.6), it is explained by one compact extra dimension whose size is proportional to the gauge coupling.

This discussion can be readily generalized to non-Abelian gauge fields. In this case, the zero mode $A_{+y}^{a(0)}$ becomes a complex scalar in the adjoint representation with the index $a$. In what follows, we will use the matrix notation for non-Abelian gauge fields,

$$A_M \equiv A_M^A T^A, \tag{5.27}$$

$$F_{MN} \equiv \partial_M A_N - \partial_N A_M - i[A_M, A_N]. \tag{5.28}$$

## Compactification on $S^1/\mathbb{Z}_2$ Orbifold

As in the case of scalar field, a discrete symmetry associated with the orbifold action forces the fields to be eigenstates of the associated action, and only the zero mode in the even eigenstates survives.

To be concrete, consider the $S^1/\mathbb{Z}_2$ orbifold again by the identification $\theta$ : $y \sim -y$. Because a non-Abelian gauge field (5.27) transforms as the adjoint representation, we associate the point group action $\theta$ with the symmetry action $P$ (now a matrix) in the group space as

$$A_\mu(x, -y) = P A_\mu(x, y) P^{-1}, \tag{5.29}$$

but also note that

$$A_y(x, -y) = -P A_y(x, y) P^{-1}, \tag{5.30}$$

$$\Phi(x, -y) = P \Phi(x, y). \tag{5.31}$$

These symmetries are correlated, because under which the Lagrangian should be invariant. Taking care of the covariant derivative, the transformation property of the $A_y$ should be the same as that of $\partial_y$, while that of $A_\mu$ is the same as that of $\partial_\mu$. Hence, we have the minus sign in (5.30).

As an example consider gauge symmetry breaking of SU(3) down to SU(2) × U(1). The generators of SU(3) are the Gell-Mann matrices $\lambda^a$

$$\lambda_1 = \begin{pmatrix} 0 & 1 & 0 \\ 1 & 0 & 0 \\ 0 & 0 & 0 \end{pmatrix}, \ \lambda_2 = \begin{pmatrix} 0 & -i & 0 \\ i & 0 & 0 \\ 0 & 0 & 0 \end{pmatrix}, \ \lambda_3 = \begin{pmatrix} 1 & 0 & 0 \\ 0 & -1 & 0 \\ 0 & 0 & 0 \end{pmatrix},$$

$$\lambda_4 = \begin{pmatrix} 0 & 0 & 1 \\ 0 & 0 & 0 \\ 1 & 0 & 0 \end{pmatrix}, \ \lambda_5 = \begin{pmatrix} 0 & 0 & -i \\ 0 & 0 & 0 \\ i & 0 & 0 \end{pmatrix}, \ \lambda_6 = \begin{pmatrix} 0 & 0 & 0 \\ 0 & 0 & 1 \\ 0 & 1 & 0 \end{pmatrix}, \tag{5.32}$$

$$\lambda_7 = \begin{pmatrix} 0 & 0 & 0 \\ 0 & 0 & -i \\ 0 & i & 0 \end{pmatrix}, \ \lambda_8 = \frac{1}{\sqrt{3}} \begin{pmatrix} 1 & 0 & 0 \\ 0 & 1 & 0 \\ 0 & 0 & -2 \end{pmatrix}.$$

If we take $P$ such that

$$\lambda^a \rightarrow P\lambda^a P^{-1} = +\lambda^a, \quad a = 1, 2, 3 \text{ and } 8,$$
$$\lambda^{\hat{a}} \rightarrow P\lambda^{\hat{a}} P^{-1} = -\lambda^{\hat{a}}, \quad \hat{a} = 4, 5, 6, 7. \quad (5.33)$$

The generators $\lambda_1, \lambda_2,$ and $\lambda_3$ belong to the SU(2) subgroup and $\lambda_8$ belongs to U(1). The corresponding gauge fields $A_{+\mu}^{(0)a}\lambda^a/2$ are expanded by even functions and have zero modes, providing generators of the unbroken symmetry SU(2) × U(1) in the low-energy theory. This is achieved if we take

$$P = \pm \text{diag}(1, 1, -1), \quad (5.34)$$

which differ by overall sign. Restating the relation (5.33), the unbroken generators satisfy

$$[\lambda^a, P] = 0, \quad \text{(no summation)} \quad (5.35)$$

as in the breaking by an adjoint Higgs field.

Consider a scalar $\Phi$ in the fundamental representation **3**. Under SU(3) → SU(2) × U(1), it branches into **2** and **1**. Under $\theta$, this $\Phi$ transforms as (5.31). For the choice of $+$ sign, the doublet **2** component is invariant and survives. For $-$ sign, the singlet **1** survives.

This is explicit symmetry breaking. From (5.25) we see that, if an element of $\pm P$ is odd (even), the corresponding field $A_\mu (A_y)$, the corresponding tower is completely removed. We also have a tower of states from $A_y^{(n)}$ including the zero modes. Each combination $(A_y^{4(n)} + i A_y^{5(n)}, A_y^{6(n)} + i A_5^{7(n)})$ transforms as $(\mathbf{2}, \mathbf{1})$ under the SU(2) × U(1).

This SU(3) model had been a candidate for unification of the electroweak symmetry SU(2) × U(1) but ruled out by wrong prediction of unified coupling. However, if we consider extra dimensions, this is possibly revived [17, 23].

### Symmetry Restoration at a Fixed Point

Another point to note is the symmetry behavior at the fixed point $y = 0$ and $\pi R$. The operator $P$ breaks gauge symmetry. The matrix valued gauge field $A_M(x, y)$ has the following expansion in terms of small $y$

$$A_M(x, y) = A_M(x) + f_{MN}(x)y^M + O(y^2), \quad (5.36)$$

where $f_{MN}$ is simply a function of $x$. At a fixed point with a finite $y$, $y = \pi R$, $A_M(x, y)$ is not invariant under the change $y \rightarrow -y$, i.e. under $P$. But at the fixed point $y = 0$, $A_M(x, y) = A_M(x)$ and the transformation under $P$ leaves the SU(3) symmetry intact. In a field theoretic orbifold, this is a general feature: at the origin $y = 0$ the gauge symmetry is unbroken. On the other hand, at the fixed point with a finite $y$ the gauge symmetry is a broken one if $P$ breaks the symmetry.

**Fig. 5.2** The fundamental region of Fig. 5.1. Symmetries at the fixed points are shown

It is schematically shown in Fig. 5.2. The effective gauge symmetry below the KK compactification is the common union of the symmetries at the fixed points(SU(3) and $G_{SM}$). In the above example, the common intersection is the SU(2)× U(1). For the bulk, one may consider the bulk symmetry as SU(3) since the symmetry is broken only by the boundary conditions. But, the massless gauge bosons in the bulk do not form a complete SU(3) multiplet due to the boundary conditions, and hence one can consider the bulk symmetry as SU(2)× U(1). This example shows that one may find the effective 4D gauge symmetry by studying the massless gauge bosons in the bulk or by picking up the common union of the symmetries respected at the fixed points.

### 5.1.3  Fermions

**Five Dimensional Spinors**
We briefly review the properties of five dimensional spinors along the line discussed in Sect. 4.1.2. The 5D gamma matrices are defined in terms of four dimensional ones $\Gamma^\mu = \gamma^\mu$

$$\Gamma^0 = \begin{pmatrix} 0 & \mathbf{1}_2 \\ \mathbf{1}_2 & 0 \end{pmatrix}, \quad \Gamma^i = \begin{pmatrix} 0 & \sigma^i \\ -\sigma^i & 0 \end{pmatrix}, \quad i = 1, 2, 3, \tag{5.37}$$

with Pauli matrices $\sigma^i$ and unit matrix $\mathbf{1}_2$. The product of 4D gamma matrices

$$\Gamma^y = \Gamma^0 \Gamma^1 \Gamma^2 \Gamma^3 = \begin{pmatrix} i\mathbf{1}_2 & 0 \\ 0 & -i\mathbf{1}_2 \end{pmatrix} \tag{5.38}$$

provides the fifth, because they satisfy the anti-commutation relation

$$\{\Gamma^M, \Gamma^N\} = 2\eta^{MN} \mathbf{1}_5, \quad M, N = 0, 1, 2, 3, 4, \tag{5.39}$$

and we have four component spinor. The minimal spinor in five dimension has eight real components, which is presented in Table 4.1. The action is

$$S = \int d^5x \overline{\Psi}(x, y) \left[ i\Gamma^M (\partial_M - iA_M) - m(y) \right] \Psi(x, y), \tag{5.40}$$

where we allowed $y$ dependence of mass $m$ for a later use. There are two Lorentz invariant bilinears: The Dirac mass term $\overline{\Psi}\Psi$ and the symplectic Majorana mass term $\Psi_1^\top C_5 \Psi_2$ with the five dimensional charge conjugation matrix $C_5 = \Gamma^0 \Gamma^2 \Gamma^4$.

## Mode Expansion

It is important to note that the fifth gamma matrix is proportional to the four dimensional chirality operator defined in (4.9)

$$\Gamma^y = -i\gamma^5. \tag{5.41}$$

The four dimensional mass is the eigenvalue of the Dirac operator $i\Gamma^y$ in the fifth direction. Its eigenstate should be that of the chirality operator at the same time. So we decompose the KK modes into the $\gamma^5$ eigenstates

$$\gamma^5 \Psi_L^{(n)} = -\Psi_L^{(n)},$$
$$\gamma^5 \Psi_R^{(n)} = \Psi_R^{(n)}. \tag{5.42}$$

They become Weyl fermions in four dimensions. This sign difference should be compensated by different harmonic functions $f_L^{(n)}(y)$, $f_R^{(n)}(y)$ satisfying

$$\left( \frac{d}{dy} + m(y) \right) f_L^{(n)}(y) = -M_n f_R^{(n)}(y), \tag{5.43}$$

$$\left( \frac{d}{dy} - m(y) \right) f_R^{(n)}(y) = +M_n f_L^{(n)}(y), \tag{5.44}$$

$$\frac{1}{L} \int_0^L dy f_{L,R}^{(n)}(y) f_{L,R}^{(m)}(y) = \delta_{m+n,0}. \tag{5.45}$$

As in (5.5), the last integration is done for $f_L f_L$ or $f_R f_R$; That between $f_L$ and $f_R$ is zero. In fact, there is supersymmetry relating the basis functions with those in the bosonic case. If we act $(d/dy - m(y))$ operator on both sides of Eq. (5.43), we obtain the quadratic relation (5.4) used in the bosonic case.

Consider the case without bulk mass $m(y) = 0$. Both $f_L^{(n)}(y)$ and $f_R^{(n)}(y)$ are solutions to the equation (5.4). So we take the set (5.6) for $f_L^{(n)}$. We relate them to $f_R^{(n)}$ using the relations (5.43) and (5.44). So we have

$$\left\{ f_{L+}^{(0)}(y) = -f_{R+}^{(0)}(y) = 1, \ \ f_{L+}^{(n)}(y) = -f_{R+}^{(n)}(y) = \sqrt{2}\cos\frac{ny}{R}, \right.$$

$$\left. f_{L-}^{(n)}(y) = f_{R-}^{(n)}(y) = \sqrt{2}\sin\frac{ny}{R}, n = 1, 2, \dots \right\}, \tag{5.46}$$

giving $M_n = n/R$. Note also that if the KK mass is zero, this does not apply, because the $y$-differentiation makes a constant vanish. Thus we have the mode expansion

$$\Psi(x^\mu, y) = \sum_{n=0}^{\infty} \left( \Psi_{L+}^{(n)}(x^\mu) - \Psi_{R+}^{(n)}(x^\mu) \right) f_{L+}^{(n)}(y) + \sum_{n=1}^{\infty} \left( \Psi_{L-}^{(n)}(x^\mu) + \Psi_{R-}^{(n)}(x^\mu) \right) f_{L-}^{(n)}(y).$$
(5.47)

We have a tower of four dimensional Dirac spinors with masses $M_n = n/R$, $n = 0, 1, 2, \ldots$. Like in the case of the complex scalar, we have a pair of states for each chirality and each $n$. In Eq. (5.48), the zero mode has both left- and right-handed Weyl fermions; If charged, they should have the same representation thus the zero mode content is parity symmetric and is *not chiral* in the simple toroidal compactification.

Plugging them to the action (5.40) and integrating over $y$, we have the four dimensional action

$$S = \int d^4x \sum_{n=0}^{\infty} \left[ \sum_{\chi=L,R} \sum_{\alpha=+,-} \overline{\psi}_{\chi\alpha}^{(n)}(x^\mu) i\gamma^\mu \partial_\mu \psi_{\chi\alpha}^{(n)}(x^\mu) \right.$$

$$\left. -2M_n \left( \overline{\psi}_{L+}^{(n)}(x^\mu) \psi_{R-}^{(n)}(x^\mu) + \overline{\psi}_{R+}^{(n)}(x^\mu) \psi_{L-}^{(n)}(x^\mu) \right) \right],$$
(5.48)

with the usual redefinition $\psi_{L/R,\pm}^{(n)}(x^\mu) = \sqrt{2\pi R} \Psi_{L/R,\pm}^{(n)}(x^\mu)$ and the gamma matrix relations (5.37). Be warned that the zero modes exist only for positive parity states, $\psi_{L+}^{(0)}(x^\mu)$ and $\psi_{R+}^{(0)}(x^\mu)$.

As is well known, the mass term can be formed by fermions of different chirality. What is new here is also the mass terms are possible only between the fields with the opposite parity. Since the charge conjugate in four dimension exchanges chirality and charge, we may write the mass term as

$$-M_n \left( \overline{\psi}_{L+}^{(n)}(x^\mu) \psi_{R-}^{(n)}(x^\mu) + \overline{\psi}_{R+}^{(n)}(x^\mu) \psi_{L-}^{(n)}(x^\mu) \overline{\psi}_{L-}^{(n)}(x^\mu) \psi_{R+}^{(n)}(x^\mu) + \overline{\psi}_{R-}^{(n)}(x^\mu) \psi_{L+}^{(n)}(x^\mu) \right).$$

Finally, we may redefine four dimensional Dirac spinors

$$\psi^{(n)} = \begin{pmatrix} \psi_{L+}^{(n)} \\ \psi_{R-}^{(n)} \end{pmatrix}, \quad \psi'^{(n)} = \begin{pmatrix} \psi_{L-}^{(n)} \\ \psi_{R+}^{(n)} \end{pmatrix},$$

and we may rewrite the action (5.48 ) suggesting the decoupling

$$S = \int d^4x \sum_{n=0}^{\infty} \left[ \overline{\psi}^{(n)}(x^\mu) \left( i\gamma^\mu \partial_\mu - M_n \right) \psi^{(n)}(x^\mu) + \overline{\psi}'^{(n)}(x^\mu) \left( i\gamma^\mu \partial_\mu - M_n \right) \psi'^{(n)}(x^\mu) \right].$$

### $\mathbb{Z}_2$ **Action on Fermions**

Consider now the dimensional reduction of the fermions on the $S^1/\mathbb{Z}_2$ orbifold. The point group action is $\theta : y \to -y$, while leaving $x^\mu$ invariant. Consider the kinetic term first. For a Dirac spinor $\Psi(x, y)$, the inverted one $\Psi(x^\mu, -y)$ does not satisfy the Dirac equation. We should rearrange the components as $M\Psi(x, y)$ so that the newly obtained spinor satisfies the Dirac equation. This translates to finding a matrix $M$ satisfying

$$M^\dagger \Gamma^0 \Gamma^y M^{-1} = -\Gamma^0 \Gamma^y, \quad M^\dagger \Gamma^0 \Gamma^\mu M^{-1} = \Gamma^0 \Gamma^\mu. \tag{5.49}$$

We see that

$$M = -\eta i \Gamma^y = \eta \gamma^5 \tag{5.50}$$

satisfies the condition (5.49). We may allow a parity $\eta = \pm 1$. It is no coincidence that the transformation matrix is the *4D chirality operator.*

Therefore, under the $\mathbb{Z}_2$ action, the associated projection condition leads to

$$\Psi(x, -y) = \eta \gamma^5 \Psi(x, y). \tag{5.51}$$

When we extend this to non-Abelian symmetries, the set of phase $\eta$ becomes a projection matrix $P$ as before

$$A_\mu(x, -y) = P A_\mu(x, y) P^{-1}, \tag{5.52}$$

$$\Psi(x, -y) = P \gamma^5 \Psi(x, y). \tag{5.53}$$

It follows that the *constant* mass term is not allowed by $\mathbb{Z}_2$, although the Lorentz invariance allows it. It is because under the inversion of $y$, the Dirac mass term acquires a minus sign,

$$\overline{\Psi}(x, -y)\Psi(x, -y) \to -\overline{\Psi}(x, y)\gamma^{5\dagger}\gamma^5\Psi(x, y) = -\overline{\Psi}(x, y)\Psi(x, y), \tag{5.54}$$

where we used the fact $\gamma^{5\dagger}\Gamma^0 = -\Gamma^0 \gamma^{5\dagger}$. Instead, we can use a kink type mass

$$m(-y) = -m(y) \tag{5.55}$$

which can arise from a soliton background of scalar field.

The solution to the fifth component of the Dirac equation at the orbifold fixed points $y = 0, \pi R$ for *normalizable* fermion wave functions

$$f^{(0)}_{L,R}(y) = N_{L,R} \exp\left(\mp \int_0^y dz\, m(z)\right), \tag{5.56}$$

where $L, R$ are, respectively, correlated to $-, +$. Thus only the left-handed fermion can have normalizable zero mode $N_R = 0$. The zero mode has a *definite chirality.*

For example, if the background scalar has a linear profile $m(y) = 2a^2 y$ to have a constant energy density, the left-handed zero mode becomes

$$\psi_L = \frac{\sqrt{a}}{(\pi/2)^{1/4}} e^{-a^2 y^2}. \tag{5.57}$$

Then the fermion wave function is *localized* at $y = 0$.

**Orbifold Leads to Chiral Theory**

As before, consider the $S^1/\mathbb{Z}_2$ orbifold under identification by (5.14). The field is forced to be an eigenstate of the action (5.14), and the eigenvalues are opposite for L- and R-handed Weyl fields. Choosing $\eta = +1$, Eq. (5.42) restricts the left-handed spinor to be the even eigenstate. Therefore, we have

$$\Psi(x, y) = \Psi_L^{(0)}(x) + \sum_{n=1}^{\infty} \Psi_L^{(n)}(x) \cos\frac{ny}{R} + \sum_{n=1}^{\infty} \Psi_R^{(n)}(x) \sin\frac{ny}{R}. \tag{5.58}$$

Comparing with the simple toroidal compactification result (5.47), we have zero mode only for the left-handed spinor. Therefore, we obtain a *chiral theory* from the orbifold condition. For $\eta = -1$ we have a right-handed zero mode.

### 5.1.4   Graviton

Finally, we review the original idea of Kaluza and Klein. Consider five dimensional action

$$S = -\frac{1}{2\kappa_5^2} \int d^5 x \sqrt{\det \hat{G}} R^{(5)}, \tag{5.59}$$

where $R^{(5)}$ is the Ricci tensor made of the five dimensional metric $\hat{G}_{MN}$. We compactify the $y$-direction on the circle of radius $R$ as in (5.2). The metric $\hat{G}_{MN}$ can is decomposed as

$$\hat{G}_{MN} = \begin{pmatrix} G_{\mu\nu} + A_\mu A_\nu & A_\mu \\ A_\nu & G_{44} \end{pmatrix}, \tag{5.60}$$

giving the metric

$$ds^2 = \hat{G}_{MN} dx^M dx^N = G_{\mu\nu} dx^\mu dx^\nu + G_{44}(dx^4 + A_\mu dx^\mu)^2. \tag{5.61}$$

Note that $\hat{G}_{\mu\nu}$ is different from $G_{\mu\nu}$. Now we observe that there is the position-dependent translational symmetry in the metric,

$$x'^4 = x^4 + \lambda(x^\mu), \qquad (5.62)$$

$$A'_\mu = A_\mu - \partial_\mu \lambda(x^\mu). \qquad (5.63)$$

That is, the isometry of the interspace became the four dimensional gauge symmetry. In this case we have U(1) gauge symmetry. Plugging the metric with $G_{44} = e^{-2\sigma}$, we have the four dimensional effective action

$$S = -\frac{2\pi R}{2\kappa_5^2} \int d^4x \, e^{-\sigma} \left( R + \frac{1}{2e^{2\sigma}} F_{\mu\nu} F^{\mu\nu} + 2\partial_\mu \sigma \partial^\mu \sigma \right). \qquad (5.64)$$

We have four dimensional gravity plus U(1) gauge theory and scalar theory.

Compactification of many dimensions yield as many Abelian gauge bosons and a generalized isometry to (5.62) might even lead to a non-Abelian gauge group, whose possibility will not be pursued here.

What happens to a five dimensional scalar field $\phi(x^M)$? We perform the decomposition (5.9). For simplicity we take the limit $\tilde{G}_{44} = 1$ and $G_{\mu\nu} = \eta_{\mu\nu}$. Plugging the metric (5.61) into the equation of motion $\partial_M \partial^M \phi = 0$, we observe that a "covariant derivative" is appearing,

$$S = \int d^4x \sum_{n=-\infty}^{\infty} \left( \left( \partial_\mu + i\frac{n}{R} A_\mu \right) \phi^{(n)*} \left( \partial^\mu - i\frac{n}{R} A^\mu \right) \phi^{(n)} - \frac{n^2}{R^2} \phi^{(n)*} \phi^{(n)} \right).$$

$$(5.65)$$

The lesson we learn from this example is that the momentum $n/R$ in the compact dimension plays the both role of *charge and mass*. Compactification of more dimensions leads to as many Abelian gauge fields $A_\mu^n = G_{\mu n}$. We will see shortly that, because of the stringy effect, in the string theory we have more gauge fields carrying momentum as internal quantum number like $n$; therefore they act like charged boson $W^\pm$ in the non-Abelian gauge group.

Evidently, the remaining $(D-1)$ dimensional part of the Einstein–Hilbert action still has the general covariance and hence gives rise to the Einstein gravity. This is what Kaluza and Klein originally obtained [24,25] as a unification of the U(1) gauge theory and gravity. The main difficulty for the KK theory to be a realistic model for particle physics is that it is hard to fit $m/R$ as charge and mass at the same time. In string theory, this can be overcome because there is universal shift of the mass in the form of zero point energy. Thus we can obtain charged massless fields. Masses of elementary particles can be given by Higgs mechanism.

## 5.2    Realistic GUT Models

GUTs in extra dimensions may shed light on some difficult issues of GUTs when
they are compactified to a 4D SM. In this section, we discuss some attempts on
extra dimensional GUTs. The well-known GUT groups SU(5) and SO(10) will be
discussed in Sect. 5.2.1 for 5D and Sect. 5.2.2 for 6D, respectively, not because these
groups depend on the number of dimensions but because we want to introduce two
GUTs in an economical way. The dimension restricts the GUT group representations
only for cancelling the gauge anomalies.

### 5.2.1    SU(5) GUT in Five Dimension

We study the SU(5) GUT model in Ref. [2] to solve doublet-triplet splitting
problem.

**A More $\mathbb{Z}_2$ Action**
We compactify the fifth direction on the $S^1/(\mathbb{Z}_2 \times \mathbb{Z}_2')$ orbifold, studied in Sect. 3.2.
It is obtained by modding out $S^1$ by two $\mathbb{Z}_2$ point group actions

$$g : y \to -y, \quad h : y \to \pi R - y. \tag{5.66}$$

We may also define $y' = \pi R/2 - y$ and consider an equivalent action $h : y' \to -y'$.
This situation is drawn in Fig. 5.3.

For each action, we may associate boundary conditions. For a bulk complex
scalar field $\phi(x, y)$, we have

$$\begin{aligned}
g &: \phi(x, y) \to \eta \phi(x, -y), \\
h &: \phi(x, y') \to \eta' \phi(x, -y').
\end{aligned} \tag{5.67}$$

**Fig. 5.3** $S^1/(\mathbb{Z}_2 \times \mathbb{Z}_2')$
orbifold. Fixed points at
$y = 0, \frac{\pi R}{2}$ are denoted by
bullets. The thick arc is the
fundamental region

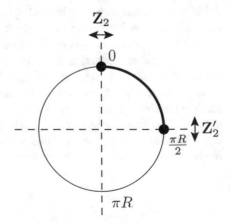

We call $\eta$ and $\eta'$ the parities of the field $\phi(x, y)$ under the $\mathbb{Z}_2 \times \mathbb{Z}'_2$ actions. This is extended to fermions as

$$g : \Psi(x, -y) = \eta \gamma^5 \Psi(x, y), \quad \eta = \pm 1,$$
$$h : \Psi(x, \pi R - y) = \eta' \gamma^5 \Psi(x, y), \quad \eta' = \pm 1. \tag{5.68}$$

If it belongs to the same hypermultiplet, it should have the same parities.

We consider the eigenstates under these parities $(\eta, \eta')$. The complete set of functions satisfying (5.43) and (5.43) are

$$\left\{ f_{++}^{(2n)}(y) = \cos \frac{2ny}{R}, \quad f_{+-}^{(2n+1)}(y) = \cos \frac{(2n+1)y}{R}, \right.$$
$$\left. f_{-+}^{(2n+1)}(y) = \sin \frac{(2n+1)y}{R}, \quad f_{--}^{(2n+2)}(y) = \sin \frac{(2n+2)y}{R}, n = 0, 1, \dots \right\}. \tag{5.69}$$

We have mode expansions

$$\phi_{++}(x^\mu, y) = \sum_{n=0}^{\infty} \Phi_{++}^{(2n)}(x^\mu) \cos \frac{2ny}{R}, \tag{5.70}$$

$$\phi_{+-}(x^\mu, y) = \sum_{n=0}^{\infty} \Phi_{+-}^{(2n+1)}(x^\mu) \cos \frac{(2n+1)y}{R}, \tag{5.71}$$

$$\phi_{-+}(x^\mu, y) = \sum_{n=0}^{\infty} \Phi_{-+}^{(2n+1)}(x^\mu) \sin \frac{(2n+1)y}{R}, \tag{5.72}$$

$$\phi_{--}(x^\mu, y) = \sum_{n=0}^{\infty} \Phi_{--}^{(2n+2)}(x^\mu) \sin \frac{(2n+2)y}{R}. \tag{5.73}$$

Note that only $\phi_{++}$ can have a zero mode, allowing a massless 4D field in the effective low-energy theory. It survived both of projections.

A technical note on compactification. The functions in (5.69) are not mutually orthogonal if we integrate their product over the fundamental region $[0, \pi R/2]$. We define the modes from the integration over the whole circle $[0, 2\pi R)$. For example, the $(++)$ mode comes from the expansion

$$\phi_{++}^{(2n)}(x^\mu) \equiv \frac{1}{\pi R} \int_0^{2\pi R} dy \cos \frac{2ny}{R} \Phi(x^\mu, y), \tag{5.74}$$

having the mass $M_n = 2|n|/R$. Note that the mass $M_n$ are different for distinct eigenfunctions.

**Model**

The 5D $\mathcal{N} = 1$ supersymmetry was reviewed in the previous chapter. Here, we introduce the minimal content of the supersymmetric SU(5) GUT in 5D [26].

- One vector multiplet $V$ in the bulk, transforming as **24** of SU(5)

$$V = \{(A_\mu, \lambda^2), (A_4 + i\Sigma, \lambda^1)\}, \tag{5.75}$$

  for gauge bosons.
- Two hypermultiplets $H^{(1)}$, $H^{(2)}$ in the bulk for Higgs bosons. They transform as **5** and $\overline{\mathbf{5}}$

$$
\begin{aligned}
H^{(1)} &= \{H_{\mathbf{5}} \equiv (H_1^{(1)}, \psi_L^{(1)}), \hat{H}_{\overline{\mathbf{5}}} = (H_2^{(1)}, \overline{\psi}_R^{(1)})\} \\
H^{(2)} &= \{\hat{H}_{\mathbf{5}} \equiv (H_1^{(2)}, \psi_L^{(2)}), H_{\overline{\mathbf{5}}} = (H_2^{(2)}, \overline{\psi}_R^{(2)})\}.
\end{aligned}
\tag{5.76}
$$

- Three generations of matter multiplets $\Phi_{\overline{\mathbf{5}}} + \Phi_{\mathbf{10}}$ at the fixed point $y = 0$, transforming as $\overline{\mathbf{5}} + \mathbf{10}$.

Note that the gauge symmetry at the fixed point $y = 0$ is SU(5) as commented before. The projection associated with orbifold imposes boundary conditions and hence some bulk fields are projected out at low energy. We will design such projections to remove unwanted fields. The matter fields located at the fixed point(s) remain intact. The gauge-invariant action is given by

$$S = \int d^5 x \mathcal{L}^{(5)} + \frac{1}{2} \int d^5 x \delta(y) \mathcal{L}^{(4)}, \tag{5.77}$$

where the Lagrangian for the Yang–Mills and Higgs in the bulk ,

$$
\begin{aligned}
\mathcal{L}^{(5)} &= \mathcal{L}_{YM}^{(5)} + \mathcal{L}_H^{(5)}, \\
\mathcal{L}_{YM}^{(5)} &= -\frac{1}{2g_5^2}\mathrm{Tr}F_{MN}^2 + \mathrm{Tr}|D_M\Sigma|^2 + \mathrm{Tr}(i\bar{\lambda}_i\Gamma^M D_M\lambda^i) - \mathrm{Tr}(\bar{\lambda}_i[\Sigma, \lambda^i]), \\
\mathcal{L}_H^{(5)} &= |D_M H_i^{(s)}|^2 + i\overline{\psi}_{(s)}\Gamma^M D_M\psi^{(s)} - (i\sqrt{2}\overline{\psi}_{(s)}\lambda^i H_i^{(s)} + \text{h.c.}) \\
&\quad - \overline{\psi}_{(s)}\Sigma\psi^{(s)} - H_{(s)}^{\dagger i}\Sigma^2 H_i^{(s)} - \frac{1}{2}g_5^2\sum_{m,A}(H_{(s)}^{\dagger i}(\sigma^m)_i^j T^A H_j^{(s)})^2, \tag{5.78}
\end{aligned}
$$

and that for matter fields at the fixed point $y = 0$,

$$
\begin{aligned}
\mathcal{L}^{(4)} &\equiv \sum_{3 \text{ families}} \int d^2\bar{\theta} d^2\theta \left(\Psi_{\overline{\mathbf{5}}}^\dagger e^{2V^A T^A}\Psi_{\overline{\mathbf{5}}} + \Psi_{\mathbf{10}}^\dagger e^{2V^A T^A}\Psi_{\overline{\mathbf{10}}}\right) \\
&\quad + \sum_{3 \text{ families}} \int d^2\theta \left(y_u H_{\mathbf{5}}\Phi_{\mathbf{10}}\Phi_{\mathbf{10}} + \hat{y}_u \hat{H}_{\mathbf{5}}\Phi_{\mathbf{10}}\Phi_{\mathbf{10}}\right.
\end{aligned}
$$

$$+y_d H_{\bar{5}} \Phi_{\bar{5}} \Phi_{10} + \hat{y}_d \hat{H}_{\bar{5}} \Phi_{\bar{5}} \Phi_{10} \Big) + \text{h.c.},$$

where $\lambda^i = (\lambda^i_L, \epsilon^{ij} \bar{\lambda}_{Lj})^T$, $\mathscr{D}_M$ is the covariant derivative (5.20) and $g_5$ is a 5D gauge coupling constant, $\sigma^m$ are Pauli matrices, the $T^A$ are SU(5) generators, $V^A T^A$ is an SU(5) vector multiplet.

**Projections**

To break symmetry, we associate the actions (5.66) with projectors

$$g \to P, \quad h \to P'. \tag{5.79}$$

It can be checked that the two actions $g$ and $h$ commute; thus we require that the associated actions $P$ and $P'$ also commute. This is achieved if we take both $P$ and $P'$ diagonal

$$P = \text{diag}(1, 1, 1, 1, 1), \tag{5.80}$$

$$P' = \text{diag}(-1, -1, -1, 1, 1). \tag{5.81}$$

The SU(5) gauge symmetry is broken down to that of the standard model (SM) gauge group, $G_{\text{SM}} = \text{SU}(3) \times \text{SU}(2) \times \text{U}(1)$. As seen in (5.33) the boundary condition acts differently on the SU(5) generators $T^A (A = 1, 2, \ldots, 24)$,

$$P' T^A P'^{-1} = T^a, \quad T^A \in G_{\text{SM}} \tag{5.82}$$

$$P' T^{\hat{A}} P'^{-1} = -T^{\hat{A}}, \quad T^{\hat{A}} \in \text{SU}(5)/G_{\text{SM}}. \tag{5.83}$$

Now consider supersymmetry. As studied before, once we determine a projection $P$ for one field, the property for the rest fields are completely fixed by the invariance of the action under gauge and supersymmetry [27, 28]. The SU(2)$_R$ symmetry exchanges $\psi_L$ and $-\psi_R$, and does $\lambda_L$ and $-\lambda_R$. So the projection acts as

$$h(x, -y) = P h(x, y)$$

$$\hat{h}(x, -y) = -P \hat{h}(x, y)$$

$$v_v(x, -y) = P v_v(x, y) P^{-1} \tag{5.84}$$

$$v_c(x, -y) = -P v_c(x, y) P^{-1},$$

in notation of (4.42) and (4.43). Therefore, all component fields in $\mathscr{N} = 2$ multiplets have definite transformation properties which include the parity.

Componentwise they have the form

$$\phi_1(x, -y) = P\phi_1(x, y)$$

$$\psi_L(x, -y) = P\psi_L(x, y)$$

$$\phi_2(x, -y) = -P\phi_2(x, y)$$

$$\psi_R(x, -y) = -P\psi_R(x, y)$$

$$A_\mu(x, -y) = PA_\mu(x, y)P^{-1} \qquad\qquad (5.85)$$

$$\lambda_2(x, -y) = P\lambda_2(x, y)P^{-1}$$

$$A_5(x, -y) = -PA_5(x, y)P^{-1}$$

$$\Sigma(x, -y) = -P\Sigma(x, y)P^{-1}$$

$$\lambda_1(x, -y) = -P\lambda_1(x, y)P^{-1}.$$

The same holds true for $P'$.

Note that two supercharges of $\mathcal{N} = 2$ have the opposite chirality. Seen in both equations above, any projector $P$ acts differently on each supercharge because of the relation (5.42): it inevitably *breaks a half of supersymmetry*. In four dimensions, the vector multiplet always survives regardless of the overall sign. The chiral multiplets with $+1$ eigenvalue survive. Overall, we have $\mathcal{N} = 1$ supersymmetry. It is explicit symmetry breaking: there is no asymptotic limit of symmetry restoration.

In Table 5.1, we list the parity assignments and the mass spectrum of the KK modes of the bulk fields. Each Higgs multiplet in $H_5(\hat{H}_{\bar{5}}, \hat{H}_5, H_{\bar{5}})$ is divided into the SU(3)-color triplet $H_C(\hat{H}_{\bar{C}}, \hat{H}_C, H_{\bar{C}})$ and the SU(2)-weak doublet $H_u(\hat{H}_d, \hat{H}_u, H_d)$. Note that only $H_u$ and $H_d$ have zero modes. All the color triplet fields have masses of order the KK scale, $\sim 1/R$. Thus the doublet-triplet splitting problem of SU(5) is nicely resolved by assigning the boundary conditions given in Eq. (5.85).

Let us scrutinize the roles of two projections in (5.79). Why do we need two discrete symmetries? As seen in Sect. 5.1.2, $P'$ alone might break gauge symmetry down to SM gauge group $G_{SM}$, as seen in the $\mathbb{Z}'_2$ parity in Table 5.1. It also breaks a half of the supersymmetries because it chooses only fermions of one chirality in each supermultiplet, for example, choosing $\lambda^{2a}$ but not $\lambda^{2\hat{a}}$ [8]. Clearly this is not sufficient, because the unwanted fields still persist, for example, charged scalars $A_\mu^{\hat{a}}(\mathbf{3}, \mathbf{2})$ and triplet Higgs $\hat{H}_C(\bar{\mathbf{3}}, \mathbf{1})$ and so on, which are not observed and possibly mediate rapid proton decay. Therefore, we need another projection $P$ to make them heavy and solve the doublet-triplet splitting problem

As discussed at the end of Sect. 4.2.1, each projection $P$ or $P'$ breaks a half of $\mathcal{N} = 2$ supersymmetries down to $\mathcal{N} = 1$. Their common intersection is also $\mathcal{N} = 1$. It is also observed that the full SU(5) group remains intact at the origin $y = 0$, as seen before, because the projection is not acted at this point. However,

**Table 5.1**   The $\mathbb{Z}_2 \times \mathbb{Z}'_2$ parities and the KK masses, in units of $\frac{1}{R}$, of the orbifolded SU(5) bulk fields

| 4D fields | Quantum numbers | $\mathbb{Z}_2 \times \mathbb{Z}'_2$ | Mass |
|---|---|---|---|
| $A_\mu^{a(2n)}$, $\lambda^{2a(2n)}$ | $(\mathbf{8}, \mathbf{1}) + (\mathbf{1}, \mathbf{3}) + (\mathbf{1}, \mathbf{1})$ | $(+, +)$ | $2n$ |
| $A_\mu^{\hat{a}(2n+1)}$, $\lambda^{2\hat{a}(2n+1)}$ | $(\mathbf{3}, \mathbf{2}) + (\bar{\mathbf{3}}, \mathbf{2})$ | $(+, -)$ | $2n + 1$ |
| $A_5^{a(2n+2)}$, $\Sigma^{a(2n+2)}$, $\lambda^{1a(2n+2)}$ | $(\mathbf{8}, \mathbf{1}) + (\mathbf{1}, \mathbf{3}) + (\mathbf{1}, \mathbf{1})$ | $(-, -)$ | $2n + 2$ |
| $A_5^{\hat{a}(2n+1)}$, $\Sigma^{\hat{a}(2n+1)}$, $\lambda^{1\hat{a}(2n+1)}$ | $(\mathbf{3}, \mathbf{2}) + (\bar{\mathbf{3}}, \mathbf{2})$ | $(-, +)$ | $2n + 1$ |
| $H_C^{(2n+1)}$ | $(\mathbf{3}, \mathbf{1})$ | $(+, -)$ | $2n + 1$ |
| $H_u^{(2n)}$ | $(\mathbf{1}, \mathbf{2})$ | $(+, +)$ | $2n$ |
| $\hat{H}_{\bar{C}}^{(2n+1)}$ | $(\bar{\mathbf{3}}, \mathbf{1})$ | $(-, +)$ | $2n + 1$ |
| $\hat{H}_d^{(2n+2)}$ | $(\mathbf{1}, \mathbf{2})$ | $(-, -)$ | $2n + 2$ |
| $\hat{H}_C^{(2n+1)}$ | $(\mathbf{3}, \mathbf{1})$ | $(-, +)$ | $2n + 1$ |
| $\hat{H}_u^{(2n)}$ | $(\mathbf{1}, \mathbf{2})$ | $(-, -)$ | $2n + 2$ |
| $H_{\bar{C}}^{(2n+1)}$ | $(\bar{\mathbf{3}}, \mathbf{1})$ | $(+, -)$ | $2n + 1$ |
| $H_d^{(2n)}$ | $(\mathbf{1}, \mathbf{2})$ | $(+, +)$ | $2n$ |

in the bulk there is only $G_{SM}$ and the dimensional reduction yields only this gauge group.

## 5.2.2   SO(10) GUT in Six Dimension

We discuss the next simplest GUT SO(10) in 6D [7, 9, 18]. One interesting feature of 6D SO(10) is that each fixed point respects different gauge groups. In this case the low-energy effective theory is the common intersection of the groups respected at each fixed point. This is the generalization of the symmetry breaking we discussed in the previous subsection and has more complex structure. This kind of the common intersection as the gauge group appears in string orbifold also.

### SO(10) GUT

Group theoretically, the SO(10) has some merits over the SU(5) except not being the minimal one. Firstly, the fifteen chiral fields are put in a single representation **16** together with an SU(5) singlet neutrino and realizes our *theme of unification*. Second, since it contains an SU(5) singlet neutrino in the spinor **16** of SO(10), it is possible to introduce small Majorana neutrino masses through the see-saw mechanism [29, 30]. Of course, one can introduce SU(5) singlets in the SU(5) GUT and introduce a similar see-saw mechanism, but in the SO(10) GUT the see-saw neutrino mass is related to other couplings dictated from the SO(10) symmetry. Third, because the top and bottom quarks are put in the same representation **16**, it is possible to relate their masses, i.e. the so-called top-bottom unification is possible. Thus, it seems that the SO(10) GUT has its own merit to study [31–33].

To break SO(10) down to $G_{SM}(\times U(1))$ just by orbifolding, we need to go beyond five dimension for the following reasons. With a single projection $P$ we can break SO(10) down to one of its maximal groups only, thus cannot directly go to $G_{SM}$. We need at least two projections. Also, if we want the doublet-triplet splitting from orbifold, one needs more projections. Thus we need at least three projections. As seen in Sect. 3.2, the discrete group on a circle can be at most $\mathbb{Z}_2 \times \mathbb{Z}_2$. Moreover, if we use the second $\mathbb{Z}_2$ action to break gauge group further, there remain unwanted massless fields from $A_5$ components of the vector multiplet [18]. These do not form a complete representation of GUT groups. Thus the gauge coupling unification may not be accomplished.

In addition, two extra dimensional compactification shows the essential features of compactification in still higher dimensions. In the following chapters on the string compactification, we encounter more internal dimensions to be compactified. These features include the localization of gauge groups and matter spectra at fixed points due to the presence of Wilson lines.

### Subgroups of SO(10)

Before presenting a full orbifold model, let us recapitulate the group theoretical aspects of SO(10). As pointed out in Sec. 2.2, the interesting rank 5 subgroups of SO(10) are

   (i)  $SU(5) \times U(1)$, where $SU(5)$ is the Georgi–Glashow (GG) group [26]
  (ii)  $SU(4)_c \times SU(2)_L \times SU(2)_R$, the Pati–Salam (PS) group [34]
 (iii)  $SU(5)' \times U(1)_X$, the flipped SU(5) (flipped SU(5)) group [35].[2]

Of course, the differences arise in the way of embedding matter contents. In Table 5.2 the sixteen chiral fields are classified under these three cases. All these groups can be obtained when one nontrivial $\mathbb{Z}_2$ boundary conditions are imposed.

Since $G_{SM}$ is a subgroup of each Cases (i), (ii), and (iii), $G_{SM}$ can be a common union of them. Consider for example, the quark doublet $q$ and the lepton doublet $l$. Both of these complete the PS representation $(\mathbf{4}, \mathbf{2}, \mathbf{1})$ under the PS group. However, they belong to two different representations under the GG-SU(5). So, we must split $\mathbf{10}_F$ and $\overline{\mathbf{5}}_F$ so that $q$ and $l$ themselves become a complete representation. It is most easily achieved from chopping off $\overline{\mathbf{5}}_F$ so that $l$ is split. Then, $\mathbf{10}_F$ is also split to produce $q$. When we chop off $\overline{\mathbf{5}}_F$ into $\mathbf{3} \oplus \mathbf{2}$, the unbroken group is $G_{SM}$. For the part of the PS group, the fourth color is separated from the remaining three colors to produce $q$ and $l$. This means that the common intersection of the $SU(5)_{GG} \times U(1)$ and the PS group is $G_{SM} \times U(1)$. In this way, one can confirm that the common

---

[2]The flipped SU(5) is the flipped SU(5) in field theory models. In string compactifications in Chaps. 13 and 14, we will use the word anti-SU(5) [36] instead of flipped SU(5), to stress the needed antisymmetric representations.

**Table 5.2** The chiral fields are L-handed

| $G_{SM}$ | Fields | GG-SU(5) | PS 422 | Flipped SU(5) |
|---|---|---|---|---|
| $q$ | $(3, 2)$ | $\mathbf{10}_F$ | $(4, 2, 1)$ | $\mathbf{10}_F$ |
| $u^c$ | $(\bar{3}, 1)$ | $\mathbf{10}_F$ | $(4, \bar{1}, 2)$ | $\bar{\mathbf{5}}_F$ |
| $d^c$ | $(\bar{3}, 1)$ | $\bar{\mathbf{5}}_F$ | $(4, \bar{1}, 2)$ | $\mathbf{10}_F$ |
| $l$ | $(1, 2)$ | $\bar{\mathbf{5}}_F$ | $(4, 2, 1)$ | $\bar{\mathbf{5}}_F$ |
| $e^c$ | $(1, 1)$ | $\mathbf{10}_F$ | $(4, \bar{1}, 2)$ | $\mathbf{1}_F$ |
| $N$ | $(1, 1)$ | $\mathbf{1}_F$ | $(4, \bar{1}, 2)$ | $\mathbf{10}_F$ |

Note that the discrete $\mathbb{Z}_2$ element of $SU(2)_R$ exchanges up and down type quarks and leptons, thus this relates GG-SU(5) and flipped SU(5)

intersection of any two columns of Table 5.2 contains $G_{SM}$. If we considered the PS group and the flipped SU(5) group, then the common intersection is again $G_{SM} \times U(1)$. Similarly, the common intersection of $SU(5)_{GG} \times U(1)$ and the flipped SU(5) is $G_{SM} \times U(1)$.

The subgroup structure of the SO(10) can be understood more clearly when we classify the 45 generators $T^a$ [9, 18]. They are represented by *imaginary and antisymmetric (thus Hermitian) $10 \times 10$ matrices*. To deal with SU(5) subgroup, the standard convention is to embed U($n$) group into SO($2n$). Then, it is convenient to write these imaginary and antisymmetric generators as direct products of $2 \times 2$ and $5 \times 5$ matrices, giving

$$SO(10): \quad \mathbf{1}_2 \otimes A_5, \ \sigma^1 \otimes A_5, \ \sigma^2 \otimes S_5, \ \sigma^3 \otimes A_5 . \tag{5.86}$$

Here $\mathbf{1}_2$ and $\sigma_{1,2,3}$ are the $2 \times 2$ unit matrix and the Pauli matrices; $S_n$ and $A_n$ are real and symmetric $n \times n$ matrices, and imaginary and antisymmetric $n \times n$ matrices, respectively. It is easily checked that the surviving generators are fifteen $S_5$s and ten $A_5$s. The U(5) subgroup of SO(10) is then generated by

$$U(5): \quad \mathbf{1}_2 \otimes A_5, \ \sigma^2 \otimes S_5 \tag{5.87}$$

whose total number is 25 the number of U(5) generators. Excluding U(1) generator $\sigma^2 \otimes \mathbf{1}_5$, the rest forms the generators of SU(5), which are traceless under this convention.

It is useful to denote the matrix in the 3-2-3-2 block highlighting the $2 \times 2$ block structure [33]. Writing our $2 \times 2$ $\sigma$ space in the $\{i, 5+i\}$ coordinates, $10 \times 10$ SO(10) generators can be written as

$$M = \begin{pmatrix} A + C & B + S \\ B - S & A - C \end{pmatrix}_{10 \times 10} \tag{5.88}$$

where $A$, $B$, and $C$ are antisymmetric and $S$ is symmetric. For example, we have

$$\mathbf{1}_2 \otimes A_5 = \begin{pmatrix} A_3 & A_X & 0 & 0 \\ -A_X^\top & A_2 & 0 & 0 \\ 0 & 0 & A_3 & A_X \\ 0 & 0 & -A_X^\top & A_2 \end{pmatrix},$$

$$\sigma^1 \otimes B_5 = \begin{pmatrix} 0 & 0 & B_3 & B_X \\ 0 & 0 & -B_X^\top & B_2 \\ B_3 & B_X & 0 & 0 \\ -B_X^\top & B_2 & 0 & 0 \end{pmatrix}.$$

The unitary transformation is given by

$$U = \frac{1}{\sqrt{2}} \begin{pmatrix} \mathbf{1}_5 & i\mathbf{1}_5 \\ i\mathbf{1}_5 & \mathbf{1}_5 \end{pmatrix}. \tag{5.89}$$

Under the unitary transformation, $M$ transforms to

$$M' = \begin{pmatrix} A - iS & B - iC \\ B + iC & A + iS \end{pmatrix}_{10 \times 10}. \tag{5.90}$$

In this case, the 24 SU(5) generators are $A$ and traceless $S$. The U(1) generator of U(5) is

$$I_{\mathrm{SU}(5)} = \begin{pmatrix} \mathbf{1}_5 & 0 \\ 0 & -\mathbf{1}_5 \end{pmatrix}_{10 \times 10} \tag{5.91}$$

which belongs to $M'$ up to a phase.

**Projections**  With one of the following projections $P_i$ we can break SO(10) to one of them[3]

$$P_{\mathrm{GG}} \equiv \sigma^2 \otimes \mathbf{1}_5, \tag{5.92}$$

$$P_{\mathrm{F}} \equiv \sigma^2 \otimes \mathrm{diag}(1, 1, 1, -1, -1), \tag{5.93}$$

$$P_{\mathrm{PS}} \equiv \mathbf{1}_2 \otimes \mathrm{diag}(1, 1, 1, -1, -1). \tag{5.94}$$

---

[3]In Ref. [18], $P_{\mathrm{GG}}$, $P_{\mathrm{F}}$, and $P_{\mathrm{PS}}$ are represented as $P_2$, $P_3$, and $P_4$, respectively.

In the above notation, we may also write as

$$P_{\text{GG}} = \begin{pmatrix} 0 & 0 & -i\mathbf{1}_3 & 0 \\ 0 & 0 & 0 & -i\mathbf{1}_2 \\ i\mathbf{1}_3 & 0 & 0 & 0 \\ 0 & i\mathbf{1}_2 & 0 & 0 \end{pmatrix},$$

and so on. The unbroken generators satisfy the condition (5.35)

$$[T^a, P] = 0. \tag{5.95}$$

We assigned the name of the projectors according to the resulting subgroup. In what follows we treat each subgroup separately.

**Georgi–Glashow SU(5)**  The above choice of SU(5) embedding is the Georgi–Glashow type, because in the conventional basis, the hypercharge generator $Y \propto T^{24}$ is diagonal and its eigenvalues are proportional to that of $\bar{\mathbf{5}}$ of GG-SU(5) embedding.

By embedding the SM gauge group into this U(5), we can divide the $5 \times 5$ matrix further by choosing the first three indices $1, 2, 3$ for the SU(3)$_c$ and the last two indices $4, 5$ for the SU(2)$_L$. Then, $A_3$, $S_3$, $A_2$, and $S_2$ contain the SM group generators. The total number of these are 13 out of which the identity generator is not belonging to the SM gauge group. The remaining 12 generators are those of the SM. Now, let us denote the left-over pieces of $A_5$ and $S_5$ as $A_X$ and $S_X$. Then, the generators of the Georgi–Glashow SU(5)$_{\text{GG}} \times$U(1) subgroup can be grouped as

$$\text{SU(5)}_{\text{GG}} \times \text{U(1)} : \frac{\mathbf{1}_2 \otimes A_3, \ \mathbf{1}_2 \otimes A_2, \ \mathbf{1}_2 \otimes A_X}{\sigma^2 \otimes S_3, \ \sigma^2 \otimes S_2, \ \sigma^2 \otimes S_X}. \tag{5.96}$$

The total number of generators in Eq. (5.96) is 25.

**Pati–Salam SU(4)$\times$SU(2)$\times$SU(2)**  The Pati–Salam group is obtained as

$$\text{SO(10)} \ \rightarrow \ \text{SO(6)} \times \text{SO(4)} \simeq \text{SU(4)} \times \text{SU(2)} \times \text{SU(2)}.$$

It means that the SO(10) generators are partitioned into blocks of dimensions 6 and 4 and the diagonal blocks survived, i.e. the SO(6)$\times$SO(4) generators remain, which is equivalent to the PS group. The $P_{\text{PS}}$ of (5.94) can be denoted as $\mathbf{1}_6 \otimes (-\mathbf{1}_4)$ and yield the ones we want, by a similar relation as (5.33). The resulting generators of the Pati–Salam group SU(4)$_c \times$SU(2)$_L \times$SU(2)$_R$ are given by

$$\text{PS 422} : \frac{(\mathbf{1}_2, \sigma^1, \sigma^3) \otimes A_3, \ \sigma^2 \otimes S_3}{(\mathbf{1}_2, \sigma^1, \sigma^3) \otimes A_2, \ \sigma^2 \otimes S_2,} \tag{5.97}$$

where the first line lists the SU(4) generators and the second line lists the SU(2)×SU(2) generators. Here the total number of generators is 21.

**Flipped SU(5)′×U(1)** Consider the SU(2)$_R$ rotation generated by $T_{3R}$ of the PS group. Rotating by angle $\pi$ flips the signs of the last two entries in $P_F$. Thus, we have $P_F = P_{PS} P_{GG}$. This rotation flips the representations of SU(5)$_{GG}$ are flipped to those of the flipped SU(5):

$$e_L^c \leftrightarrow N$$
$$d_L^c \leftrightarrow u_L^c$$

from which the name "flipped" has been coined.

If the U(1) part is not the identity as given in Eq. (5.93), then traceless $Q_{em}$ in SO(10) must contain the U(1)$_X$ piece in the flipped SU(5), i.e. $U(1)_X \in$ flipped SU(5). Here, the commutators of $\sigma^2 \otimes \mathrm{diag}(1, 1, 1, -1, -1)$ with the generators $\sigma^1 \otimes A_X$ and $\sigma^3 \otimes A_X$ are nonvanishing but put again in the set. Note that there are two nonvanishing factors $[\sigma^2, \sigma_{1,3}]$ and $[\mathrm{diag}(1, 1, 1, -1, -1), A_X]$. If we assign one, for a nonvanishing commutator, we end up with $+$ for the two commutator factors; thus the flipped SU(5) gauge bosons carry the $+$ parity. The surviving generators are the following [34, 36].

$$\mathrm{SU(5)}_F \times \mathrm{U(1)}_X' : \frac{\mathbf{1}_2 \otimes A_3, \ \mathbf{1}_2 \otimes A_2, \ \sigma^1 \otimes A_X}{\sigma^2 \otimes S_3, \ \sigma^2 \otimes S_2, \ \sigma^3 \otimes A_X}. \tag{5.98}$$

Here also, the total number of generators is 25.

**SM as the Common Intersection** It can be clearly seen from the above decompositions that the intersection of any combination of two of the GG-SU(5), flipped SU(5), and PS 422, is $G_{SM} \times$ U(1), whose common generators are

$$\begin{aligned} \mathbf{1}_2 \otimes A_3, \ \mathbf{1}_2 \otimes A_2, \\ \sigma^2 \otimes S_3, \ \sigma^2 \otimes S_2, \end{aligned} \tag{5.99}$$

which are underlined in Eqs. (5.96, 5.98, and 5.97). The common intersection of these SO(10) subgroups is shown schematically in Fig. 5.4. The U(1) generator in the common intersection is $\sigma^2 \otimes \mathbf{1}_5$.

Because the Cartan subalgebra always commutes with all the projectors, as in (5.95), we need another method to reduce the rank 5 of SO(10) down to the rank 4 of the SM. One way is to employ the Higgs doublet. For example, 5D SO(10) models can be considered even if one $\mathbb{Z}_2$ is used for the rank preserving breaking SO(10)→SU(4)×SU(2)×SU(2) and eventually to $G_{SM}$ is realized by the VEV of Higgs $\langle(\mathbf{4}, \bar{\mathbf{1}}, \mathbf{2})\rangle$.

Another method is to employ continuous Wilson lines, where the orbifold actions do not commute and hence the projections do not commute. Therefore, some Cartan

**Fig. 5.4** The common intersection of SO(10) subgroups

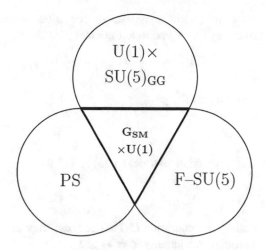

subalgebra do not remain invariant so that the rank is reduced. This is explained in Sect. 12.4.2.

## $T^2/\mathbb{Z}_2$ Orbifold and Gauge Symmetries at Fixed Points

We compactify two dimensions $y^1$, $y^2$ on $T^2/\mathbb{Z}_2$ orbifold considered in Sect. 3.3. The torus $T^2$ are made by identification

$$y^m \sim y^m + e_i^m. \tag{5.100}$$

The point group is a $\mathbb{Z}_2$ generated by

$$\theta : (y^1, y^2) \to (-y^1, -y^2). \tag{5.101}$$

The 6D $\mathcal{N} = 1$ SO(10) gauge multiplet can be decomposed into 4D $\mathcal{N} = 1$ SUSY multiplets as in (4.42): a vector multiplet $V$ and chiral adjoint multiplet $\Phi$. The bulk action is given by

$$S = \int d^6 x \left\{ \frac{1}{4g^2} \text{Tr} \left[ \int d^2\theta \, \mathcal{W}^\alpha \mathcal{W}_\alpha + \text{h.c.} \right] \tag{5.102}$$

$$+ \int d^4\theta \frac{1}{g^2} \text{Tr} \left[ (\sqrt{2}\partial^\dagger + \Phi^\dagger)e^{-V}(-\sqrt{2}\partial + \Phi)e^V + \partial^\dagger e^{-V}\partial e^V \right] \right\},$$

where $V = V^a T^a$, $\Phi = \Phi^a T^a$, and $\partial = \partial_5 - i\partial_6$.

We associate translations $e_i$ with the phase phases $T_i$. That is,

$$V(\vec{y} + e_i) = T_i V(\vec{y}) T_i^{-1}, \tag{5.103}$$

$$\Phi(\vec{y} + e_i) = T_i \Phi(\vec{y}) T_i^{-1}. \tag{5.104}$$

Now, we require a consistency condition. Successive transportation by $e_1$, $e_2$, $-e_1$, and $-e_2$ move a point to the same point,

$$V(y) = T_2^{-1} T_1^{-1} T_2 T_1 V(y) T_1^{-1} T_2^{-1} T_1 T_2, \qquad (5.105)$$

thus we have

$$T_2 T_1 = T_1 T_2. \qquad (5.106)$$

The projection matrices (5.93) and (5.94)

$$T_1 = P_{GG}, \quad T_2 = P_F. \qquad (5.107)$$

satisfy the condition (5.106) because they are all diagonal. With the action $\theta$ we associate the identity $Z = \mathbf{1}_2 \otimes \mathbf{1}_5$,

$$V(-\vec{y}) = Z V(\vec{y}) Z^{-1}, \qquad (5.108)$$

$$\Phi(-\vec{y}) = -Z \Phi(\vec{y}) Z. \qquad (5.109)$$

Note that $Z$ does not belong to the SO(10) generators (5.86) since it is real and symmetric, breaking the symmetry.

At the fixed points on the orbifold, certain gauge transformation parameters are forced to vanish. Remember that the 5D $\mathbf{Z}_2$ example in Eq. (5.36) at the fixed point $y = 0$ leaves the SU(5) symmetry intact. But at the fixed point $y = \pi R$, the $\mathbf{Z}_2$ transformation leaves only the SM group $G_{SM}$ invariant.

This observation is also applicable in the present 6D case. The matter contents and interactions located at the fixed points O, A, B, and C of Fig. 5.5 respect different gauge symmetries. At O, the full SO(10) gauge symmetry is respected.

Recall that in (3.30) and (3.47), we have labelled fixed points by space group actions

$$O : (\theta, 0), \quad A : (\theta, e^1), \quad B : (\theta, e^2), \quad C : (\theta, e^1 + e^2). \qquad (5.110)$$

Considering the fixed point $A$, under action $\theta$ we are forced to move along $e^1$ to come back to the original point A in the covering space. The vector field transforms as

$$\theta : V(\pi R, 0) = V(-\pi R, 0)$$

$$e^1 : V(\pi R, 0) = P_{GG} V(-\pi R, 0) P_{GG}^{-1},$$

where $T_1 = P_{GG}$. From (5.35), generators commuting with $P_{GG}$ survive and the others are projected out; thus the gauge symmetry at $A$ is SU(5)$_{GG} \times$U(1). Similarly, at the point $B$ we have the projector $P_F$ and preserve SU(5)$_F \times$ U(1)$_X$. At $C$, one

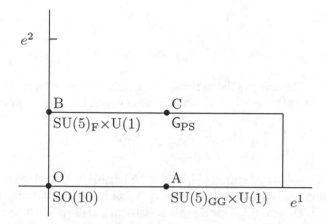

**Fig. 5.5** The $T^2/\mathbb{Z}_2$ orbifold in the $(y_1, y_2)$ plane. The orbifold fixed points are denoted by bullets and localized groups at each fixed point are indicated. The fundamental region is the rectangle

must apply both $e^1$ and $e^2$ translations after $\mathbb{Z}_2$ action. The projector is $T_2 T_1 = P_F P_{GG} = P_{PS}$ from Eqs. (5.92) and (5.93). The gauge symmetry is the PS group $SU(4)_c \times SU(2)_L \times SU(2)_R$.

From Eqs. (5.108 and 5.109) one notes that the 4D $\mathcal{N} = 1$ supersymmetry is preserved at each fixed point.

We could have taken $T_1 = P_{PS}$ and $T_2 = P_{GG}$. Then, the fixed point A preserves $SU(4)_c \times SU(2)_L \times SU(2)_R$, and the fixed point B preserves $SU(5)_{GG} \times U(1)$. The unbroken gauge group is the $G_{SM} \times U(1)$ as before. This is so because geometrically the four fixed points are equivalent and the different choice of projectors is equivalent to renaming the fixed points.

## 5.3    Local Anomalies at Fixed Points

We noted that an effective 4D theory obtained by compactifying the 5D theory on a circle is parity symmetric and anomaly free. Therefore, possible obstruction of gauge symmetry can be present only at fixed points. In this respect, the U(1) gauge anomaly in 5D theories compactified on an $S^1/\mathbb{Z}_2$ orbifold was first discussed in Ref. [14]. They considered a single bulk fermion with unit charge and imposed chiral boundary conditions. The anomaly—defined as the five dimensional divergence of the current—lives entirely on the orbifold fixed points,

$$\partial_M J^M = \frac{1}{2} \left[ \delta(y) + \delta(y - \pi R) \right] \mathcal{Q}(x, y), \qquad (5.111)$$

where $J^M$ is the 5D fermionic current

$$J^M = \overline{\psi} \Gamma^M \Psi, \qquad (5.112)$$

and

$$\mathcal{Q}(x, y) = \frac{1}{16\pi^2} F_{\mu\nu}(x, y) \widetilde{F}^{\mu\nu}(x, y), \qquad (5.113)$$

is the 4D chiral anomaly in the external gauge field $A_M(x, y)$ in (5.19). We noted that the effective 4D anomaly is absent when the localized anomalies cancel

$$\int dy \partial_M J^M = 0 \implies \mathcal{Q}(x, \pi R) = -\mathcal{Q}(x, 0).$$

But if any anomaly is present in 4D, it must be localized at the fixed points. So, if $\mathcal{Q}(x, 0)$ is nonzero, then the only possibility is that it is proportional to $\delta(x)$ which is not allowed. Therefore, for the $\mathbb{Z}_2$ case the absence of 4D anomaly is sufficient to ensure $\mathcal{Q}(x, 0) = 0$ and hence the consistency of the higher dimensional orbifold theory.

For the orbifold $S^1/(\mathbb{Z}_2 \times \mathbb{Z}_2')$ defined in (5.67), this phenomenon does not persist. Despite the fact that the orbifold projections remove both fermionic zero modes, gauge anomalies localized at the fixed points were found [15],

$$\partial_M J^M = \frac{1}{4} [\delta(y) - \delta(y - \pi R/2) + \delta(y - \pi R) - \delta(y - 3\pi R/2)] \mathcal{Q}(x, y),$$
$$(5.114)$$

for gauge fields having the odd parity as the boundary conditions. Even if the 4D effective theory is anomaly free because anomalies cancel after integration over the fifth dimension, it is possible that the gauge invariance is broken at the fixed points, spoiling the consistency of the 5D theory. Thus, the full 5D anomaly structure must be checked for the models with $S^1/(\mathbb{Z}_2 \times \mathbb{Z}_2')$. The anomaly structure of the Abelian gauge theory with arbitrary boundary condition on 5D orbifold is analyzed in Refs. [16]. For the non-Abelian gauge theory, Ref. [17] discussed the anomaly structure and the cure by Chern–Simons terms in detail.

Now we calculate (5.114). Take the five dimensional space $R^4 \times S^1/(\mathbb{Z}_2 \times \mathbb{Z}_2')$. Consider a 5D Dirac fermion in it, coupled with a U(1) external gauge field. The action of this U(1) gauge theory is the sum of (5.40) and (5.19). The two orbifold projections act on the spacetime points as

$$g : y \to -y, \quad h : y \to \pi R - y, \qquad (5.115)$$

as in (5.67). For the 5D spinor, the following boundary condition is imposed:

$$g : \Psi(x, -y) = \eta \gamma^5 \Psi(x, y), \quad \eta = \pm 1$$
$$h : \Psi(x, \pi R - y) = \eta' \gamma^5 \Psi(x, y), \quad \eta' = \pm 1. \qquad (5.116)$$

We consider the eigenstates under these parities $(\eta, \eta')$. The $\gamma^5$ matrix in Eq. (5.116) distinguishes the parities of two 4D Weyl spinors in $\Psi$ and allows only one Weyl spinor for having zero mode (the + parity). Notice that a 5D bulk mass term $m\Psi^\dagger\Gamma^0\Psi$ is forbidden unless $m$ has a nontrivial profile in the bulk with parities $(--)$.

The complete set of functions satisfying the boundary conditions

$$f_{++} = \sqrt{2}\cos\frac{2ny}{R}, \quad f_{+-} = \sqrt{2}\cos\frac{(2n+1)y}{R},$$

$$f_{-+} = \sqrt{2}\sin\frac{(2n+1)y}{R}, \quad f_{--} = \sqrt{2}\sin\frac{(2n+2)y}{R}, \tag{5.117}$$

where $n = 0, 1, \ldots$ for all,

Even though the 5D current (5.112) is classically conserved, it may have a local anomalous divergence at quantum level. One can rewrite the action (5.40) as a collection of 4D massive Dirac fermions by expanding $\Psi$ in terms of the complete set formed by the solutions of free Dirac equation in 5D as in (5.47). The KK modes $f_{L,R}^{(n)}$ are those in Eqs. (5.70–5.73).

Using the orthogonality of the KK modes, the action (5.40) can be written as

$$S = \int d^4x \left[ \sum_n \overline{\psi}^{(n)} \left( i\gamma^\mu\partial_\mu - M_n \right) \psi^{(n)} \right.$$

$$\left. - \sum_{n,m} \left( j_{Lmn}^\mu A_{\mu mn}^L + j_{Rmn}^\mu A_{\mu mn}^R - i j_{5mn} A_{5mn} \right), \right] \tag{5.118}$$

where $M_n$ are the mass eigenvalue in (5.43) and (5.44). We define a tower of four dimensional gauge fields

$$A_{\mu mn}^L = \frac{1}{\pi R} \int_0^L dy f_{+-}^{(m)}(y) f_{+-}^{(n)}(y) A_\mu(x, y), \tag{5.119}$$

$$A_{\mu mn}^R = \frac{1}{\pi R} \int_0^L dy f_{-+}^{(m)}(y) f_{-+}^{(n)}(y) A_\mu(x, y), \tag{5.120}$$

$$A_{ymn} = \frac{1}{\pi R} \int_0^L dy f_L^{(m)}(y) f_R^{(n)}(y) A_y(x, y), \tag{5.121}$$

where $L, R$ are correlated. We also have chiral and mixed currents

$$j_{Lmn}^\mu = \overline{\psi}_L^{(m)} \gamma^\mu \psi_L^{(n)} \tag{5.122}$$

$$j_{Rmn}^\mu = \overline{\psi}_R^{(m)} \gamma^\mu \psi_R^{(n)} \tag{5.123}$$

$$j_{ymn} = \overline{\psi}_L^{(m)} \psi_R^{(n)} - \overline{\psi}_R^{(m)} \psi_L^{(m)}. \tag{5.124}$$

Using the symmetry (5.24), we may choose the gauge $A_5 = 0$. Using the well-known result for the 4D anomalous divergence of chiral current in 4D gives the relations [14]

$$\partial_\mu j^\mu_{Lmn} = i\left[\overline{\psi}^{(m)}_L M_m \psi^{(n)}_L - \overline{\psi}^{(m)}_R M_n \psi^{(n)}_R\right] - \frac{1}{16\pi^2}\sum_{k=1}^{\infty} F^L_{\mu\nu mk}\tilde{F}^{L\mu\nu}_{kn} \qquad (5.125)$$

$$\partial_\mu j^\mu_{Rmn} = i\left[\overline{\psi}^{(m)}_R M_m \psi^{(n)}_R - \overline{\psi}^{(m)}_L M_n \psi^{(n)}_L\right] + \frac{1}{16\pi^2}\sum_{k=1}^{\infty} F^R_{\mu\nu mk}\tilde{F}^{R\mu\nu}_{kn}, \qquad (5.126)$$

where $\mathscr{F}^{L,R}_{\mu mn}$ are the field strengths of $A^{L,R}_{\mu mn}$. On the other hand, the 5D current can be expanded in terms of the 4D currents

$$J^\mu(x, y) = \sum_{m,n}\left[f^{(m)}_R(y)f^{(n)}_R(y)j^\mu_{Rmn}(x) + f^{(m)}_L(y)f^{(n)}_L(y)j^\mu_{Lmn}(x)\right]. \qquad (5.127)$$

$$J^y(x, y) = -i\sum_{m,n}f^{(m)}_L(y)f^{(n)}_R(y)j^y_{mn}(x). \qquad (5.128)$$

Noticing that at the classical level, using (5.43) and (5.44),

$$\partial_y J^y = \sum_{mn} i\left[M_n\left(f^{(m)}_{+-}f^{(n)}_{+-} - f^{(m)}_{-+}f^{(n)}_{-+}\right)\overline{\psi}^{(m)}\left(-\gamma^5\right)\psi^{(n)}\right] \qquad (5.129)$$

the divergence of the 5D current can be expressed by

$$\partial_M J^M = \frac{1}{2}\mathscr{Q}\sum_n\left(f^{(n)}_{+-}(y)f^{(n)}_{+-}(y) - f^{(n)}_{-+}(y)f^{(n)}_{-+}(y)\right), \qquad (5.130)$$

with $\mathscr{Q}$ defined in (5.113) It should be stressed that in deriving (5.130) one tacitly assumed that Eq. (5.129) is still valid at the quantum level and all the quantum effects are encoded in Eqs. (5.125) and (5.126). To discuss the complete anomaly structure, one should also consider the parity anomaly.

The sum over the KK modes in (5.130) can be computed using the completeness property of the mode functions [14]. For the case of $(++)$ parity, consider

$$\pi R\Delta(y, y') \equiv \sum_{n=1}^{\infty} f^{(n)}_L(y)f^{(n)}_L(y') - \sum_{n=1}^{\infty} f^{(n)}_R(y)f^{(n)}_R(y'), \qquad (5.131)$$

which reduces to the last factor of Eq. (5.130) if we set $y = y'$. Here we defined $f^{(n)}_{++}(y)$, $f^{(m)}_{--}(y)$ in Eq. (5.69). Because the $f^{(m)}_L(y)$ functions are odd while the $f^{(m)}_R(y)$ are even, we can write

$$\Delta(y, -y') = \sum_{m\geq 0} f^{(m)}_R(y)f^{(m)}_R(y') + \sum_{m>0} f^{(m)}_L(y)f(y')^{(m)}_L. \qquad (5.132)$$

Since $f_R^{(m)}(y)$ and $f_L^{(m)}(y)$ form a complete set of functions with periodic boundary conditions in the interval $[0, \pi R)$, (although they are still normalized in the interval $[0, 2\pi R]$). The sum of the eigenfunctions in the complete set is a delta function

$$\Delta(y, -y') = \frac{1}{2} \sum_{n=-\infty}^{\infty} \delta(y - y' - n\pi R), \tag{5.133}$$

which leads to the expression, by setting $y' = -y$,

$$\Delta(y, y) = \frac{1}{2} \sum_{n=-\infty}^{\infty} \delta(2y - n\pi R) = \sum_{n=-\infty}^{\infty} \frac{1}{4R} \delta(y/R - n\pi/2). \tag{5.134}$$

Using this method for the case of $\mathbb{Z}_2 \times \mathbb{Z}_2'$ orbifold, various choices of the fermion parity lead to

$$\Delta^{(++)}(y) = -\Delta^{(--)}(y) = \frac{1}{4R} \sum_{n=-\infty}^{\infty} \delta(y/R - n\pi/2), \tag{5.135}$$

$$\Delta^{(+-)}(y) = -\Delta^{(-+)}(y) = \frac{1}{4R} \sum_{n=-\infty}^{\infty} (-1)^n \delta(y/R - n\pi/2), \tag{5.136}$$

where $+$ and $-$ denote the parities of $\mathbb{Z}_2$ and $\mathbb{Z}_2'$. In particular, if the fermions have opposite parities $(+, -)$ and $(-, +)$, one recovers Eq. (5.114).

With a similar calculation, one find for the case of $S^1/\mathbb{Z}_2$,

$$f^{(+)}(y) = -f^{(-)}(y) = \frac{1}{2R} \sum_{n=-\infty}^{+\infty} \delta(y/R - n\pi), \tag{5.137}$$

which reproduces Eq. (5.111).

For a consistent quantum theory, there should not appear any fixed point anomaly. In this regard, note that the above fixed point anomalies can be always cancelled by adding appropriate Chern–Simons terms in the bulk and some fermions at the fixed point(s) [17, 37].

## Bulk and Local Anomalies in Six Dimension

The automatic cancellation of bulk anomaly in 5D is not maintained in 6D. In even dimensions, chiral anomalies can be present. The gauge anomalies in even dimensions were studied by many groups, notably by Frampton and Kephart [38]. In 6D, the chiral anomaly arises from box diagrams. So, the 4D intuition that $\overline{N}$ of $SU(N)$ carries $-1$ unit of the anomaly of $N$ does not work in 6D. In 6D, unlike in 4D, orthogonal groups can have gauge anomalies. Usually, in even dimensions the $\mathcal{N} = 1$ theory can be made anomaly free by adjusting bulk matter contents by hand. Although it is possible to remove the SO(10) gauge anomalies in 6D in this way [9], it is desirable to have some guidance for cancelling the gauge anomaly. For an effective 4D theory, the orbifolding has been used in this chapter. One method to remove the gauge anomalies is to consider parity symmetric $\mathcal{N} = 2$ SUSY theory [9]. Even if the 4D gauge anomaly is cancelled in this way, the local anomaly cancellation has to be checked carefully as we will discuss later. Another method is to form a complete representation of higher rank gauge group for which the anomaly cancellation is guaranteed. In 6D, exceptional groups can be used for an automatic absence of gauge anomalies as $SO(4N+2)$ groups with complex fermion representations do in 4D [39].

For the localized anomaly we have the same result as in five dimensions [40]: the anomaly freedom implies the cancellation of localized anomalies at the fixed points.

## Exercises

▶ **Exercise 5.1**  Check the signs in (5.3).

▶ **Exercise 5.2**  Another convenient set of eigenfunctions satisfying (5.4) and (5.5) is

$$f^{(n)} = e^{iny/R}, \quad M_n^2 = \frac{n^2}{R^2}, \quad n = 0, \pm 1, \pm 2, \ldots.$$  (5.138)

(1) Explain that the negative mode numbers $n$ are allowed.
(2) Expand the modes

$$\Phi(x, y) = \sum_{n=-\infty}^{\infty} \Phi^{(n)}(x^\mu) e^{iny/R},$$  (5.139)

and use the completeness relation (5.5) and plug them back to the original action (5.3) to obtain

$$S = \int d^4x \sum_{n=-\infty}^{\infty} \left( \partial_\mu \phi^{(n)*} \partial^\mu \phi^{(n)} - \frac{n^2}{R^2} \phi^{(n)*} \phi^{(n)} \right),$$

$$= \int d^4x \left[ \partial_\mu \phi^{(0)*} \partial^\mu \phi^{(0)} + \sum_{n=1}^{\infty} \left( \partial_\mu \phi^{(n)*} \partial^\mu \phi^{(n)} - \frac{n^2}{R^2} \phi^{(n)*} \phi^{(n)} \right) \right.$$

$$\left. + \sum_{n=1}^{\infty} \left( \partial_\mu \phi^{(-n)*} \partial^\mu \phi^{(-n)} - \frac{n^2}{R^2} \phi^{(-n)*} \phi^{(-n)} \right) \right].$$  (5.140)

(3) Show that this is the same result as (5.10). Also it is more convenient to show the Sherck–Schwarz relations (5.12) and ((5.13).

▶ **Exercise 5.3** Show that, if the scalar $\Phi(x^\mu, y)$ were real, we should have $\Phi^{(n)}(x^\mu) = \Phi^{(-n)}(x^\mu)$ and count the states with $n \geq 0$. Using this, expand the gauge field

$$A_\mu(x, y) = \sum_{n=-\infty}^{\infty} A_\mu^{(n)}(x^\mu)e^{iny/R},$$

and show that it gives the same result as (5.26).

# References

1. L.J. Dixon, J.A. Harvey, C. Vafa, E. Witten, Strings on orbifolds. 2. Nucl. Phys. B **274**, 285–314 (1986)
2. Y. Kawamura, Gauge symmetry breaking from extra space $S^1/Z_2$. Prog. Theor. Phys. **103**, 613–619 (2000)
3. G. Altarelli, F. Feruglio, SU(5) grand unification in extra dimensions and proton decay. Phys. Lett. **B511**, 257–264 (2001)
4. L.J. Hall, Y. Nomura, Gauge unification in higher dimensions. Phys. Rev. **D64**, 055003 (2001)
5. R. Barbieri, L.J. Hall, Y. Nomura, Softly broken supersymmetric desert from orbifold compactification. Phys. Rev. **D66**, 045025 (2002).
6. Y. Kawamura, Split multiplets, coupling unification and extra dimension. Prog. Theor. Phys. **105**, 691–696 (2001)
7. T. Asaka, W. Buchmuller, L. Covi, Gauge unification in six-dimensions. Phys. Lett. **B523**, 199–204 (2001)
8. A. Hebecker, J. March-Russell, The structure of GUT breaking by orbifolding. Nucl. Phys. **B625**, 128–150 (2002)
9. L.J. Hall, Y. Nomura, T. Okui, D. Tucker-Smith, SO(10) unified theories in six-dimensions. Phys. Rev. **D65**, 035008 (2002)
10. H.-D. Kim, J.E. Kim, H.M. Lee, Top–bottom mass hierarchy, $s - \mu$ puzzle and gauge coupling unification with split multiplets. Eur. Phys. J. **C24**, 159–164 (2002)
11. K.S. Babu, S.M. Barr, B. Kyae, Family unification in five-dimensions and six-dimensions. Phys. Rev. **D65**, 115008 (2002)
12. R. Dermisek, A. Mafi, SO(10) grand unification in five-dimensions: proton decay and the mu problem. Phys. Rev. **D65**, 055002 (2002)
13. H.D. Kim, S. Raby, Unification in 5D SO(10). J. High Energy Phys. **1**, 056 (2003)
14. N. Arkani-Hamed, A.G. Cohen, H. Georgi, Anomalies on orbifolds. Phys. Lett. **B516**, 395–402 (2001)
15. C.A. Scrucca, M. Serone, L. Silvestrini, F. Zwirner, Anomalies in orbifold field theories. Phys. Lett. **B525**, 169–174 (2002)
16. L. Pilo, A. Riotto, On anomalies in orbifold theories. Phys. Lett. **B546**, 135–142 (2002)
17. H.-D. Kim, J.E. Kim, H.M. Lee, TeV scale 5-D SU(3)$_W$ unification and the fixed point anomaly cancellation with chiral split multiplets. J. High Energy Phys. **6**, 048 (2002)
18. B. Kyae, C.-A. Lee, Q. Shafi, Low-energy consequences of five-dimensional SO(10). Nucl. Phys. **B683**, 105–121 (2004)
19. J. Scherk, J.H. Schwarz, Spontaneous breaking of supersymmetry through dimensional reduction. Phys. Lett. **B82**, 60–64 (1979)

20. J. Polchinski, Dirichlet Branes and Ramond–Ramond charges. Phys. Rev. Lett. **75**, 4724–4727 (1995)
21. L. Randall, R. Sundrum, An Alternative to compactification. Phys. Rev. Lett. **83**, 4690–4693 (1999)
22. G.R. Dvali, M.A. Shifman, Domain walls in strongly coupled theories. Phys. Lett. **B396**, 64–69 (1997). [Erratum: Phys. Lett. **B407**, 452 (1997)]
23. S. Dimopoulos, D.E. Kaplan, The Weak mixing angle from an SU(3) symmetry at a TeV. Phys. Lett. **B531**, 127–134 (2002)
24. Th. Kaluza, Zum Unitätsproblem der Physik. Sitzungsber. Preuss. Akad. Wiss. Berlin (Math. Phys.) **1921**, 966–972 (1921). [Int. J. Mod. Phys. **D27**(14), 1870001 (2018)]
25. O. Klein, Quantum Theory and Five-Dimensional Theory of Relativity. (In German and English). Z. Phys. **37**, 895–906 (1926). [76 (1926)]
26. H. Georgi, S.L. Glashow, Unity of all elementary particle forces. Phys. Rev. Lett. **32**, 438–441 (1974)
27. N. Arkani-Hamed, T. Gregoire, J.G. Wacker, Higher dimensional supersymmetry in 4-D superspace. J. High Energy Phys. **3**, 055 (2002)
28. N. Marcus, A. Sagnotti, W. Siegel, Ten-dimensional supersymmetric Yang–Mills theory in terms of four-dimensional superfields. Nucl. Phys. **B224**, 159 (1983)
29. T. Yanagida, Horizontal gauge symmetry and masses of neutrinos. Conf. Proc. **C7902131**, 95–99 (1979)
30. M. Gell-Mann, P. Ramond, R. Slansky, Complex spinors and unified theories. Conf. Proc. **C790927**, 315–321 (1979)
31. H. Georgi, The state of the art—gauge theories. AIP Conf. Proc. **23**, 575–582 (1975)
32. H. Fritzsch, P. Minkowski, Unified interactions of leptons and hadrons. Ann. Phys. **93**, 193–266 (1975)
33. F. Wilczek, A. Zee, Families from spinors. Phys. Rev. **D25**, 553 (1982)
34. J.C. Pati, A. Salam, Unified lepton-hadron symmetry and a gauge theory of the basic interactions. Phys. Rev. **D8**, 1240–1251 (1973)
35. S.M. Barr, A new symmetry breaking pattern for SO(10) and proton decay. Phys. Lett. **B112**, 219–222 (1982)
36. J.P. Derendinger, J.E. Kim, D.V. Nanopoulos, Anti-SU(5). Phys. Lett. **B139**, 170–176 (1984)
37. C.G. Callan, Jr., J.A. Harvey, Anomalies and fermion zero modes on strings and domain walls. Nucl. Phys. **B250**, 427–436 (1985)
38. P.H. Frampton, T.W. Kephart, The analysis of anomalies in higher space-time dimensions. Phys. Rev. **D28**, 1010 (1983)
39. T. Asaka, W. Buchmuller, L. Covi, Exceptional coset spaces and unification in six-dimensions. Phys. Lett. **B540**, 295–300 (2002)
40. T. Asaka, W. Buchmuller, L. Covi, Bulk and brane anomalies in six-dimensions. Nucl. Phys. **B648**, 231–253 (2003)

# Quantization of Strings

**6**

Starting from this chapter, we discuss string theory for particle physics. Our emphasis will be, starting from string theory, understanding low-energy physics described by the Standard Model (SM). As summarized in Chap. 2, the SM is a chiral theory for fermions, and hence obtaining a chiral spectrum from string theory is of utmost importance. This goes with string theory with fermions, i.e., superstring rather than bosonic string. Superstring is written in ten spacetime dimensions(10D) but the effective quantum field theory of the SM is in four spacetime dimensions(4D), and the extra six dimensions(6D) must be cleverly hidden from the 4D observers. For hiding these extra dimensions, we follow the compactification scheme via orbifolds introduced in Chap. 3 and applied to quantum field theory in Chap. 5. It will be exploited fully in string theory in the subsequent chapters. In this chapter, we introduce basics in string theory.

There are excellent books on string theory [1–4]. In this spirit, we attempt to excerpt the key formulae of string theory in this chapter which will be used in the subsequent chapters for constructing 4D string models.

## 6.1   Bosonic String

Besides its own interesting points, bosonic string is a building block of all string theories. We discuss quantization and modular invariance of the partition function that can be easily generalized to superstrings.

© Springer Nature Switzerland AG 2020                                                    129
K.-S. Choi, J. E. Kim, *Quarks and Leptons From Orbifolded Superstring*,
Lecture Notes in Physics 954, https://doi.org/10.1007/978-3-030-54005-0_6

## 6.1.1   Action and Its Invariance Properties

We begin with an example of relativistic point particle. One successful form for the action with the Poincaré invariance is

$$S = m \int d\tau \sqrt{\dot{X}^\mu \dot{X}_\mu}, \qquad (6.1)$$

where the dot denotes $d/d\tau$. We parametrized $\tau$ along the world line. Physically there should be no dependence on $\tau$ thus it has reparametrization invariance $X'_\mu(\tau'(\tau)) = X_\mu(\tau)$ and we can easily check indeed it is. Varying (6.1) with respect to $\delta X^\mu$ gives

$$\delta S = m \int d\tau \; \dot{u}_\mu \delta X^\mu,$$

where $u^\mu = \dot{X}^\mu / \sqrt{\dot{X}^\nu \dot{X}_\nu}$ is the $D$-velocity. The equation of motion is $\dot{u}^\mu = 0$ which is a free particle motion. We can identify $m$ as the particle mass.

Consider another form of the action, by introducing world-line metric $e(\tau)$,

$$S' = \frac{1}{2} \int d\tau \left( \frac{1}{e} \dot{X}^\mu \dot{X}_\mu + em^2 \right) \qquad (6.2)$$

which has the Poincaré invariance. The invariant volume element in the world line is $e(\tau) d\tau$. Variation of (6.2) with respect to $e(\tau)$ relates

$$e^2 = \frac{\dot{X}^\mu \dot{X}_\mu}{m^2}$$

and the action $S'$ of (6.1) reduces to the original action $S$ of (6.2). We feel much better for the latter form because it is quadratic in $X^\mu$ and successfully describes the massless limit.

Let us extend the above discussion to the string. Strings propagate in spacetime manifolds, sweeping out an area. The world volume swept out by a string is a two dimensional surface for whose coordinates we introduce $\tau$ and $\sigma$. The spatial extent of the string is defined to, in a suitable definition of the parameter $\sigma$,

$$0 \le \sigma \le \pi. \qquad (6.3)$$

We will treat bosonic degrees $X^\mu$, $(\mu = 0, 1, 2, \cdots, D-1)$ as $D$ displacements defined on a worldsheet parametrized by $\tau$ and $\sigma$. On this worldsheet, we introduce a metric $h_{\alpha\beta}(\tau, \sigma)$. Thus, we introduce the following action of type (6.2) for the relativistic string:

$$S = \frac{T}{2} \int d\tau d\sigma \sqrt{-h}\, h^{\alpha\beta} \eta^{\mu\nu} \partial_\alpha X_\mu \partial_\beta X_\nu, \qquad (6.4)$$

where $h = \det h_{\alpha\beta}$ and $T$ is string tension having dimension of mass squared. Here, we can write a worldsheet action in analogy with the relativistic point particle action in the world line, and one can write Eq. (6.4) blindly and start from there. The action (6.4) has the following symmetries:

1. *General covariance* (worldsheet) which implies that the action is invariant under the reparametrization of coordinates

$$\tau \rightarrow \tau'(\tau, \sigma), \sigma \rightarrow \sigma'(\tau, \sigma).$$

   Therefore,

$$X'^{\mu}(\tau', \sigma') = X^{\mu}(\tau, \sigma),$$

$$h'_{\gamma\delta}(\tau', \sigma') = \frac{\partial \sigma^{\alpha}}{\partial \sigma'^{\gamma}} \frac{\partial \sigma^{\beta}}{\partial \sigma'^{\delta}} h_{\alpha\beta}(\tau, \sigma). \tag{6.5}$$

   Due to the reparametrization invariance on the worldsheet, we can deal with two dimensional general relativity.
2. *Weyl invariance* (worldsheet) under the local scaling,

$$X'^{\mu}(\tau, \sigma) = X^{\mu}(\tau, \sigma)$$

$$h'_{\alpha\beta}(\tau, \sigma) = e^{\Lambda(\tau, \sigma)} h_{\alpha\beta}(\tau, \sigma), \tag{6.6}$$

   which holds only in two dimensions.
3. *Poincaré invariance* (target space),

$$X'^{\mu} = \Lambda^{\mu}{}_{\nu} X^{\nu} + a^{\mu}. \tag{6.7}$$

   Therefore, we have a unitary representation under Poincarè algebra and the state is labelled by momentum, spin, and internal quantum numbers.

The standard way of quantizing the system having gauge symmetry is to fix the gauge and use the gauge condition as a constraint. In this regard, one can remind the Coulomb gauge condition $\nabla \cdot \mathbf{A} = 0$ in Maxwell's theory.

There are three degrees of freedom for the symmetric tensor $h_{\alpha\beta}$. Using reparametrization invariance, we can fix $h_{\alpha\beta}$ as

$$h_{\alpha\beta} = e^{\varphi} \eta_{\alpha\beta}, \tag{6.8}$$

called the *conformal gauge*. Plugging (6.8) into (6.4), the $\varphi$ dependence disappears due to the two dimensional nature of the worldsheet. The resulting action is

$$S = \frac{T}{2} \int d^2\sigma \left( \dot{X}^2 - X'^2 \right) = \int d^2\sigma \mathscr{L}, \tag{6.9}$$

where dot and prime represent derivative with respect to $\tau$ and $\sigma$, respectively. This makes our later analysis easy although the following considerations make it nontrivial:

- We should note that either the scale invariance or the reparametrization invariance is broken in quantum theory. Fortunately, there is no anomaly in a critical dimension (for example $D = 26$ in the bosonic string case).
- Riemann and Roch theorem tells us about fixing of $h_{\alpha\beta}$. It is always possible to fix gauge (6.8) in the tree Feynman diagram (topologically it is reduced to a sphere). However, in the loop diagrams there remain non-fixable degrees in general. The one-loop case will be discussed in Sect. 6.1.4.
- Even if we succeed in fixing gauge $h_{\alpha\beta}$, there is a residual symmetry, which will be discussed in the following subsection.

The string equations are obtained by varying with respect to $X^\mu$. Besides the surface term we have

$$\partial_\alpha \partial^\alpha X^\mu = \left( \frac{\partial^2}{\partial \tau^2} - \frac{\partial^2}{\partial \sigma^2} \right) X^\mu = 0. \qquad (6.10)$$

As is familiar, the solution of this partial differential equation is separated into two independent ones

$$X^\mu(\tau, \sigma) = X_R^\mu(\tau - \sigma) + X_L^\mu(\tau + \sigma). \qquad (6.11)$$

Therefore, it is useful to introduce the light-cone variables $\sigma^\pm = \tau \pm \sigma$. Then, Eq. (6.10) becomes $\partial_+ X_R^\mu = \partial_- X_L^\mu = 0$; thus we have chiral scalars in two dimensions. This contributes to gravitational and Weyl anomalies in two dimensions.

This is complemented by the constraint, that the variation with respect to $h^{\alpha\beta}$ vanishes,

$$T_{\alpha\beta} \equiv \frac{2}{T\sqrt{-h}} \frac{\delta S}{\delta h^{\alpha\beta}} \bigg|_{h_{\alpha\beta}=\eta_{\alpha\beta}}$$
$$= -\partial_\alpha X^\mu \partial_\beta X_\mu + \frac{1}{2} \eta_{\alpha\beta} \partial^\gamma X^\mu \partial_\gamma X_\mu = 0, \qquad (6.12)$$

which defines the *worldsheet energy-momentum tensor*. It corresponds to the Gauss law constraint $\nabla \cdot \mathbf{E} = 0$ in Maxwell's theory, originating from $A_0 = 0$ which in turn corresponds to Eq. (6.8).

Being symmetric tensor, this $T_{\alpha\beta}$ seems to have three independent constraints. However, the scaling invariance (6.6) is translated to the traceless condition of the energy-momentum tensor

$$\mathrm{Tr}\, T_{\alpha\beta} = \eta^{\alpha\beta} T_{\alpha\beta} = T_{00} - T_{11} = 0, \tag{6.13}$$

and only two constraints are independent. In view of (6.11), it is convenient to define

$$T_{++} \equiv \frac{1}{2}(T_{00} + T_{01}) = -\frac{1}{4}(\dot{X} + X')^2 = -\partial_+ X_L^\mu \partial_+ X_{\mu L},$$

$$T_{--} \equiv \frac{1}{2}(T_{00} - T_{01}) = -\frac{1}{4}(\dot{X} - X')^2 = -\partial_- X_R^\mu \partial_- X_{\mu R}. \tag{6.14}$$

The constraint equation (6.12) becomes

$$T_{++} = T_{--} = 0. \tag{6.15}$$

As in the quantization of gauge theories, we use these as constraint equations on physical states.

The boundary condition is to abolish the surface term contributions,

$$T \int d\tau \frac{\partial X_\mu}{\partial \sigma} \delta X^\mu \Big|_{\sigma=0}^{\pi} = 0. \tag{6.16}$$

## Closed Strings

The following boundary condition satisfying Eq. (6.16) specifies the *closed string*:

$$X^\mu(\tau, \sigma + \pi) = X^\mu(\tau, \sigma). \tag{6.17}$$

(In fact, we also need $h_{\alpha\beta}(\tau, \sigma + \pi) = h_{\alpha\beta}(\tau, \sigma)$.) Let us try to solve the equation of motion, consistent with this condition. We have mode expansion

$$X_R^\mu(\tau - \sigma) = \frac{1}{2}x^\mu + \alpha' p^\mu(\tau - \sigma) + i\sqrt{\frac{\alpha'}{2}} \sum_{n \neq 0} \frac{1}{n}\alpha_n^\mu e^{-2in(\tau-\sigma)}, \tag{6.18}$$

$$X_L^\mu(\tau + \sigma) = \frac{1}{2}x^\mu + \alpha' p^\mu(\tau + \sigma) + i\sqrt{\frac{\alpha'}{2}} \sum_{n \neq 0} \frac{1}{n}\tilde{\alpha}_n^\mu e^{-2in(\tau+\sigma)}, \tag{6.19}$$

where the index $n$ runs over all the integers except 0. It is useful to define the Regge slope $\alpha'$ and the string length scale $\ell_s$ as

$$\alpha' \equiv \frac{1}{2\pi T} \equiv \frac{1}{2}\ell_s^2. \tag{6.20}$$

This is the only free parameter of the string theory and will be related to experimentally observable quantities in Chap. 11. What will be more useful is the (anti-)holomorphic derivatives

$$\partial_- X_R^\mu(\tau - \sigma) = \ell_s \sum_{n=-\infty}^{\infty} \alpha_n^\mu e^{-2in(\tau-\sigma)}, \tag{6.21}$$

$$\partial_+ X_L^\mu(\tau + \sigma) = \ell_s \sum_{n=-\infty}^{\infty} \tilde{\alpha}_n^\mu e^{-2in(\tau+\sigma)}. \tag{6.22}$$

Here we include the zero modes, which are naturally defined as

$$\alpha_0^\mu = \tilde{\alpha}_0^\mu = \sqrt{\frac{\alpha'}{2}} p^\mu. \tag{6.23}$$

Now let us quantize $X^\mu$. First we treat them as operators. The conjugate momentum to $X^\mu$ is

$$\Pi^\mu(\tau, \sigma) = \frac{\delta \mathcal{L}}{\delta \dot{X}_\mu} = T \dot{X}^\mu. \tag{6.24}$$

The first quantization of string leads to the following equal $\tau$ commutators:

$$[\Pi^\mu(\tau, \sigma), X^\nu(\tau, \sigma')] = T[\dot{X}^\mu(\tau, \sigma), X^\nu(\tau, \sigma')] = i\delta(\sigma - \sigma')\eta^{\mu\nu} \tag{6.25}$$

with others vanishing. They lead to the following commutators for the center-of-mass variables and oscillator mode operators:

$$[x^\mu, p^\nu] = -i\eta^{\mu\nu}, \tag{6.26}$$

$$[\alpha_m^\mu, \alpha_n^{\nu\dagger}] = -m\delta_{mn}\eta^{\mu\nu}, \tag{6.27}$$

$$[\tilde{\alpha}_m^\mu, \tilde{\alpha}_n^{\nu\dagger}] = -m\delta_{mn}\eta^{\mu\nu}, \tag{6.28}$$

with the other commutators vanishing. Using the Hermiticity of $X^\mu$, operators $\alpha, \tilde{\alpha}$ with negative indices are related to positive indices by

$$(\alpha_n^\mu)^\dagger = \alpha_{-n}^\mu, \quad (\tilde{\alpha}_n^\mu)^\dagger = \tilde{\alpha}_{-n}^\mu. \tag{6.29}$$

The worldsheet Hamiltonian becomes

$$H = \int_0^\pi d\sigma(\dot{X} \cdot \Pi - \mathcal{L})$$

$$= \sum_{n\neq 0}(\alpha_{-n} \cdot \alpha_n + \tilde{\alpha}_{-n} \cdot \tilde{\alpha}_n) + \alpha' p \cdot p + \tilde{c} + c, \tag{6.30}$$

where the dot product means the sum over $\mu$ and the worldsheet Lagrangian density $cL$ is defined in (6.9). Since $\alpha_n$ and $\alpha_{-n}$ do not commute, we have the ordering problem. We reserved normal ordering constant $c$ and $\tilde{c}$. We also have the momentum generating translation along the $\sigma$ direction

$$
P = \int_0^\pi d\sigma \, X' \cdot \Pi
$$
$$
= \sum_{n\neq 0} \tilde{\alpha}_{-n} \cdot \tilde{\alpha}_n - \sum_{n\neq 0} \alpha_{-n} \cdot \alpha_n + \tilde{c} - c. \tag{6.31}
$$

In fact we have an unwanted state of negative norm

$$
\|\alpha^0_{-m}|0; p\rangle\|^2 < 0
$$

due to the opposite sign in the commutation relation along the time direction.[1] This negative norm state is pathological and we should decouple these states. It turns out that such unphysical degrees are decoupled when the constraint is appropriately imposed, as we encounter in the Gupta–Bleuler formalism in QED [5, 6]. The remaining part of this section is devoted to removing such an ambiguity.

## Open Strings

The open string has the Neumann boundary condition

$$
\frac{\partial X^\mu}{\partial \sigma} = 0, \quad \text{at } \sigma = 0, \pi. \tag{6.32}
$$

Technically, we can make an open string by $X_R(\sigma + \pi) = X_L(\sigma)$, $X_L(\sigma + \pi) = X_R(\sigma)$. Then we can use the above closed string formula by setting $\alpha^\mu = \tilde{\alpha}^\mu$. We have

$$
X^\mu = x^\mu + 2\alpha' p^\mu \tau + i\sqrt{2\alpha'} \sum_{n\neq 0} \frac{1}{n} \alpha_n^\mu e^{-in\tau} \cos(n\sigma). \tag{6.33}
$$

It has the same quantization

$$
[x^\mu, p^\nu] = -i\eta^{\mu\nu}, \quad [\alpha_m^\mu, \alpha_n^\nu] = -m\delta_{m+n,0}\,\eta^{\mu\nu}. \tag{6.34}
$$

The Hamiltonian becomes

$$
H = -\frac{1}{2} \sum_{n\neq 0} \alpha_{-n} \cdot \alpha_n - \alpha' p \cdot p. \tag{6.35}
$$

---

[1] Because of the mostly negative metric convention, we define the norm of a state $\||\phi\rangle\|^2$ as $-\langle\phi|\phi\rangle$.

Traditionally, it has been the end of the story. Because of $\delta X^\mu$ in the boundary term (6.16), the Dirichlet boundary condition

$$X^\mu(\sigma = 0) = X_0^\mu, \quad X^\mu(\sigma = \pi) = X_\pi^\mu, \tag{6.36}$$

where RHSs are constant vectors, is also allowed for some directions. Although it breaks the Lorentz symmetry, we have only that symmetry in four dimensions. So we can allow for such boundary condition for the extra dimensions.

## 6.1.2  Conformal Symmetry and Virasoro Algebra

In this section we will consider the so-called old covariant quantization. In the worldsheet point of view, string theory is a two dimensional theory, which has an infinite-dimensional local conformal symmetry, generated by the Virasoro algebra [7]. Thanks to this symmetry, we may have a quantum theory of string since many ambiguities disappear.

### Residual Symmetry

We have discussed in the covariant gauge where the worldsheet metric was chosen as $h_{\alpha\beta} = \eta_{\alpha\beta}$. But still there remains a gauge freedom. We can further fix the gauge, exhausting all the gauge freedom, as in the unitary gauge in field theory. This fixing guarantees that we do not expect any unphysical states.

This residual gauge freedom can be seen as follows. After choosing a covariant gauge, any combined reparametrization of Eqs. (6.5) chosen in the following way

$$\partial^\alpha \xi^\beta + \partial^\beta \xi^\alpha = -\Lambda \eta^{\alpha\beta} \tag{6.37}$$

is consistent with the Weyl scaling (6.6). Under this, the angle between two vectors is preserved and thus it is called *conformal symmetry*. It seems to be peculiar, but it is consistent with the infinite number of conserved quantities in the covariant gauge, following from $\partial_- T_{++} = 0$. This is always satisfied when we transform coordinates such that

$$\sigma^+ \to \tilde\sigma^+(\sigma^+), \quad \sigma^- \to \tilde\sigma^-(\sigma^-). \tag{6.38}$$

It turns out that the generators for this local transformation are the energy–momentum tensors $T_{++}$, $T_{--}$, respectively, in (6.14). Since it is local, it is useful to expand it in terms of the Fourier components, by substituting the mode expansions (6.18, 6.19),

$$L_m = \frac{T}{2} \int_0^\pi d\sigma_- e^{2im\sigma_-} T_{--} = -\frac{1}{2} \sum_{n=-\infty}^\infty \alpha_{m-n} \cdot \alpha_n, \quad (m \neq 0), \tag{6.39}$$

$$\tilde L_m = \frac{T}{2} \int_0^\pi d\sigma_+ e^{2im\sigma_+} T_{++} = -\frac{1}{2} \sum_{n=-\infty}^\infty \tilde\alpha_{m-n}^\mu \tilde\alpha_{\mu n}, \quad (m \neq 0) \tag{6.40}$$

which are called Virasoro operators [8]. We have defined the Virasoro operators for $m \neq 0$. For $m = 0$, we have again the ordering problem as in (6.30). We *define* $L_0, \tilde{L}_0$ as normal ordered products $: :$,

$$L_0 \equiv -\frac{1}{2} \sum_{n=-\infty}^{\infty} : \alpha_{-n} \cdot \alpha_n := -\frac{1}{2}\alpha_0 \cdot \alpha_0 - \sum_{n=1}^{\infty} \alpha_{-n} \cdot \alpha_n$$

$$\tilde{L}_0 \equiv -\frac{1}{2} \sum_{n=-\infty}^{\infty} : \tilde{\alpha}_{-n} \cdot \tilde{\alpha}_n := -\frac{1}{2}\tilde{\alpha}_0 \cdot \tilde{\alpha}_0 - \sum_{n=1}^{\infty} \tilde{\alpha}_{-n} \cdot \tilde{\alpha}_n.$$

(6.41)

This definition of $L_0$ and $\tilde{L}_0$ differs by a normal ordering constant from the one having the ambiguity by using (6.39) and (6.40) with a brute force substitution $m = 0$. These constants are the same as $\tilde{c}$ and $c$ defined in the Hamiltonian (6.30) and the momentum (6.31)

$$H = \tilde{L}_0 + \tilde{c} + L_0 + c, \tag{6.42}$$

$$P = \tilde{L}_0 + \tilde{c} - L_0 - c. \tag{6.43}$$

**Virasoro Algebra**

Using the commutation relation (6.39) for $m + n \neq 0$, we obtain

$$[L_m, L_n] = \frac{1}{4}\left[ \sum_{p=-\infty}^{\infty} \alpha_{m-p} \cdot \alpha_p, \sum_{q=-\infty}^{\infty} \alpha_{n-q} \cdot \alpha_q \right] = (m-n)L_{m+n}.$$

For $n = -m \neq 0$, we may allow a central term as $[L_m, L_{-m}] = 2mL_0 + A(m)\delta_{m+n,0}$, catching the "anomaly," that is, the quantum effect from the zero point energies. Using Jacobi identity, one can show that [1]

$$A(m) = \frac{D}{12}m(m^2 - 1).$$

Here $D = \eta_{\mu\nu}\eta^{\mu\nu}$ counts the number of worldsheet bosons.

This is the Virasoro algebra [8], whose general form is

$$[L_m, L_n] = (m-n)L_{m+n} + \frac{C}{12}m(m^2 - 1)\delta_{m+n,0}, \tag{6.44}$$

where $C$, the $D$ above, is the *central charge*. We have the same algebra for the left-moving operators $\tilde{L}_m$, which are independent from the right-moving operators $[L_m, \tilde{L}_n] = 0$.

**Physical States, Level Matching, and Mass Shell Condition**

Now, we apply the constraint Eqs. (6.15). In quantum mechanics, we might treat them as operators and impose on physical states $|\varphi\rangle$ in the Hilbert space

$$T_{++}|\varphi\rangle = 0, \quad T_{--}|\varphi\rangle = 0 \quad \text{(too strong)}. \tag{6.45}$$

However, this is a too strong condition with which no nontrivial physical states survive. Instead, we employ a milder constraint, as in the Gupta–Bleuler quantization [5, 6]. Namely, impose vanishing conditions only for non-negative components

$$L_m|\varphi\rangle = \tilde{L}_m|\varphi\rangle = 0, \quad \text{for } m > 0$$
$$(L_0 + c)|\varphi\rangle = (\tilde{L}_0 + \tilde{c})|\varphi\rangle = 0. \tag{6.46}$$

The mass shell condition follows from the last condition, since $L_0$ contains $p^\mu$ (6.23)

$$M^2 = p \cdot p = \frac{2}{\alpha'}\alpha_0 \cdot \alpha_0 = \frac{2}{\alpha'}\tilde{\alpha}_0 \cdot \tilde{\alpha}_0. \tag{6.47}$$

This means, we need matching between the left and right movers

$$L_0 + c = \tilde{L}_0 + \tilde{c}, \tag{6.48}$$

at the operator level.

Putting the mass shell condition (6.47) to the defining relation (6.41), we have

$$\frac{1}{4}\alpha' M^2 = \sum_{n=1}^{\infty} \alpha_{-n} \cdot \alpha_n + c = \sum_{n=1}^{\infty} \tilde{\alpha}_{-n} \cdot \tilde{\alpha}_n + \tilde{c}. \tag{6.49}$$

It is convenient to define the oscillator number

$$\tilde{N}^\alpha = \sum_{n=1}^{\infty} \tilde{\alpha}_{-n} \cdot \tilde{\alpha}_n$$
$$N^\alpha = \sum_{n=1}^{\infty} \alpha_{-n} \cdot \alpha_n. \tag{6.50}$$

The level matching condition (6.48) becomes $\tilde{N} = N$. Since it relates the left and right movers, we may separate the mass condition of open string theory into the form

$$M^2 \equiv M_L^2 + M_R^2, \tag{6.51}$$

$$M_L^2 = M_R^2, \tag{6.52}$$

$$\frac{1}{2}\alpha' M_L^2 \equiv \sum_{n=1}^{\infty} \alpha_{-n} \cdot \alpha_n + c = N + c, \tag{6.53}$$

$$\frac{1}{2}\alpha' M_R^2 \equiv \sum_{n=1}^{\infty} \tilde{\alpha}_{-n} \cdot \tilde{\alpha}_n + \tilde{c} = \tilde{N} + \tilde{c}. \tag{6.54}$$

For the open string, we put $\alpha_n = \tilde{\alpha}_n$ in (6.51) and obtain

$$\alpha' M^2 = \sum_{n=0}^{\infty} \alpha^i_{-n} \alpha^i_n + \frac{1}{2} c. \tag{6.55}$$

### 6.1.3 Light-Cone Gauge

Here we will consider quantization in the light-cone frame. Although we discard the manifest Lorentz invariance(although not lost) of the theory, we can easily obtain the spectrum explicitly. Due to the residual gauge freedom, we can solve the constraint equation (6.14) explicitly in this light-cone frame, to leave only physical degrees.

In the previous subsection, we observed that there is a residual gauge freedom (6.38). Under it, $\tau = \frac{1}{2}(\sigma^+ + \sigma^-)$ transforms as

$$\tilde{\tau} = \frac{1}{2}[\tilde{\sigma}^+(\tau + \sigma) + \tilde{\sigma}^-(\tau - \sigma)]. \tag{6.56}$$

For $\tilde{\tau}$, this is nothing but a general solution (6.11) of the wave equation (6.10), $(\partial_\sigma^2 - \partial_\tau^2)\tilde{\tau} = 0$. So any linear combination of string coordinates $X^\mu$ can be a solution up to a multiplicative and an additive constant.

What choice will be the most useful one? We will show that, by a suitable choice we can solve the constraint equation (6.14) fully to leave only the physical (transverse) degrees of freedom. It is done by introducing the following light-cone coordinates:

$$X^\pm \equiv \frac{1}{\sqrt{2}}(X^0 \pm X^{D-1}). \tag{6.57}$$

In this coordinate,[2] $\eta_{ij} = \text{diag}(-1, -1, \ldots, -1)$ with $(i, j = 1, 2, \ldots, D-2)$, $\eta_{+-} = 1$, and $\eta_{-+} = 1$, thus the inner products look as

$$v \cdot w = v^+ w^- + v^- w^+ - v^i w^i. \tag{6.58}$$

---

[2]We will denote the transverse coordinates by roman indices $i, j$. The remaining two coordinates can be called $X^\pm$, or $\tilde{\tau}, \tilde{\sigma}$.

Now, using the gauge freedom let us set $\tilde{\tau}$ to a spacetime direction as $\tilde{\tau} = X^+/p^+ +$ (constant). In other words, we choose

$$X^+(\tau, \sigma) = x^+ + \ell_s p^+ \tilde{\tau}, \tag{6.59}$$

and $p^+$ is interpreted as the conjugate momentum to $x^+$. Stated differently, *the oscillator coefficients $\alpha_n^+$ are chosen to be zero for $n \neq 0$*. It amounts to making the worldsheet time direction and $X^+$ coincident in the infinite $p^+$ limit; thus this gauge choice is called the *light-cone gauge*. From (6.14) the constraint equations $(\dot{X} \pm X')^2 = 0$ becomes

$$\dot{X}^- \pm X'^- = \frac{1}{2p^+\ell_s^2} \sum_i (\dot{X}^i \pm X''^i)^2. \tag{6.60}$$

Thus, in the light-cone gauge $X^i$ are the only dynamical degrees.

We are left to specify the closed or open string boundary conditions. Here we will consider the closed string case, since the open string case is obtained therefrom by the doubling trick. We have mode expansion for $X^-$,

$$X^- = x^- + p^-\tau + i\ell_s \sum_{n \neq 0} \left( \frac{1}{n}\alpha_n^- e^{-in(\tau-\sigma)} + \frac{1}{n}\tilde{\alpha}_n^- e^{-in(\tau+\sigma)} \right). \tag{6.61}$$

Inserting this into (6.60) and comparing both sides term by term, we can solve $\alpha_n^-, \tilde{\alpha}_n^-$ in terms of $\alpha_n^i, \tilde{\alpha}_n^i$

$$\alpha_n^- = \frac{1}{2p^+\ell_s} \left( \sum_{i=1}^{D-2} \sum_{m=-\infty}^{\infty} :\alpha_{n-m}^i \alpha_m^i: + :\tilde{\alpha}_{n-m}^i \tilde{\alpha}_m^i: + (c + \tilde{c})\delta_{n0} \right). \tag{6.62}$$

In particular, the identification $p^- = \alpha_0^-$ gives the mass shell condition,

$$\frac{1}{2}\alpha' M^2 = \frac{1}{2}\alpha' p^2 = \frac{1}{2}\alpha' \left( 2p^+ p^- - \sum_i p^i p^i \right)$$

$$= \sum_{i=1}^{D-2} \sum_{n=1}^{\infty} \left( \alpha_{-n}^i \alpha_n^i + \tilde{\alpha}_{-n}^i \tilde{\alpha}_n^i \right) + c + \tilde{c}, \tag{6.63}$$

where we used (6.62) in the second equation. We have the same result as (6.51), provided the following is met. First, we agree that the oscillators are only excited along the physical directions with the index $i$. Second, we interpret $c$ and $\tilde{c}$ be the vacuum constants (6.39) and (6.40). For each $n$, we have $[\alpha_n^i, \alpha_{-n}^i] = n$ and

$[\tilde{\alpha}_n^i, \tilde{\alpha}_{-n}^i] = n$. Since we have $(D - 2)$ physical degrees,

$$c + \tilde{c} = (D - 2) \sum_{n=1}^{\infty} n = (D - 2)\zeta(-1) = -\frac{D - 2}{12}. \tag{6.64}$$

The formally divergent sum is calculated by an analytic continuation of the Riemann zeta function $\zeta(s) = \sum_{n=1}^{\infty} n^s$ to give [9]

$$\zeta(-1) = -\frac{1}{12}. \tag{6.65}$$

We will justify this prescription by requiring the modular invariance shortly.

Let us discuss the spectrum. Consider first open string. From the mode expansion, we construct a tower of physical states. Because of the Poincaré invariance in the target space, we have the unitary representation in terms of mass, spins, etc. With oscillators, we define the ground state that is annihilated by all the oscillator annihilation operators,

$$
\begin{aligned}
p^i|0; k\rangle &= k^i|0; k\rangle, \\
\alpha_m^i|0; k\rangle &= 0, \quad m > 0,
\end{aligned}
\tag{6.66}
$$

where $k^\mu$ is the eigenvalue of the momentum operator $p^\mu$.

Note that we have $c = -(D - 2)/24$ from (6.64). The condition for vanishing Lorentz anomaly fixes $D = 26$ by "no-ghost" theorem [10, 11], thus $c = -1$. Here we just present a heuristic argument. In the light-cone gauge of open strings, all string excitations are created by operating transverse oscillators $\alpha_{-n}^i$ on the vacuum. The ground state $|0\rangle$ is a tachyon since the eigenvalue of $M^2$ is $c < 0$. As known in the Higgs potential, this signals that this vacuum is not the true vacuum, which requires a study of string field theory. Here, we ignore this tachyon problem. Then consider the first excited state

$$\alpha_{-1}^i|0\rangle. \tag{6.67}$$

It has $(D-2)$ components $(i = 1, 2, \ldots, D-2)$ and is a vector representation of the transverse rotation group $SO(D-2)$. Its mass is $\frac{1}{4}\alpha' M^2 = 1 + c = 1 - (D-2)/24$. If we require the Lorentz symmetry in $D$ dimensions, it should be massless. Therefore, we have

$$D = 26, \quad c = -1.$$

For the closed string, we define the ground state as the tensor product

$$
\begin{aligned}
&|0\rangle_L \otimes |0\rangle_R, \\
&\tilde{\alpha}_m^i|0\rangle_L = \alpha_m^i|0\rangle_R = 0, \quad m > 0.
\end{aligned}
\tag{6.68}
$$

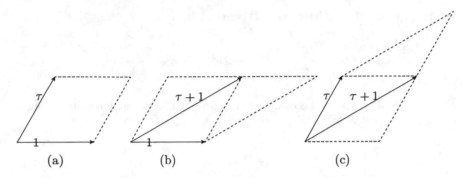

**Fig. 6.1** (a) $\tau$ defines a torus bounded by the parallelogram. The subsequent figures show those modified by $\mathscr{T}$ (b), and by $\mathscr{T}\mathscr{S}\mathscr{T}$ (c) which is conformally equivalent to the one by $\mathscr{S}$

According to (6.53, 6.54), the first excited states are massless states

$$\tilde{\alpha}^i_{-1}|0\rangle_L \otimes \alpha^j_{-1}|0\rangle_R. \tag{6.69}$$

By the same argument, and the level matching condition (6.52), we have $D = 26$ and $c = \tilde{c} = -1$. We can decompose this as symmetric traceless, trace, and antisymmetric parts. They provide graviton $G_{ij}$, dilaton $\phi$, and antisymmetric tensor $B_{ij}$. They are the massless representation of the Lorentz group SO(1, 25).

Note that although we will not employ the covariant quantization, it is proven equivalent to the light-cone quantization [1]. Therefore, without confusion we will conveniently use the covariant Lorentz indices $\mu$, $\nu$ instead of the physical degree indices $i$, $j$.

### 6.1.4   Partition Function and Modular Invariance

The most important consistency condition that we will inspect every time later is the so-called *modular invariance*, emerging from one-loop amplitude. Let us calculate the partition function, or Euclidianized one-loop vacuum-to-vacuum amplitude,

$$\mathscr{Z} = \int [Dh][DX]e^{-S_E(h,X)}. \tag{6.70}$$

Its Feynman diagram is torus. This torus is described by one complex number $\tau = \tau_1 + i\tau_2$, called the modular parameter,[3] or in the case of torus, *complex structure*. It is made by identifying the opposing sides of a parallelogram parametrized by 1 and $\tau$, as shown in Fig. 6.1a. Without loss of generality we can fix $\tau_2 > 0$. For example, $\tau = i$ corresponds to a rectangle.

---

[3]The modular parameter should not be confused with the worldsheet coordinate $\tau = \sigma^0$.

The problem arises from the measure $[Dh]$ in (6.70). We cannot maintain the covariant gauge $h_{\alpha\beta} = \eta_{\alpha\beta}$ globally *maintaining the original periodicity* [12]: the residual degree of freedom is embedded in $\tau$ so that at best we may have the form

$$ds^2 = |d\sigma^1 + \tau d\sigma^2|^2. \tag{6.71}$$

Such non-fixable parameter(s) is called *modulus* and its number is determined by topology.

The string loop amplitude should be invariant for the equivalent torus. Its equivalence class is generated by reparametrization of $\tau$, belonging to the modular transformation $\mathrm{PSL}(2, \mathbb{Z})$,

$$\tau \rightarrow \frac{a\tau + b}{c\tau + d}, \quad a, b, c, d \in \mathbb{Z}, \quad ad - bc = 1. \tag{6.72}$$

It is generated by two elements,

$$\mathscr{T} : \tau \rightarrow \tau + 1, \quad \mathscr{S} : \tau \rightarrow -1/\tau. \tag{6.73}$$

They satisfy the relations $\mathscr{S}^2 = (\mathscr{S}\mathscr{T})^3 = 1$. Under $\mathscr{T}$ the torus is deformed as Fig. 6.1b. The role of $\mathscr{S}$ is not transparent, but the $\mathscr{T}\mathscr{S}\mathscr{T}$ generates the transformation $\tau \rightarrow \tau/(\tau + 1)$ which corresponds to Fig. 6.1c, which is equivalent to $\mathscr{S}$ up to rescaling. Note that it exchanges worldsheet time $\sigma^0$ and space $\sigma^1$ direction.

As in the Faddeev–Popov gauge fixing in gauge theories, we should divide the measure by the redundancy. We should restrict the integration region to the *fundamental region* $\mathbb{C}^1/\mathrm{PSL}(2, \mathbb{Z})$,

$$|\tau| > 1, \quad |\tau_1| < \tfrac{1}{2}, \quad \tau_2 > 0 \tag{6.74}$$

as shown in Fig. 6.2. This is an orbifold having fixed points at $\tau = i$ and $\tau = e^{2\pi i/3}$. With the field theory analogy, the loop amplitude has the ultraviolet diverges when the region goes to $\mathrm{Im}\,\tau \rightarrow 0$ [12], which is absent due to the nature of the string being an extended object. In the point particle limit, where the torus shrinks to a circle, this becomes a loop diagram, which is responsible for the chiral and gravitational anomalies. It turns out that its low-energy limit is anomaly free since anomaly comes from the failure of its regularization.

The resulting one-loop amplitude has the form

$$\mathscr{Z} = V \int \frac{d\tau d\bar{\tau}}{(\mathrm{Im}\,\tau)^2} \mathscr{Z}(\tau, \bar{\tau}). \tag{6.75}$$

We will not consider constant volume factor $V$. We form such field theory on a torus [12] by a field theory on a circle, evolving the Euclidian time by $2\pi\tau_2$ and translating in the $\sigma$ direction by $2\pi\tau_1$, and then gluing the ends together. They are generated

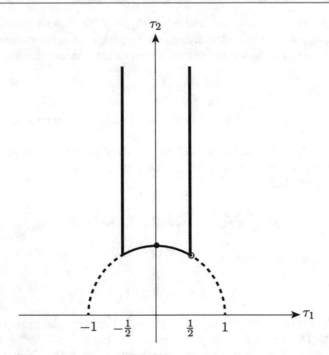

**Fig. 6.2** The fundamental region is bounded by the bold lines and arc. Two lines are identified. Two fixed points are shown as •($i$) and $\circ(e^{2\pi i/3})$

by the worldsheet Hamiltonian (6.42) and the momentum (6.43), respectively. Thus, we have

$$\mathscr{Z}(\tau,\bar\tau) = \mathrm{Tr}\left[\exp(2\pi i\tau_1 P - 2\pi\tau_2 H)\right] \tag{6.76}$$

$$= \mathrm{Tr}\left[q^{\tilde L_0+\tilde c}\bar q^{L_0+c}\right], \tag{6.77}$$

where

$$q = \exp(2\pi i\tau). \tag{6.78}$$

An immediate consequence is that invariance under $\mathscr{T}$ requires that the worldsheet momentum is zero and this is nothing but the level matching condition (6.48) in most general form.

Let us calculate it for the bosonic string case. From (6.41), $\tilde L_0$ and $L_0$ are separated into continuous (center of momentum) and discrete (oscillator) parts. In the light-cone gauge we have 24 physical degrees of freedom; For each degree of freedom, it is

$$\mathscr{Z}(\tau,\bar\tau) \equiv \int\frac{dk}{2\pi}e^{-\pi\tau_2 k^2}\prod_{n=1}^{\infty}\sum_{N_n,\tilde N_n=0}^{\infty}q^{n\tilde N_n+\frac{n}{2}}\bar q^{nN_n+\frac{n}{2}}, \tag{6.79}$$

where $n\tilde{N}_n = \tilde{\alpha}^i_{-n}\tilde{\alpha}^i_n$ and $nN_n = \alpha^i_{-n}\alpha^i_n$ for fixed $i$ (no summation). The continuous part is the Gaussian integral

$$\int \frac{dk}{2\pi} e^{-\pi\tau_2 k^2} = (4\pi^2\tau_2)^{-1/2}. \tag{6.80}$$

The oscillator part is a geometric sum. For the $q$-sum, we have

$$\prod_{n=1}^{\infty}\sum_{\tilde{N}_n=0}^{\infty} q^{\frac{n}{2}+n\tilde{N}_n} = \prod_{n=1}^{\infty} q^{\frac{n}{2}} \sum_{\tilde{N}_n=0}^{\infty} q^{n\tilde{N}_n} \tag{6.81}$$

$$= q^{-\frac{1}{24}} \prod_{n=1}^{\infty}(1 + q^n + q^{2n} + \ldots) \tag{6.82}$$

$$= q^{-\frac{1}{24}} \prod_{n=1}^{\infty}(1 - q^n)^{-1} \tag{6.83}$$

$$\equiv \eta(\tau)^{-1}, \tag{6.84}$$

which is the definition of Dedekind eta function. Note that the zeta function regularization (6.65) provides consistent modular invariance. Performing $q$-expansion, we see that each coefficient of $q^E$ counts the number of states with the energy $E$.

The right mover is equivalent to the left one, except the antiholomorphic argument $\tau \to \bar{\tau}$. With 24 identical such physical degrees of freedom, we have finally

$$\mathscr{Z}(\tau, \bar{\tau}) = [\mathscr{L}(\tau, \bar{\tau})]^{24} = \left[(4\pi^2\tau_2)^{-1/2}|\eta(\tau)|^{-2}\right]^{24}. \tag{6.85}$$

One can verify that this is invariant under the modular transformations (exercise),

$$\mathscr{T} : \eta(\tau + 1) = e^{i\pi/12}\eta(\tau) \tag{6.86}$$

$$\mathscr{S} : \eta(-1/\tau) = (-i\tau)^{1/2}\eta(\tau). \tag{6.87}$$

## 6.2 Superstring

Now let us introduce supersymmetry in the string, which leads to *superstring theory*. Many important features of superstring rely on supersymmetry. Historically, before the linear realization of spacetime supersymmetry in four spacetime dimensions, supersymmetry was introduced in string theory first in two dimensional worldsheet [13, 14].

For our purpose, it will suffice to restrict our discussion to the Neveu–Schwarz–Ramond (NSR) formalism. It has only the worldsheet supersymmetry manifestly, but indeed has the spacetime supersymmetry also when we introduce an appropriate projection. As in the previous section, here we will consider the light-cone quantization also.

### 6.2.1  Worldsheet Action

In addition to the bosonic fields $X^\mu(\tau, \sigma)$, $\mu = 0, 1, \ldots, D-1$, let us also introduce the same number of fermionic fields $\Psi^\mu(\tau, \sigma)$ in the worldsheet. We choose two-component Majorana fermions. The action

$$S = \frac{1}{4\pi\alpha'} \int d^2\sigma \left( \partial_\alpha X^\mu \partial^\alpha X_\mu + i \overline{\Psi}^\mu \rho^\alpha \partial_\alpha \Psi_\mu \right) \tag{6.88}$$

is invariant under the two dimensional global supersymmetry in the worldsheet,

$$\delta X^\mu = \bar{\xi} \Psi^\mu, \quad \delta \Psi^\mu = -i\rho^\alpha \partial_\alpha X^\mu \xi. \tag{6.89}$$

The two dimensional gamma matrices are taken as

$$\rho^0 = \begin{pmatrix} 0 & -i \\ i & 0 \end{pmatrix}, \quad \rho^1 = \begin{pmatrix} 0 & i \\ i & 0 \end{pmatrix}, \tag{6.90}$$

for which the algebra is

$$\{\rho^\alpha, \rho^\beta\} = 2\eta^{\alpha\beta} \mathbf{1}_2, \tag{6.91}$$

where $\mathbf{1}_2$ is the unit matrix. Since we chose purely imaginary gamma matrices, the Dirac operator is real and so are the components of Majorana spinors $\Psi^* = \Psi$. In (6.88), we formed a Lorentz scalar with $\overline{\Psi} \equiv \Psi^\dagger \rho^0$.

We see that the fermion $\Psi^\mu$ has the same vector index as the boson $X^\mu$ and has the same $SO(D-1, 1)$ Poincaré invariance. It seems strange that the fermion transforms as a vector, defying the spin-statistics theorem. However, from the worldsheet point of view, the vector transformation property is the symmetry of target space and hence it is merely an internal symmetry. Without violating the spin-statistics theorem, it is a spin half fermion on the worldsheet but interestingly we will see that it can be either boson or fermion in the spacetime.

For our purpose, there is no need to construct an explicit action with reparametrization invariance or local supersymmetry. It turns out [2] that the action (6.88) is understood as gauge fixed form of local supersymmetric one, as in the bosonic case. Thus we only need to take care of additional constraint. The Euler-Lagrange equations are

$$\partial_\alpha \partial^\alpha X^\mu = 0, \quad i\rho^\alpha \partial_\alpha \Psi^\mu = 0. \tag{6.92}$$

They are to be supplemented by the constraints: the vanishing worldsheet energy-momentum tensor

$$T_{\alpha\beta} = -\partial_\alpha X^\mu \partial_\beta X_\mu - \frac{i}{4}\bar\Psi^\mu(\rho_\alpha \partial_\beta + \rho_\beta \partial_\alpha)\Psi_\mu$$
$$+ \frac{1}{2}\eta_{\alpha\beta}\left(\partial_\gamma X^\mu \partial^\gamma X_\mu + \frac{i}{2}\bar\Psi^\mu \rho^\gamma \partial_\gamma \Psi_\mu\right), \tag{6.93}$$

and the vanishing worldsheet supercurrent

$$J_\alpha = \frac{1}{2}\rho^\beta \rho_\alpha \Psi^\mu \partial_\beta X_\mu, \tag{6.94}$$

whose spatial integral gives supersymmetry generator $Q_\alpha$. They are superpartners of each other, which is natural because two successive supersymmetry transformations generate a translation which is generated by the energy-momentum tensor.

It is convenient to rewrite the fermionic part of the action (6.88) as

$$S_F = \frac{i}{2\pi\alpha'}\int d^2\sigma(\Psi^\mu_L \partial_- \Psi_{\mu L} + \Psi^\mu_R \partial_+ \Psi_{\mu R}) \tag{6.95}$$

with

$$\Psi^\mu = \begin{pmatrix} \Psi^\mu_R(\tau - \sigma) \\ \Psi^\mu_L(\tau + \sigma) \end{pmatrix}. \tag{6.96}$$

This is possible because in two dimensions, a Majorana fermion is further decomposed to the left and right handed Majorana–Weyl fermions $\Psi_L, \Psi_R$ satisfying

$$\partial_- \Psi^\mu_L = 0, \quad \partial_+ \Psi^\mu_R = 0. \tag{6.97}$$

The boundary conditions come from the vanishing boundary term

$$\left[\Psi_L \delta \Psi_L - \Psi_R \delta \Psi_R\right]^\pi_{\sigma=0} = 0. \tag{6.98}$$

For closed string, since it has two $\Psi$s with one variation, the surface term goes away if the solution is periodic or antiperiodic. So we may have

$$\Psi^\mu_L(\tau, \sigma + \pi) = (-1)^\nu \Psi^\mu_L(\tau, \sigma),$$
$$\Psi^\mu_R(\tau, \sigma + \pi) = (-1)^\nu \Psi^\mu_R(\tau, \sigma). \tag{6.99}$$

The periodic boundary condition $\nu = 0$ is called the *Ramond (R) boundary condition*, and the antiperiodic boundary condition $\nu = \frac{1}{2}$ is called the *Neveu-Schwarz (NS) boundary condition*. Therefore, we have four sets of possible boundary conditions. We will see that the spacetime supersymmetry requires that the theory should contain all the four sectors.

For the open string, we have

$$\Psi_L^\mu(\tau, 0) = \Psi_R^\mu(\tau, 0),$$
$$\Psi_L^\mu(\tau, \pi) = (-1)^\nu \Psi_R^\mu(\tau, \pi). \tag{6.100}$$

Due to chiral symmetry, we may fix the relative signs in the first line.

### Super-Virasoro Algebra

For the closed strings, the boundary conditions R or NS are imposed independently
for the left and right movers $\Psi_L$ and $\Psi_R$. The mode expansion goes as

$$\Psi_R^\mu = \sum_{r \in \mathbb{Z}+\nu} \psi_r^\mu e^{-2ir(\tau-\sigma)},$$
$$\Psi_L^\mu = \sum_{r \in \mathbb{Z}+\nu} \tilde{\psi}_r^\mu e^{-2ir(\tau+\sigma)}, \tag{6.101}$$

where $\nu = 0$ for R and $\nu = 1/2$ for NS.

From the action (6.95), the canonical momenta corresponding to $\Psi_{L,R}$ are
$\frac{i}{2\pi}\Psi_{L,R}$, whence the canonical quantization of the fermionic degrees,

$$\{\Psi_L^\mu(\tau, \sigma), \Psi_L^\nu(\tau, \sigma')\} = \{\Psi_R^\mu(\tau, \sigma), \Psi_R^\nu(\tau, \sigma')\} = -2\pi\delta(\sigma - \sigma')\eta^{\mu\nu}, \tag{6.102}$$

$$\{\Psi_L^\mu(\tau, \sigma), \Psi_R^\nu(\tau, \sigma')\} = 0, \tag{6.103}$$

leads to the following anti-commutators for the oscillators

$$\{\psi_r^\mu, \psi_s^\nu\} = \{\tilde{\psi}_r^\mu, \tilde{\psi}_s^\nu\} = -\delta_{r+s,0}\,\eta^{\mu\nu} \tag{6.104}$$

with the anti-commutator of left and right movers vanishing. Since the left and right-
moving states are identical, we will deal with the right movers in what follows. The
expressions for left movers are easily obtained when we replace $(\Psi_L, \tilde{\psi}_r, \tau - \sigma)$
with $(\Psi_R, \psi_r, \tau + \sigma)$. For $r = s = 0$, the $\psi_0^\mu$ has the commutation relation

$$\{\psi_0^\mu, \psi_0^\nu\} = -\eta^{\mu\nu}. \tag{6.105}$$

This is the 10D gamma matrices if we define $\Gamma^\mu$ as

$$\Gamma^\mu = i\sqrt{2}\psi_0^\mu. \tag{6.106}$$

Thus, the R sector corresponds to the fermionic sector in 10D spacetime. On the
other hand, the NS sector corresponds to the bosonic sector in 10D spacetime.

Here also, we will consider the closed strings only. We have seen that the bosonic
Virasoro algebra was introduced from the constraint equations. In the fermionic

case also, the super-Virasoro algebra arises from constraint equations, $T_{\alpha\beta} = 0$ and $J^\alpha = 0$. They are

$$T_{++} = -\partial_+ X_L \cdot \partial_+ X_L - \frac{i}{2}\Psi_L \cdot \partial_+ \Psi_L = 0, \tag{6.107}$$

$$T_{--} = -\partial_- X_R \cdot \partial_- X_R - \frac{i}{2}\Psi_R \cdot \partial_- \Psi_R = 0, \tag{6.108}$$

$$J_+ = \Psi_L \cdot \partial_+ X_L = 0 \tag{6.109}$$

$$J_- = \Psi_R \cdot \partial_- X_R = 0. \tag{6.110}$$

Here and in the sequel, the dot product means contraction of the spacetime index $\mu$. As in the bosonic case, the Virasoro operators are defined as, for $m \neq 0$,

$$\begin{aligned}
L_m &\equiv \frac{1}{\pi}\int_0^\pi d\sigma\, e^{2im\sigma} T_{--} \\
&= -\frac{1}{2}\sum_{n\in\mathbb{Z}} \alpha_{m-n} \cdot \alpha_n + \frac{1}{2}\sum_{r\in\mathbb{Z}+\nu}\left(\frac{m}{2}-r\right)\psi_{m-r}\cdot\psi_r.
\end{aligned} \tag{6.111}$$

Again, we need the normal ordering for the $m = 0$ case as

$$L_0 \equiv -\frac{1}{2}\sum_{n\in\mathbb{Z}} :\alpha_{-n}\cdot\alpha_n: -\frac{1}{2}\sum_{r\in\mathbb{Z}+\nu} r :\psi_{-r}\cdot\psi_r: +c \tag{6.112}$$

with the zero point energy $c$. Similarly, the $\tilde{L}_m$ operators are defined from $T_{++}$. For the supercurrent, the Fourier components of the right movers are

$$G_r = \frac{\sqrt{2}}{\pi}\int_0^\pi d\sigma\, e^{2ir\sigma} J_- = \sum_{n\in\mathbb{Z}} \alpha_{-n}\cdot\psi_{r+n}, \quad r \in \mathbb{Z}+\nu. \tag{6.113}$$

Also, a similar generators $\tilde{G}_r$ are obtained from $J_-$ for the left mover.

These generate the super-Virasoro algebra that can be written with $G$s and $L$s. For the NS sector we have

$$[L_m, L_n] = (m-n)L_{m+n} + \frac{D}{8}m(m^2-1)\delta_{m+n,0}, \tag{6.114}$$

$$[L_m, G_r] = \left(\frac{m}{2}-r\right)G_{m+r}, \tag{6.115}$$

$$\{G_r, G_s\} = 2L_{r+s} + \frac{D}{2}\left(r^2 - \frac{1}{4}\right)\delta_{r+s,0}. \tag{6.116}$$

For the R sector,

$$[L_m, L_n] = (m - n)L_{m+n} + \frac{D}{8}m^3\delta_{m+n,0} \tag{6.117}$$

$$[L_m, G_n] = \left(\frac{m}{2} - n\right)G_{m+n}, \tag{6.118}$$

$$\{G_m, G_n\} = 2L_{m+n} + \frac{D}{2}m^2\delta_{m+n,0}. \tag{6.119}$$

All the same relations hold for the left mover operators. We see that the $L_n$ has the same Virasoro algebra (6.44) except the central charge $C = 3D/2$. It is accounted for by an antiholomorphic boson $\partial_- X_R$ and Majorana–Weyl fermion $\Psi_R$, contributing 1 and $\frac{1}{2}$ for each dimension, respectively.

For the physical states, we require the constraint equations

$$\text{NS}: \quad (L_n + \delta_{n0}c_{\text{NS}})|\phi\rangle = 0, \; n \geq 0; \quad G_r|\phi\rangle = 0, \; r > 0$$

$$\text{R}: \quad (L_n + \delta_{n0}c_{\text{R}})|\phi\rangle = 0, \; n \geq 0; \quad F_m|\phi\rangle = 0, \; m \geq 0, \tag{6.120}$$

where $c_{\text{NS}}$ and $c_{\text{R}}$ are the zero point energy to be determined shortly. Similar relations hold for the left movers.

## 6.2.2   Light-Cone Gauge

To draw physical degrees only, we take the light-cone gauge as done in the bosonic case. Taking the light-cone coordinate, we choose the same form for $X^+$ as given in (6.59),

$$X^+(\sigma, \tau) = x^+ + p^+\tilde{\tau}.$$

All the steps are the same as presented throughout (6.18–6.24). By supersymmetry transformation (6.89) we have $\delta X^+ = \bar{\xi}\psi^+ = 0$ and fix

$$\psi^+ = 0. \tag{6.121}$$

Then we solve the constraint equations,

$$p^+\partial_+ X_L^- = \partial_+ X_L^i \partial_+ X_L^i + \frac{i}{2}\Psi_L^i\partial_+\Psi_L^i, \tag{6.122}$$

$$p^+\partial_- X_R^- = \partial_- X_R^i \partial_- X_R^i + \frac{i}{2}\Psi_R^i\partial_-\Psi_R^i, \tag{6.123}$$

$$\frac{1}{2}p^+\Psi_L^- = X_L^i\partial_+\Psi_L^i, \tag{6.124}$$

$$\frac{1}{2}p^+\Psi_R^- = X_R^i\partial_+\Psi_R^i, \tag{6.125}$$

for $X^-$ and $\Psi^-$ in terms of $X^i$ and $\Psi^i$s. Using the same method as done in the bosonic case, we obtain the oscillators $\alpha^-$ in terms of $\alpha^i_n$, $\psi^i_n$ as

$$\alpha_n^- = \alpha_n^-(X) + \frac{1}{2p^+} \sum_{i=1}^{D-2} \sum_{r=-\infty}^{\infty} \left(r - \frac{1}{2}n\right) : \psi^i_{n-r}\psi^i_r : + \frac{c}{2p^+}\delta_{n,0}, \qquad (6.126)$$

where the first term on the RHS is from the bosonic part of the action (6.62). Also for the fermionic degrees, we have

$$\psi_r^- = \frac{1}{p^+} \sum_{i=1}^{D-2} \sum_{s \in \mathbb{Z}+\nu} \alpha^i_{r-s}\psi^i_s. \qquad (6.127)$$

As in the bosonic case, we want to check the hidden Lorentz invariance for the generators,

$$M^{\mu\nu} = \int_0^\pi d\sigma \left[(X^\mu P^\nu - X^\nu P^\mu) + (\Psi^\mu P_\Psi^\nu - \Psi^\nu P_\Psi^\mu)\right] \qquad (6.128)$$

which should satisfy the Lorentz algebra,

$$[M^{\mu\nu}, M^{\rho\sigma}] = i(\eta^{\mu\rho}M^{\nu\sigma} + \eta^{\nu\sigma}M^{\mu\rho} - \eta^{\mu\sigma}M^{\nu\rho} - \eta^{\nu\rho}M^{\mu\sigma}). \qquad (6.129)$$

The additional pieces are provided by the fermionic superpartners. Thus, the generator for the Lorentz group has another piece $K$,

$$M^{\mu\nu} = -M^{\nu\mu} = M_0^{\mu\nu} + K^{\mu\nu}, \qquad (6.130)$$

where $M_0^{-i}$ is the part contributed by bosonic coordinate only. The mixed commutator $[M, K]$ should vanish. For the transverse degrees $M^{ij}$, it is straightforward to verify the Lorentz algebra. The check for the Lorentz algebra in the light-cone gauge is needed for those involving the lightlike coordinates. Thus, we find

$$[M^{-i}, M^{-j}] = \frac{-1}{(p^+)^2} \sum_{i=1}^{\infty} (\alpha^i_{-n}\alpha^j_n - \alpha^j_{-n}\alpha^i_n)(\Delta_n - n), \qquad (6.131)$$

where

$$\Delta_n = n\left(\frac{D-2}{8}\right) - \frac{n}{2}\left(c_{NS} + \frac{D-2}{16}\right), \qquad (6.132)$$

where $a_{NS}$ is the normal ordering constant in the NS sector. Since the Lorentz algebra (6.129) gives $[M^{-i}, M^{-j}] = 0$, we need $\Delta_n = n$ for Lorentz symmetry, i.e., for a consistent NSR string we must require

$$D = 10, \quad c_{NS} = -\frac{1}{2}. \tag{6.133}$$

This provides another proof of relations among Lorentz symmetry, conformal anomaly, and consistent spacetime dimension $D$. Recall that for each physical boson we have the contribution $-\frac{1}{24}$ to $c$ by (6.64). In the NS sector, we have $v = \frac{1}{2}$ and each fermion has $-\frac{1}{48}$. With the bosonic oscillator we have $-\frac{1}{16}$ to have the result (6.133). So we have the relation (6.133). In the Ramond sector, we have $v = 0$. By supersymmetry, we have the (negative) same contribution $+\frac{1}{24}$ for each fermion in the R sector. There is cancellation

$$c_R = 0. \tag{6.134}$$

Finally, we get the mass shell conditions following the same method of bosonic string:

$$M^2 = M_R^2 + M_L^2, \tag{6.135}$$

supplemented by the level matching condition

$$M_R^2 = M_L^2. \tag{6.136}$$

With the oscillator numbers

$$N^{\psi} \equiv \sum_{r \in \mathbb{Z}+v}^{\infty} r \psi_{-r}^i \psi_r^i, \tag{6.137}$$

$$\tilde{N}^{\psi} \equiv \sum_{r \in \mathbb{Z}+v}^{\infty} r \tilde{\psi}_{-r}^i \tilde{\psi}_r^i, \tag{6.138}$$

and the bosonic ones in (6.50), we obtain

$$\frac{1}{2} \alpha' M_R^2 = N^{\alpha} + N^{\psi} + c, \tag{6.139}$$

$$\frac{1}{2} \alpha' M_L^2 = \tilde{N}^{\alpha} + \tilde{N}^{\psi} + \tilde{c}. \tag{6.140}$$

The difference between NS and R sectors is the following. First, we have, respectively, $v = 0$ and $v = \frac{1}{2}$ in the oscillator numbers (6.137 ) and (6.138). And, the zero point energies $c$ in (6.139) and (6.140) are different.

### 6.2.3 Spectrum and GSO Projection

The states are constructed by piling up oscillators in the Fock space. We will consider a single right mover, which is exactly the same as the left mover.

**NS Sector**

Consider the NS sector $\nu = \frac{1}{2}$ in (6.99), in which the oscillator number is half-integer. The lowest state is the ground state $|0\rangle_{NS}$ defined as

$$\alpha_m^i |0\rangle_{NS} = \psi_r^i |0\rangle_{NS} = 0, \quad m, r > 0. \tag{6.141}$$

From the mass shell condition (6.139), it is tachyonic since the mass is $\frac{1}{8} M^2 = -\frac{1}{2}$, which is undesirable. We may project out this state if we introduce a $G$-parity,

$$G = (-1)^F, \quad F = \sum_{r=1/2}^{\infty} \psi_{-r}^i \psi_r^i \tag{6.142}$$

with fermion oscillator number $F$, and mod out the non-invariant state. Acting on the "ground state," we have

$$G|0\rangle = -|0\rangle, \tag{6.143}$$

so that it is projected out. The first excited states are

$$\psi_{-1/2}^i |0\rangle, \tag{6.144}$$

which are massless. They carry the transversal SO(8) vector index $i$, and therefore form a massless representation of SO(1,9). This means, the NS states become *bosonic* states in the spacetime sense. Applying a few more oscillators, we immediately meet the following problem. The Fock space has a tower of states,

$$|\varphi\rangle = \psi_{-r_1}^{i_1} \psi_{-r_2}^{i_2} \cdots \psi_{-r_n}^{i_n} |0\rangle. \tag{6.145}$$

For $\psi_{-r_k}^{i_k}$ satisfying the worldsheet fermionic relation (6.104), exchanging two fields yields extra factor $(-1)$, which is not desirable for statistics of spacetime bosons. So again, we require the even number of excitations $n \in 2\mathbb{Z}$.

This projection in the Fock space has been formulated by Gliozzi et al. [15] and is known as *the GSO projection*,

$$\frac{1 - (-1)^F}{2}, \tag{6.146}$$

with which we can project out massive and higher spin states as well.

## R Sector

In the R sector $v = 0$, the ground state is defined as that annihilated by all the positive operators

$$\alpha_m^i |0\rangle_R = \psi_m^i |0\rangle_R = 0, \quad m > 0. \tag{6.147}$$

The zero point energy $c_R = 0$; thus the lowest state is massless and this sector is tachyon-free. The commutation relation (6.104) $\{\psi_0^\mu, \psi_0^\nu\} = -\eta^{\mu\nu}$ implies that $\psi_0^\mu$ behave as gamma matrices (6.106). In the light-cone gauge, defining

$$\psi_0^{i\pm} = \frac{1}{\sqrt{2}}(\psi_0^{2i} \pm i\psi_0^{2i+1}), \quad i = 1, 2, 3, 4, \tag{6.148}$$

they satisfy the algebra of harmonic oscillators

$$\{\psi_0^{i+}, \psi_0^{j-}\} = \delta^{ij}, \tag{6.149}$$

with other (anti)commutators vanishing. We have spinorial representation that we saw in the previous chapter. Therefore, each represents the creation $(+)$ and annihilation $(-)$ operator of the spin-$\frac{1}{2}$ system. Thus, we can build up the spinorial state denoted as

$$|s_1 \, s_2 \, s_3 \, s_4\rangle, \tag{6.150}$$

where each $s_i$ can assume $\pm\frac{1}{2}$. We define the ground state of the algebra[4] (6.149), by one annihilated by all the annihilation operators

$$\psi_0^{i-} |0\rangle_R = 0, \quad i = 1, 2, 3, 4. \tag{6.151}$$

With this vacuum we can make the following 16 massless states:

$$\begin{aligned}
|0\rangle_R &= |-\tfrac{1}{2} - \tfrac{1}{2} - \tfrac{1}{2} - \tfrac{1}{2}\rangle \\
\psi_0^{i+} |0\rangle_R &= |\tfrac{1}{2} - \tfrac{1}{2} - \tfrac{1}{2} - \tfrac{1}{2}\rangle \\
\psi_0^{i_1+} \psi_0^{i_2+} |0\rangle_R &= |\tfrac{1}{2} \tfrac{1}{2} - \tfrac{1}{2} - \tfrac{1}{2}\rangle \\
\psi_0^{i_1+} \psi_0^{i_2+} \psi_0^{i_3+} |0\rangle_R &= |\tfrac{1}{2} \tfrac{1}{2} \tfrac{1}{2} - \tfrac{1}{2}\rangle \\
\psi_0^{i_1+} \psi_0^{i_2+} \psi_0^{i_3+} \psi_0^{i_4+} |0\rangle_R &= |\tfrac{1}{2} \tfrac{1}{2} \tfrac{1}{2} \tfrac{1}{2}\rangle,
\end{aligned} \tag{6.152}$$

---

[4]This Fock space is different from the one constructed by the algebra (6.104).

where the underline means possible permutations. They make up spacetime fermions. Especially, we can define the chirality operator

$$\Gamma = \psi_1 \psi_2 \cdots \psi_8 = 2^4 s_1 s_2 s_3 s_4, \tag{6.153}$$

which gives $\Gamma = +1$ or $\Gamma = -1$, respectively, for even or odd numbers of $-\frac{1}{2}$ in $s_1 s_2 s_3 s_4$. Accordingly such states are called spinorial $\mathbf{8}_s$ of SO(8) for $\Gamma = +1$ and conjugate sponorial $\mathbf{8}_c$ of SO(8) for $\Gamma = -1$.

Then, we encounter a problem if we require the spacetime supersymmetry, since we have too many R states compared to those in the NS sector. To match, we introduce another GSO projector in this R sector

$$\frac{1 + \Gamma(-1)^F}{2}, \tag{6.154}$$

with fermion number in the R sector given as, similarly to (6.142),

$$F = \sum_{m=1}^{\infty} \psi^i_{-m} \psi^i_m, \tag{6.155}$$

or in the $s$ basis,

$$\Gamma = \sum_{i=1}^{4} s_i = 0, \quad \mathrm{mod}\ 2. \tag{6.156}$$

We can choose $\Gamma$, which can be called the chirality, as we wish. We can check that we have the same number of fermions in the R sector as bosons in the NS sector. Furthermore, every state matches at the full massive level.

## Partition Function

We may understand the necessity of the GSO projection, by calculating partition function for superstring. The procedure is the same as before, discussed in Sect. 6.1.4. The alteration is that the generators $L_0$ and $\tilde{L}_0$ of (6.112) now contain fermionic parts in addition, and we have eight physical degrees of freedom.

Consider first the NS sector, defined by the boundary condition (6.99). The fermionic oscillator can be either occupied or unoccupied, so the traces become

$$\mathrm{Tr}\, q^{N^\psi} = \prod_{i=1}^{8} \prod_{r=1/2}^{\infty} \sum_{N_r=0}^{1} q^{r/2+rN_r} = \prod_{i=1}^{8} q^{1/16} \prod_{n=1}^{\infty} (1 + q^{n-\frac{1}{2}}) \tag{6.157}$$

$$= q^{1/2} \prod_{n=1}^{\infty} (1 + q^{n-1/2})^8, \tag{6.158}$$

$$\mathrm{Tr}(-1)^F q^{N^\psi} = \prod_{r=1/2}^{\infty} \mathrm{Tr}(-1)^F q^{r\psi_{-r}\cdot\psi_r} = \prod_{n=1}^{\infty} (1 - q^{n-1/2})^8, \tag{6.159}$$

where we renamed the dummy variable as $r + \frac{1}{2} = n$. The zero point energy will be calculated in the next chapter. We also make use of the calculation in the bosonic part $\mathrm{Tr}\, q^{N^\alpha}$ in (6.81). For the NS sector, we have

$$
\mathscr{Z}_{NS}(\tau) = \mathrm{Tr}\, \frac{1 - (-1)^F}{2} q^{N^\alpha + N^\psi}
$$

$$
= \frac{1}{2} q^{-1/2} \prod_{n=1}^{\infty} (1 - q^n)^{-8} \left[ \prod_{n=1}^{\infty} (1 + q^{n-1/2})^8 - \prod_{n=1}^{\infty} (1 - q^{n-1/2})^8, \right]
$$

where

$$
N = \sum_{n=1}^{\infty} \alpha_{-n} \cdot \alpha_n + \sum_{r=\frac{1}{2}}^{\infty} \sum_{i=1}^{8} N_r, \quad N_r \equiv \psi_{-r}^i \psi_r^i. \tag{6.160}
$$

Also, we included the vacuum energy of the NS state (6.141). Here and in the sequel we omit the continuous part (6.80).

Similarly, we have the R sector, defined in (6.98), partition function,

$$
\mathscr{Z}_R(\tau) = 16\, \mathrm{Tr}\, \frac{1 + \Gamma(-1)^F}{2} q^N, \tag{6.161}
$$

where

$$
N = \sum_{n=1}^{\infty} (\alpha_{-n} \cdot \alpha_n + n\psi_{-n} \cdot \psi_n). \tag{6.162}
$$

The multiplicity 16 reflects the degeneracy of the ground states (6.152). After the GSO projection, it becomes 8. For massive states, the chirality operator $\Gamma$ commutes with $\psi_n^i$ operators with $n \neq 0$, and the corresponding pairs of states are cancelled out. Using the results given above, we obtain

$$
\mathscr{Z}_R(\tau) = 8 \prod_{n=1}^{\infty} (1 - q^n)^{-8} (1 + q^n)^8. \tag{6.163}
$$

These functions can be rewritten in terms of modular forms

$$
\mathscr{Z}_R(\tau) = \frac{\vartheta \begin{bmatrix} 1/2 \\ 0 \end{bmatrix}^4}{2\eta^{12}} + \frac{\vartheta \begin{bmatrix} 1/2 \\ 1/2 \end{bmatrix}^4}{2\eta^{12}}, \tag{6.164}
$$

$$
\mathscr{Z}_{NS}(\tau) = \frac{\vartheta \begin{bmatrix} 0 \\ 0 \end{bmatrix}^4}{2\eta^{12}} - \frac{\vartheta \begin{bmatrix} 0 \\ 1/2 \end{bmatrix}^4}{2\eta^{12}}. \tag{6.165}
$$

Here, we define Jacobi theta function as

$$\vartheta \begin{bmatrix} \alpha \\ \beta \end{bmatrix} (\tau) = \eta(\tau) e^{2\pi i \alpha \beta} q^{\alpha^2/2 - 1/24} \prod_{n=1}^{\infty} (1 + q^{n+\alpha-1/2} e^{2\pi i \beta})(1 + q^{n-\alpha-1/2} e^{-2\pi i \beta}),$$

(6.166)

and Dedekind eta function is defined in (6.84). We can make $q$-expansion

$$\vartheta \begin{bmatrix} 1/2 \\ 1/2 \end{bmatrix} (\tau) = 0,$$

(6.167)

$$\vartheta \begin{bmatrix} 1/2 \\ 0 \end{bmatrix} (\tau) = 2q^{1/4}(1 + q^2 + q^6 + q^{12} + q^{20} + \cdots),$$

(6.168)

$$\vartheta \begin{bmatrix} 0 \\ 0 \end{bmatrix} (\tau) = 1 + 2q + 2q^4 + 2q^9 + 2q^{16} + \cdots,$$

(6.169)

$$\vartheta \begin{bmatrix} 0 \\ 1/2 \end{bmatrix} (\tau) = 1 - 2q + 2q^4 - 2q^9 + 2q^{16} + \cdots.$$

(6.170)

The factor 2 in (6.168) and hence the factor 8 in (6.161) naturally come from the factor $(1 + q^{n-1}) = 2$ for $n = 1$. Miraculously, we have $\mathscr{Z}_{NS} = \mathscr{Z}_R$ as an "abstruse identity" by Jacobi [9]

$$\vartheta \begin{bmatrix} 1/2 \\ 0 \end{bmatrix}^4 (\tau) - \vartheta \begin{bmatrix} 1/2 \\ 0 \end{bmatrix}^4 (\tau) - \vartheta \begin{bmatrix} 0 \\ 1/2 \end{bmatrix}^4 (\tau) = 0,$$

(6.171)

which can be checked term by term in the above expansion. It means that, for given energy, the number of states in the R and NS sectors are the same. This relation is a necessary condition for the existence of spacetime supersymmetry. We construct explicit string models with the GSO projection, having the supersymmetric spectrum.

## Spin Structure

The theta function is a modular function. Its useful property is well-defined transformation

$$\mathscr{T} : \vartheta \begin{bmatrix} \alpha \\ \beta \end{bmatrix} (\tau + 1) = e^{i\pi(\alpha^2 - \alpha)} \vartheta \begin{bmatrix} \alpha \\ \alpha + \beta - \frac{1}{2} \end{bmatrix} (\tau),$$

(6.172)

$$\mathscr{S} : \vartheta \begin{bmatrix} \alpha \\ \beta \end{bmatrix} (-1/\tau) = (-i\tau)^{1/2} e^{2\pi i \alpha \beta} \vartheta \begin{bmatrix} \beta \\ -\alpha \end{bmatrix} (\tau).$$

(6.173)

This means

$$\mathscr{T} : \quad \vartheta [^{1/2}_0] \to e^{\pi i/4} \vartheta [^{1/2}_0], \quad \vartheta [^0_0] \leftrightarrow e^{\pi i/4} \vartheta [^0_{1/2}]$$

$$\mathscr{S} : \quad \vartheta [^0_0]/\eta \to \vartheta [^0_0]/\eta, \quad \vartheta [^0_{1/2}]/\eta \leftrightarrow \vartheta [^{1/2}_0]/\eta.$$

(6.174)

With antiholomorphic right movers, the whole partition function is made invariant. Therefore, we *cannot* make a nontrivial modular invariant theory *with only one of four sectors*. The consistency forces modular invariance.

We can reinterpret the above result as follows (again, we neglect the continuous part). We decompose the partition functions (6.164) and (6.165) as follows:

$$\mathscr{Z}_{NS} = \mathscr{Z}_{NS}^{NS} + \mathscr{Z}_{NS}^{R}, \quad \mathscr{Z}_{R} = \mathscr{Z}_{R}^{NS} + \mathscr{Z}_{R}^{R}, \tag{6.175}$$

where

$$\mathscr{Z}_{NS}^{NS} \equiv \operatorname{Tr} q^{L_0(NS)} = \frac{\vartheta[^0_0]^4}{2\eta^{12}}, \tag{6.176}$$

$$\mathscr{Z}_{R}^{NS} \equiv \operatorname{Tr} q^{L_0(R)} = \frac{\vartheta[^{1/2}_0]^4}{2\eta^{12}}, \tag{6.177}$$

$$\mathscr{Z}_{NS}^{R} \equiv \operatorname{Tr}(-1)^F q^{L_0(NS)} = \frac{\vartheta[^0_{1/2}]^4}{2\eta^{12}}, \tag{6.178}$$

$$\mathscr{Z}_{R}^{R} \equiv \operatorname{Tr}(-1)^F q^{L_0(R)} = \frac{\vartheta[^{1/2}_{1/2}]^4}{2\eta^{12}} = 0, \tag{6.179}$$

where $L_0(NS, R)$ is the Virasoro operator constructed with the NS and R oscillators. $F$ in Eq. (6.177) is the NS sector $F$ number (6.142) and $F$ in Eq. (6.179) is the R sector $F$ number (6.155).

Their transformation properties are summarized in Fig. 6.3. We have the right number of degrees of freedom that cancels the factor arising from $\mathscr{T}$ transformation. For $\mathscr{S}$ transformation, the nontrivial $\tau$ dependence of $\vartheta(\tau)$ is the same as that of $\eta(\tau)$. Thus, by dividing $\eta(\tau)$ we expect some linear combination of these give a modular invariant partition function.

As discussed below (6.73), the $\mathscr{S}$ transformation exchanges the two boundary conditions along the worldsheet directions $\sigma^0$ and $\sigma^1$. In view of Eq. (6.174), it indicates that we may assign R or NS boundary conditions in the $\tau$ *direction*. For instance, the partition function $\mathscr{Z}_{R}^{NS}$ in (6.177) is defined by the R boundary condition but is transformed into $\mathscr{Z}_{NS}^{R}$ defined by NS condition. This means, in fact $\mathscr{Z}_{R}^{NS}$ in (6.177) has the NS boundary condition along the $\tau$ direction.

The GSO projection has an effect of taking into account both boundary conditions, which is referred to as *spin structure* [16]. This is required when we consider loop amplitudes. The partition function corresponds to the one-loop amplitude. The

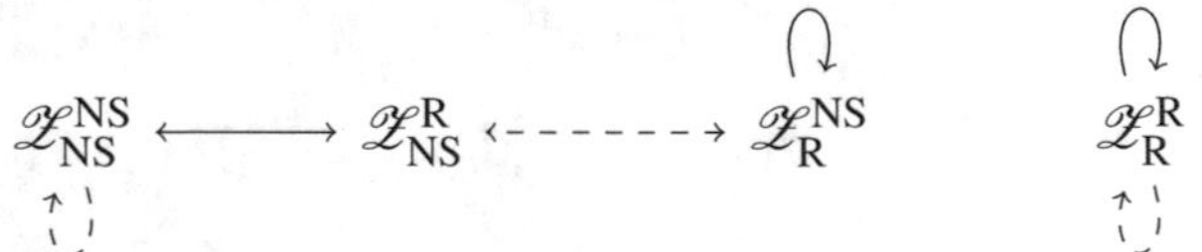

**Fig. 6.3** Modular transformation of spin structures. Solid line denotes $\mathscr{T}$ and dashed $\mathscr{S}$

$$\mathscr{Z}_{NS}^{NS} \longleftrightarrow \mathscr{Z}_{NS}^{R} \dashleftarrow\dashrightarrow \mathscr{Z}_{R}^{NS} \qquad \mathscr{Z}_{R}^{R}$$

**Fig. 6.4** When a fermion is transported around a cycle in the non-simply connected Feynman diagram, it acquires an ambiguous phase, which is remedied by the GSO projection

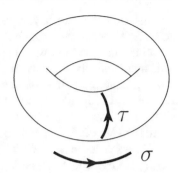

corresponding Feynman diagrams are not simply connected (having genera), and there is a subtlety in defining boundary conditions for fermions living on them. The situation is depicted in Fig. 6.4. When rotated once around a closed cycle, a boson remains invariant, i.e., with no extra phase. However, a fermion acquires an ambiguous phase, where there are two possibilities. The GSO projection projects out consistently one of them [2].

### 6.2.4 Superstring Theories

Now we are ready to define superstring theories. Let us investigate massless states, formed by the following left and right movers. Recall that when we perform the GSO projection in the R sector (6.154) for each mover, we have two choices of chirality $\Gamma$. Since only the relative chirality matters, we have the following two choices:

$$\text{Type IIA:} \quad (\mathbf{8}_v \oplus \mathbf{8}_c) \otimes (\mathbf{8}_v \oplus \mathbf{8}_s),$$

$$\text{Type IIB:} \quad (\mathbf{8}_v \oplus \mathbf{8}_c) \otimes (\mathbf{8}_v \oplus \mathbf{8}_c).$$

We expand the tensor product as in (4.40). What they have in common in both theories are the products

$$\text{NSNS}: \mathbf{8}_v \otimes \mathbf{8}_v = \mathbf{35}_v + \mathbf{28} + \mathbf{1}, \tag{6.180}$$

$$\text{NSR}: \mathbf{8}_c \otimes \mathbf{8}_c = \mathbf{56}_s + \mathbf{8}_s, \tag{6.181}$$

which correspond to graviton $G_{\mu\nu}$, antisymmetric tensor $B_{\mu\nu}$ and dilaton $\phi$, and their superpartners gravitino $\psi_\mu$ and dilatino $\lambda$. The selection of chirality in the Ramond sector of either mover leads to different spectrum. First consider choosing the different chirality $\Gamma_L = 1$, $\Gamma_R = -1$. From (4.40), the resulting spectrum is

$$\text{RR}: \mathbf{8}_c \otimes \mathbf{8}_s = \mathbf{56}_v + \mathbf{8}_v, \tag{6.182}$$

$$\text{RNS}: \mathbf{8}_c \otimes \mathbf{8}_v = \mathbf{56}_c + \mathbf{8}_c. \tag{6.183}$$

In the RR sector, we have a vector $C_\mu$ and rank three antisymmetric tensor field $C_{\mu\nu\rho}$. The RNS sector completes the supersymmetric multiplet: another set of dilatino and gravitino with opposite chirality. Therefore the theory is non-chiral and automatically anomaly free. This field content is that of type IIA supergravity, so we name the corresponding string theory as *Type IIA string theory*. We have $\mathcal{N} = (1, 1)$ supersymmetry in ten dimensions generated by 32 real supercharges. The name comes from the supergravity theory having the same spectrum, namely type IIA ($N = 2$, 32 supercharges) supergravity.

If we select the same chirality $\Gamma_L = \Gamma_R = 1$, from (4.41), we obtain *Type IIB string theory*, whose massless spectrum is

$$\text{RR} : \mathbf{8}_s \otimes \mathbf{8}_s = \mathbf{35}_c + \mathbf{28} + \mathbf{1}, \tag{6.184}$$

$$\text{RNS} : \mathbf{8}_s \otimes \mathbf{8}_v = \mathbf{56}_s + \mathbf{8}_s. \tag{6.185}$$

The RR antisymmetric tensors are scalar $C_0$, rank-two $C_{\mu\nu}$ tensors plus a rank four self-dual tensor $C^{SD}_{\mu\nu\rho\sigma}$ self-dual. The theory is now chiral. Not only we have two gravitinos of the same chirality but also the self-dual tensor is chiral. We can check that their anomaly cancels. The resulting supergravity is IIB and has $\mathcal{N} = (2, 0)$ supersymmetry in ten dimensions.

We have also ten dimensional string theories with lower supersymmetry, type I and two heterotic strings, which will be discussed below.

## 6.3   Heterotic String

The later part of this book will be devoted to compactification of heterotic string theory [17] which we introduce here. It describes the gauge group directly by assigning charge along the string. Such charge comes from the momentum-winding from compactifying higher dimensions. The consistent theory leads to the gauge group as either SO(32) or $E_8 \times E_8$ with the dimension 496, which is expected after the work of Green and Schwarz [18]. They showed that in the ten dimensional supergravity theories with these gauge groups the gravitational and gauge anomalies are miraculously cancelled. Soon afterward, the heterotic string theory was found [17].

### 6.3.1   Non-Abelian Gauge Symmetry

We now discuss how non-Abelian gauge symmetry naturally arises in the string theory. We may obtain ten dimensional theory plus gauge theory of rank sixteen by compactifying the sixteen left-moving bosonic degrees on the torus. In Sect. 5.1.4, we reviewed Kaluza–Klein theory, which explains the origin of gauge theory from gravity. On top of this, stringy behavior may enhance the gauge symmetry.

## Toroidal Compactification of String

We first compactify one dimension of 26D bosonic string theory as in (5.2). Let us call this direction $x^{25}$ and we identify

$$x^{25} \sim x^{25} + 2\pi R. \tag{6.186}$$

As in the point particle case (5.11), the wave function $e^{ip^{25}x^{25}}$ (in the center-of-mass frame) should be single-valued, obeying the same periodicity of (6.186). It leads quantization of the momentum $p^{25}$.

$$p^{25} = \frac{m}{R}, \quad m = \text{integer}. \tag{6.187}$$

A characteristic feature of a closed string is that its winding around torus is

$$X^{25}(\tau, \sigma + \pi) = X^{25}(\tau, \sigma) + 2\pi L^{25}. \tag{6.188}$$

The winding modes are also quantized, viz. (6.186),

$$L = nR, \quad n = \text{integer}. \tag{6.189}$$

A negative $n$ corresponds to the reverse direction of winding. Winding numbers are good quantum numbers preserved by string interactions.

The mode expansion obeying these conditions reads

$$X^{25}(\sigma, \tau) = x^{25} + 2\alpha' p^{25}\tau + 2L\sigma$$
$$+ i\sqrt{\frac{\alpha'}{2}} \sum_{n \neq 0} \frac{1}{n} \left[ \alpha_n^{25} e^{-2in(\tau-\sigma)} + \tilde{\alpha}_n^{25} e^{-2in(\tau+\sigma)} \right]. \tag{6.190}$$

It is decomposed to the right and left movers

$$X_R^{25} = x_R^{25} + \sqrt{2\alpha'} p_R^{25}(\tau - \sigma) + i\sqrt{\frac{\alpha'}{2}} \sum_{n \neq 0} \frac{1}{n} \alpha_n^{25} e^{-2in(\tau-\sigma)}, \tag{6.191}$$

$$X_L^{25} = x_L^{25} + \sqrt{2\alpha'} p_L^{25}(\tau + \sigma) + i\sqrt{\frac{\alpha'}{2}} \sum_{n \neq 0} \frac{1}{n} \tilde{\alpha}_n^{25} e^{-2in(\tau+\sigma)}, \tag{6.192}$$

where

$$x^{25} = x_L^{25} + x_R^{25},$$
$$\sqrt{2\alpha'} p_R^{25} = \alpha' p^{25} - L = \alpha' \frac{m}{R} - nR, \tag{6.193}$$
$$\sqrt{2\alpha'} p_L^{25} = \alpha' p^{25} + L = \alpha' \frac{m}{R} + nR.$$

Observe that $p_L^{25}$ and $p_R^{25}$ are dimensionless and contain the contribution from the winding. Now the above equations change the mass shell conditions such as (6.53) and (6.54) to

$$\frac{1}{2}\alpha' M_L^2 = \frac{1}{2}\left(p_L^{25}\right)^2 + \tilde{N} - 1,$$
$$\frac{1}{2}\alpha' M_R^2 = \frac{1}{2}\left(p_R^{25}\right)^2 + N - 1. \tag{6.194}$$

They are supplemented by the level matching condition (6.52),

$$\frac{1}{2}M^2 = M_L^2 = M_R^2, \tag{6.195}$$

which is equivalent to

$$N - \tilde{N} = \frac{1}{2}(p_L^{25})^2 - \frac{1}{2}(p_R^{25})^2 = mn. \tag{6.196}$$

Let us denote the string ground state as

$$|k_L\rangle_L \otimes |k_R\rangle_R, \tag{6.197}$$

where $|k_{L,R}\rangle_{L,R}$ are annihilated by all the positive oscillators and the eigenstates of the momentum operators $\sqrt{2}p_L|k_L\rangle_L = k_L|k_L\rangle_L$, $\sqrt{2}p_R|k_R\rangle_R = k_R|k_R\rangle_R$. At generic value of $R$, only the massless states are those with $\tilde{N} = N = 1, n = m = 0$,

$$\tilde{\alpha}_{-1}^\mu|0\rangle_L \otimes \alpha_{-1}^\nu|0\rangle_R.$$

They correspond to graviton, dilaton, and antisymmetric tensor. In particular, the states

$$\tilde{\alpha}_{-1}^\mu|0\rangle_L \otimes \alpha_{-1}^{25}|0\rangle_R$$
$$\tilde{\alpha}_{-1}^{25}|0\rangle_L \otimes \alpha_{-1}^\mu|0\rangle_R \tag{6.198}$$

correspond to two U(1) gauge bosons. It has the right vector index and they correspond to the KK states of $G_{\mu\nu}$ and $B_{\mu\nu}$ as in (5.61). In the field theory case we expected one gauge field $A_\mu$ from $G_{\mu\nu}$, but in the string case there is an additional one from $B_{\mu\nu}$.

This shows how string theory evades the problem of the Kaluza–Klein (KK) theory. As discussed in Sect. 5.1.4, the KK theory relates the mass and charge of a particle, predicting heavy mass of the compactification scale. In string theory, the zero point energy coming from the normal ordering is negative and exactly cancelled by the energy of the first excitations, giving rise to gauge fields. They are massless in the effective field theory and their masses are given by Higgs mechanism.

**Gauge Symmetry Enhancement**

Exchanging the momentum $m$ in (6.187) and the winding $n$ in (6.189) gives the same spectrum if we also invert the radius

$$m \leftrightarrow n, \quad R \leftrightarrow \frac{\alpha'}{R}. \tag{6.199}$$

This is called $T$-duality and believed to the property of the string theory. It means that physics is the same for a moving string in a large radius and winding string in a small radius.

For a critical radius

$$R = \sqrt{\alpha'}, \tag{6.200}$$

the dimensionless momenta (6.193) are integrally quantized as

$$\sqrt{2} p_R = m - n, \quad \sqrt{2} p_L = m + n, \tag{6.201}$$

(the factor $\sqrt{2}$ is the length of the root vector of the SU(2).) so that we have four *more* massless states

$$\begin{aligned}
&\tilde{\alpha}^\mu_{-1} |0\rangle_L \otimes |2\rangle_R, \quad m = 1, n = -1, \\
&\tilde{\alpha}^\mu_{-1} |0\rangle_L \otimes |-2\rangle_R, \quad m = -1, n = 1, \\
&|2\rangle_L \otimes \alpha^\mu_{-1} |0\rangle_R, \quad m = 1, n = 1, \\
&|-2\rangle_L \otimes \alpha^\mu_{-1} |0\rangle_R, \quad m = -1, n = -1.
\end{aligned} \tag{6.202}$$

All of them carry the vector indices, so that they become gauge bosons. We find that the states in (6.198) are Cartan subalgebra and the states in (6.202) have the quantum number of ladder operators. Thus the gauge group is enhanced to SU(2) × SU(2). Note that these even quantized momenta $p_L$, $p_R$ can satisfy the mass shell condition (6.196). We also have charged matter states $m = \pm 2, n = \tilde{N} = N = 0$, and $m = \tilde{N} = N = 0, n = \pm 2$, carrying no Lorentz index.

## 6.3.2 Compactifying Several Dimensions

Generalizing the above result, let us compactify the extra $d$ dimensional space on a torus $T^d = \mathbb{R}^d / \Gamma$, as discussed in Sect. 3.1.1. Here $\Gamma$ is the lattice as in (3.7)

$$\Gamma = \left\{ \sum_{I=1}^{d} n_I e^I \, \middle| \, n_I \in \mathbb{Z} \right\}. \tag{6.203}$$

The nontrivial geometry is described by the basis vectors $e^I = (e_M^I)$, $M = 1, 2, \ldots, d$. Defining their inverse vectors $e_I$ such that $e_M^I e_J^M = \delta_J^I$, we may understand the torus geometry using the orthogonal coordinates in $\mathbb{R}^d$,

$$x^M \sim x^M + 2\pi R e_I^M.$$

The description becomes simpler in the torus coordinate $x^I = x^M e_M^I$,

$$x^I \sim x^I + 2\pi R, \quad I = 1, 2, \ldots, d. \tag{6.204}$$

So, the winding (6.188) is naturally extended as

$$X^I(\tau, \sigma + \pi) = X^I(\tau, \sigma) + 2\pi L^I. \tag{6.205}$$

The vector $L^I$ on a lattice $\Gamma$ is an integer

$$L^I = n^I R, \quad n^I = \text{integer}. \tag{6.206}$$

Also, the momentum $p^I$ is quantized for the same reason of the single valuedness of $e^{ip \cdot x}$. Being conjugate to $x$, the momentum lattice is spanned in the dual lattice

$$\Gamma^\vee = \left\{ \sum n_I e^{\vee I} \ \middle| \ \sum_{I=1}^d e_M^I e_N^{\vee I} = \delta_{MN}, n_I \in \mathbb{Z}, e^I \in \Gamma \right\} \tag{6.207}$$

with basis vectors $e^{\vee I} = (e_M^{\vee I})$. Then, $p^I$ are quantized in the dual lattice $\Gamma^\vee$

$$p^I = \frac{m^I}{R}, \quad m^I = \text{integer}. \tag{6.208}$$

We have similar mode expansions as (6.191), and (6.192) and mass conditions as (6.194). We know how to quantize the string in the orthonormal space. So we come back to this space with the indices $M, N, \ldots$ from the lattice space with the indices $I, J, \ldots$. The quantization condition (6.26) reads

$$[x^M, p^N] = i\delta^{MN}. \tag{6.209}$$

We write the center-of-mass coordinates and momentum as in (6.193) as

$$x^M \equiv x_L^M + x_R^M,$$

$$\sqrt{\alpha'} p_R^M \equiv \alpha' p^M - L^M = \alpha' p^I e_I^{\vee M} - L^I e_I^M = \alpha' \frac{m^I}{R} e_I^{\vee M} - n^I R e_I^M,$$

$$\sqrt{\alpha'} p_L^M \equiv \alpha' p^M + L^M = \alpha' p^I e_I^{\vee M} + L^I e_I^M = \alpha' \frac{m^I}{R} e_I^{\vee M} + n^I R e_I^M.$$

$$\tag{6.210}$$

Naturally, we require that the left-moving momentum $p_L^M$ and the center of mass $x_L^M$ are independent hence commute with right-moving ones. This implies the following commutation relation:

$$\left[ x_L^M, p_L^N \right] = \left[ x_R^M, p_R^N \right] = \frac{i}{2} \delta^{MN}. \tag{6.211}$$

### 6.3.3 Heterotic String

As the word heterotic ("hybrid vigor") implies, the *heterotic string* has different worldsheet theories for the left and right movers. This possibility arises because left and right movers of the closed string behave independently, as in (6.11). Conventionally we define the heterotic string as

a closed string with $(\mathcal{N}_L, \mathcal{N}_R) = (0, 1)$ worldsheet supersymmetries on the left and right movers, respectively.

Which one is called left is a matter of convention. The left mover is that of 26 dimensional bosonic string and the right mover is that of the 10 dimensional superstring. Also we may think of other heterotic strings with different number of worldsheet supersymmetries [4].

We have the action for the heterotic string as

$$S = \frac{1}{4\pi \alpha'} \int d^2\sigma \left( \partial_\alpha X_\mu \partial^\alpha X^\mu + \partial_\alpha X_I \partial^\alpha X^I + i \Psi_R^\mu \partial_+ \Psi_{\mu R} \right). \tag{6.212}$$

Here, the indices run $\mu = 0, 1, \ldots, 9$ and $I = 1, \ldots, 16$. Alternatively, we may unify $X^\mu$ and $X^I$ as $X^M$ and let the index run as $M = 0, \ldots, 26$. We have only the R-moving Majorana–Weyl fermions $\Psi_{\mu R}$ defined in (6.95).

With the extra sixteen bosonic degrees only in the *left-moving* sector, the absence of right movers in $X^I$ make things nontrivial. For this, we treat the right movers as auxiliary fields and eliminate them by imposing constraints,

$$(\partial_\tau - \partial_\sigma) X^I = 0. \tag{6.213}$$

It is hard to think of momentum and winding. Also it is difficult to quantize.[5]

$$p_R^I = 0. \tag{6.214}$$

---

[5] Using Dirac bracket [19] we may obtain we may quantize the momentum and winding.

So we use the result of the previous section with $d = 16$ and impose the constraint (6.214). Then, the momentum and the winding are forced to be identified; thus the massless momentum becomes

$$p_L^I = 2p^I \equiv P^I, \tag{6.215}$$

with the usual commutation relation (6.209).

For proper quantization, the length-squared of this vector $P^2 \equiv P \cdot P$ should be integer quantized. In terms of the basis vectors, all the inner products in (6.206) should be integers. If the metric is integral, we call also the lattice *integral*. If we further want the level matching condition, the lattice should be even. If all the diagonal elements $G_{II}$ are even the matrix is symmetric $G_{IJ} = G_{JI}$. Finally, we apply the constraint condition of (6.214), so that

$$\alpha' \sum_{I=1}^{16} \frac{m^I}{R} e_I^{\vee M} = \sum_{I=1}^{16} n_I R e_I^M. \tag{6.216}$$

This condition is so restrictive that it is only satisfied by *self-dual* lattice $\Gamma = \Gamma^\vee$ with the critical radii $R = \sqrt{\alpha'}$ for all $I$. We expect gauge symmetry will be enhanced beyond $U(1)^{16}$. Of course, finite radius is meaningful only when the dimensions are *compact*.

We will see shortly that even and self-dual *Euclidian* lattices exist when the number of dimensions is a multiple of 8. We define lattices $\Gamma_k$ as

$$\Gamma_k = \left\{ (n_1, n_2, \ldots, n_k), (n_1 + \tfrac{1}{2}, n_2 + \tfrac{1}{2}, \ldots, n_k + \tfrac{1}{2}) \Big| \sum n_i \in 2\mathbb{Z} \right\}, \tag{6.217}$$

with integers $n_i$. They are weight lattices of the $SO(2k)$. They are decomposed into four sublattices, or conjugacy classes, as displayed in Table 6.1.

In eight dimensions, there is a unique even and self-dual lattice $\Gamma_8$. This is called $E_8$ lattice because it is coincident with $\Lambda_{E_8}$ in (3.63). In sixteen dimensions, we have two such lattices

$$\Gamma_8 \times \Gamma_8, \quad \Gamma_{16}. \tag{6.218}$$

The former is a direct product of the two and can be called $E_8 \times E_8$ lattice. The latter is called $\text{Spin}(32)/\mathbb{Z}_2$ lattice. The $\text{Spin}(32)$ is the double covering group of

**Table 6.1** The four conjugacy classes of $SO(2n)$ group

| Symbol | Vector | Constraint |
|---|---|---|
| 0 | $(n_1, \ldots, n_n)$ | $\sum n_i$ even |
| v | $(n_1, \ldots, n_n)$ | $\sum n_i$ odd |
| s | $(n_1 + \tfrac{1}{2}, \ldots, n_n + \tfrac{1}{2})$ | $\sum n_i$ even |
| c | $(n_1 + \tfrac{1}{2}, \ldots, n_n + \tfrac{1}{2})$ | $\sum n_i$ odd |

SO(32) and its lattice contains all four conjugacy classes in Table 6.1. We mod out the state by $\sum n_i$ even condition, so that the Spin(32)/$\mathbb{Z}_2$ lattice contains even classes 0 and s. In this book, without confusion we just stick to the name SO(32). In 24 dimensions, there are 24 lattices classified by Niemeier [20].

## Spectrum and Current Algebra

As before, let us choose the light-cone gauge. For the left mover, we use the 26D bosonic string compactified on 16D torus,

$$\frac{1}{2}\alpha' M_L^2 = \frac{1}{2}\sum_{M=1}^{16}(P^M)^2 + \tilde{N} - 1, \tag{6.219}$$

where

$$\tilde{N} = \sum_{n=1}^{\infty}\left(\sum_{i=1}^{8}\tilde{\alpha}_{-n}^{i}\tilde{\alpha}_{n}^{i} + \sum_{M=1}^{16}\tilde{\alpha}_{-n}^{M}\tilde{\alpha}_{n}^{M}\right), \tag{6.220}$$

and sixteen dimensional vector $(P^I)$ belongs to one of the lattices (6.218). We labelled 16 bosons $I = \mu - 9 = 1, \ldots, 16$. We construct Fock states on the ground state.

The right mover is that of ten dimensional superstring, discussed in Sect. 6.2. The mass shell condition is (6.139)

$$\frac{1}{2}\alpha' M_R^2 = N^{\alpha} + N^{\psi} + c, \tag{6.221}$$

where

$$\text{NS}: \nu = \frac{1}{2}, \quad c = -\frac{1}{2}, \tag{6.222}$$

$$\text{R}: \nu = 0, \quad c = 0. \tag{6.223}$$

A complete state is the tensor product of the left and right-moving states. The mass of string is given as

$$M^2 = M_L^2 + M_R^2, \tag{6.224}$$

with

$$M_L^2 = M_R^2. \tag{6.225}$$

Let us find out the lowest lying states. From (6.219), it seems to be the ground state with $P^2 = 0, \tilde{N} = 0$

$$|0\rangle_L \otimes |0\rangle_R$$

to have $\frac{1}{2}\alpha' M_L^2 = -1$. However, this state is projected out by (6.241). Also there is no right-moving state from (6.221) satisfying the level matching condition (6.225). Thus, this state is projected out, and this theory is tachyon-free.

The right movers allow the following lowest mass states:

$$\psi_{-1/2}^{\mu}|0\rangle_R \ (\text{NS}), \quad \psi_0|0\rangle_R \ (\text{R}). \tag{6.226}$$

Noting the spacetime index, the NS states provide bosonic states and the R states provide fermionic states, forming together superpartners. Combined with massless left movers, they give rise to low-energy fields. We will present bosonic states:

- $\tilde{N} = 1$ and $P^2 = 0$ of the following kinds:

$$\tilde{\alpha}_{-1}^{\mu}|0\rangle_L \otimes \psi_{-1/2}^{\nu}|0\rangle_R. \tag{6.227}$$

They make up graviton $G^{\mu\nu}$, dilaton $\phi$, and the antisymmetric tensor $B^{\mu\nu}$.
- For $\tilde{N} = 1$ and $P^2 = 0$, there are states carrying internal index $I$

$$\tilde{\alpha}_{-1}^{I}|0\rangle_L \otimes \psi_{-1/2}^{\mu}|0\rangle_R, \tag{6.228}$$

which provide sixteen U(1) generators $A^{\mu I}$ of the Cartan subalgebra $H^I$. They are vectors with the single spacetime index $\mu$ in 10D.
- $\tilde{N} = 0$ and $P^2 = 2$ states of the form

$$|P\rangle_L \otimes \psi_{-1/2}^{\mu}|0\rangle_R, \tag{6.229}$$

where $P$ belonging to the lattice (6.218). For $\Gamma_8$, the possible forms of $P$ are

$$\begin{aligned}
&\left(\underline{\pm 1 \ \pm 1\, 0\, 0\, 0\, 0\, 0\, 0}\right)(0\, 0\, 0\, 0\, 0\, 0\, 0\, 0), \\
&\left(\left[\tfrac{1}{2}\, \tfrac{1}{2}\, \tfrac{1}{2}\, \tfrac{1}{2}\, \tfrac{1}{2}\, \tfrac{1}{2}\, \tfrac{1}{2}\, \tfrac{1}{2}\right]\right)(0\, 0\, 0\, 0\, 0\, 0\, 0\, 0), \\
&(0\, 0\, 0\, 0\, 0\, 0\, 0\, 0)\left(\underline{\pm 1 \ \pm 1\, 0\, \cdots\, 0}\right), \\
&(0\, 0\, 0\, 0\, 0\, 0\, 0\, 0)\left(\left[\tfrac{1}{2}\, \tfrac{1}{2}\, \tfrac{1}{2}\, \tfrac{1}{2}\, \tfrac{1}{2}\, \tfrac{1}{2}\, \tfrac{1}{2}\, \tfrac{1}{2}\right]\right).
\end{aligned} \tag{6.230}$$

The underline means permutations of the corresponding entries and the square bracket means even flips of signs. They are root vectors of $E_8 \times E_8$. We know that $E_8$ roots **248** can be made by giving a suitable commutation relation between the adjoint **120** and the spinorial **128** of SO(16)

$$\mathbf{248} = \mathbf{120} + \mathbf{128}. \tag{6.231}$$

Indeed some sum of the two roots in (6.230) are contained the same set thus the algebra is closed and make a single representation. For $\Gamma_{16}$, $P$ vectors are

$$(\pm 1 \ \pm 1 \ 0 \ \cdots \ 0) \tag{6.232}$$

with 16 entries. They are the adjoint **496** of $SO(32)$.

The effective theory made out of these fields is the minimal ten dimensional supergravity coupled to the corresponding Yang–Mills gauge group [21, 22]. We will study the resulting effective action in more detail in Sect. 11.1.

**Current Algebra**

The massless spectrum is not sufficient to show the group property, although they are indeed the roots of the algebra. This is because they do not show the relations among them. It turns out that the vertex operators have the exact properties we want in this regard. We claim [23–25] that the vertex operators of the massless roots and weights considered above are

$$H^I(z) = \dot{X}_L^I(z), \tag{6.233}$$

$$E^P(z) = c_P : \exp[2i P \cdot X_L(z)] :, \tag{6.234}$$

where $P$ is now interpreted as the root vectors and $c_P$ is $\pm 1$ determined by the commutation relations among them. As a simple check, assume "zero modes" of the above operators

$$H_0^I(z) = P^I, \quad E_0^P =: e^{2i P \cdot X_L(\tau + \sigma)} : . \tag{6.235}$$

Then by the commutation relations coming from (6.211), we have

$$[H_0^I, H_0^J] = 0,$$
$$[H_0^I, E_0^P] = P^I E_0^P. \tag{6.236}$$

This is nothing but the relations between generators of a simple Lie algebra, between the Cartan subalgebra and ladder operators.

## 6.3.4  Bosonization and Fermionization

In two dimensional field theory there is equivalence between bosons and fermions, with which we may describe the above string theories in alternative forms. This relation is clearly seen by the partition function.

## Partition Function

In the heterotic string theory, the modular invariance condition determines the gauge group structure. Here, we require the modular invariance for the gauge degrees of freedom. The oscillator part is decoupled and given by eta function. Thus the partition function contains the contribution from the dimensionless momenta (6.193)

$$
\mathcal{Z}(\tau, \bar{\tau}) = |\eta(\tau)|^{-2d} \sum_{p\in\Lambda^*} \sum_{L\in\Lambda} q^{\frac{1}{2}p_L^2} \bar{q}^{\frac{1}{2}p_R^2}
$$

$$
= |\eta(\tau)|^{-2d} \sum_{p\in\Lambda^*} \sum_{L\in\Lambda} \exp\left[-\pi\tau_2\left(\frac{\alpha' p^2}{R^2} + \frac{L^2 R^2}{\alpha'}\right) + 2\pi i\tau_1 p\cdot L\right],
$$

(6.237)

where the square of the vector is calculated by the metric in (6.206). It is the Cartan matrix, presented in Table 12.3.

Under $\mathcal{T}$, the function (6.237) acquires the phase $2\pi i(\frac{1}{2}p_L^2 - \frac{1}{2}p_R^2)$. Thus it is invariant if

$$
p_L \cdot p_L - p_R \cdot p_R = \text{even}.
$$

(6.238)

We also call the corresponding lattice even lattice. To see the invariance under $\mathcal{S}$, we use the Poisson resummation formula

$$
\sum_{L\in\Gamma} e^{-\pi a(L+p)^2} e^{2\pi i Y\cdot(L+U)} = (\text{vol}\Gamma)^{-1} a^{-D/2} \sum_{p\in\Gamma} e^{-\pi(p+Y)^2/a} e^{-2\pi i p\cdot U},
$$

(6.239)

where $U$ and $Y$ are some vectors and vol$\Gamma$ is the volume of the unit cell in $\Lambda$. Since the momentum and winding are exchanged and the radii are inverted, we need the critical radii. The resulting lattice is the *self-dual* lattice

$$
\Gamma = \Gamma^\vee.
$$

(6.240)

Therefore, for the modular invariance, we need *the even and self-dual lattice*.

Let us construct the $\Gamma_8$, in (6.217) partition function. Observing the lattice structure, we can sum over the lattice by introducing a projection,

$$
\mathcal{Z}^{E_8}(\tau) = \frac{1}{\eta(\tau)^8} \sum_{P\in\Gamma_8} q^{P^2/2}
$$

(6.241)

$$
= \frac{1}{\eta(\tau)^8}\left(\sum_{\{n_j\in\mathbb{Z}\}} \frac{1+(-1)^{\sum n_j}}{2} q^{\frac{1}{2}\sum n_j^2} + \sum_{\{r_j\in\mathbb{Z}+\frac{1}{2}\}} \frac{1+(-1)^{\sum r_j}}{2} q^{\frac{1}{2}\sum r_j^2}\right),
$$

where $j$ runs from 1 to 8 so that the sum in the second line runs over $n_j \in \mathbb{Z}, r_j \in \mathbb{Z} + \frac{1}{2}$ for all $j$. In the fermionic construction, this projection corresponds to the GSO projection. Observing the regular pattern, it can be reexpressed in a fancy form,

$$\mathscr{Z}^{E_8}(\tau) = \frac{1}{2\eta(\tau)^8} \left( \vartheta \begin{bmatrix} 0 \\ 0 \end{bmatrix}^8 (\tau) + \vartheta \begin{bmatrix} 0 \\ 1/2 \end{bmatrix}^8 (\tau) + \vartheta \begin{bmatrix} 1/2 \\ 0 \end{bmatrix}^8 (\tau) + i\vartheta \begin{bmatrix} 1/2 \\ 1/2 \end{bmatrix}^8 (\tau) \right),$$

$$(6.242)$$

where we have included $i\vartheta[\begin{smallmatrix} 1/2 \\ 1/2 \end{smallmatrix}] = 0$ by taking into account of the last term in (6.241). Here, the Jacobi theta function (6.166) admits alternative definition

$$\vartheta \begin{bmatrix} \alpha \\ \beta \end{bmatrix} (\tau) = \sum_{n \in \mathbb{Z}} q^{(n+\alpha)^2/2} e^{2\pi i (n+\alpha)\beta}.$$

We can check that $\mathscr{Z}^{E_8}$ is invariant under modular transformations. The $\mathscr{T}$ transformation gives rise to a nontrivial phase $-i\pi/4$ when $\alpha = 1/2$, so only the theta function with the power $8k$ with integer $k$ leaves the partition function invariant. This is the reason why we have even and self-dual Euclidian lattices in $8k$ dimensions. Under the $\mathscr{S}$, the term shuffles each other as in (6.174).

The complete $E_8 \times E_8$ partition function is the square of the single one

$$\mathscr{Z}^{\Gamma_8 \times \Gamma_8} = (\mathscr{Z}^{E_8})^2.$$

It can be generalized to the SO(32) case also. Observe that the eight dimensional lattice is formed by (6.241) whose indices $n_j$ run over from 1 to 8. Running from 1 to 16, we have a sixteen dimensional lattice. Remarkably, it is known that there is only one modular invariant function of sixteen dimensions [9, 26]; therefore the $E_8 \times E_8$ and SO(32) partition functions are the same

$$\mathscr{Z}^{\Gamma_8 \times \Gamma_8} = \mathscr{Z}^{\Gamma_{16}}.$$

$$(6.243)$$

This partitioning corresponds to giving boundary conditions simultaneously to some parts, as we have constructed the original heterotic string theory.

**Bosonic Description of Superstring**

Using the method for obtaining the partition function for $\Gamma_8$ in (6.241), we can rewrite the partition functions of the right-moving superstring (6.164) and (6.165) as

$$\mathscr{Z}_R(\tau) = \frac{1}{\eta^{12}(\tau)} \sum_{s \in \Lambda_0} q^{s^2/2}, \qquad (6.244)$$

$$\mathscr{Z}_{NS}(\tau) = \frac{1}{\eta^{12}(\tau)} \sum_{s \in \Lambda_s} q^{s^2/2}, \qquad (6.245)$$

where $\Lambda_0$ and $\Lambda_s$ are the lattices consisting of the conjugacy classes in Table 6.1, respectively. The $\eta^{-12}$ part is the partition function of 12 worldsheet bosons as in (6.85). Before we had 8 bosons and 8 fermions in the worldsheet, which means that 8 fermion is replaced by 4 bosons. Note that they carry 4-component momentum-winding $s$, which makes sense when the corresponding dimensions are compact. In two compact dimensions, two Majorana–Weyl fermions are equivalent to one chiral boson.

The zero point energy can be read off from $\eta^{-12}$, giving $c = 12 \cdot \left(-\frac{1}{24}\right) = -\frac{1}{2}$, therefore the right mover mass is

$$\frac{1}{2}\alpha' M_R^2 = \frac{s^2}{2} - \frac{1}{2} = 0. \tag{6.246}$$

In later calculations, we will employ these bosonized right movers, because it transparently shows massless spectrum and its chirality. All the massless vectors $s$ satisfy the mass condition $s^2 = 1$. They are the three fundamental weights of SO(8)

$$\mathbf{8}_{\mathrm{v}} : \pm(\underline{1\ 0\ 0\ 0})$$

$$\mathbf{8}_{\mathrm{s}} : ([+ + ++]) \tag{6.247}$$

$$\mathbf{8}_{\mathrm{c}} : ([- + ++])$$

where $+, -$ means $\frac{1}{2}, -\frac{1}{2}$, respectively, the underline means permutations and the square bracket means possible even flips of signs. All of these can be transformed into each other by the *triality* relation. In the spacetime, an $\mathbf{8}_{\mathrm{v}}$ is the spacetime boson and one among $\mathbf{8}_{\mathrm{s}}$ and $\mathbf{8}_{\mathrm{c}}$ is chosen by the GSO projection to be its partner as the spacetime fermion.

Note that (6.244) and (6.245) take the same form in the bosonic description. In particular, the GSO projections for NS (6.146) and R (6.154) sectors are unified in (6.241). These spinorial representations are unified to a momentum vector; $\mathbf{8}_{\mathrm{v}}$ is NS and $\mathbf{8}_{\mathrm{s,c}}$ is R states and we can show that there exists one-to-one correspondence between $\mathbf{8}_{\mathrm{v}}$ and $\mathbf{8}_{\mathrm{s,c}}$, due to supersymmetry.

### Fermionic Construction of the Heterotic String

Some readers may feel unsatisfactory in that we have introduced different spacetime dimensions for the left and right movers. Alternatively, we may introduce 32 fermions instead of 16 bosons in *ten dimensions* as we have seen just before. The mode expansion is the same as (6.101),

$$\lambda^I = \sum_{r \in \mathbb{Z} + v} \lambda_r^I e^{-2ir(\tau + \sigma)} \tag{6.248}$$

with $v = 0, 1/2$ for periodic and antiperiodic boundary conditions, respectively. The quantization and the mass conditions are the same.

In fact, it is an equivalent choice, since in two (worldsheet) dimensions, two holomorphic fermions play the role of one holomorphic boson. Their conformal dimension, correlation functions, and spectrum are the same. These 32 real fermions $\lambda^I$ have an O(32) internal symmetry. As we have done to fermionic strings, we introduce the GSO projection (6.146 and 6.154) on these fermions also via the operator

$$(-1)^F, \quad F = \sum_{r=1/2}^{\infty} \lambda^I_{-r} \lambda^I_r. \tag{6.249}$$

States with the odd number of fermionic oscillators are projected out. Actually this projection has a counterpart in the bosonic string theory. It is equivalent to our choosing of ($\sum P_I$ = even only) in (6.217).

Since there are two kinds of boundary conditions on each fermions, we have essentially two choices for possible theories. One is that we can assign the same boundary conditions on all of them. This leads to the SO(32) gauge group. Here we briefly present the other less straightforward case yielding the $E_8 \times E_8$ group. Divide the 32 fermions into two partitions, say $n$ and $(32 - n)$ fermions for the R and NS fermions. We know that for each degree of freedom, the zero point energy is $\frac{1}{24}$ and $\frac{1}{48}$ for the R and NS fermions, respectively. With 24 bosonic degrees of freedom, we have the following total zero point energy $-a$ with

$$a_{\text{NSNS}} = \frac{8}{24} + \frac{n}{48} + \frac{32-n}{48} = 1, \tag{6.250}$$

$$a_{\text{NSR}} = \frac{8}{24} + \frac{n}{48} - \frac{32-n}{24} = \frac{n}{16} - 1, \tag{6.251}$$

$$a_{\text{RNS}} = \frac{8}{24} - \frac{n}{24} + \frac{32-n}{48} = 1 - \frac{n}{16}, \tag{6.252}$$

$$a_{\text{RR}} = \frac{8}{24} - \frac{n}{24} - \frac{32-n}{24} = -1, \tag{6.253}$$

where $\frac{8}{24}$ is from the eight bosonic coordinates of 10D. (Note that considering the right mover, we specify three boundary conditions on a state.) For the NSR or RNS sectors to have massless states, we require $n = 16$, otherwise it reduces to a theory with the same boundary conditions for all the movers.

First, noting that $-a_{\text{NSNS}} = -1$, we can find massless states in the NSNS states with

$$\lambda^I_{-1/2} \lambda^J_{-1/2} |0\rangle, \tag{6.254}$$

which split into the following three cases

$$I = 1, \ldots, 16, \quad J = 1, \ldots, 16 \quad (\mathbf{120}, \mathbf{1})$$

$$I = 1, \ldots, 16, \quad J = 17, \ldots, 32 \quad (\mathbf{16}, \mathbf{16})$$

$$I = 17, \ldots, 32, \quad J = 17, \ldots, 32 \quad (\mathbf{1}, \mathbf{120}).$$

We should apply the GSO projection. Since we have partitioned the fermions into two sectors, we will see that a reasonable choice is to use an independent projection for each sector

$$(-1)^{F_1}, \quad (-1)^{F_2}. \tag{6.255}$$

In this NSNS case, both partitions are in the NS sector; thus the GSO projection (6.146) is applied. The $(\mathbf{16}, \mathbf{16})$ has an odd fermion number since the number operator $F_A$, $A = 1, 2$ acts on the first 16 indices and the last 16 indices separately, and hence is projected out. Note that the Cartan subalgebra is contained in (6.254).

In addition, we have $a_{\mathrm{RNS}} = 0$ in the RNS sector. These become spinorial massless states, in view of (6.152) with eight $s_i$s,

$$|s_1 s_2 \cdots s_8\rangle, \quad \text{spinorial made of } \lambda_0^I, \quad I = 1, \ldots, 16. \tag{6.256}$$

They lead to the states

$$(\mathbf{256}, \mathbf{1}) = (\mathbf{128_s}, \mathbf{1}) + (\mathbf{128_c}, \mathbf{1}).$$

Again the GSO projection should be applied.

$$(-1)^F, \quad F = \sum s_i \mod 1.$$

The fermions are partitioned into sixteen R and sixteen NS, therefore (6.154) and (6.146) are separately used. They project out one, say $(\mathbf{128_c}, \mathbf{1})$, and there remains $(\mathbf{128_s}, \mathbf{1})$. By a similar reasoning applied to the NSR sector, we obtain $(\mathbf{1}, \mathbf{128_s})$. They all belong to one $E_8 \times E_8$ multiplet

$$(\mathbf{248}, \mathbf{1}) + (\mathbf{1}, \mathbf{248}). \tag{6.257}$$

The currents (6.233) and (6.234) are generated by

$$T^a = T_{IJ}^a \lambda^I \lambda^J \tag{6.258}$$

where $a = 1, \ldots, d$ is the adjoint index.

## 6.4   Open Strings

Finally, we discuss open strings. With the discovery of Dirichlet brane, the most interesting physics comes from the open string sector [27].

### 6.4.1   Charged Open Strings

We briefly discuss another possibility of describing group degrees of freedom by open strings. Recall that the heterotic string has uniform charge distribution along the closed string, whose local relation is given by (6.236).

Alternatively, we can assign charges on both ends of an open string. This was the original idea of introducing string in 1960s attempted for describing strong interactions. We assign a new degree of freedom at each end labeled by indices $i$ and $j$ as shown in Fig. 6.5a, called *the Chan–Paton factors*. As a representation, we introduce an $n \times n$ matrix $\lambda_{ij}^a$. A string state with momentum $k$ can carry an adjoint index $a$,

$$|k, a\rangle = \sum_{i,j=1}^{n} |k, ij\rangle \lambda_{ij}^a. \tag{6.259}$$

In the open string Feynman diagram, in and out states have definite Chan–Paton factors. For example, the four-point tree level diagram has ends with the same indices, as shown in Fig. 6.5b.

Summing up all the possible states, the amplitude is proportional to the trace of the products $\lambda^a$,

$$\lambda_{ij}^1 \lambda_{jk}^2 \lambda_{kl}^3 \lambda_{li}^4 = \text{Tr}\, \lambda^1 \lambda^2 \lambda^3 \lambda^4.$$

Generally, the amplitude is invariant under a global U($n$) transformation

$$\lambda^a \to U\lambda^a U^{-1}$$

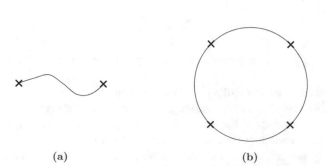

(a)                                      (b)

**Fig. 6.5** (a) Assigning the charges on both ends. (b) The four point tree level diagram

under which the endpoint($\times$) transforms as a fundamental–anti-fundamental representation, $\mathbf{n}$ and $\bar{\mathbf{n}}$. This U($n$) is the worldsheet global symmetry. Now in the target space, this is coordinate($X^i$) dependent and hence is escalated to a *local symmetry*. This is in fact the general argument in string theory: in string theory every symmetry is a local symmetry, i.e., there is no global symmetry. Similarly, when we consider unoriented strings we can also describe SO($n$) and Sp($2n$) groups.

## 6.4.2  Dirichlet Brane

The $T$-duality in (6.199) exists and the open string theory possesses such duality as well [28].

$$(p_L, p_R) \leftrightarrow (p_L, p_R). \tag{6.260}$$

This means

$$(X_L^M, X_R^M) \leftrightarrow (X_L^M, -X_R^M), \quad (\psi_L^M, \psi_R^M) \leftrightarrow (\psi^M, -\psi_R^M). \tag{6.261}$$

In the dual space, the Neumann boundary condition (6.32) becomes the Dirichlet boundary condition: the open string endpoints should be placed in some specific hypersurfaces. This object at the hypersurfaces is understood as *Dirichlet brane*, or D-brane in short. The location of D-branes is the eigenvalue of the Chan–Paton factor.

On it the open string dynamics can be understood alternatively as the dynamics of D-branes.

In the beginning, Type II theory was defined as a theory of closed strings. However, inserting D-branes is also consistent: away from the brane the massless spectrum contains closed and oriented strings only. However, the D-brane is an object for an open string to be attached: near the brane open strings emerge. In the presence of D-branes, only "unoriented" half of the $\mathcal{N} = 8$ supersymmetries $\tilde{Q} + Q$ of Type II is conserved. In the $T$-dual picture without the brane, where the D-brane is present,

$$\tilde{Q}_\alpha + P Q_\alpha, \tag{6.262}$$

where $P = \prod P_m$ is a parity action and the product is over the real dualized dimension.

The D-brane is a Bogomol'nyi–Prasad–Sommerfield (BPS) state. Supersymmetry guarantees some stable configurations, even nonperturbertively. With one brane, a string with both ends at the brane, the resulting massless state is charged under U(1). BPS state guarantees the stability when more than one branes are coincident. Alternatively we can check that there is no force between parallel D-branes; by the worldsheet supersymmetry NS–NS force(containing gravitational attraction) is cancelled by the R–R force. Now there is a symmetry enhancement: a string can

**Fig. 6.6** String stretched between two D-branes, describing $U(1) \times U(1)$. An open string ending on the same brane gives massless gauge boson. An open string stretched between different branes has charges $\pm 1, \mp 1$ giving rise to massive $W^\pm$ bosons. If they become coincident, the gauge symmetry is enhanced to $U(2)$, making $W$ bosons massless. We can generalize it to $U(n)$ with $n$ slices

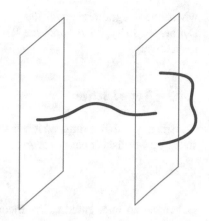

end at *different branes*, whose 4D positions are the same, i.e., the same $x^\mu$ as explained in the caption of Fig. 6.6. It is like a charged boson $W^\pm$. With $n$ coincident branes, the resulting symmetry is extended to $U(n)$. An open string ends between two different D-branes, whose ends have different orientations, so we can assign opposite charges. This corresponds to the charge of $W^\pm$ bosons.

Its dynamics is described by Dirac–Born–Infeld action,

$$S = -T_p \int d^{p+1}\xi \, \mathrm{Tr} \, e^{-\Phi} \sqrt{- \det(G_{ab} + B_{ab} + 2\pi\alpha' F_{ab})}. \tag{6.263}$$

Expanding to the quadratic order in $\alpha' F_{MN}$, it reduces to $(p+1)$-dimensional Yang–Mills action,

$$S_p = -\frac{T_p (2\pi\alpha')^2}{4g_s} \int d^{p+1}x \, \mathrm{Tr} \, F_{MN} F^{MN} \tag{6.264}$$

with a potential of transverse scalar degrees. Here $T_p$ and $g_s$ are the tension and Type II string coupling fixed by the vacuum expectation value (VEV) of dilaton. Therefore, the YM coupling is

$$g_{p+1}^2 = g_s T_p^{-1} (2\pi\alpha')^{-2}. \tag{6.265}$$

Every BPS object has a conserved charge, seen by the supersymmetry commutation relations. We have seen in Sect. 6.2.4 Type II string theories have various RR tensors. Like a vector field sourced by a charge carried by point particle, they are sourced by membranes. Type IIA and Type IIB theories contain $(p+1)$-form fields $C_{\mu_1\cdots\mu_{p+1}}$ with $p+1$ being odd and even, respectively. They couple to the D-brane as

$$\int C_{\mu_0\cdots\mu_p} dx^{\mu_0} dx^{\mu_1} \ldots dx^{\mu_p},$$

where the integration is over the $p + 1$ dimensional world volume spanned by the $Dp$ brane. So Type IIA and Type IIB theories have $Dp$-brane with $p$ even and odd, respectively.

### 6.4.3 Type I String

Consider type IIB string, which is parity symmetric. We may mod out the string states by worldsheet parity reversal

$$\Omega : \sigma \to -\sigma \tag{6.266}$$

and make an unoriented string theory. This is orbifold action on the worldsheet Hence projects out non-invariant states. The open string arises in the twisted sector under the parity reversal (6.266). Exchanging the left and right movers corresponds to exchanging two indices of the rank 2 fields; thus the antisymmetric tensor $B_{\mu\nu}$ is projected out. The fundamental string should be source to this tensor but in this time it is missing. It decays, with the lifetime inversely proportional to the string coupling constant. In the weakly coupled limit, this open string is long lived. The graviton $G_{\mu\nu}$ and the dilaton $\phi$ are invariant. However, in the RR sector, anticommuting worldsheet fields multiply an additional minus sign when exchanged, thus the RR tensor of rank two only $C_{\mu\nu}$ survives the projection. This should be sourced by another open string. This open string is the D1-brane. Thus the field content is again that of $\mathcal{N} = 1$ supergravity in ten dimensions.

Calculation of one-loop diagram shows potential divergence from tadpole diagram and it is cancelled by 16 spacetime filling D-branes or D9-branes. This $\Omega$ also projects out U($n$) gauge group and makes it real, and reflects the D-branes with respect to the end of the interval. Thus we have SO(32) gauge group in the bulk. This defines *Type I string theory*. We have seen embedding U($n$) in SO($2n$) in Sect. 5.2.2. Unoriented string can end between a D-brane and its mirror brane, filling in the multiplets.

In the open string sector, massless degrees of freedom are gauge bosons, described by the Chan–Paton factors attached at both ends. Consistency of one-loop diagrams imposes a certain condition called the Ramond–Ramond tadpole cancellation [29]. It fixes the gauge group completely as SO(32).

### 6.4.4 Duality of Strings

Two string theories, type IIA and type IIB, are $T$-dual to each other. They are different guises of a single, unified theory.

It is known that type IIA supergravity is obtained from dimensional reduction of the eleven dimensional supergravity on a circle. The size of the circle is proportional to the IIA string dilaton $e^{\phi}$. The latter contains the rank three antisymmetric tensor field $C_{MNP}$, whose source is $(2 + 1)$-dimensional brane. We also have a higher

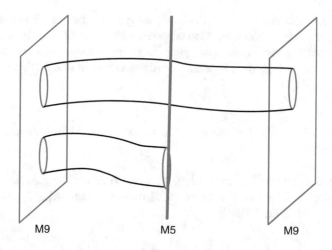

**Fig. 6.7** M2-branes stretched between two M9-branes, in the heterotic M-theory picture. Shrinking it along radial direction gives an open string shown in Fig. 6.6. Taking the zero interval length shrinks it along the longitudinal direction, making the heterotic string. An M2-brane ending between M9 and M5-branes is called E-string. An M2-brane ending between two M5-branes is called M-string

dimensional theory whose compactification on the same circle gives the IIA string theory. It is called M-theory and that 2-brane is promoted to M2-brane. By Hodge duality, we have magnetic source for this 3-form field, which are M5-branes. Various D-branes can be understood as being obtained from dimensional reduction of M2 and M5-branes [4]. The string coupling is proportional to $e^{\phi}$, where $\phi$ is the type IIA dilaton, so we may define M-theory as strongly coupling limit of IIA string. An M2-brane can end on two M5-branes, providing the source for anti-self-dual tensor field $B_{\mu\nu}^{-}$ in the six dimensional supersymmetry.

Compactifying M-theory on an interval, or $S^1/\mathbb{Z}_2$ orbifold discussed in 3.2, we obtain strongly coupled heterotic string [30, 31]. The size of the extra dimension is again proportional to the heterotic string dilaton $e^{\phi}$ (Fig. 6.7).

We have seen at Sect. 5.3 the anomaly is localized at the ends of the interval. The ten dimensional gravitational anomaly is equally localized at each end. Thus each has a 9-brane, which we call M9-brane, harboring $E_8$ worldvolume gauge theory. The $E_8 \times E_8$ heterotic string is charged under the both, hence understood as M2-brane stretched between these two M9-branes (Fig. 6.7, the upper brane).

M5-branes become 5-branes in the heterotic string, providing magnetic source for the antisymmetric tensor field $B_{\mu\nu} = C_{\mu\nu11'}$, where we labelled the M-theory direction as another "11th" direction different from that of IIA theory. We can calculate a variant of partition function, elliptic genus, including the effect of the M5-branes. Without background gauge field, the modular invariance of the elliptic genus gives that we need 24 M5-branes [32]. Likewise we have E-string that is M2-brane stretched between the M9-brane and M5-brane (Fig. 6.7, the lower brane). This is the source for the tensor multiplet in the six dimensional $(1, 0)$ supersymmetry, whose scalar component parametrizes the distance between the M9 and M5-branes. We may understand dynamics of string theory as a low-energy limit of M-theory. Its moduli space is schematically drawn as in Fig. 2.10.

## Exercises

▶ **Exercise 6.1** Verify that the partition function (6.85) is invariant under the modular group.

▶ **Exercise 6.2** To avoid overcounting, we need to define the fundamental region (6.74) precisely. Show that the fundamental region of the $SL(2, \mathbb{Z})$ modulus $\tau$ is one satisfying all of the following

$$|\tau| > 1, \quad -\frac{1}{2} < \tau_1 \le \frac{1}{2}, \quad -\frac{1}{2} < \tau_1 < 0 \Rightarrow |\tau| > 1, \quad \tau_2 > 0.$$

▶ **Exercise 6.3** By introducing orthonormal coordinates as in (6.206), express the momentum states (6.202) and the mass condition (6.194) and the level matching condition (6.196).

## References

1. M.B. Green, J.H. Schwarz, E. Witten, *Superstring Theory. Vol. 1: Introduction*. Cambridge Monographs on Mathematical Physics, vol. 7 (Cambridge University Press, Cambridge, 1988)
2. M.B. Green, J.H. Schwarz, E. Witten, *Superstring Theory. Vol. 2: Loop Amplitudes, Anomalies and Phenomenology*, vol. 7 (Cambridge University Press, Cambridge, 1988)
3. J. Polchinski, *String Theory. Vol. 1: An Introduction to the Bosonic String*. Cambridge Monographs on Mathematical Physics (Cambridge University Press, Cambridge, 12 2007)
4. J. Polchinski, *String Theory. Vol. 2: Superstring Theory and Beyond*. Cambridge Monographs on Mathematical Physics, vol. 12 (Cambridge University Press, Cambridge, 2007)
5. S.N. Gupta, Theory of longitudinal photons in quantum electrodynamics. Proc. Phys. Soc. **A63**, 681–691 (1950)
6. K. Bleuler, A New method of treatment of the longitudinal and scalar photons. Helv. Phys. Acta **23**, 567–586 (1950)
7. A.A. Belavin, A.M. Polyakov, A.B. Zamolodchikov, Infinite conformal symmetry in two-dimensional quantum field theory. Nucl. Phys. **B241**, 333–380 (1984)
8. M.S. Virasoro, Subsidiary conditions and ghosts in dual resonance models. Phys. Rev. **D1**, 2933–2936 (1970)
9. E.T. Whittaker, G.N. Watson, *A Course of Modern Analysis* (Cambridge University Press, Cambridge, 1927, 1996)
10. R.C. Brower, Spectrum generating algebra and no ghost theorem for the dual model. Phys. Rev. **D6**, 1655–1662 (1972)
11. P. Goddard, C.B. Thorn, Compatibility of the dual Pomeron with unitarity and the absence of ghosts in the dual resonance model. Phys. Lett. **40B**, 235–238 (1972)
12. J. Polchinski, Evaluation of the one loop string path integral. Commun. Math. Phys. **104**, 37 (1986)
13. P. Ramond, Dual theory for free fermions. Phys. Rev. **D3**, 2415–2418 (1971)

14. A. Neveu, J.H. Schwarz, Factorizable dual model of pions. Nucl. Phys. **B31**, 86–112 (1971)
15. F. Gliozzi, J. Scherk, D.I. Olive, Supersymmetry, supergravity theories and the dual spinor model. Nucl. Phys. **B122**, 253–290 (1977)
16. N. Seiberg, E. Witten, Spin structures in string theory. Nucl. Phys. **B276**, 272 (1986)
17. D.J. Gross, J.A. Harvey, E.J. Martinec, R. Rohm, The heterotic string. Phys. Rev. Lett. **54**, 502–505 (1985)
18. M.B. Green, J.H. Schwarz, Anomaly cancellation in supersymmetric $D = 10$ gauge theory and superstring theory. Phys. Lett. **149B**, 117–122 (1984)
19. K.S. Narain, M.H. Sarmadi, E. Witten, A note on toroidal compactification of heterotic string theory. Nucl. Phys. **B279**, 369–379 (1987)
20. H.V. Niemeier, J. Number Theory **5**, 142 (1973). Definite Quadratische Formen der Dimension 24 und Diskcriminante 1
21. W. Nahm, Supersymmetries and their representations. Nucl. Phys. **B135**, 149 (1978)
22. E. Cremmer, B. Julia, J. Scherk, Supergravity theory in eleven-dimensions. Phys. Lett. **B76**, 409–412 (1978) [25 (1978)]
23. I.B. Frenkel, V.G. Kac, Basic representations of affine Lie algebras and dual resonance models. Invent. Math. **62**, 23–66 (1980)
24. G. Segal, Unitarity Representations of some infinite dimensional groups. Commun. Math. Phys. **80**, 301–342 (1981)
25. P. Goddard, D.I. Olive, Kac-Moody and Virasoro algebras in relation to quantum physics. Int. J. Mod. Phys. **A1**, 303 (1986)
26. D. Mumford, *Projective Invariants of Projective Structures and Applications* (1963)
27. C. Angelantonj, A. Sagnotti, Open strings. Phys. Rep. **371**, 1–150 (2002). [Erratum: Phys. Rep. **376**(6), 407 (2003)]
28. J. Polchinski, Dirichlet Branes and Ramond–Ramond charges. Phys. Rev. Lett. **75**, 4724–4727 (1995)
29. J. Polchinski, Y. Cai, Consistency of open superstring theories. Nucl. Phys. **B296**, 91–128 (1988)
30. P. Horava, E. Witten, Heterotic and type I string dynamics from eleven-dimensions. Nucl. Phys. B **460**, 506–524 (1996)
31. P. Horava, E. Witten, Eleven-dimensional supergravity on a manifold with boundary. Nucl. Phys. B **475**, 94–114 (1996)
32. K.-S. Choi, S.-J. Rey, Elliptic genus, anomaly cancellation and heterotic M-theory (2017). e-print:1710.07627

# Strings on Orbifolds

**7**

Heterotic string possesses very rich symmetries. It naturally describes SO(32) and $E_8 \times E_8$ gauge group, by uniform charge on the closed string. Also it has sixteen real ($\mathcal{N} = 4$ in four dimension) supersymmetries. We would say that it also has enough number of spacetime dimensions, 10, to explain our spacetime. However, these symmetries are too large from the phenomenological point of view, by the criteria discussed in Chap. 2. In weakly coupled string theories, it is better if a big gauge group is given already so that the standard model (SM) gauge group SU(3) × SU(2) × U(1) can be embedded there.

By compactifying down to four dimension and relating such symmetries with spacetime properties, we can break them. Here we will deal with the easiest and intuitive compactification scheme, the orbifold compactification, such that the resulting four dimensional theory has gauge symmetry of the standard model and only four ($\mathcal{N} = 1$) supersymmetries. Alternatively, on the way to the SM we may meet the grand unified theory (GUT). In this process, the chiral nature of the SM fermions should result, which is one of the most mysterious puzzles of the Planck-scale physics.

In this chapter, we quantize the string on orbifold and calculate the spectrum. We associate point group action with shift vector. We will turn on Wilson lines whose shift vectors are associated with translational elements. We will mainly focus on prime orbifold. More formal construction as well as discussion on nonprime orbifold will be done in the next chapter. There are many good treatments on this topic [1,2], as well as the original paper [3,4].

## 7.1 Twisted String

A closed string is characterized by the boundary condition

$$X^\mu(\sigma + \pi)^\mu = X^\mu(\sigma).$$

© Springer Nature Switzerland AG 2020
K.-S. Choi, J. E. Kim, *Quarks and Leptons From Orbifolded Superstring*,
Lecture Notes in Physics 954, https://doi.org/10.1007/978-3-030-54005-0_7

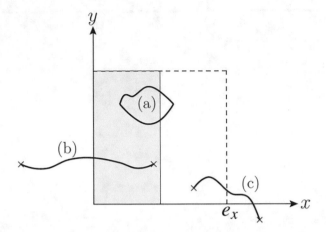

**Fig. 7.1** Strings on $T^2/\mathbb{Z}_2$ orbifold, defined by an identification $(x, y) \sim (-x, -y)$. The shaded region depicts the fundamental region and dotted border bounds the covering torus. (**a**) An untwisted closed string. (**b**) A closed string winding around torus. (**c**) A twisted string localized at the fixed point $e_x$. It is associated with the space group action $(-1, 2e_x)$

A normal closed string is drawn in Fig. 7.1a. Contrary to the point particle, a string has spatial extent so that it may behave differently in a topologically nontrivial space. A closed string on a torus, defined in Eq. (3.6),

$$T^6 = \mathbb{R}^6/\Lambda$$

can wind a non-contractable cycle as we have seen in (6.188). The boundary condition becomes

$$X(\sigma + \pi)^i = X(\sigma)^i + 2\pi R L^i,$$

where the index $i$ is for the orthogonal coordinates and $L^i$ are the winding numbers forming vector in $\Lambda$, as in Eq. (6.205). It is shown in Fig. 7.1b.

We may further mod out the torus to have orbifold

$$T^6/\mathbb{Z}_N \equiv T^6/\overline{\mathsf{P}},$$

as we have done in Chap. 3. The point group $\overline{\mathsf{P}}$ is generated by rotations $\theta$, which is a cyclic group $\mathbb{Z}_N$. We have identification

$$z^a \sim \theta^a{}_b z^b \tag{7.1}$$

up to a lattice translation. From now on, we will suppress the spacetime indices. $\theta$ is diagonalized and we call the exponents of the components twist vector $\phi \equiv (\phi_1, \phi_2, \phi_3)$. The generalization to arbitrary even dimensions is straightforward.

On this orbifold, we have a closed string modulo an action $(\theta^k, v) \in \mathsf{S}$ in (3.12) since the space is identified under $\mathsf{S}$,

$$Z(\sigma + \pi) = \theta^k Z(\sigma) + v. \tag{7.2}$$

We call such string twisted string. If none of the eigenvalues of $\theta^k$ are 1, there is one-to-one correspondence between a space group element $(\theta^k, v)$ and a fixed point

$$f = \left(1 - \theta^k\right)^{-1} v. \tag{7.3}$$

We learned in Chap. 3 that the twisted string is labelled by conjugacy class $[(\theta^k, v)]$. That is, all the states labelled by

$$(\omega, u) \left(\theta^k, v\right) (\omega, u)^{-1}, \quad (\omega, u) \in \mathsf{S}, \tag{7.4}$$

should be treated equivalent.

With the defining element $\theta$, we can broadly classify the twisted sectors using $\theta^k, k = 1, \ldots, (N - 1)$, calling each set, respectively, the $k$th twisted sector. This situation is depicted in Fig. 7.1. We call $k = 0$ sector and string, respectively, the *untwisted sector* and *untwisted string*.

A string in the untwisted sector freely moves in the bulk, whereas a string in the twisted sector is localized at the fixed point.

## 7.2   Mode Expansion and Quantization

We now perform mode expansion and quantize twisted strings. The modification is that the mode number is fractional on the orbifold to satisfy the boundary condition (7.2).

### 7.2.1   Bosonic Left and Right Movers

From (7.2), the worldsheet boson in the $k$th twisted sector acquires the phase

$$Z^a(\sigma + \pi) = e^{2\pi i k \phi^a} Z^a(\sigma) + v^a.$$

We take components of the twist $k\phi^a = k^a/N$. In this book, we have the convention $0 < k^a/N < 1$ in the first twisted sector. If the component of the twist vector does not lie on this interval, we may add an appropriate integer. If $k\phi^a$ is an integer, the mode expansion is the same as that of the untwisted string, viz. Eqs. (6.18) and (6.19). This happens on the fixed torus; the translational symmetry is recovered and thus the string component can have momentum as well.

The following mode expansions have the desired phases:

$$
Z_R^a(\tau - \sigma) = f_R^a + \frac{i}{2} \sum_{n \in \mathbb{Z}} \frac{\alpha_{n-k^a/N}^a}{n - k^a/N} e^{-2i(n-k^a/N)(\tau-\sigma)},
$$

$$
Z_L^a(\tau + \sigma) = f_L^a + \frac{i}{2} \sum_{n \in \mathbb{Z}} \frac{\tilde{\alpha}_{n+k^a/N}^a}{n + k^a/N} e^{-2i(n+k^a/N)(\tau+\sigma)},
$$

(7.5)

where we used the convention $\alpha' = \frac{1}{2}$. Using $Z^{\bar{a}} = [Z^a]^*$, we also have

$$
Z_R^{\bar{a}}(\tau - \sigma) = f_R^{\bar{a}} + \frac{i}{2} \sum_{m \in \mathbb{Z}} \frac{\alpha_{n+k^a/N}^{\bar{a}}}{n + k^a/N} e^{-2i(n+k^a/N)(\tau-\sigma)},
$$

$$
Z_L^{\bar{a}}(\tau + \sigma) = f_L^{\bar{a}} + \frac{i}{2} \sum_{m \in \mathbb{Z}} \frac{\tilde{\alpha}_{n-k^a/N}^{\bar{a}}}{n - k^a/N} e^{-2i(n-k^a/N)(\tau+\sigma)}.
$$

(7.6)

Here, the center of mass is at the *fixed point* $f^a = f_L^a + f_R^a$, obtained in (7.3). The twisted string does not have a momentum.

We quantize the string by assigning commutation relations of the oscillators

$$
\left[ \tilde{\alpha}_{n+k^a/N}^a, \tilde{\alpha}_{-m-k^a/N}^{\bar{b}} \right] = \left( n + \frac{k^a}{N} \right) \delta^{ab} \delta_{mn},
$$

(7.7)

$$
\left[ \alpha_{n-k^a/N}^a, \alpha_{-m+k^a/N}^{\bar{b}} \right] = \left( n - \frac{k^a}{N} \right) \delta^{ab} \delta_{mn}.
$$

(7.8)

Therefore we have creation operators

$$
\tilde{\alpha}_{-n-k^a/N}^{\bar{a}}(n \geq 0), \alpha_{-n+k^a/N}^{\bar{a}}(n \geq 1), \alpha_{-n-k^a/N}^a(n \geq 0), \tilde{\alpha}_{-n+k^a/N}^a(n \geq 1).
$$

Also we define a ground state $|\sigma_k\rangle \equiv |\sigma_k\rangle_L \otimes |\sigma_k\rangle_R$ as one annihilated by all the annihilation operators, for all $a$,

$$
\begin{aligned}
\alpha_{n-k^a/N}^a |\sigma_k\rangle_R &= 0, & n &\geq 1, \\
\tilde{\alpha}_{n+k^a/N}^a |\sigma_k\rangle_L &= 0, & n &\geq 0, \\
\alpha_{n+k^a/N}^{\bar{a}} |\sigma_k\rangle_R &= 0, & n &\geq 0, \\
\tilde{\alpha}_{n-k^a/N}^{\bar{a}} |\sigma_k\rangle_L &= 0, & n &\geq 1.
\end{aligned}
$$

(7.9)

We may understand more properties of these ground states in Sect. 6.3.

The fractional mode numbers make the following things nontrivial.

1. The number operator

$$\tilde{N} = \sum_{a=1}^{4} \tilde{N}_a + \sum_{a=1}^{4} \tilde{N}_{\bar{a}} + \sum_{I}^{16} \tilde{N}_I, \tag{7.10}$$

where each

$$\tilde{N}_a = \sum_{n+k^a/N>0} \tilde{\alpha}^{\bar{a}}_{-n-k^a/N} \tilde{\alpha}^{a}_{n+k^a/N},$$

$$\tilde{N}_{\bar{a}} = \sum_{n-k^a/N>0} \tilde{\alpha}^{a}_{-n+k^a/N} \tilde{\alpha}^{\bar{a}}_{n-k^a/N},$$

$$\tilde{N}_I = \sum_{n=1}^{\infty} \tilde{\alpha}^{I}_{-n} \tilde{\alpha}^{I}_{n},$$

can have fractional eigenvalue. For example, an excited state $\tilde{\alpha}^1_{-1/3}|\sigma^1_1\rangle_L$ has a fractional oscillator number $\tilde{N} = \frac{1}{3}$.

2. When we rewrite (7.5) componentwise, we see that, under the point group action $\theta$, each oscillator transforms as

$$\alpha^a_{n-k^a/N} \to e^{-2\pi i k^a/N} \alpha^a_{n-k^a/N},$$

$$\tilde{\alpha}^a_{n+k^a/N} \to e^{2\pi i k^a/N} \tilde{\alpha}^a_{n+k^a/N},$$

$$\alpha^{\bar{a}}_{n+k^a/N} \to e^{2\pi i k^a/N} \alpha^{\bar{a}}_{n+k^a/N}, \tag{7.11}$$

$$\tilde{\alpha}^{\bar{a}}_{n-k^a/N} \to e^{-2\pi i k^a/N} \tilde{\alpha}^{\bar{a}}_{n-k^a/N}.$$

Later it will be necessary to use a vector $\rho_L$ $(\rho_R)$, whose $a$-th component counts the number of the excited oscillators along $a$-th directions in the left (right) mover. We have a convention that a holomorphic component has $+1$ and antiholomorphic $-1$. For instance, the state with oscillator excitations has the vector

$$|\varphi\rangle_L \equiv \left(\tilde{\alpha}^1_{-2/3}\right)^2 \tilde{\alpha}^2_{-1/3} \tilde{\alpha}^{\bar{3}}_{-1/3} |\sigma_k\rangle_L : \rho_L = (2, 1, -1).$$

Then, the above phase can be easily represented as

$$|\varphi\rangle_L \to e^{2\pi i \rho_L \cdot \phi} |\varphi\rangle_L. \tag{7.12}$$

We can similarly define $\rho_R$ from the right-moving oscillators. Also the ground state acquires a phase

$$| \prod \sigma_k \rangle_L \to e^{-\pi i k \phi^2} | \prod \sigma_k \rangle_L \tag{7.13}$$

that we will prove in the next chapter.

3. The zero point energy $\tilde{c}$, the sum of the normal ordering constant, is changed. To obtain it, we need a generalized Riemann zeta function that is regularized as

$$f(\eta) = \frac{1}{2} \sum_{n=0}^{\infty} (n + \eta) = -\frac{1}{24} + \frac{1}{4} \eta(1 - \eta), \tag{7.14}$$

for each *real, bosonic* degree of freedom with $0 \le \eta \le 1$. We will derive it in Appendix A. For the bosonic left movers, we have the zero point energy

$$\tilde{c} = 2 \sum_{a=0}^{3} f(\phi_a) + \sum_{I=1}^{16} f(0). \tag{7.15}$$

The factor 2 in the first term comes from the definition (3.68), so that we rotated pairwise. We verify that the untwisted string has $\tilde{c} = -1$. The fermionic states have the contributions with the same magnitude but with the opposite sign.

## 7.2.2 Fermionic Right Movers

For the superstring right movers, we have the fermionic degrees $\psi^M$. Spacetime supersymmetry requires that the phases of $X_R^M$ and $\psi^M$ are the same. We need

$$\psi^a(\sigma + \pi) = e^{2\pi i (k\phi_a + v)} \psi^a(\sigma), \tag{7.16}$$

where $v = 0$ for R and $v = \frac{1}{2}$ for NS states. As before, we consider the compact dimensions with the holomorphic and antiholomorphic indices $a$ and $\bar{a}$, respectively. However, this traditional distinction becomes meaningless due to the orbifold twist. What is previously called the NS state is just shifted further from the R state by the extra phase $-1$.

The mode expansions are, using the same convention as before,

$$\psi_R^a = \sum_{r \in \mathbb{Z} + v} \psi_{r+k^a/N}^a e^{-2i(r+k^a/N)(\tau - \sigma)},$$

$$\psi_R^{\bar{a}} = \sum_{r \in \mathbb{Z} + v} \psi_{r-k^a/N}^{\bar{a}} e^{-2i(r-k^a/N)(\tau - \sigma)}. \tag{7.17}$$

Thus, we have the following modified anti-commutation relations:

$$\left\{ \psi^a_{r+k^a/N}, \psi^{\bar b}_{-r-k^a/N} \right\} = \left( r + \tfrac{k}{N} \right) \delta^{ab} \delta_{mn},$$

$$(7.18)$$

$$\left\{ \psi^a_{r-k^a/N}, \psi^{\bar b}_{-r+k^a/N} \right\} = \left( r - \tfrac{k}{N} \right) \delta^{ab} \delta_{mn}.$$

$$(7.19)$$

We may define the ground states in the same way as the bosonic ones. The oscillator numbers now have the bosonic part which is similar to (7.10). The oscillator numbers of the fermionic part is

$$N_F = \sum_{a=1}^{4} \sum_{r+k^a/N>0} \left( r + k^a/N \right) \psi^{\bar a}_{-r-k^a/N} \psi^a_{r+k^a/N}$$

$$+ \sum_{a=1}^{4} \sum_{r-k^a/N>0} \left( r - k^a/N \right) \psi^a_{-r+k^a/N} \psi^{\bar a}_{n-k^a/N},$$

$$(7.20)$$

where $r \in \mathbb{Z} + v$. Consequently, we have the following modified zero point energies,

$$c = 2 \sum_{a=1}^{4} f(\phi_a) - 2 \sum_{a=1}^{4} f(\phi_a + v).$$

$$(7.21)$$

For the untwisted superstring, $c = -\tfrac{1}{2}$ and 0 (where $\phi_i = 0$) for the NS and R states, respectively.

For a practical calculation, we will use the bosonized description for the right movers. Recall that the right mover of heterotic string is the same as that of Type II string, as we have seen in Sect. 6.3.3. Also we replaced eight fermions by four bosons $s^a$, $a = 0, 1, 2, 3$, so that we have 12 bosons in total. We have the spinorial $\mathbf{8}_s$ in the R sector and the vector $\mathbf{8}_v$ in the NS sector.

For the twisted string, we have the right movers

$$|s + k\phi\rangle_R,$$

$$(7.22)$$

where $N^{(k)} = \sum_a N^{(k)}_a$ and

$$R : s = ([+ + + +]),$$
$$NS : s = (\pm 1\ 0\ 0\ 0).$$

The twist does not affect the oscillators of the bosonized fermions but shifts the momentum as $s \to s + k\phi$. So we have the zero point energy for the right movers,

$$c = 4f(0) + 2\sum_{a=0}^{3} f(k\phi_a). \tag{7.23}$$

This is tabulated in Appendix A. Comparing between Eqs. (7.15) and (7.20), we observe

$$c = \tilde{c} - \frac{1}{2}. \tag{7.24}$$

The resulting mass shell condition becomes (7.38).

The spacetime Lorentz representation is still encoded in the $s$ vector, which is now unified. Under the point group action $\theta^k$, each state transforms as

$$\left(\prod \alpha\right) |s + k\phi\rangle_R \to e^{-2\pi i[(s+k\phi+\rho_R)\cdot\phi - \frac{k}{2}\phi^2]} \left(\prod \alpha\right) |s + k\phi\rangle_R, \tag{7.25}$$

where $\prod \alpha$ is defined in the notion of Eq. (7.13). This is essentially the same as (7.33), including the phase from the ground state. The overall minus sign in the exponent reflects that the state is right moving. There is also an offset phase $\frac{k}{2}\phi^2$ from the ground state. Note that this phase is exactly the same as that of the bosonic one, so there is a complete cancellation.

### 7.2.3  Shifting

We have another way to make an orbifold, by *translational* elements that we may call *shifting*. We formed a torus $T^d$ by modding out $\mathbb{R}^d$ by this symmetry encoded in the lattice. We may consider a *further* action on the torus

$$\mathscr{V}: \quad x^i \to x^i + 2\pi v^i, \tag{7.26}$$

where each component $v^i$ is an integer multiple of $R/N$, where we took the coordinate so that $2\pi R$ is the periodicity in all the directions. Certainly, it is an action of order $N$ and defines an orbifold $T^d/\mathbb{Z}_N$, as seen in Sect. 3.2. The operation (7.26) is free, and there is no fixed point.

Apart from the usual untwisted string, there is a twisted string up to the shift (7.26)

$$X^i(\sigma + \pi) = X^i(\sigma) + 2\pi R L^i + 2\pi R v^i, \tag{7.27}$$

with no summation over $i$. Although this is equivalent to (7.26), we have displayed some possible contribution from string winding $L^i$. This is a generalization of winding: it completely closes the string on the *modded torus* along the smaller circumference $2\pi/N$. Formally, the mode expansion is same as (6.191, 6.192) with $L$ replaced by $L + v$,

$$X^i = x^i + p^i \tau + 2 \left( L^i + v^i \right) \sigma + \frac{i}{2} \sum \left( \frac{1}{n} \tilde{\alpha}_n^i e^{-2in(\tau+\sigma)} + \frac{1}{n} \alpha_n^i e^{-2in(\tau-\sigma)} \right),$$

and therefore the same are quantization, mass condition, and so on. For the defining element (7.26), we have a number of twisted sectors shifted by $kv$ with $k = 1, \ldots, N - 1$.

We are mainly interested in the case where the shift (7.26) acts on the current algebra described by extra sixteen bosons $X_L^I$. We have *only left movers*, so we should act the shift in an *asymmetric* manner

$$X_L^I(\sigma + \pi) = X_L^I(\sigma) + \pi R L^I + \pi R k V^I. \tag{7.28}$$

Here, we use uppercase letter for those variables of the current algebra. The missing factor 2 in the last term is due to the fact that $X^I$ are only the left movers, which hence has no geometric interpretation. To implement this situation, we formally assume the presence of right movers as done before. Then, the mode expansion goes like

$$X_R^I = x_R^I + \left( \frac{1}{2} P^I - L^I \right) (\tau - \sigma) + \frac{i}{2} \sum \frac{1}{n} \alpha_n^I e^{-2in(\tau-\sigma)}, \tag{7.29}$$

$$X_L^I = x_L^I + \left( \frac{1}{2} P^I + L^I + k V^I \right) (\tau + \sigma) + \frac{i}{2} \sum \frac{1}{n} \tilde{\alpha}_n^I e^{-2in(\tau+\sigma)}. \tag{7.30}$$

Again we impose the "absence of right mover" constraint as in (6.214) $\frac{1}{2} P - L = 0$. Like the original heterotic string, this forces the current algebra to be non-Abelian so that we need a critical radius $R = 1/\sqrt{\alpha'} = \sqrt{2}$. Then, the mode expansion becomes

$$X_L^I = x_L^I + \left( P^I + k V^I \right) (\tau + \sigma) + \frac{i}{2} \sum \frac{1}{n} \tilde{\alpha}_n^I e^{-2in(\tau+\sigma)}, \tag{7.31}$$

with the new momentum

$$P \rightarrow P + kV. \tag{7.32}$$

Combining with the spacetime degrees of freedom, the mass shell condition becomes (7.37).

By $\mathcal{V}$, a state $|P + kV\rangle_L$ transforms as

$$|P + kV\rangle_L \rightarrow e^{2\pi i[(P+kV)\cdot V - \frac{1}{2}(kV)\cdot V]}|P + kV\rangle_L. \tag{7.33}$$

There is a vacuum phase $-\frac{k}{2} V^2$, since we have associated the twist with the shift, so we have essentially the same contribution as (7.13).

## 7.3    Embedding Gauge Group

To obtain the four dimensional world, we should make six dimensions out of ten compact and small. We compactify heterotic string on the six dimensional orbifold. Then, the modular invariance condition becomes nontrivial and relates this compact space with the internal gauge symmetries, resulting in some projections. This will provide boundary conditions to break the gauge group and obtain a chiral theory.

### 7.3.1   Associating Shift

We associate the action $\theta$ on $T^6$ with a translation by a *shift vector* $V$ on the group lattice $T_{G,L}^{16}$,

$$\theta \longrightarrow (1, V). \tag{7.34}$$

This is an *embedding* of the orbifold action in the group space. The resulting total orbifold is

$$\mathcal{O} = T_{G,L}^{16}/\mathcal{V} \times T^6/\overline{\mathsf{P}}, \tag{7.35}$$

where $\mathcal{V}$ is the shifting (7.26) and $\overline{\mathsf{P}}$ is the generalized point group action. It is not the direct product of two sets of tori, but the actions $\theta$ and $(1, V)$ are done simultaneously (Fig. 7.2).

This is the simplest setup in the sense that we use only the translational modding $\mathcal{V}$ on the gauge lattice and use the symmetric modding $\overline{\mathsf{P}}$ on the left and the right

**Fig. 7.2** We associate point action $\theta$ with lattice shift $V$ in the group space, that is, by doing them simultaneously. Then, the momentum state $|P\rangle$ has a different boundary condition and breaks the group

movers. We may also mod out the $T_{G,L}^{16}$ by a point group action. Shifting on the gauge lattice in (7.28) is quite general due the following theorem [5]:

Every inner automorphism of finite order can always be represented by a shift vector.

We will study automorphisms in Sect. 12.4. The lattice defining these tori is the group lattice. A point group action takes one weight to another and thus is an inner automorphism. Since the $E_8$ group has only inner automorphisms, all of its subgroups can be obtained by shift vectors. If we have more than one embeddings, we cannot simultaneously transform all the embeddings into the shifts.

With $\theta$, $V$ defining the orbifold $\mathcal{O}$, we have twisted sectors in which we associate $\theta^k$ with $kV$. This leads us to the quantization discussed in Sect. 7.2. We have the complete states by combining left and right movers

$$\left(\prod \tilde{\alpha}\right) |P + kV\rangle_L, \otimes \left(\prod \alpha\right) |s + k\phi\rangle_R, \tag{7.36}$$

with simplified notation for the oscillator excitations. We have the mass shell condition

$$\tilde{L}_0(kV, k\phi) + \tilde{c}^{(k)} = \frac{(P + kV)^2}{2} + \tilde{N}^{(k)} + \tilde{c}^{(k)} = \frac{1}{2}\alpha' M_L^2, \tag{7.37}$$

$$L_0(k\phi) + c^{(k)} = \frac{(s + k\phi)^2}{2} + N^{(k)} + c^{(k)} = \frac{1}{2}\alpha' M_R^2, \tag{7.38}$$

$$\frac{1}{2}M^2 = M_L^2 = M_R^2, \tag{7.39}$$

where $\tilde{c}^{(k)}$ and $c^{(k)}$ depend on the twist, given in Eqs. (7.15) and (7.21). The total states satisfy the mass shell condition $M^2 = 0$ in (7.37). We have such twisted sectors associated with $k\phi$ and $kV$ with $k = 1, 2, \ldots, N-1$ as well as the untwisted sector $k = 0$ (or the twisted sector $k = N$). Since string mass is proportional to the length stretched, the lowest mass state is shrunk and localized on the fixed point $f$.

Since we have embedded the point group action into the action in the current algebra, we need to keep invariant states under these actions. We have

$$\left(\prod \tilde{\alpha}\right) |P + kV\rangle_L \to e^{2\pi i[(P+kV)\cdot V + \rho_L \cdot \phi - \frac{k}{2}(V^2 - \phi^2)]} \left(\prod \tilde{\alpha}\right) |P + kV\rangle_L, \tag{7.40}$$

$$\left(\prod \alpha\right) |s + k\phi\rangle_R \to e^{-2\pi i(s+k\phi+\rho_R)\cdot\phi} \left(\prod \alpha\right) |s + k\phi\rangle_R. \tag{7.41}$$

Here, $\rho_L$ and $\rho_R$ vectors parametrize the oscillators (7.11), and we have vacuum phase.

## 7.3.2   Modular Invariance

The consistency condition of the heterotic string theory comes from the modular invariance of the partition function. It is invariant under the transformation of the modular parameter $\tau = \tau_1 + i\tau_2$ in (6.72), generated by

$$\mathcal{T}: \tau \to \tau + 1, \quad \mathcal{S}: \tau \to -1/\tau, \tag{7.42}$$

as in (6.73).[1]

In the construction of heterotic string, modular invariance on the current algebra direction constrains the group to $E_8 \times E_8$ or $SO(32)$. The spacetime part is *independently* modular invariant.

By orbifolding, each part loses its own modular invariance; however, we pair them up, by relating as (7.34), to make the *whole theory* invariant. This introduces projection condition and twisted sectors. Here, we sketch the procedure for obtaining constraints, postponing the proof to the next chapter.

### $\mathcal{S}$ Invariance

A twisted string has the boundary condition (7.2), or in terms of a space group element $h$,

$$Z(\tau, \sigma + \pi) = h Z(\tau, \sigma). \tag{7.43}$$

Orbifolding introduces a projection. We have seen, in Sect. 6.2.3, that we should consider spin structure for a well-defined spin for worldsheet fermions on a torus. It led us to introduce the GSO projection. The R and NS sectors are further divided into two subsectors with extra boundary conditions. The boundary condition (6.99) formally looks the same as a $\mathbb{Z}_2$ boundary condition from orbifolding. We may generalize it to $\mathbb{Z}_N$ orbifolds. This phase can be understood as the eigenvalue of the space group action $g \in \mathsf{S}$, which we will show in the next chapter. This is equivalent to assigning another boundary condition

$$Z(\tau + 2\pi\tau_2, \sigma + \pi\tau_1) = g Z(\tau, \sigma). \tag{7.44}$$

We say that such a field belongs to the $(g, h)$-twisted sector. Thus, even in the untwisted sector, we should also specify the boundary conditions in the $\tau$ direction. Also, once we introduce orbifold projection, we also need twisted sectors.

Under general finite transformation $\tau \to \frac{a\tau+b}{c\tau+d}$ of (6.72), the boundary condition goes like

$$(h, g) \to \left( h^d g^c, h^b g^a \right). \tag{7.45}$$

---

[1]The reader would not confuse the notation with that for the worldsheet time direction.

In particular,

$$\mathcal{T} : (h, g) \rightarrow (h, hg), \tag{7.46}$$

$$\mathcal{S} : (h, g) \rightarrow \left(g, h^{-1}\right). \tag{7.47}$$

This will shuffle various $(g, h)$-twisted sectors.

### $\mathcal{T}$ Invariance

The condition from the $\mathcal{T}$ invariance provides a guideline for choosing the shift vector $V$ for a given orbifold geometry described by $\phi$.

The real axis is identified with $\sigma$, so that the $\mathcal{T}$ generate a shift in the $\sigma$ direction whose operator is generated by the worldsheet momentum in (6.43). The exponentiated form is

$$U(\sigma) = \exp\left[2i\left(\tilde{L}_0 + \tilde{c} - L_0 - c\right)\sigma\right]. \tag{7.48}$$

Since the closed string has no preferred origin, these levels should match modulo an integer

$$\tilde{L}_0 + \tilde{c} - L_0 - c = 0 \mod 1. \tag{7.49}$$

Using $\tilde{L}_0 + \tilde{c}$ from (7.37) and $L_0 + c$ from (7.38), it is easy to see that the oscillator number N is a multiple of $\frac{1}{N}$, and the zero point energy and $(P+V)^2$ are multiples of $\frac{1}{N^2}$. In general, it is not possible to make sum of them integer to satisfy the relation (7.49). Thus, a necessary condition is imposed to make the $\frac{1}{N^2}$ dependence to be an integer multiple of $1/N$,

$$(P + V)^2 - (s + \phi)^2 = 0 \mod \frac{2}{N}.$$

Since we are in the even and self-dual lattice, $P^2 + 2P \cdot V$ is a multiple of $2/N$. The same holds for $1 - s^2 - 2s \cdot \phi$. Thus, it leads to

$$N \sum_{a=1}^{3} \phi_a^2 = N \sum_{I=1}^{16} V_I^2 \mod 2. \tag{7.50}$$

In effect, we have only considered the transformation $\mathcal{T}^N$, taking an $(h, g)$-twisted sector to $(h, h^N g)$, to be invariant.

We required the condition of definite orders (3.75)

$$N \sum_{a=1}^{3} \phi_a = 0 \mod 2. \tag{7.51}$$

The shift vector is compatible to the point group if it has the same or less order, similar to (3.75)

$$N \sum_{I=1}^{16} V_I = 0 \mod 2, \tag{7.52}$$

where the modulo 2 condition is reserved for spinorial states, e.g. in the $E_8 \times E_8'$ theory.

Once these conditions (for the first twisted sector) are satisfied, so are the relations for the $k$th twisted sectors $kV, k\phi$ for any integer $k$. The condition for the modular invariance is a necessary condition. We show that this is also the sufficient condition in the next chapter.

### Hilbert Space on Orbifold

In reality, the gauge symmetry is broken, and we may obtain matter states. This will give rise to a more general string theory in four dimensions. In particular, the unbroken gauge boson comes from the untwisted sector. The massless condition is $P^2 = 2$. The solutions $P$ here become root vectors. Gauge transform adds them to weights of the matter. During this, the state should not acquire a nontrivial phase, so that the root vectors should not carry such a phase in (7.33)

$$P \cdot V = 0 \mod 1. \tag{7.53}$$

For the right mover, since the space group action does not affect the untwisted components, we always have singlets

$$R : (\pm 1\, 0\, 0\, 0), \quad NS : \pm(+, +, +, +).$$

For the entire state to be invariant, the gauge current left movers should combine with multiplicity 1. So, we always have a vector multiplet.

The rest of the states which do not satisfy the condition (7.53) become *matter states*. In the untwisted sector, we have $P \cdot V \neq 0 \mod 1$ with $P^2$. This means that the matter states are obtained from the branching of the previously adjoint representation. In twisted sectors, the solutions $P + kV$ from (7.37) become weight vectors as quantum numbers carried by matter fields. Combined with the right mover that we see next, a complete state surviving projection is what we observe in low energy. We have introduced twisted sectors and required invariance under the generalized GSO projection. Note that a state may not be necessarily invariant under the projection, but a linear combination of states at different fixed points may be invariant. We will examine such states in the next chapter.

## 7.4   The Standard Embedding

The most famous example employs the $T^6/\mathbb{Z}_3$ orbifold, discussed in Sect. 3.5. It is specified by a twist vector (3.68)

$$\phi = \left(\tfrac{2}{3}\ \tfrac{1}{3}\ \tfrac{1}{3}\right).$$

We choose *the standard embedding*, that is, the orbifold action that is associated with the shift

$$V = \phi \tag{7.54}$$

with other degrees is not touched. That is, the shift vector is specified by

$$V = \left(\tfrac{2}{3}\ \tfrac{1}{3}\ \tfrac{1}{3}\ 0\,0\,0\,0\,0\right)(0\,0\,0\,0\,0\,0\,0\,0). \tag{7.55}$$

This satisfies modular invariance condition (7.50) by construction.

### 7.4.1   Untwisted Sector

The untwisted states are a part of the original states invariant under the point group action $\theta$ and associated shift $V$. It is not necessary that each left and right mover itself be invariant, but only the combined state should be invariant. By the structure, we see that (7.55) leaves the second $E_8$ intact; thus, our concern is the first $E_8$. Through this example, we will display the vectors and the representation for the first $E_8$.

#### Left Mover
The left mover in the untwisted sector has zero point energies $\tilde{c}^{(0)} = -1$. From (7.37), we see that massless states satisfy either

$$|P\rangle_L,\ P^2 = 2, \quad \text{or} \quad \tilde{\alpha}^M_{-1}|P\rangle_L,\ P^2 = 0.$$

Depending on the orbifold projection, these are divided into the left movers for gauge and matter fields.

**Gauge Fields**  The massless states satisfying $P \cdot V =$ integer of (7.53) include[2]

$$\pm (0\,1\,-1\,0\,0\,0\,0\,0)$$

$$\pm (1\,\underline{1\,0}\,0\,0\,0\,0\,0)$$

and permutations of the underlined elements. With the two Cartan generators, linear combinations of $\alpha^I_{-1}|0\rangle$, $I = 1, 2, 3$, they form the root vectors **8** of SU(3). We may choose the *simple roots* as $(-1\,-1\,0\,0\,0\,0\,0\,0)$ and $(0\,1\,-1\,0\,0\,0\,0\,0)$, since we can generate all the roots from them. The inner product shows that indeed the group is SU(3).

The remaining vector satisfying the condition (7.53) is the roots of $E_6$.

$$(0\,0\,0\,\underline{\pm 1}\,\underline{\pm 1}\,0\,0\,0)$$

$$\pm (-++[++++-])$$

where the square brackets denote even sign flips. Here, the plus and minus represents $+\frac{1}{2}$ and $-\frac{1}{2}$, respectively. We can check that, together with the remaining six Cartan generators, they form the **78** of $E_6$.

Therefore, the unbroken gauge group is SU(3) $\times$ $E_6$ $\times$ $E'_8$.

**Matter Fields**  Now, we come to matter representation. States satisfying $P \cdot V = \frac{1}{3}$ are

$$
\begin{aligned}
&(+\,\underline{+\,-}\,[+\,+\,+\,+\,-]) \\
&(-\,-\,-\,[+\,+\,+\,+\,-]) \\
&(0\,\underline{1\,0}\,\underline{\pm 1}\,0\,0\,0\,0) \\
&(-1\,0\,0\,\underline{\pm 1}\,0\,0\,0\,0) \\
&(1\,\underline{-1}\,0\,0\,0\,0\,0\,0) \\
&(0\,-1\,-1\,0\,0\,0\,0\,0).
\end{aligned}
\tag{7.56}
$$

Under a $\mathbb{Z}_3$ rotation, these acquires a phase $\alpha \equiv e^{2\pi i/3}$. We can check that with an aid from the root vectors these form the weight vectors $(\mathbf{3}, \mathbf{27})$ of SU(3) $\times$ $E_6$.

We have other states from $P \cdot V = \frac{2}{3}$ modulo 1. Their weight vectors are precisely $-P$ for the above states: $-P \cdot V = -\frac{1}{3} = \frac{2}{3}$ mod 1. They acquire phase $\alpha^2$. The resulting states are $(\overline{\mathbf{3}}, \overline{\mathbf{27}})$. *They become the $\mathscr{CPT}$ conjugates to the above states.*

---

[2]In the following vectors, if we flip the signs of the first entries we may easily notice regular patterns. Nevertheless, we employ this convention because the form manifestly show the group theoretical origin of the standard root and weight vectors, shown in Table 12.1.

**Right Mover**

The right movers are supersymmetric, and the lowest modes come from the worldsheet fermions. As in the previous chapter, we bosonize these fermions. In this description, all the bosons are periodic, and we have vanishing zero point energy. Thus, all the states $|s\rangle_R$ satisfy $s^2 = 1$. Under the point group action, the state acquires a phase

$$|s\rangle_R \to e^{-2\pi i s \cdot \phi}|s\rangle_R, \tag{7.57}$$

where $\phi = \frac{1}{3}(0;\ 2\ 1\ 1)$.

In the R sector, the weights $s$ are spinorial representation (6.247).

$$\mathbf{3} : (+ - \underline{+-}),\ (+ + ++),\ \sim \alpha^1, \tag{7.58}$$

$$\mathbf{\bar{3}} : (- + \underline{+-}),\ (- - --),\ \sim \alpha^2, \tag{7.59}$$

$$\mathbf{1} + \mathbf{1} : (+ + --),\ (- - ++),\ \sim \alpha^0. \tag{7.60}$$

We have the NS sector, whose states have integral entries

$$\mathbf{3}_{NS} : (0\ 1\ 0\ 0),\ (0\ 0\ \underline{-1}\ 0),\ \sim \alpha^1, \tag{7.61}$$

$$\mathbf{\bar{3}}_{NS} : (0\ -1\ 0\ 0),\ (0\ 0\ \underline{1}\ 0),\ \sim \alpha^2, \tag{7.62}$$

$$\mathbf{1}_{NS} + \mathbf{1}_{NS} : (1\ 0\ 0\ 0),\ (-1\ 0\ 0\ 0),\ \sim \alpha^0. \tag{7.63}$$

We see that the two sectors are connected by the supercharge

$$s_R + r = s_{NS}, \tag{7.64}$$

where $r$ is the $R$-vector defined in (3.72)

$$r = \left(-\tfrac{1}{2};\ \tfrac{1}{2}\ -\tfrac{1}{2}\ -\tfrac{1}{2}\right). \tag{7.65}$$

It is the relation between the spacetime bosons and the fermions.

**Combined States and Chirality**

The complete state is made by combining the left and right movers. We have vector multiplet

$$|\mathbf{8}, \mathbf{1}\rangle_L \otimes |\mathbf{1} + \mathbf{1}\rangle_R, \quad |\mathbf{1}, \mathbf{78}\rangle_L \otimes |\mathbf{1} + \mathbf{1}\rangle_R, \quad \alpha^I_{-1}|0\rangle_L \otimes |\mathbf{1} + \mathbf{1}\rangle_R. \tag{7.66}$$

The R states with half-integral entries describes gaugino and fermions. The NS states with integral entries describe gauge bosons and fermions

$$P \in (\mathbf{8}, \mathbf{1}) + (\mathbf{1}, \mathbf{78}) \text{ of SU(3)} \times E_8, \quad s \in \mathbf{1} + \mathbf{1}. \tag{7.67}$$

In (7.57), $\phi$ acts on the last three components $s_1, s_2, s_3$ of $s$, while the noncompact component $s_0$ remains untouched. For R-states, we interpret $s_0$ *as the four dimensional helicity*. Our convention is that $s_0 = +\frac{1}{2}, -\frac{1}{2}$ are, respectively, right- and left-handed.

For the matter, we have $\mathbb{Z}_3$ invariant states

$$|3, 27\rangle_L \otimes |\overline{3}\rangle_R, \quad |\overline{3}, \overline{27}\rangle_L \otimes |3\rangle_R, \tag{7.68}$$

$$|3, 27\rangle_L \otimes |\overline{3}_{NS}\rangle_R, \quad |\overline{3}, \overline{27}\rangle_L \otimes |3_{NS}\rangle_R. \tag{7.69}$$

Note that the state $|\overline{3}\rangle_R$ has weight vectors (7.59). Hence, the $|3, 27\rangle_L \otimes |\overline{3}\rangle_R$ multiplet has four dimensional helicity $s_0 = -\frac{1}{2}$, which is left-handed in our convention. It is not complete state under Lorentz group, and the antiparticle is missing. We can easily see that the number of *real* degrees of freedom for the $|3, 27\rangle_L$ state is the same as the number of vectors $P$, which is $3 \times 27$. We need complex representation but half of the degrees is missing. The rest of the state comes from the state $|\overline{3}, \overline{27}\rangle_L \otimes |3\rangle_R$. It is the $\mathscr{C}\mathscr{P}\mathscr{T}$ conjugate to the previous state, having the opposite helicity, time flow, and complex conjugate representation.

So the two states in (7.68) form the complete representation. The resulting spectrum is *chiral*. They have the same *chirality*, although $\mathscr{C}\mathscr{P}\mathscr{T}$ conjugates have the opposite *helicities*. We say we have a *left-handed* Weyl fermion in $(3, 27)_-$. (It is not to be confused with string left movers.) In four dimensions, we may instead call it right-handed with the complex conjugate charge $(\overline{3}, \overline{27})_+$.

Likewise, the two states in (7.69) are $\mathscr{C}\mathscr{P}\mathscr{T}$ conjugates to each other, forming the complete complex scalar. They form the chiral multiplet.

Counting all the states, we have the SU(3) × E$_8$ vector multiplet and

$$3(3, 27)_-$$

as $\mathscr{N} = 1$ chiral multiplet in the untwisted sector. The multiplicity three in the untwisted sector matter is due to this triplet, due to $|\overline{3}\rangle_R$ and $|3\rangle_R$ in Eq. (7.68).

We also have massless states from the remaining oscillators $\tilde{\alpha}_{-1}^M$. We can decompose index $M = (\mu, a, \bar{a})$. From (7.11), we see that the invariant states are

$$\alpha_{-1}^a |0\rangle_L \otimes |s\rangle_R,$$

$$a = 1, \bar{2}, \bar{3}, \quad s \in \overline{3}, \overline{3}_{NS}, \tag{7.70}$$

$$a = \bar{1}, 2, 3, \quad s \in 3, 3_{NS}.$$

They are moduli fields $G_{a\bar{b}}, G_{\bar{a}b}, B_{a\bar{b}}, B_{\bar{a}b}$. We will see later that they are naturally combined. Those with noncompact four dimensional index $\mu$

$$\alpha_{-1}^\mu |0\rangle_L \otimes |s\rangle_R, \quad s \in \mathbf{1} + \mathbf{1}, \mathbf{1}_{NS} + \mathbf{1}'_{NS}. \tag{7.71}$$

are not affected, forming four dimensional graviton and antisymmetric tensor field. The dilaton is inherited from that of ten dimensions with the scaling from the internal volume.

### 7.4.2 Twisted Sector

In the twisted sector, as discussed in Eq. (7.10) and below, there are some modifications to oscillator numbers, phases, and zero point energies.

**Left Mover**

The zero point energy is shifted by $f(\eta)$ of Eq. (7.14) for each real bosonic degree of freedom. With twist $\phi = (0; \frac{2}{3}, \frac{1}{3}, \frac{1}{3})$ on the *complex* degrees, we have vacuum energy from Eq. (7.15)

$$\tilde{c} = 16 f(0) + 2 f(0) + 4 f\left(\tfrac{1}{3}\right) + 2 f\left(\tfrac{2}{3}\right) = -\tfrac{2}{3}, \tag{7.72}$$

and the level matching condition becomes

$$\tfrac{1}{4} M_L^2 = \tfrac{1}{2}(P + V)^2 + \tilde{N} - \tfrac{2}{3}. \tag{7.73}$$

We need massless state $M_L^2 = 0$.

- Without oscillator $\tilde{N} = 0$, the massless states $|P + V\rangle$ are

$$\left(\tfrac{2}{3} - \tfrac{2}{3} - \tfrac{2}{3} 0^5\right),$$

$$\left(-\tfrac{1}{3} 0 0 \pm 1 0^4\right),$$

$$\left(\tfrac{1}{3} - \tfrac{1}{3} - \tfrac{1}{3}[+ + + + -]\right).$$

They transform as $(\mathbf{1}, \mathbf{27})$ by the above root vectors. The state acquires a GSO phase

$$e^{2\pi i (P+V)\cdot V} = 1, \tag{7.74}$$

up to the phase form the twisted vacuum.

- From the mode expansion of the twisted states (7.6), there are fractional oscillators $\tilde{\alpha}^a_{-1/3}$ with $\tilde{N} = \frac{1}{3}$. So, there are additional massless states

$$\tilde{\alpha}^a_{-1/3}|P + V\rangle_L, \quad a = 1, \bar{2}, \bar{3}, \tag{7.75}$$

with $P + V$ satisfying $\frac{1}{2}(P + V)^2 = \frac{1}{3}$,

$$\left(\frac{2}{3} \; \frac{1}{3} \; \frac{1}{3} \; 0^5\right),$$

$$\left(-\frac{1}{3} \; -\frac{2}{3} \; \frac{1}{3} \; 0^5\right),$$

which form $(\bar{\mathbf{3}}, \mathbf{1})$. Since $a$ assumes $1, \bar{2}, \bar{3}$, they transforms as a triplet under the holonomy group $\mathbb{Z}_3$; thus, the multiplicity is 3. The state acquire phases $\alpha$ from $\tilde{\alpha}_{-1/3}$ and $\alpha^2 = e^{2\pi i(P+V)\cdot V}$ from $|P + V\rangle_L$, so that the state (7.75) is invariant. Again, we neglected the vacuum phase.

**Right Mover**

We have the same twist on the right movers. Again, we will use the bosonized right movers, among which only the original eight bosons get twisted, while the four bosonized fermions have no twists (which are reflected in the weight vectors). Accordingly the zero point energy is

$$c = 6f(0) + 2f\left(\tfrac{2}{3}\right) + 4f\left(\tfrac{1}{3}\right) = -\tfrac{1}{6}.$$

One may check that it is always satisfied $c = \tilde{c} + 1/2$. Resorting to the mass shell condition

$$\frac{M_R^2}{4} = \frac{(s + \phi)^2}{2} + N - \frac{1}{6}, \tag{7.76}$$

we find two massless states with $N = 0$

$$s + \phi = \left(-\tfrac{1}{2} \; \tfrac{1}{6} \; -\tfrac{1}{6} \; -\tfrac{1}{6}\right), \left(0 \; -\tfrac{1}{3} \; \tfrac{1}{3} \; \tfrac{1}{3}\right).$$

They correspond to (twisted) R and NS states, respectively, in the sense that $s$ is belonging to the conjugacy classes $\mathbf{8}_s$ and $\mathbf{8}_v$ in (6.247). In the spacetime, they are superpartners to the others, as it should be. We verify that the phase

$$e^{-2\pi i(s+\phi)\cdot\phi} = 1 \tag{7.77}$$

saves the state which survives the GSO projection. The resulting spectrum has the negative (left-handed) chirality, with untwisted matter defined from $P \cdot V = \frac{1}{3}$.

In the above calculation, the phases for left and right movers vanished independently, but this is not necessary in general. Only the *combined state* should have vanishing phase. Moreover, there might be a nontrivial "vacuum phase" that cannot be calculated by operator method here, which we will meet in Eq. (8.61) in Chap. 8.

The second twisted sector is obtained by the shift vector $2V$. It is equivalent to $-V$ up to a lattice translation. From the relation $(P - V)^2 = (-P + V)^2$, the state vectors in this twisted sector $P - V$ are the minus of the first twisted sector. The right movers are twisted by $2\phi \simeq -\phi$, which provides the opposite (positive) helicity states. They are the $\mathscr{CPT}$ conjugates with the opposite helicities and the complex conjugate of the states in the first twisted state. However, it is nontrivial to check that the total phase is invariant (see Exercise).

(We present an example of how the massive oscillators are formed. The lowest massive oscillator states are in the form

$$
\begin{aligned}
\alpha^a_{-2/3}|0\rangle, \quad & a = \bar{1}, 2, 3, \quad \mathbf{3}, \\
\alpha^a_{-1/3}\alpha^b_{-1/3}|0\rangle, \quad & a, b = 1, \bar{2}, \bar{3}, \quad \mathbf{6},
\end{aligned}
\tag{7.78}
$$

and have the oscillator number $\tilde{N} = \frac{2}{3}$. The holonomy quantum number is the representation of the point group $\mathbb{Z}_3$.)

### 7.4.3  Need Improvement

Summing up, we have obtained $SU(3) \times E_6$ gauge group with the matter fields

$$
3(\mathbf{3}, \mathbf{27})
$$

in the untwisted sector and

$$
27(\mathbf{1}, \mathbf{27}) + 81(\bar{\mathbf{3}}, \mathbf{1})
$$

in the first and the twisted sector. They are all four dimensional $\mathscr{N} = 1$ supergravity multiplets, with left-handed fermions. We also have moduli fields. In the $E_6$ GUT, each $\mathbf{27}$ contains one complete family, so we have 36 chiral families.

It is a nontrivial check for the anomaly cancellation of the $SU(3)$ part between chiral fermions from the untwisted and twisted sectors. There are $27 \times 3 = 81$ $\mathbf{3}$s in the untwisted sector and 81 $\bar{\mathbf{3}}$s in the twisted sector. $E_6$ is anomaly free. Thus, there is no gauge anomaly. In addition, there is no gravitational anomaly. Shortly we will see that in the presence of more than one shift vector, there still does not exist any anomaly. This is a generic feature of the orbifold theory with modular invariance, although there are only proofs [6–8] for the case of toroidal compactification on self-dual lattices. In some models, at most there exists an anomalous $U(1)$ which is cancelled by the generalized Green–Schwarz mechanism [9].

In most cases, the number of generations is a multiple of three, which originates either (1) from the right movers forming a $\mathbf{3}$ under the point group or (2) from the number of fixed points 27. Thus, the $\mathbb{Z}_3$ orbifold is a natural candidate giving three generations. The model presented above needs a further symmetry breaking because:

1. There are too many generations, 36 copies of **27** of $E_6$.
2. The gauge group $SU(3) \times E_6$ is still too big, predicting unobserved gauge bosons and charged matter fields. One may be satisfied at this stage as having obtained a GUT model. However, we do not have appropriate representations to break $E_6$ down to the SM gauge group.
3. All the twisted sectors have the same spectrum, because we have no way of distinguishing them. If we can distinguish them in some way, they give different spectra at different fixed points. Then, we may have different Yukawa couplings for different family members and thus can provide the observed SM mass hierarchy.

In the following section, we introduce Wilson lines. They introduce more shift vectors breaking the gauge group further. Also Wilson lines distinguish fixed points for them to have different spectra.

## 7.5    Wilson Lines

A Wilson line is a background gauge field that cannot be gauged away due to the non-simply connected topology of the space under consideration. It is characterized by a loop in the space that cannot be shrunk into a point. A torus has such non-simply connected circles that are inherited by a toroidal orbifold. We may associate their translations defining the torus with more shift vectors and break symmetries further.

### 7.5.1    Shifts Associated with Translations

Consider a constant gauge field $A_i(x) = A_i$. *Locally*, it is a pure gauge

$$A_i(x) = -iU^{-1}(x)\frac{\partial U(x)}{\partial x^i}, \quad U(x) = \exp\left(i A_i x^i\right), \tag{7.79}$$

and can be eliminated by gauge transformation $U(x)$

$$A_i(x) \rightarrow A_i(x) + iU^{-1}(x)\frac{\partial U(x)}{\partial x^i} = 0. \tag{7.80}$$

*Globally*, it is not possible to gauge away a constant gauge field, if the geometry is non-simply connected. We consider circles on the torus, which partially define the orbifold. The gauge parameter $U(x)$ does not obey periodicity,

$$U\left(x^i + 2\pi R\right) = U\left(x^i\right)\exp(2\pi i A_i R) \neq U\left(x^i\right), \tag{7.81}$$

which means that the ungauging cannot be done uniformly over the entire circles, unless $A_i$ is an integer multiple of $1/R$.

Alternatively, we force the gauge field to vanish by the passive transformation for the charged fields

$$\psi(x) \to U^{-1}(x)\psi(x) = e^{-iA_ix^i}\psi(x). \tag{7.82}$$

Then, it is unavoidable that the wave function $\psi(x)$ acquires a phase $e^{-2\pi iA_iR}$ once we circle around the $x^i$ direction. This phase in Eq. (7.82) is interpreted as the Scherk–Schwarz phase in Eq. (5.11).

For general (non-constant) gauge field, the gauge-invariant object of this effect is called the *Wilson line*,

$$W_i = \exp\left[i \oint A_i \cdot H dx^i\right] \quad \text{no summation over } i, \tag{7.83}$$

where the dot product means

$$A_i \cdot H \equiv \sum_{I=1}^{16} A_i^I H^I.$$

This leads us to define *the Wilson line shift vectors*

$$a_i^I \equiv \frac{1}{2\pi} \oint A_i^I dx^i \quad \text{no summation.} \tag{7.84}$$

Like the shift vector (7.28), each $a_i$ is a 16-component vector with the gauge index $I$ suppressed. The index $i = 1, \ldots, 6$ runs over the lattice direction defining the torus. The relation (7.81) shows that each component of $a_i^I$ is defined modulo 1.

Let us now associate the Wilson line with such a translation. In effect, the state acquires a phase when we move around $e_i$

$$|P\rangle \to W_i|P\rangle = \exp(2\pi ia_i \cdot P)|P\rangle. \tag{7.85}$$

This is nothing but a shift vector associated with translation

$$(1, e_i) \longrightarrow a_i. \tag{7.86}$$

To be concrete, it is realized as in (3.10),

$$X^m(\sigma + \pi) = X^m(\sigma) + 2\pi Re_i^m, \quad i = 1, \ldots, 6, \tag{7.87}$$

with $e_i^m$ being the $m$th component of the basis vector labelled by $i$. The label means that the vector $e_i$ spans the $i$th lattice and hence torus. Coming back to the original point by (6.205), we associate the effect of Wilson line with the following in the group space

$$X_L^I(\sigma + \pi) = X_L^I(\sigma) + \pi R a_i^I \quad I = 10, \ldots, 25,$$

as in (7.28). This gives rise to the shift in the momentum space

$$P^I \to P^I + 2\pi R a_i^I. \tag{7.88}$$

From now on, again we will suppress the index $I$.

This restricts the boundary condition of the wave function and thus is used to break the gauge group. It is the well-known Hosotani mechanism [10]. Here, we associate the translation with the additional shift vector $a_i$. The shifting does not touch the Cartan subalgebra, so the total rank is not changed.

### Compatibility of Wilson Lines

Wilson lines should be compatible with the lattice. For example, we can show it for the $T^6/\mathbb{Z}_3$ case as

$$\theta e_1 = e_2 \quad \text{implies} \quad a_1 = a_2, \tag{7.89}$$

$$\theta e_2 = -e_1 - e_2 \quad \text{implies} \quad a_2 = -a_1 - a_2, \tag{7.90}$$

where we may relax the equalities between $a_i$'s up to lattice translation. Combining them, we obtain[3]

$$3a_2 = 3a_1 \in \Lambda.$$

That is, we have one independent Wilson line, and it is discrete of order 3. The rest of directions have similar conditions. Denoting the same vector up to lattice translation using the sign "$\approx$," we may denote the conditions as

$$3a_3 \approx 3a_4 \approx 0, \quad 3a_5 \approx 3a_6 \approx 0.$$

There is no more relation relating, e.g. $a_1$ and $a_3$. So, there can exist three independent Wilson lines on $T^6/\mathbb{Z}_3$ orbifold.

This guarantees that each conjugacy class has well-defined shift

$$\left[ \left( \theta^k, v \right) \right] = \left\{ \left( \theta^k, \left( 1 - \theta^k \right) u + \theta^l v \right), \left( \theta^l, u \right) \in \mathsf{S} \right\}, \tag{7.91}$$

---

[3]It is possible that $3a_2 - 3a_1 \in \Lambda$, but there is no observable effect here.

because the Wilson lines along $u$, $v$ directions are, respectively, the same as those of $\theta^k u$, $\theta^l v$ directions. Thus, the actions (7.28) and (7.86) commute and the association (7.92) is well-defined.

We will generalize it to nonprime $N$ in the next chapter. In general, the compatibility also depends on the choice of the lattice. The order of Wilson lines is in general not the same as that of the point group.

## 7.5.2 The Combination: Local Twists at the Fixed Points

We have a combined homomorphism of (7.28) and (7.86) to have

$$\left(\theta^k, \sum_{i=1}^{6} m_i e_i\right) \longrightarrow kV + \sum_{i=1}^{6} m_i a_i, \tag{7.92}$$

with the shift in the momentum space

$$P \rightarrow P + kV + \sum_{i=1}^{6} m_i a_i. \tag{7.93}$$

The conjugacy classes for different fixed points, in (3.30), have different translational parts in general. So far, there is no way to distinguish a specific fixed point. Now, Wilson lines associated with the translations affect the spectrum. The fixed points are *not equivalent and can be distinguished*, improving Condition 3.

How can we obtain the mass shell condition and projection? Recall that each fixed point is parametrized by the space group element $(\theta^k, m_i e^i)$. The converse is true: for every space group element, we have independent fixed points if we consider reducible elements. Thus, the mass shell condition becomes

$$\frac{1}{2}\alpha' M_L^2 = \frac{(P + kV + \sum_i m_i a_i)^2}{2} + \tilde{N} + \tilde{c}^{(k)}, \tag{7.94}$$

noting that $m_i$ parametrize the specific fixed point. We suppress the current algebra index, so $P$, $V$, $a_i$ are 16 dimensional vectors. Again, $\tilde{c}^{(k)}$ is the zero point energy from the internal field oscillators; thus, it remains the same as that without the Wilson line.

Under the space group action, also the state acquires the phase. To fully understand it, we need the complete partition function. Alternatively, we consider the following local picture.

Around each fixed point $(\theta^j, m_i e^i)$, the situation is like we have a noncompact orbifold

$$\mathbb{R}^6/S \simeq \mathbb{R}^6/\mathbb{Z}_n, \quad S = \left\{\left(\theta^j, m_i e^i\right)^k\right\}. \tag{7.95}$$

That is, the point group is generated by $\theta^j$ with the effective order $n$ such that $(\theta^j)^n = 1$ and $k = 0, 1, \ldots, n-1$.

We associate the point group action with the "local shift vector"

$$\left(\theta^j, 0\right) \to V' = kV + \sum_i m_i a_i. \tag{7.96}$$

We have the localized state at this fixed point satisfying the same mass shell condition (7.94) and obtain the phase by replacing $kV \to kV'$

$$\exp 2\pi i \left[ \left(P + kV + \sum_i m_i a_i\right) \cdot \left(kV + \sum_i m_i a_i\right) - (s + k\phi + \rho_R - \rho_L) \cdot \phi \right.$$

$$\left. - \frac{1}{2}\left(kV + \sum_i m_i a_i\right) \cdot \left(kV + \sum_i m_i a_i\right) + \frac{1}{2}k\phi \cdot \phi \right]. \tag{7.97}$$

In the vicinity of this fixed point, we have non-localized "untwisted sector." We have the projection playing the role of (7.40) from (7.97)

$$P \cdot \left(kV + \sum_{i=1}^{6} m_i a_i\right) = \text{integer.} \tag{7.98}$$

With this in mind, we modify the modular invariance condition (7.50) to

$$\phi^2 = \left(kV + \sum m_i a_i\right)^2, \qquad \text{mod } 2/N, \tag{7.99}$$

where the vector indices are suppressed.

Note that all the space group elements $(\theta^j, m_i e^i)^k$ refer to the same fixed point. Conversely, the fixed point $(\theta^j, m_i e^i)$ may be equivalent to another fixed point in the lower-order twisted sector. For instance, in the $T^6/\mathbb{Z}_4$ orbifold we may have $(\theta^2, e_1 + e_2) = (\theta, e_1)$. Then, we regard the state from $(\theta^2, e_1 + e_2)$ as second twisted sector and do not overcount the state. This justifies the point group in (7.95).

### 7.5.3  Projection Conditions in the Bulk

Let us come back to the global geometry $T^6/\mathbb{Z}_N$. The corresponding orbifold theory can be obtained by patching all those theories on the local geometries (7.95) together.

**Gauge Group** The unbroken gauge symmetry in the bulk geometry, that is, in the untwisted sector, is the common intersection of the "untwisted sectors" of the above local models. Imposing the projection condition (7.98) for all $i$ gives the following equivalent conditions for the root vectors $P$:

$$P \cdot V = \text{integer},$$
$$P \cdot a_i = \text{integer}, \quad \text{for all } i. \tag{7.100}$$

Therefore, Wilson lines $a_i$ *further break gauge symmetry from* $V$, improving Condition 2 [11].

**Untwisted Matter** In Sect. 7.4, we have considered in detail the matter fields for the case without Wilson lines. With Wilson lines, any $V + \sum_i m_i a_i$ is a local shift vector. Therefore, for the untwisted matter we require for the weight vectors $P$:

$$P \cdot V \neq \text{integer},$$
$$P \cdot a_i = \text{integer}, \quad \text{for all } i. \tag{7.101}$$

Thus, the matter fields also receive *further projection* from the Wilson lines. It branches the representation of the original theory with $V$ into those of the new theory with $a_i$s, improving Condition 1.

**Modular Invariance** We conclude with another form of the modular invariance equivalent to (7.99)

$$\phi^2 - V^2 = 0 \quad \text{mod } 2/N,$$
$$a_i^2 = 0 \quad \text{mod } 2/N,$$
$$2V \cdot a_i = 0 \quad \text{mod } 2/N, \quad \text{for all } i, \tag{7.102}$$
$$2a_i \cdot a_j = 0 \quad \text{mod } 2/N, \quad \text{for all } i, j \text{ with } i \neq j.$$

This is convenient because independent objects are restricted.

### 7.5.4 $\mathbb{Z}_3$ Example

We construct a toy SU(5) model on a $T^6/\mathbb{Z}_3$ orbifold. With the point group action $\theta$, we associate a shift vector

$$V = \left(-\tfrac{2}{3} \; \tfrac{1}{3} \; \tfrac{1}{3} \; \tfrac{1}{3} \; \tfrac{1}{3} \; 0 \; 0 \; 0\right) \left(-\tfrac{2}{3} \; 0 \; 0 \; 0 \; 0 \; 0 \; 0 \; 0\right)'. \tag{7.103}$$

Although this vector $V$ preserves the SU(5), this shift vector yields the unbroken gauge group SU(9) $\times$ SO(14)$'$ $\times$ U(1)$'$, due to gauge symmetry enhancement that made SO(16) into $E_8$.[4] We need a nonzero entry for the second $E_8'$ direction to satisfy the modular invariance condition.

To break the gauge symmetry further, we turn to a Wilson line along, say, the $e_1$ direction along the first torus, with a shift

$$a_1 = \left(0\,0\,0\,0\,0\,0\,0\,\tfrac{2}{3}\right)\left(0\,\tfrac{1}{3}\,\tfrac{1}{3}\,0\,0\,0\,0\,0\right)' \qquad (7.104)$$

This should be compatible with the point group, $\theta e_1 = e_2$, so we need $a_2 = a_1$. We can verify that these shift vectors satisfy the modular invariance condition (7.102).

The resulting group is

$$\mathrm{SU}(5) \times \mathrm{SU}(2)_1 \times \mathrm{SU}(2)_2 \times \mathrm{U}(1)^2 \times \left[\mathrm{SU}(2) \times \mathrm{SO}(10) \times \mathrm{U}(1)^2\right]',$$

identified by the gauge fields satisfying the conditions in (7.100). We can obtain matter spectrum in the untwisted sector (U) using the rule (7.101). The resulting spectrum is listed on the first row of Table 7.1.

We have three inequivalent twisted sectors by $V$, $V + a_1$, $V - a_1$, which we call, respectively, T0, T1, T2. Each has nine equivalent fixed points. The center of mass of twisted string is at the fixed point. There are three fixed points $\circ$, $\bullet$, $\times$ in the first torus, depicted in Fig. 7.3. By the space group element, they return to the original point, so we can parametrize the fixed point. The origin $\circ$ remains invariant and is parametrized by $(\theta, 0)$. The point $\times$ needs $e_2$ translation after the point group action

$$X_\times^i(\pi) = (\theta X_\times)^i(0) + e_2. \qquad (7.105)$$

So, we associate $(\theta, e_2)$ with the fixed point $\times$. Similarly, we may associate the point $\bullet$ with the element $(\theta, e_1 + e_2)$. The corresponding shift vector is $V + a_1 + a_2$. In this $\mathbb{Z}_3$ case, it is identical to $V + 2a_1 \simeq V - a_1$.

The mass shell condition (7.94) of the twisted sector spectrum is

$$\frac{M_L^2}{4} = \frac{(P + V + m_1 a_1)^2}{2} + \tilde{N} - \frac{3}{2}, \quad m_1 = 0, 1, -1.$$

The zero point energy only depends on the twist $\theta$, not the translations along $e_1$ and $e_2$. All the states have $\tilde{N} = 0$ and survive the generalized GSO projection condition (7.97). They are again tabulated in Table 7.1.

All the matter fields are left-handed, by appropriate charge conjugation. With the help of some singlet fields with nonzero VEVs, this example can yield three complete generations of $\overline{\mathbf{10}}$ and $\mathbf{5}$. But we lack matter field with the adjoint representation $\mathbf{24}$ for breaking SU(5) down to the SM.

---

[4]We will sometimes put a prime $(')$ to distinguish the original group as $E_8 \times E_8'$.

**Table 7.1** Spectrum of the SU(5) model

| Sector | State vector | Representation |
|---|---|---|
| Gauge | $(\underline{1 - 1\, 0^3}\, 0^3)(0^8)$ | $(\mathbf{24}, \mathbf{1}, \mathbf{1}; \mathbf{1}, \mathbf{1})_{0,0,0,0}$ |
| | $\pm(0^5\, 1^2\, 0)(0^8)$ | $(\mathbf{1}, \mathbf{3}, \mathbf{1}; \mathbf{1}, \mathbf{1})_{0,0,0,0}$ |
| | $(0^5\, \underline{1 - 1}\, 0)(0^8)$ | $(\mathbf{1}, \mathbf{1}, \mathbf{3}; \mathbf{1}, \mathbf{1})_{0,0,0,0}$ |
| | $(0^8)(0\, \underline{1 - 1}\, 0^5)$ | $(\mathbf{1}, \mathbf{1}, \mathbf{1}; \mathbf{3}, \mathbf{1})_{0,0,0,0}$ |
| | $(0^8)(0^3\, \pm 1\, \pm 1\, 0^3)$ | $(\mathbf{1}, \mathbf{1}, \mathbf{1}; \mathbf{1}, \mathbf{45})_{0,0,0,0}$ |
| U0 | $(\underline{1\, 0^4\, \pm 1\, 0\, 0})(0^8)$ | $3(\mathbf{5}, \mathbf{2}, \mathbf{2}; \mathbf{1}, \mathbf{1})_{1,0,0,0}$ |
| | $(\underline{-1^2\, 0^3}\, 0^3)(0^8)$ | $3(\overline{\mathbf{10}}, \mathbf{1}, \mathbf{1}; \mathbf{1}, \mathbf{1})_{-2,0,0,0}$ |
| | $(0^8)(\underline{-1\, 0\, 0\, \pm 1\, 0^4})$ | $3(\mathbf{1}, \mathbf{1}, \mathbf{1}; \mathbf{10}, \mathbf{1})_{0,0,0,-1}$ |
| | $(0^8)(-\frac{1}{2}\, \frac{1}{2}\, -\frac{1}{2}[\frac{1}{2}^4\, -\frac{1}{2}])$ | $3(\mathbf{1}, \mathbf{1}, \mathbf{1}; \mathbf{2}, \mathbf{16})_{0,0,0,\frac{1}{2}}$ |
| T0 | $(-\frac{2}{3}\, \frac{1}{3}^4\, 0^3)(0 - \frac{2}{3}\, 0^6)$ | $9(\overline{\mathbf{5}}, \mathbf{1}, \mathbf{1}; \mathbf{1}, \mathbf{1})_{\frac{2}{3},0,0,\frac{2}{3}}$ |
| V | $(-\frac{1}{6}^5\, [\frac{1}{2}\, \frac{1}{2}]\, \frac{1}{2})(-\frac{2}{3}\, 0^7)$ | $9(\mathbf{1}, \mathbf{2}, \mathbf{1}; \mathbf{1}, \mathbf{1})_{-\frac{6}{5},0,-\frac{1}{2},-\frac{2}{3}}$ |
| | $(-\frac{1}{6}^5\, -\frac{1}{2}\, \frac{1}{2}\, -\frac{1}{2})(-\frac{2}{3}\, 0^7)$ | $9(\mathbf{1}, \mathbf{1}, \mathbf{2}; \mathbf{1}, \mathbf{1})_{-\frac{6}{5},0,\frac{1}{2},-\frac{2}{3}}$ |
| T1 | $(\frac{1}{3}^5\, 0^2\, -\frac{1}{3})(-\frac{2}{3}\, -\frac{1}{3}^2\, 0^5)$ | $9(\mathbf{1}, \mathbf{1}, \mathbf{1}; \mathbf{2}, \mathbf{1})_{\frac{5}{3},-\frac{1}{3},-\frac{1}{3},-\frac{1}{3}}$ |
| $V + a_1$ | $(\frac{1}{3}^5\, 0^2\, -\frac{1}{3})(-\frac{2}{3}\, \frac{1}{3}^2\, 0^5)$ | $9(\mathbf{1}, \mathbf{1}, \mathbf{1}; \mathbf{1}, \mathbf{1})_{\frac{5}{3},\frac{2}{3},-\frac{1}{3},\frac{2}{3}}$ |
| | $(-\frac{1}{6}^5\, \frac{1}{2}\, -\frac{1}{2}\, \frac{1}{6})(\frac{1}{3}\, \frac{1}{3}\, -\frac{2}{3}\, 0^5)$ | $9(\mathbf{1}, \mathbf{2}, \mathbf{1}; \mathbf{2}, \mathbf{1})_{-\frac{5}{6},-\frac{1}{3},\frac{1}{6},-\frac{1}{3}}$ |
| | $(-\frac{1}{6}^5\, \frac{1}{2}\, -\frac{1}{2}\, \frac{1}{6})(-\frac{2}{3}\, \frac{1}{3}^2\, 0^5)$ | $9(\mathbf{1}, \mathbf{2}, \mathbf{1}; \mathbf{1}, \mathbf{1})_{-\frac{5}{6},\frac{2}{3},\frac{1}{6},\frac{2}{3}}$ |
| T2 | $(\frac{1}{3}^5\, 0^2\, \frac{1}{3})(\frac{1}{3}\, \underline{-\frac{1}{3}}\, -\frac{2}{3}\, 0^5)$ | $9(\mathbf{1}, \mathbf{1}, \mathbf{1}; \mathbf{2}, \mathbf{1})_{\frac{5}{3},\frac{1}{3},\frac{1}{3},-\frac{1}{3}}$ |
| $V - a_1$ | $(\frac{1}{3}^5\, 0^2\, \frac{1}{3})(-\frac{2}{3}\, -\frac{1}{3}\, -\frac{1}{3}\, 0^5)$ | $9(\mathbf{1}, \mathbf{1}, \mathbf{1}; \mathbf{1}, \mathbf{1})_{\frac{5}{3},-\frac{2}{3},\frac{1}{3},\frac{2}{3}}$ |
| | $(-\frac{1}{6}^5\, [\frac{1}{2}\, \frac{1}{2}]\, -\frac{1}{6})(\frac{1}{3}\, \underline{-\frac{1}{3}}\, \frac{2}{3}\, 0^5)$ | $9(\mathbf{1}, \mathbf{1}, \mathbf{2}; \mathbf{2}, \mathbf{1})_{-\frac{5}{6},\frac{1}{3},-\frac{1}{6},\frac{1}{3}}$ |
| | $(-\frac{1}{6}^5\, [\frac{1}{2}\, \frac{1}{2}]\, -\frac{1}{6})(-\frac{2}{3}\, -\frac{1}{3}\, -\frac{1}{3}\, 0^5)$ | $9(\mathbf{1}, \mathbf{1}, \mathbf{2}; \mathbf{1}, \mathbf{1})_{-\frac{5}{6},-\frac{2}{3},-\frac{1}{6},-\frac{2}{3}}$ |

The underline means permutations and the square bracket [ ] means permutations up to even numbers of minus sign changes. They are all left-handed in four dimensions

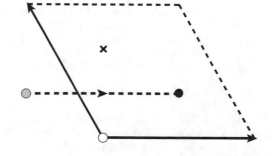

**Fig. 7.3** We may associate a translational element with a Wilson line shift. Taking the closed path, the localized string is affected by the Wilson line along this direction

It is nontrivial check that we have no $SU(5)$ anomaly: the contribution from the twelve **5** is cancelled by those from three $\overline{\textbf{10}}$ from U and nine $\overline{\textbf{5}}$ from T0. We have four U(1) groups. There is a potential anomalous U(1) among them that can be cancelled by the Green–Schwarz mechanism. We will come back to this in Chap. 13.

In the above model, we have obtained three generations of quarks and leptons from the untwisted sector, due to the multiplicity coming from the right mover. In $\mathbb{Z}_3$ orbifold, we have another natural way to have three generations. They are from as many fixed points, if we turn on Wilson lines along two independent directions, say in $e_1$ and $e_3$.

## Exercises

▶ **Exercise 7.1**  Obtain the modular invariance condition (7.50) from the fermionic construction.

▶ **Exercise 7.2**  In the standard embedding in Sect. 7.4.1, we may choose simple roots of SU(3) differently. What other choices are possible? Are they all equivalent? If not, how many inequivalent choices are?

▶ **Exercise 7.3**  We may choose the complex structure $\tau = e^{2\pi i/6}$ for the $A_2$ lattice.

(a)  Show that this is also compatible to the point group action $\theta = e^{2\pi i/3}$.
(b)  Obtain the condition for the Wilson lines associated with this lattice vectors.

▶ **Exercise 7.4**  Show that the second twisted sector of the standard embedding has automatically invariant phase, if the $\mathscr{C}\mathscr{P}\mathscr{T}$ conjugate states in the first twisted sector are invariant. Discuss that we need modular invariance condition for this.

## References

1. L.E. Ibanez, *The Search for a Standard Model SU(3) × SU(2) × U(1) Superstring: An Introduction to Orbifold Constructions* (1987). No. CERN-TH-4769-87
2. A. Font, L.E. Ibanez, H.P. Nilles, F. Quevedo, Degenerate orbifolds. Nucl. Phys. **B307**, 109–129 (1988). [Erratum: Nucl. Phys. **B310**, 764 (1988)]
3. L.J. Dixon, J.A. Harvey, C. Vafa, E. Witten, Strings on orbifolds. Nucl. Phys. **B261**, 678–686 (1985)
4. L.J. Dixon, J.A. Harvey, C. Vafa, E. Witten, Strings on Orbifolds (II) Nucl. Phys. **B274**, 285–314 (1986)
5. J. Fuchs, C. Schweigert, *Symmetries, Lie Algebras and Representations: A Graduate Course for Physicists* (Cambridge University, Cambridge, 2003)
6. A.N. Schellekens, N.P. Warner, Anomalies and modular invariance in string theory. Phys. Lett. **B177**, 317–323 (1986)
7. A.N. Schellekens, N.P. Warner, Anomaly cancellation and selfdual lattices. Phys. Lett. **B181**, 339–343 (1986)
8. A.N. Schellekens, N.P. Warner, Anomalies, characters and strings. Nucl. Phys. **B287**, 317 (1987)

9. M.B. Green, J.H. Schwarz, Anomaly cancellation in supersymmetric D=10 gauge theory and superstring theory. Phys. Lett. **B149**, 117–122 (1984)
10. Y. Hosotani, Dynamical mass generation by compact extra dimensions. Phys. Lett. **B126**, 309–313 (1983)
11. L.E. Ibanez, H.P. Nilles, F. Quevedo, Orbifolds and Wilson lines. Phys. Lett. **B187**, 25–32 (1987)

# Formal Construction

**8**

Modular invariance of the partition function is an important guiding principle of orbifold construction. In the construction of heterotic string in Sect. 6.3.3, we have required *separate* modular invariance in the sixteen internal and the ten dimensional spacetime dimensions. This yielded $E_8 \times E_8$ and $SO(32)$ heterotic strings. If we relax this to be invariance in the *entire 26 dimensions*, we may have a more general theory. One fruitful result was the discovery of $SO(16) \times SO(16)$ heterotic string without supersymmetry [1], but its applicability is much more profound, leading us to a huge class of orbifold models. Furthermore, allowing different orbifolding in the right mover, we can have asymmetric orbifolds.

In this chapter we define a more formal string theory on orbifold. Again the modular invariance is the only guiding principle. We have been focusing on the $T^6/\mathbb{Z}_N$ orbifold with prime $N$, but we can generalize the discussion to non-prime one. The explicit partition function will provide the proof on the orbifold rules presented in the previous chapter. Phases of twisted ground states can be obtained. Using the modular forms we can construct theories that do not admit geometric interpretation. Also we apply this to threshold correction. Many references are available [2–5].

## 8.1 The String Hilbert Space

We also extend the discussion to more general cases, non-prime orbifolds and factor orbifolds: $\mathbb{Z}_N$ with a non-prime $N$ and $\mathbb{Z}_N \times \mathbb{Z}_M$.

### 8.1.1 Considering Modular Invariance

Now we are ready to define orbifold Hilbert spaces and partition functions [6]. Consider a complexified worldsheet boson $Z$. The geometry of orbifold requires

© Springer Nature Switzerland AG 2020

K.-S. Choi, J. E. Kim, *Quarks and Leptons From Orbifolded Superstring*,
Lecture Notes in Physics 954, https://doi.org/10.1007/978-3-030-54005-0_8

for a string to be closed up to space group action $h = (\theta, v) \in S$,

$$Z(\sigma + \pi) = hZ(\sigma) = \theta Z(\sigma) + v. \tag{8.1}$$

It appears that, for every $h$ there is an independent Hilbert space. However, there is redundancy. Because $S$ is non-Abelian group, some different elements are regarded as equivalent. Consider an additional action $g \in S$ on (8.1),

$$gZ(\sigma + \pi) = ghZ(\sigma) = \left(ghg^{-1}\right) gZ(\sigma). \tag{8.2}$$

Since $Z$ and $gZ$ are identified, the twisted state specified by $ghg^{-1}$ should be equivalent to the one specified by $h$. The reader may recall that this is the notion of conjugacy class $[h]$, discussed in Sect. 3.1.3. Therefore, the Hilbert space is specified by the conjugacy class

$$\mathcal{H}_{[h]} = \left\{ Z \mid Z(\sigma + \pi) = hZ(\sigma), h \sim ghg^{-1}, g, h \in S \right\}. \tag{8.3}$$

Given a twisted string by $h$, we keep the invariant states under $g \in S$. They should commute with $h$, otherwise the invariance does not make sense. A set of such commuting elements to $h$ are called centralizer $C(h)$. We are led to define a *generalized GSO projector* making linear combination and an invariant state

$$P_g = \frac{1}{|C(h)|} \sum_{g \in C(h)} g, \tag{8.4}$$

where $|C(h)|$ is the order of the centralizer. By summation, we mean the linear superposition of the quantum states.

Summing up, a modular invariant theory is formed in the following way. We want to consider a given worldsheet theory compactified on orbifold defined by a space group element $h \in S$.

1. Include the $h$-twisted strings for all $h \in S$ as in (8.1). The corresponding conjugacy class defines $h$-twisted sector.
2. For each element $g \in C(h)$, keep only the $g$-invariant states by projection (8.4).

Via the twisted mode expansion (7.5), we have Virasoro generators $\tilde{L}_0(h)$ and $L_0(h)$ made of the fields twisted by $h$. The resulting partition function has the form

$$\mathscr{Z}(\tau, \bar{\tau}) = \sum_{h \in S} \mathscr{Z}_h(\tau, \bar{\tau}) \tag{8.5}$$

$$= \sum_{h \in S} \left[ \frac{1}{|C(h)|} \sum_{g \in S} \text{Tr} \left( g q^{\tilde{L}_0(h) + \tilde{c}} \bar{q}^{L_0(h) + c} \right) \right], \tag{8.6}$$

where

$$q = e^{2\pi i \tau} \tag{8.7}$$

is the modular parameter defined in (6.77). The $g$ gives rise to a phase. This procedure is equivalent to introducing another twisted boundary conditions for the other worldsheet direction as in (7.44)

$$Z\left(\sigma^1 + 2\pi\tau_1, \sigma^0 + 2\pi\tau_2\right) = gZ(\sigma, \tau).$$

Later we will call this sector as $(h, g)$-twisted sector.

For the $\mathbb{Z}_2$ orbifold with the representation $g = -1$, this is similar to giving boundary condition for the worldsheet fermions. The latter gives rise to the spin structure which is projected by GSO projector, which is formally the same as the operator $P_g$. We also have seen that, in the fermionic construction of the current algebra, the condition for even lattice has the same form as the GSO projection. Hence we may call the projection (8.59) *generalized GSO projection*.

Under the modular transformations, the partition function becomes

$$\mathcal{T} : \mathscr{Z}_{(h,g)} \to \mathscr{Z}_{(h,hg)}, \tag{8.8}$$

$$\mathcal{S} : \mathscr{Z}_{(h,g)} \to \mathscr{Z}_{(g,h^{-1})}. \tag{8.9}$$

Thus we can calculate the entire partition function from the chain of modular transformation. Since this chain is closed due to the finiteness of the action of $\mathsf{S}$, by construction, the total partition function $\mathscr{Z}(\tau, \bar{\tau})$ is invariant under the modular transformation. The $h = 1$ case is our definition of the untwisted sector. The summation is illustrated in Fig. 8.1.

In this book, we consider a toroidal orbifold, that is, a torus modded out by a point group $\theta = e^{2\pi i \phi_a J_a} \in \bar{\mathsf{P}}$. Since a point group is Abelian, we have independent Hilbert space for each $\theta^k \in \bar{\mathsf{P}}$ and all the elements belong to the centralizer

**Fig. 8.1** Summing over the spin structure is extended. The phase ambiguity is resolved by the generalized GSO projection $g$

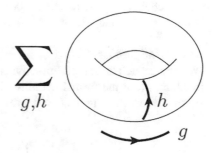

$|C(\theta^k)| = |\mathbb{Z}_N| = N$. Then we mod out the spectrum by the generalized GSO projector

$$P_{\theta^l} = \frac{1}{N} \sum_{l=1}^{N-1} \theta^l. \tag{8.10}$$

The above partition function takes the form

$$\mathscr{Z}(\tau, \bar{\tau}) = \sum_{k=0}^{N-1} \left[ \frac{1}{N} \sum_{l=0}^{N-1} \text{Tr} \left( \theta^l q^{\tilde{L}_0(h)+\tilde{c}} \bar{q}^{L_0(h)+c} \right) \right] \tag{8.11}$$

$$= \frac{1}{N} \sum_{k=0}^{N-1} \sum_{l=0}^{N-1} \mathscr{Z}_{(\theta^k,\theta^l)}(\tau, \bar{\tau}) \tag{8.12}$$

$$= \sum_{k=0}^{N-1} \mathscr{Z}_{\theta^k}(\tau, \bar{\tau}). \tag{8.13}$$

## 8.2    Building Blocks of Partition Functions

With the above tools we may construct a modular invariant partition function.

### 8.2.1    Bosonic String

In this section we will calculate the partition function, which contains all the necessary information on the spectrum. The generalized GSO projection condition will be derived, which contains the vacuum phase in nonstandard embedding. Still, the Jacobi elliptic function is a useful tool [7, 8] for studying such a property. First we consider the simplest modular invariant theory of bosonic string on orbifolds as a toy model. The same building block will be used when we construct the heterotic string partition function.

**Untwisted Sector**
We first calculate the partition function that only includes bosonic degrees of freedom. From the formal definition in the previous section, the untwisted sector states are contained in the $k = 0$ term in (8.13)

$$\mathscr{Z}^X_{\theta^0}(\tau, \bar{\tau}) = \frac{1}{N} \sum_{l=0}^{N-1} \mathscr{Z}^X_{(\theta^0,\theta^l)}(\tau, \bar{\tau}). \tag{8.14}$$

For this, we use the untwisted generator $L_0(1)$ and we project onto a $\theta^l$ invariant state

$$\mathscr{Z}^X_{(\theta^0,\theta^l)}(\tau,\bar{\tau}) = \mathrm{Tr}\left(\theta^l q^{\tilde{L}_0(1)+\tilde{c}} \bar{q}^{L_0(1)+c}\right). \tag{8.15}$$

First, consider one complex dimension $z^a$ and the states generated by the operators from its mode expansion. We have a unique ground state $|0\rangle$ with the zero point energy $-\frac{1}{12}$. If we act an oscillator along this direction, the next ones are $H = -\frac{1}{12} + 1$ states $\tilde{\alpha}^a_{-1}|0\rangle$ and $\tilde{\alpha}^{\bar{a}}_{-1}|0\rangle$, whose $\theta^l$ eigenvalues are, respectively, $e^{2\pi i l\phi_a}$ and $e^{-2\pi i l\phi_a}$, from (7.11). In this way we can construct the states and the corresponding terms in the partition function[1]

$$\mathscr{Z}^a(\tau) = q^{-\frac{2}{24}}\left(1 + q e^{2\pi i l\phi_a} + q e^{-2\pi i l\phi_a} + \dots\right) \tag{8.16}$$

$$= q^{-\frac{2}{24}} \prod_{n=1}^{\infty}\left(1 - q^n e^{2\pi i l\phi_a}\right)^{-1}\left(1 - q^n e^{-2\pi i l\phi_a}\right)^{-1}. \tag{8.17}$$

We can collect these into a modular form, that is, a combination of the $\vartheta$-function in (6.166) and $\eta$-function in (6.84). We use the relation

$$\frac{1}{\eta}\vartheta\begin{bmatrix}1/2\\1/2+l\phi_a\end{bmatrix} = e^{\pi i/2} e^{\pi i l\phi_a} q^{\frac{1}{12}} \prod_{n=1}^{\infty}\left(1 - q^n e^{2\pi i l\phi_a}\right)\left(1 - q^{n-1} e^{-2\pi i l\phi_a}\right)$$

$$= -2\sin(l\pi\phi_a) q^{\frac{1}{12}} \prod_{n=1}^{\infty}\left(1 - q^n e^{2\pi i l\phi_a}\right)\left(1 - q^n e^{-2\pi i l\phi_a}\right). \tag{8.18}$$

In the last line, we factored out the factor containing $q^{n-1}$ to obtain

$$\mathscr{Z}^a(\tau) = -2\sin(l\phi_a\pi) \frac{\eta(\tau)}{\vartheta\begin{bmatrix}\frac{1}{2}\\\frac{1}{2}+l\phi_a\end{bmatrix}(\tau)}. \tag{8.19}$$

---

[1] We denote the holomorphic partition function as $\mathscr{Z}$, antiholomorphic as $\overline{\mathscr{Z}}$ and combined $\mathscr{Z}$.

The contribution for the right mover is just the complex conjugation, $\overline{\mathscr{Z}^{\bar{a}}}(\bar{\tau}) = [\mathscr{Z}^a(\tau)]^*$. Gathering all for the number $d/2$ of complex dimensions, we have

$$
\mathscr{L}^X_{(\theta^0,\theta^l)}(\tau,\bar{\tau}) = \prod_{a=0}^{d/2-1} \mathscr{Z}^a(\tau)\overline{\mathscr{Z}^{\bar{a}}}(\bar{\tau})
$$

$$
= \chi(\theta^l) \left| \prod_{a=0}^{d/2-1} \frac{\eta(\tau)}{\vartheta\begin{bmatrix}\frac{1}{2}\\\frac{1}{2}+l\phi_a\end{bmatrix}(\tau)} \right|^2 . \tag{8.20}
$$

We defined the prefactor that we call *the formal number of fixed points* under the $\theta^l$ twist

$$
\chi(\theta^l) \equiv \prod_{a=0}^{d/2-1} 4\sin^2(l\phi_a\pi), \tag{8.21}
$$

to be justified shortly. The reason we introduced this is the contrary. It is a normalization factor from the definition of theta function, to make the overall coefficient of the partition function (8.16) to be 1, because there is no degeneracy in the untwisted sector. Nevertheless, it will be related to the number of fixed points in the $\theta^l$ twisted torus.

We do not perform orbifolding in the noncompact dimensions, thus the corresponding twist vector component, say $\phi_0$ is zero. This apparently makes $\sin(l\phi_0\pi)$ and hence $\chi(\theta^l)$ vanish. However, the net partition function (8.19) does *not* depend on $\sin(l\phi_0\pi)$, as seen in Eq. (8.18). Thus it is finite in the limit $l\phi_a = 0$,

$$
\lim_{l\phi_0\to 0} \frac{-2\sin(l\pi\phi_0)\eta(\tau)}{\vartheta\begin{bmatrix}\frac{1}{2}\\\frac{1}{2}+l\phi_0\end{bmatrix}(\tau)} = \frac{1}{\eta^2(\tau)}, \tag{8.22}
$$

so that Eq. (8.19) reduces to the case without orbifolding, as discussed in Sect. 6.1.4. This argument also holds true if a certain direction is invariant under the twist $l\phi_a \equiv 0 \mod 1$. Also the uncancelled, remaining prefactor $\chi(\theta^l)$ in (8.21) automatically takes into account only the orbifolded directions. Note that this does not see whether the space is compact or not, so in the noncompact case we need to put the factor from continuous momentum $2\pi\sqrt{\tau_2}$ by hand.

We define *the number of fixed points* under the twist $\theta^l$ as

$$
\tilde{\chi}(\theta^l) = \prod_a 4\sin^2(l\phi_a\pi), \tag{8.23}
$$

where the index $a$ runs over non-fixed tori.

## Twisted Sectors

We can derive the partition functions of the twisted sectors from successive modular transformations of the untwisted ones. This does not only prove the requirement of the twisted states but also fix the normalization coefficients, which explains the multiplicities of states.

Let us apply the modular transformation. The transformation $\mathscr{S}$ takes $(1, \theta^l)$ states to those of $(\theta^l, 1)$, as in (8.9). The latter belongs to the $\theta^l$-twisted sector. Indeed

$$
\mathscr{L}^X_{(\theta^l, \theta^0)}(\tau, \bar{\tau}) = \chi\left(\theta^l\right) \left| \prod_{a=0}^{3} \frac{\eta(\tau)}{\vartheta\left[\begin{matrix} \frac{1}{2} + l\phi_a \\ \frac{1}{2} \end{matrix}\right](\tau)} \right|^2
$$

$$
= \chi\left(\theta^l\right) q^{\tilde{c}} \bar{q}^c \left| \prod_{a=0}^{3} \prod_{n=1}^{\infty} \left(1 - q^{n+l\phi_a}\right)^{-1} \left(1 - q^{n-1-l\phi_a}\right)^{-1} \right|^2.
$$

(8.24)

For convenience we fix the physical dimension $d = 8$ from now on. The definition in terms of partition function on the last line fits well with the previous one, Eq. (8.20). The zero point energy $\tilde{c} = c$ is modified to the correct value

$$
c = -\frac{4}{24} + \frac{1}{4} \sum_{a=0}^{3} (l\phi_a)(1 - l\phi_a),
$$

as in (7.15), according to the regularization (7.14) (up to subtraction by integers in $l\phi_a$). The factor $\sqrt{-i\tau}$ is common to eta and theta functions so that they cancel. Expanding the above equation further, we have

$$
\mathscr{L}^X_{(\theta^l, \theta^0)}(\tau, \bar{\tau})
$$

(8.25)

$$
= \chi\left(\theta^l\right) q^{\tilde{c}} \bar{q}^c \left(1 + q^{l\phi_a} + q^{1-l\phi_a} + \dots\right) \left(1 + \bar{q}^{1-l\phi_b} + \bar{q}^{l\phi_b} + \dots\right).
$$

Every term for any $n$ in (8.24) surves, with the overall coefficient $\chi(\theta^l)$. Therefore, we have the degeneracy factor coinciding with the number of fixed points (3.82). In the untwisted sector, this factor set the overall degeneracy to 1.

Further transformations give any twisted states with any projections. Applying $\mathscr{T}$ transformation $m$ times to the $(\theta^k, 1)$ state, we obtain

$$
\mathscr{L}^X_{(\theta^k, \theta^0)}(\tau + m, \bar{\tau} + m) = \mathscr{L}^X_{(\theta^k, \theta^{mk})}(\tau, \bar{\tau}).
$$

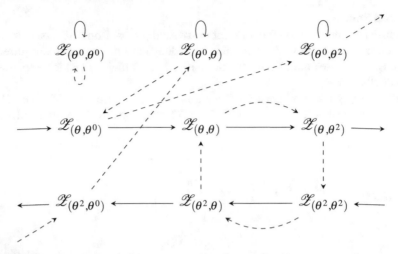

**Fig. 8.2** Modular transformation of the $\mathbb{Z}_3$ orbifold partition functions. Solid line denotes $\mathscr{T}$ and dashed $\mathscr{S}$. It is drawn on a "torus" so that, for instance, $\mathscr{T}$ takes $\mathscr{Z}_{(\theta,\theta^2)}$ to $\mathscr{Z}_{(\theta,\theta^0)}$ and $\mathscr{S}$ takes $\mathscr{Z}_{(\theta^0,\theta^2)}$ to $\mathscr{Z}_{(\theta^2,\theta^0)}$. Except $(\theta^0, \theta^0)$ sector, every sector is connected

In the particular case $k = 1$, we have $\mathscr{Z}^X_{(\theta,\theta^m)}(\tau,\bar\tau)$. We can easily see from its expansion in $q$, each field has the multiplicity $\chi(\theta)$. The complete twisted sectors and their transformations are shown in Fig. 8.2.

We can obtain the $\mathscr{Z}^X_{(\theta^2,\theta)}$ function from $\mathscr{Z}_{(\theta,1)}$ from (8.24) by successive transformations $\mathscr{S}, \mathscr{T}$,

$$\mathscr{Z}^X_{(\theta^2,\theta)}(\tau,\bar\tau) = \chi(\theta) \left| \prod_{a=0}^{3} \frac{\eta(\tau)}{\vartheta\begin{bmatrix} \frac{1}{2} + 2\phi_a \\ \frac{1}{2} + \phi_a \end{bmatrix}(\tau)} \right|^2. \tag{8.26}$$

Although the corresponding states belong to the second twisted sector, the prefactor is not $\chi(\theta^2)$, but $\chi(\theta)$ from that of the first twisted sector. The number of fixed points $\chi(\theta^2)$ is always a multiple of $\chi(\theta)$, because the latter counts the number of fixed points under the rotation of a smaller angle. Thus the latter counts the number of *simultaneous* fixed point under $\theta^2$ and $\theta$. We express this as

$$\chi(\theta) \equiv \chi\left(\theta^2, \theta\right).$$

We can show that the twisted sectors $(\theta^k, \theta^l)$ are related to the above by chain of transformations, if $k$ and $l$ are relative prime. The prefactor becomes

$$\chi(\theta) = \chi\left(\theta^k, \theta^l\right) \quad \gcd(k,l) = 1.$$

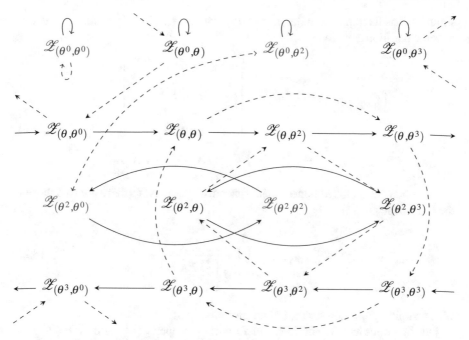

**Fig. 8.3** Modular transformation of the $\mathbb{Z}_4$ orbifold partition functions, in the same format as Fig. 8.2. There are three connected sectors, containing $(\theta^0, \theta^0)$, $(\theta^1, \theta^0)$, $(\theta^2, \theta^0)$ sharing the same coefficients $\chi(\theta^k, \theta^l)$

For example, in the $\mathbb{Z}_4$ orbifold, the component for $(\theta^2, \theta^2)$ cannot be obtained by the above chain of the modular transformations (see Fig. 8.3). It is only obtained from $\mathscr{Z}_{(\theta^0, \theta^2)}$, having the factor $\chi(\theta^2)$, by $\mathscr{S}$ and $\mathscr{T}$ transformations. Thus we have

$$\chi\left(\theta^2\right) = \chi\left(\theta^2, \theta^2\right).$$

Definitely, it is the "simultaneous" number of fixed points under $\theta^2$ and $\theta^2$.

All the twisted partition functions related by modular transformation have the same coefficients. Generalizing above, for $(\theta^k, \theta^l)$-twisted sector is related to $(\theta^0, \theta^m)$ with $m = \gcd(k, l)$. Thus $\chi(\theta^k, \theta^l) = \chi(\theta^m)$ is the number of simultaneous fixed points under $\theta^k$ and $\theta^l$ twists [3,4]. Therefore, the most general bosonic partition function is

$$
\mathscr{L}^X_{(\theta^k,\theta^l)}(\tau,\bar{\tau}) = \chi\left(\theta^k,\theta^l\right) \left| \prod_{a=0}^{3} \frac{\eta(\tau)}{\vartheta\left[\begin{array}{c} \frac{1}{2}+k\phi_a \\ \frac{1}{2}+l\phi_a \end{array}\right](\tau)} \right|^2 \tag{8.27}
$$

$$
= \chi\left(\theta^k,\theta^l\right) \mathscr{L}^X_{(\theta^k,\theta^l)}(\tau)\overline{\mathscr{L}}^X_{(\theta^k,\theta^l)}(\bar{\tau}).
$$

Note that we included noncompact spacetime degrees $X$ (there are two more in the light-cone gauge).

$$
\mathscr{L}^X_{(\theta^k,\theta^l)}(\tau) = \prod_{a=0}^{3} \frac{\eta(\tau)}{\vartheta\left[\begin{array}{c} \frac{1}{2}+k\phi_a \\ \frac{1}{2}+l\phi_a \end{array}\right](\tau)}, \tag{8.28}
$$

with a similar definition for the antiholomorphic part.

Our functions have the desired transformation property (8.8) and (8.9) thanks to

$$
\mathscr{T}: \mathscr{L}^X_{(\theta^k,\theta^l)}(\tau+1) = e^{\pi i[(k\phi)^2+\frac{1}{3}]}\mathscr{L}^X_{(\theta^k,\theta^{k+l})}(\tau), \tag{8.29}
$$

$$
\mathscr{S}: \mathscr{L}^X_{(\theta^k,\theta^l)}\left(-\frac{1}{\tau}\right) = e^{-2\pi ikl\phi^2}\mathscr{L}^X_{(\theta^l,\theta^{-k})}(\tau). \tag{8.30}
$$

Since they give rise to only phases, the complete $\mathscr{L}^X_{\theta^k}(\tau)$, containing only the combinations $|\mathscr{L}^X_{(\theta^k,\theta^l)}|^2$, is invariant.

### The Number of Simultaneously Fixed Points

Since $\theta^0$ means no rotation, all the fixed points $\chi(\theta^k)$ are by default the common fixed points, $\chi(\theta^0,\theta^k) = \chi(\theta^k)$. We have seen that

$$
\chi\left(\theta^k,\theta^0\right) = \chi\left(\theta^0,\theta^k\right) = \prod_{a} 4\sin^2(k\phi_a\pi). \tag{8.31}
$$

Under $\mathscr{S}$, the following are related

$$
\chi\left(\theta^k,\theta^l\right) = \chi\left(\theta^{N-k},\theta^l\right) = \chi\left(\theta^k,\theta^{N-l}\right) = \chi\left(\theta^{N-k},\theta^{N-l}\right), \tag{8.32}
$$

where we used $\theta^N = 1$. Thus we can limit $k$ and $l$ to lie in the range $1, \ldots, [N/2]$. In this range, we find

$$\chi\left(\theta^k, \theta^l\right) = \chi\left(\theta^m\right) = \prod_a 4 \sin^2(m\phi_a \pi), \quad m = \gcd(k, l). \tag{8.33}$$

This is again a formal expression. We have seen that, if we have fixed torus we have formally $\sin(k\phi_a \pi) = 0$ but this is to be cancelled by the same contribution from the theta function in the denominator.

## 8.2.2 Current Algebra

For the partition function for the current algebra, we also require modular form. A lesson from the first definition of theta function in (6.166) is that the lattice shift by $\frac{1}{2}$ in (6.217) is realized as a shift in the argument of the theta function. The shift vector $V$ plays the same role as the twisted sector. We assume that it is given by heterotic string before orifolding; it is $\Gamma_8 \oplus \Gamma_8$ for $E_8 \times E_8$ and $\Gamma_{16}$ for $SO(32)$. We consider a gauge group $G$ of the rank $r$. The corresponding partition function is

$$\mathscr{Z}^G_{(kV, lV)}(\tau) = \frac{1}{2} \sum_{\alpha, \beta} \prod_{I=1}^r \eta_{kl}^{\alpha\beta} \frac{1}{\eta(\tau)} \vartheta \begin{bmatrix} \alpha + kV_I \\ \beta + lV_I \end{bmatrix} (\tau). \tag{8.34}$$

Here, $\alpha$ and $\beta$ can assume 0 (R states) and $\frac{1}{2}$ (NS states), as seen in (6.176)–(6.179). First, consider the bosonic expression using the momentum $P$,

$$\mathscr{Z}^G_{(V, W)}(\tau) = \sum q^{\frac{1}{2}(P+V)^2 + \tilde{N} - \frac{16}{24}} e^{2\pi i (P+V) \cdot W}, \tag{8.35}$$

we need

$$\mathscr{Z}^G_{(V, W)}(\tau) = \frac{1}{2\eta^8} \sum_{\{n_I\}} q^{\frac{1}{2} \sum (n_I + V_I)^2} e^{2\pi i \sum (n_I + V_I) W_I} \left(1 + (-1)^{\sum n_I}\right)$$

$$+ \frac{1}{2\eta^8} \sum_{\{n_I\}} q^{\frac{1}{2} \sum (n_I + \frac{1}{2} + V_I)^2} e^{2\pi i \sum (n_I + \frac{1}{2} + V_I) W_I} \left(1 + (-1)^{\sum n_I}\right), \tag{8.36}$$

with sum over all the possible integers, $\{n_I \ (I = 1, \cdots, r)\}$.

However, in the definition of theta function,

$$\vartheta \begin{bmatrix} \alpha + V_I \\ \beta + W_I \end{bmatrix} (\tau) = q^{\frac{1}{2}(n_I + V_I + \alpha)^2} e^{(n_I + V_I + \alpha)(\beta + W_I)},$$

for each term with $\beta = \frac{1}{2}$, we have a unwanted factor

$$(-1)^{\sum(V_I+\alpha)} = e^{\pi i \sum(V_I+\alpha)} = e^{2\pi i \beta(\sum V_I+\alpha)}.$$

Since $\alpha = 1/2$ if we have the partition function as a product of $4k(k \in \mathbb{Z})$ theta functions, we have no $\alpha$ dependence. Using the last form, we can put this factor for the $\beta = 0$ case. Setting

$$\eta_{kl}^{\alpha\beta} = e^{-2\pi i \alpha\beta k V_I} \tag{8.37}$$

removes them. Also, $r$ is the rank of the gauge group: $r = 8$ for E$_8$ and $r = 16$ for Spin(32)/$\mathbb{Z}_2$. For E$_8 \times$E$_8$, we make the convention

$$\mathscr{Z}_{(kV,lV)}^G(\tau) = \left( \frac{1}{2} \sum_{\alpha,\beta} \eta_{kl}^{\alpha\beta} \prod_{I=1}^{8} \frac{1}{\eta} \vartheta \begin{bmatrix} \alpha+kV_I \\ \beta+lV_I \end{bmatrix} \right) \left( \frac{1}{2} \sum_{\alpha,\beta} \eta_{kl}^{\alpha\beta} \prod_{I=9}^{16} \frac{1}{\eta} \vartheta \begin{bmatrix} \alpha+kV_I \\ \beta+lV_I \end{bmatrix} \right) (\tau), \tag{8.38}$$

so that the partition function behave in the same way for both the gauge groups, using the same rank 16 shift vector.

The structure is best understood in the sum form of the theta function

$$\mathscr{Z}_{(kV,lV)}^G(\tau) = \sum_{\tilde{N}} \sum_{P \in \Lambda_G} q^{\frac{1}{2}(P+kV)^2 + \tilde{N} - \frac{r}{24}} e^{2\pi i(P+kV)\cdot(lV)}, \tag{8.39}$$

where $\Lambda_G$ is the root lattice of the gauge group. The function (8.34) transforms as

$$\mathscr{T} : \mathscr{Z}_{(kV,lV)}^G(\tau + 1) = e^{-\pi i((kV)^2 + \frac{1}{3})} \mathscr{Z}_{(kV,(k+l)V)}^G(\tau), \tag{8.40}$$

$$\mathscr{S} : \mathscr{Z}_{(kV,lV)}^G \left( -\frac{1}{\tau} \right) = e^{2\pi i klV^2} \mathscr{Z}_{(lV,-kV)}^G(\tau). \tag{8.41}$$

For the oscillator contribution, we used

$$\frac{\sum_P q^{\frac{1}{2}P^2}}{q^{\frac{r}{24}}(1-q)^{16}} = \sum_{\tilde{N}} \sum_P q^{\frac{1}{2}P^2 + \tilde{N} - 1}.$$

The $q^{-\frac{r}{24}}$ comes from the contribution of the untwisted zero point energy, $-\frac{1}{24}$ for each bosonic degrees of freedom. Although it is not possible to factor out the oscillator on RHS, the infinite sum can be. Each factor $(1-q)^{-1} = 1+q+q^2+\cdots$ provides the oscillator contributions in the corresponding Cartan subalgebra.

### 8.2.3  Right-Moving Fermions

There are worldsheet fermions which are superpartners of the right-moving bosons. Their partition function is obtained as in Eqs. (6.164) and (6.165), with the twist vector

$$\overline{\mathscr{Z}}^{\psi}_{(\theta^k,\theta^l)}(\bar{\tau}) = \frac{1}{2} \sum_{\alpha,\beta} e^{-2\pi i \alpha \beta} \prod_{a=0}^{3} \frac{1}{\eta^*(\bar{\tau})} \vartheta \begin{bmatrix} \alpha + k\phi_a \\ \beta + l\phi_a \end{bmatrix}^* (\bar{\tau}), \tag{8.42}$$

with the spin structures. Note the essentially same form of the group partition function (8.34). The gauge sector admits a description in the fermions. The worldsheet fermions in the right mover can also be equivalently described by bosons. The transformation can be suggestively expressed as

$$\mathscr{T} : \overline{\mathscr{Z}}^{\psi}_{(\theta^k,\theta^l)}(\bar{\tau}+1) = e^{\pi i (k^2 \phi^2 + \frac{d}{24})} \overline{\mathscr{Z}}^{\psi}_{(\theta^k,\theta^{k+l})}(\bar{\tau}), \tag{8.43}$$

$$\mathscr{S} : \overline{\mathscr{Z}}^{\psi}_{(\theta^k,\theta^l)} \left(-\frac{1}{\bar{\tau}}\right) = e^{-2\pi i k l \phi^2} \overline{\mathscr{Z}}^{\psi}_{(\theta^l,\theta^{-k})}(\bar{\tau}). \tag{8.44}$$

## 8.3  Heterotic String

Now we combine all the above partition functions to form the that of heterotic string on orbifold [9–11]. We derive the modular invariance condition and the generalized GSO projector.

### 8.3.1  The Full Partition Function of Heterotic String

The full partition function of heterotic string on orbifold is obtained by combining all the above expressions,

$$\mathscr{Z}(\tau,\bar{\tau}) = \frac{1}{N} \sum_{k=0}^{N-1} \sum_{l=0}^{N-1} \mathscr{Z}_{(\theta^k,\theta^l)}(\tau,\bar{\tau}), \tag{8.45}$$

$$\mathscr{Z}_{(\theta^k,\theta^l)}(\tau,\bar{\tau}) = \eta_{k,l} \chi\left(\theta^k,\theta^l\right) \mathscr{Z}^{G}_{(\theta^k,\theta^l)}(\tau) \mathscr{Z}^{X}_{(\theta^k,\theta^l)}(\tau) \overline{\mathscr{Z}}^{X}_{(\theta^k,\theta^l)}(\bar{\tau}) \overline{\mathscr{Z}}^{\psi}_{(\theta^k,\theta^l)}(\bar{\tau}). \tag{8.46}$$

That is, the partition function is still decomposed into those of $(\theta^k, \theta^l)$-sectors.

Under the modular transformation, we have

$$\mathscr{T} : \mathscr{Z}_{(\theta^k,\theta^l)}(\tau+1, \bar{\tau}+1) = e^{-\pi i k^2 (V^2 - \phi^2)} \mathscr{Z}_{(\theta^k,\theta^{k+l})}(\tau,\bar{\tau}), \tag{8.47}$$

$$\mathscr{S} : \mathscr{L}_{(\theta^k, \theta^l)} \left( -\frac{1}{\tau}, -\frac{1}{\bar{\tau}} \right) = e^{2\pi i k l (V^2 - \phi^2)} \mathscr{L}_{(\theta^l, \theta^{-k})}(\tau, \bar{\tau}). \tag{8.48}$$

As a result, each term of different $(k, l)$ acquires a different phase (8.48). The overall partition function is not guaranteed to be invariant. The only way to remove it is to have the phase satisfying

$$\eta_{kl} e^{-\pi i k^2 (V^2 - \phi^2)} = \eta_{k(k+l)},$$
$$\eta_{kl} e^{2\pi i k l (V^2 - \phi^2)} = \eta_{l(-k)}. \tag{8.49}$$

One can check that a simple solution is

$$\eta_{kl} = e^{-\pi i k l (V^2 - \phi^2)}. \tag{8.50}$$

This means the ground state in the $k$-th twisted sector is linear combined as

$$|\sigma_k\rangle = \sum_l e^{-\pi i k l (V^2 - \phi^2)} |\sigma_{kl}\rangle \tag{8.51}$$

and has a "vacuum phase"

$$\theta |\sigma_k\rangle = e^{\pi i k (V^2 - \phi^2)} |\sigma_k\rangle. \tag{8.52}$$

This cannot be seen in the operator formalism, discussed in Chap. 7. Especially in the case of nonstandard embedding, this phase is nontrivial.

This is a necessary condition for the shape of the partition function. This fixes the relative phases with respect to another. However, the entire partition function still acquires a nontrivial phase and we need a more condition. Successive transformations of $\mathscr{T}$ by $N$ times in (7.46) change the boundary condition to

$$\mathscr{T}^N : (h, g) \rightarrow \left( h, h^N g \right) = (h, g), \tag{8.53}$$

returning to the original one. From the phase (8.47), invariance requires for the corresponding partition function to be well-defined for arbitrary $k$, we should have

$$N \sum_{I=1}^{16} V_I^2 - N \sum_{a=0}^{3} \phi_a^2 \equiv 0, \qquad \text{mod } 2, \tag{8.54}$$

reproducing the modular invariance condition (7.50). If we fermionize the gauge coordinates, the latter condition for un-orbifolded case $N = 1$ reproduces the GSO projection. We have no nontrivial contribution under $\mathscr{S}$ transformation, because $\mathscr{S}^2$ cancels the overall phase.

This condition *relates two quantities from the different spaces,* one $V$ from the current algebra and the other $\phi$ from spacetime. Also this condition miraculously relates the anomaly cancellation from different contributions: the information on the multiplicity and the number of fixed points are contained in $\mathscr{L}^X$ and the information on the chiral representations is contained in $\mathscr{L}^G \mathscr{L}^\psi$ in Eq. (8.46).

## Expecting Asymmetric Orbifold

We have been considering symmetric orbifold, for which we applied the same twist in the left and right movers. It is evident that the bosonic part of the partition function

$$\mathscr{L}^X_{(\theta^k,\theta^l)}(\tau)\,\overline{\mathscr{L}}^X_{(\theta^k,\theta^l)}(\bar\tau) = \left| \mathscr{L}^X_{(\theta^k,\theta^l)}(\tau) \right|^2$$

does not give rise to a phase under modular transformation. However, the gauge and fermionic parts

$$\mathscr{L}^G_{(kV,lV)}(\tau)\,\overline{\mathscr{L}}^\psi_{(\theta^k,\theta^l)}(\bar\tau)$$

are not automatically invariant. The modular invariance condition (8.54) makes this and the full partition functions invariant. Thus the condition connects *holomorphic and antiholomorphic.* The $(k,l)$ twisted sectors are transformed in the same way for every degrees of freedom, $\mathscr{S}$ invariance is universal.

Note that

$$\mathscr{L}^X_{(\theta^k,\theta^l)} \quad \text{and} \quad \overline{\mathscr{L}}^\psi_{(\theta^k,\theta^l)}(\bar\tau) \tag{8.55}$$

transform exactly in the same way under modular transformation. This is because, the former has the theta functions $\vartheta\begin{bmatrix} 1/2+k\phi \\ 1/2+l\phi \end{bmatrix}$ in the denominator and the latter has its complex conjugates in the numerator.

So, alternatively, we can construct an orbifold theory by relaxing the common left and right-moving bosonic states so that they are asymmetrical.

$$\mathscr{L}_{(\theta^k,\theta^l)}(\tau,\bar\tau) = \eta_{k,l}\,\mathscr{L}^L_{(\theta^k,\theta^l)}(\tau)\,\overline{\mathscr{L}}^R_{(\theta^k,\theta^l)}(\bar\tau)$$

$$\mathscr{L}^L_{(\theta^k,\theta^l)}(\tau) = \mathscr{L}^G_{(\theta^k,\theta^l)}(\tau)\,\mathscr{L}^X_{(\theta^k,\theta^l)}(\tau) \tag{8.56}$$

$$\overline{\mathscr{L}}^R_{(\theta^k,\theta^l)}(\bar\tau) = \overline{\mathscr{L}}^X_{(\theta^k,\theta^l)}(\bar\tau)\,\overline{\mathscr{L}}^\psi_{(\theta^k,\theta^l)}(\bar\tau). \tag{8.57}$$

By requiring independent modular invariance on $\mathscr{L}^L_{(\theta^k,\theta^l)}(\tau)$ and $\overline{\mathscr{L}}^R_{(\theta^k,\theta^l)}(\bar\tau)$. In this case, we may have different twists on the different movers but we do not have geometrical interpretation of the space. We will construct asymmetric orbifold in Sect. 12.5.

## 8.3.2   Generalized GSO Projections

Due to orbifolding, a state survives if it is invariant under the space group action. Without Wilson lines, it should be invariant under $\theta$ projection. Or, a linear combination of more than one states can. In the latter case, the multiplicity of the states may not be coincident with the number of fixed points. We have constructed the partition function by inserting the projection operator. Thus, we may extract invariant states under the orbifold from the partition function.

We extract the spectrum in the $\theta^k$ twisted sector that is invariant under $\theta$. In the absence of Wilson lines, we expand the partition function into the twisted sectors

$$\mathscr{L}_{\theta^k}(\tau, \bar{\tau}) = \mathrm{Tr}\left( P_{\theta^k} q^{\tilde{L}_0(\theta^k) + \tilde{c}} \bar{q}^{L_0(\theta^k) + c} \right). \tag{8.58}$$

For each state, normalized as in (8.27), the phase sum becomes *projection operator*,

$$P_{\theta^k} = \frac{1}{N} \sum_{l=0}^{N-1} \tilde{\chi}\left(\theta^k, \theta^l\right) \Delta_{(\theta^k, \theta^l)}. \tag{8.59}$$

The phase operator is given by

$$\Delta_{(\theta^k, \theta^l)} = \left(\Delta_{\theta^k}\right)^l, \tag{8.60}$$

where

$$\Delta_{\theta^k} = \exp 2\pi i \left[ (P + kV) \cdot V - (s + k\phi + \rho_R - \rho_L) \cdot \phi - \frac{k}{2}\left(V^2 - \phi^2\right) \right] \tag{8.61}$$

and $\rho_L$ and $\rho_R$ are the oscillator numbers of the holomorphic and antiholomorphic left movers in the internal group space, respectively, defined in (7.12). We included vacuum phase Eq. (8.50) as well. We will come back to cases with Wilson lines, in the next section.

The coefficient in Eq. (8.59) is given as

$$\tilde{\chi}\left(\theta^k, \theta^l\right) = \begin{cases} \chi(\theta^k, \theta^l) & [\chi(\theta^k) \neq 0], \\ \chi(\theta^k, \theta^l) \big/ \prod_a 4\sin^2(k\phi_a\pi) & [\chi(\theta^k) = 0, l \neq 0], \\ \prod_b 4\sin^2(l\phi_b\pi) & [\chi(\theta^k) = 0, l = 0], \end{cases} \tag{8.62}$$

which is essentially the same as $\chi(\theta^k, \theta^l)$ in (8.33), as the first line shows. Depending on the twisted sector, we need a correction because there is a cancellation when $k\phi_a$ is an integer. For example, in the case of $\mathbb{Z}_4$ orbifold $2\phi = \frac{1}{2}(2\ 1\ 1)$, the component $2\phi_1$ is an integer. Under this twist the corresponding direction is intact and has a fixed torus. The corresponding factor $4\sin^2(k\phi_1\pi)$ formally vanishes, but is cancelled by the same factor in the $\vartheta$-function in the denominator, as in (8.18). It is not a coincidence that the denominator contains the same factor because it is obtained by the $\mathscr{S}$ transformation which contained exactly the same factor $\chi(\theta)$. The resulting overall factor is

$$\prod_{a=2}^{3} 4\sin^2(\phi_a\pi) = \frac{\prod_{a=1}^{3} 4\sin^2(\phi_a\pi)}{\prod_{b=1}^{1} 4\sin^2(\phi_b\pi)} = \frac{\chi\left(\theta^k,\theta^l\right)}{\prod_{b=1}^{1} 4\sin^2(\phi_b\pi)}, \qquad (8.63)$$

where $a$ does not run over the invariant directions, including $a = 1$ (it runs orver $a = 2, 3$). Thus we remove the corresponding factor as in the second line, in which the product index $a$ runs over *vanishing* factor. Recall that this resulted in $\tilde{\chi}(1,\theta^l) = 1$ in the untwisted sector. In the last line, we have the number of fixed points $\chi(\theta^k)$ with the same kind of correction, letting $b$ runs over *nonvanishing* factors. Thus we may interpret as

$\tilde{\chi}(\theta^k,\theta^l) =$ (the number of simultaneous fixed points under $\theta^k$ and $\theta^l$, on twisted tori by $\theta^k$).

Note the following asymmetry $\tilde{\chi}(\theta^k,\theta^l) \neq \tilde{\chi}(\theta^l,\theta^k)$ unlike the symmetry of $\chi(\theta^k,\theta^l)$. For the prime orbifolds $\mathbb{Z}_3$ and $\mathbb{Z}_7$, simply $\tilde{\chi}(\theta^k,\theta^l) = \chi(\theta^k,\theta^l) = \chi(\theta)$. For the first twisted sector of the general orbifold, we have $\tilde{\chi}(\theta,\theta^l) = \chi(\theta,\theta^l) = \chi(\theta)$. These numbers are most frequently used in real calculation, so we tabulated them in Appendix A.

The role of the generalized GSO projector is twofold. First, like in the GSO projector, some states satisfying the mass shell condition may not be invariant under this projector hence should be removed. Also, the multiplicity of the states are determined by its coefficients. This is nontrivial in the non-prime orbifold. For example, in the $\mathbb{Z}_4$ orbifold, the partition functions of $(\theta^2, 1)$, $(\theta^2, \theta^2)$, and $(1, \theta^2)$ are shuffled by modular transformations. Only particular linear combinations are invariant under $\theta$. We shall meet further examples in detail in the next chapter.

### 8.3.3 Partition Function of the $\mathbb{Z}_3$ Orbifold

We construct the partition function for the $\mathbb{Z}_3$ orbifold. We can see explicitly how it transforms under the modular transformations and extract the generalized GSO projection. For simplicity, we assume the standard embedding.

The partition function for the $(\theta^k, \theta^l)$-twisted sector is decomposed as

$$\mathscr{Z}_{(\theta^k,\theta^l)}(\tau,\bar{\tau}) = \mathscr{Z}^{E_8}_{(kV,lV)}(\tau)\mathscr{Z}^{E_8'}_{(kV,lV)}(\tau)\mathscr{Z}^X_{(\theta^k,\theta^l)}(\tau)\overline{\mathscr{Z}}^X_{(\theta^k,\theta^l)}(\bar{\tau})\overline{\mathscr{Z}}^\psi_{(\theta^k,\theta^l)}(\bar{\tau}),$$

where

$$\mathscr{L}^{E_8}_{(kV,lV)}(\tau) \tag{8.64}$$

$$= \sum_{\alpha,\beta} e^{-2\pi i \alpha \mathbb{I} \cdot (\beta \mathbb{I} + \phi)} \frac{\vartheta \begin{bmatrix} \alpha+kV_1 \\ \beta+\frac{2}{3}l \end{bmatrix} \vartheta \begin{bmatrix} \alpha+kV_2 \\ \beta+\frac{1}{3}l \end{bmatrix} \vartheta \begin{bmatrix} \alpha+kV_3 \\ \beta+\frac{1}{3}l \end{bmatrix} \vartheta \begin{bmatrix} \alpha+kV_4 \\ \beta \end{bmatrix} \cdots \vartheta \begin{bmatrix} \alpha+kV_8 \\ \beta \end{bmatrix}}{2\eta^8},$$

$$\mathscr{L}^{E_8'}_{(kV,lV)}(\tau) = \mathscr{L}^{E_8}_{(0,0)}(\tau), \tag{8.65}$$

$$\mathscr{L}^{X}_{(\theta^k,\theta^l)}(\tau) = \left(-\sin\frac{\pi}{3}\right)^3 \frac{\eta}{\vartheta \begin{bmatrix} \frac{1}{2}+\frac{2}{3}k \\ \frac{1}{2}+\frac{2}{3}l \end{bmatrix} \vartheta \begin{bmatrix} \frac{1}{2}+\frac{1}{3}k \\ \frac{1}{2}+\frac{1}{3}l \end{bmatrix} \vartheta \begin{bmatrix} \frac{1}{2}+\frac{1}{3}k \\ \frac{1}{2}+\frac{1}{3}l \end{bmatrix}}, \tag{8.66}$$

$$\overline{\mathscr{L}}^{X}_{(\theta^k,\theta^l)}(\bar\tau) = \left[\mathscr{L}^{X}_{(\theta^k,\theta^l)}(\tau)\right]^*, \tag{8.67}$$

$$\overline{\mathscr{L}}^{\psi}_{(\theta^k,\theta^l)}(\bar\tau) = \sum_{\alpha,\beta} e^{2\pi i \alpha \mathbb{I} \cdot (\beta \mathbb{I} + \phi)} \frac{\vartheta \begin{bmatrix} \alpha+\frac{2}{3}k \\ \beta+\frac{2}{3}l \end{bmatrix}^* \vartheta \begin{bmatrix} \alpha+\frac{1}{3}k \\ \beta+\frac{1}{3}l \end{bmatrix}^* \vartheta \begin{bmatrix} \alpha+\frac{1}{3}k \\ \beta+\frac{1}{3}l \end{bmatrix}^* \vartheta \begin{bmatrix} \alpha \\ \beta \end{bmatrix}^*}{2\eta^{*4}}, \tag{8.68}$$

where we used Eqs. (8.34) and (8.27) and $\mathbb{I} = (1, 1, 1, 1)$. The second $E_8$ is not shifted so we use the original partition function in (6.242). There is no factor $\eta_{0,1}$ so that the ground state is invariant under the twist.

## Untwisted Sector
Let us analyze the untwisted sector $k = 0$. The most comprehensible form is when we represent $\mathscr{L}^G$ as the sum and the rest as the product representations

$$\mathscr{L}_{(\theta^0,\theta^l)}(\tau) = \frac{\sum_{P \in \Gamma_8 \oplus \Gamma_8, \tilde{N}} q^{\frac{1}{2}P^2 + \tilde{N}} e^{2\pi i P \cdot (lV)}}{q \prod_{n=1}^{\infty} (1 - q^n)^2 (1 - \alpha^l q^n)^3 (1 - \alpha^{2l} q^n)^3}, \tag{8.69}$$

$$\mathscr{L}_{(\theta^0,\theta^l)}(\bar\tau) = \frac{\sum_{s,N} \bar{q}^{s^2/2} e^{-2\pi i s \cdot (l\phi)}}{\bar{q}^{1/2} \prod_n (1 - \bar{q}^n)^2 (1 - \alpha^{2l} \bar{q}^n)^3 (1 - \alpha^l \bar{q}^n)^3}, \tag{8.70}$$

where $\alpha = e^{2\pi i/3}$. Here $P$s belong to $E_8 \times E_8$ lattice $\Gamma_8 \oplus \Gamma_8$. The overall factor $q^{-1}$ comes from 24 bosonic degrees of freedom and contributes to the zero point energy. Also the overall factor $q^{-1/2}$ comes from 8 bosonic and fermionic degrees of freedom.

The numerator shows the mass shell condition with the projection and the denominator shows the oscillator contributions from the eight physical spacetime dimensions in the light-cone gauge. The factor $(-2\sin\frac{\pi}{3})^3$ is cancelled by the prefactor in $\vartheta\begin{bmatrix} 1/2 \\ 1/2+l\phi_a \end{bmatrix}$ in (8.27).

We are interested in the massless states, which are provided by $P^2 = 2, s^2 = 1$. The denominator also contribute because the leading the expansion is

$$\frac{1}{q \prod_n (1 - q^n)^2 \left(1 - \alpha^l q^n\right)^3 \left(1 - \alpha^{2l} q^n\right)^3} = q^{-1} + \left(2 + 3\alpha^l + 3\alpha^{2l}\right) + \mathcal{O}(q).$$

We can extract the generalized GSO projector

$$P_{\theta^0} = \frac{1}{3} \sum_{l=0}^{2} \left(\Delta_{(\theta^0)}\right)^l,$$

with

$$\Delta_{(\theta^0)} = e^{2\pi i[\tilde{N} + P \cdot V - s \cdot \phi]}.$$

The only way to have nonvanishing projection is $\Delta_{(\theta^0)} = 1$, then all the three terms have the unit phase giving overall multiplicity $P_{\theta^0} = 1$. Otherwise $P_{\theta^0} = \frac{1}{3}(1 + \alpha + \alpha^2) = 0$ showing that they are projected out.

The massless states are counted by the coefficients of $q^0$, which satisfy $P^2 = 2$ or $P^2 = 0$ with one excited oscillator $\tilde{N} = 1$. They are subject to the projection condition

$$P \cdot V - s \cdot \phi = 0 \mod 1. \tag{8.71}$$

Once this is satisfied, then all the $(\theta, \theta^l)$ satisfy the invariance condition.

Gauge bosons satisfy $P \cdot V \equiv 0$. To satisfy (8.71) we also need $s \cdot \phi = 0$. The right movers contributing to $q^0$ term in the expansion $\prod(1 - \bar{q}^n)^2 = (1 + 2\bar{q} + \dots)$. The contribution is $\frac{1}{3} + \frac{1}{3} + \frac{1}{3} = 1$ for each helicity.

We have also nontrivial contribution from matter states from $P \cdot V \equiv \frac{1}{3} \mod 1$ and $s \cdot \phi = \frac{1}{3} \mod 1$, so that $P_{\theta^0} = 1$. The $\mathscr{CPT}$ conjugates come from $P \cdot V \equiv \frac{2}{3}$ mod 1 and $s \cdot \phi = \frac{1}{3}$ mod 1.

The current algebra component can be expressed as a product representation,

$$\mathscr{Z}^{E_8}_{(0,lV)}(\tau) = \sum_{\alpha, \beta} q^{-\frac{1}{3}} \prod_{I=1}^{8} \prod_n \left(1 + q^{n+\alpha-\frac{1}{2}} e^{2\pi i(\beta + lV_I)}\right) \left(1 + q^{n-\alpha-\frac{1}{2}} e^{-2\pi i(\beta + lV_I)}\right). \tag{8.72}$$

Each product over $I$ is contributed from a complexified Weyl fermions, thus we have 16 fermions for each $E_8$. The spin structure then corresponds NS and R sectors. Each combines to the right mover with nontrivial phase.

**Twisted Sector**

Here, the multiplicity 27 is not cancelled so that with the right mover we have 27 "fixed points." In the asymmetric orbifold, this multiplicity is not correlated with the right movers.

Again, we represent $\mathscr{L}^G$ in the sum and the rest in the product representations

$$e^{-\pi i l V^2}\, \mathscr{L}_{(\theta,\theta^l)}(\tau) = -3\sqrt{3}\, \frac{\sum_{P,\tilde{N}} q^{(P+V)^2/2+\tilde{N}-\frac{2}{3}} e^{2\pi i l[(P+V)\cdot V - V^2/2]}}{\prod_n (1-q^n)^2 \left(1-q^{n-1/3}\alpha^{2l}\right)^3 \left(1-q^{n-2/3}\alpha^l\right)^3}$$

$$e^{\pi i l \phi^2}\, \mathscr{L}_{(\theta,\theta^l)}(\bar{\tau}) = -3\sqrt{3}\, \frac{\sum_{s,N} q^{(s+\phi)^2/2-\frac{1}{6}} e^{2\pi i l[(s+\phi)\cdot\phi - \phi^2/2]}}{\prod_n (1-\bar{q}^n)^2 \left(1-\bar{q}^{n-1/3}\alpha^l\right)^3 \left(1-\bar{q}^{n-2/3}\alpha^{2l}\right)^3}.$$

From the above equation we briefly show how to read off the twisted sector mass shell condition $\frac{1}{2}(P+V)^2+\tilde{N}-\tilde{c}$ given in (7.73), as well as the projection condition

$$P_\theta = \frac{1}{3}\sum_{l=0}^{2} 27\left(\Delta_{\theta 0}\right)^l,$$

with

$$\left(\Delta_{\theta 0}\right)^l = e^{2\pi i l\left[\tilde{N}+(P+V)\cdot V - (s+\phi)\cdot\phi - \frac{1}{2}V^2 + \frac{1}{2}\phi^2\right]}.$$

The denominator has the expansion

$$q^{-2/3}\left(1 + 3q^{1/3}\alpha + 9q^{2/3}\alpha^2 + \cdots\right).$$

Besides the zero point energy factor $q^{-2/3}$ giving $\tilde{c} = -\frac{2}{3}$, these terms correspond to 3 oscillators of $\tilde{N} = \frac{1}{3}$ and 9 oscillators of $\tilde{N} = \frac{2}{3}$, as shown in Eq. (7.78). In the product expression of the $\vartheta$ function, the power of $q$ corresponds to the oscillator number $\tilde{N}$ and their coefficient is the multiplicity times the phase. The leading term 1 is projected out since we have no left movers with the same mass.

We have the following equivalent summation expression, viz. Eq. (6.166),

$$\mathscr{L}_{(\theta,\theta^l)}(\tau) = -3\sqrt{3}\, \frac{\sum_P q^{(P+V)^2/2-\frac{3}{2}} e^{2\pi i(P+V)\cdot(lV)}}{\sum_s q^{(s+\phi)^2/2-\frac{1}{2}} e^{2\pi i(s+\phi)\cdot(l\phi)}}. \tag{8.73}$$

The bosonic left mover has a fixed vector $s = (-1,0,0,0)$. Now we can read off the phase of a state as $\exp[(P+V)\cdot lV - (s+\phi)\cdot l\phi]$. With the overall phase $\eta_{kl}$ we can obtain the vacuum phase (8.52).

## Modular Invariance

Let us check modular invariance. Under $\mathcal{T}, \mathcal{S}$, those with different spin structure and twists are shuffled. For the first, each $\vartheta[^{1/2}_{l\phi_a}]$ and $\vartheta[^{1/2}_{+1/2+l\phi_a}]$ acquires a phase $e^{\pi i/4}$, while $\vartheta[^0_\beta]$ and $\vartheta[^0_{\beta+1/2}]$ acquire no extra phases. It is cancelled due to the power 8 of the theta functions. This is not the case for the denominator giving rise to $(e^{-i\pi/4})^3$ of the theta functions is cancelled by the phase of eta functions $(e^{-\pi i/12})^{15}$.[2] Under $\mathcal{S}$, (6.173), we learn that $\vartheta/\eta$ combination does not yield nontrivial $\sqrt{-i\tau}$.

Then we see that $\mathcal{T}$ takes $(\theta^k, \theta^l)$-twisted sectors to $(\theta^k, \theta^{k+l})$ sectors, as promised. Under $\mathcal{S}$, two arguments in the theta function are interchanged, exchanging $(1, \theta^l)$ and $(\theta^{-l}, 1)$ twisted sectors. The latter is equivalent to that of $(\theta^{3-l}, 1)$. This is displayed in Fig. 8.2.

## Appendix

Modular forms are building blocks of the modular invariant partition functions. We deal with modular transformations of the theta functions with general shifts. We need to consider spin structure, which divides the twisted sectors into four subsectors.

Under $\mathcal{T}$, the theta functions transform as

$$\vartheta \begin{bmatrix} kV_I \\ lV_I \end{bmatrix} / \eta \rightarrow e^{-\pi i \left[ (kV_I)^2 + kV_I + \frac{1}{12} \right]} \vartheta \begin{bmatrix} kV_I \\ \frac{1}{2} + (k+l)V_I \end{bmatrix} / \eta, \tag{8.74}$$

$$\vartheta \begin{bmatrix} kV_I \\ \frac{1}{2} + lV_I \end{bmatrix} / \eta \rightarrow e^{-\pi i \left[ (kV_I)^2 - kV_I + \frac{1}{12} \right]} \vartheta \begin{bmatrix} kV_I \\ (k+l)V_I \end{bmatrix} / \eta, \tag{8.75}$$

$$\vartheta \begin{bmatrix} \frac{1}{2} + kV_I \\ lV_I \end{bmatrix} / \eta \rightarrow e^{-\pi i \left[ (kV_I)^2 - \frac{1}{4} + \frac{1}{12} \right]} \vartheta \begin{bmatrix} \frac{1}{2} + kV_I \\ (k+l)V_I \end{bmatrix} / \eta, \tag{8.76}$$

$$\vartheta \begin{bmatrix} \frac{1}{2} + kV_I \\ \frac{1}{2} + lV_I \end{bmatrix} / \eta \rightarrow e^{-\pi i \left[ (kV_I)^2 - \frac{1}{4} + \frac{1}{12} \right]} \vartheta \begin{bmatrix} \frac{1}{2} + kV_I \\ \frac{1}{2} + (k+l)V_I \end{bmatrix} / \eta. \tag{8.77}$$

The first is accompanied by extra transformation

$$\vartheta \begin{bmatrix} \alpha \\ \beta \end{bmatrix} = e^{-2\pi i \alpha} \vartheta \begin{bmatrix} \alpha \\ \beta+1 \end{bmatrix}$$

Under $\mathcal{S}$, we have

$$\vartheta \begin{bmatrix} kV_I \\ lV_I \end{bmatrix} / \eta \rightarrow e^{2\pi i (kV_I)(lV_I)} \vartheta \begin{bmatrix} lV_I \\ -kV_I \end{bmatrix} / \eta, \tag{8.78}$$

---

[2]In the denominator there are seven $\eta$s times extra eight $\eta$s from $\mathcal{Z}^{E_8}$.

$$\vartheta \begin{bmatrix} kV_I \\ \frac{1}{2}+lV_I \end{bmatrix} /\eta \rightarrow e^{2\pi i(kV_I)(lV_I+\frac{1}{2})} \vartheta \begin{bmatrix} \frac{1}{2}+lV_I \\ -kV_I \end{bmatrix} /\eta, \tag{8.79}$$

$$\vartheta \begin{bmatrix} \frac{1}{2}+kV_I \\ lV_I \end{bmatrix} /\eta \rightarrow e^{2\pi i(kV_I-\frac{1}{2})(lV_I)} \vartheta \begin{bmatrix} lV_I \\ \frac{1}{2}-kV_I \end{bmatrix} /\eta, \tag{8.80}$$

$$\vartheta \begin{bmatrix} \frac{1}{2}+kV_I \\ \frac{1}{2}+lV_I \end{bmatrix} /\eta \rightarrow e^{2\pi i(kV_I-\frac{1}{2})(lV_I+\frac{1}{2})} \vartheta \begin{bmatrix} \frac{1}{2}+lV_I \\ \frac{1}{2}-kV_I \end{bmatrix} /\eta. \tag{8.81}$$

So we understand how the states with different spin structures are shuffled.

The third and the fourth transforms exactly same as the bosonic. To have uniform transformations yielding the same phase, the number of products should be a multiple of four. This is satisfied by rank $8k$ gauge current algebra, including $\Gamma_8$ and $\Gamma_{16}$. For spacetime fermions, one complexified fermion is counted by the above function so we need $16k$ dimensions.

## References

1. L. Alvarez-Gaume, P.H. Ginsparg, G.W. Moore, C. Vafa, An O(16) x O(16) Heterotic string. Phys. Lett. **B171**, 155–162 (1986)
2. A. Font, L.E. Ibanez, H.P. Nilles, F. Quevedo, Degenerate orbifolds. Nucl. Phys. **B307**, 109–129 (1988). [Erratum: Nucl. Phys. **B310**,764 (1988)]
3. L.E. Ibanez, J. Mas, H.-P. Nilles, F. Quevedo, Heterotic strings in symmetric and asymmetric orbifold backgrounds. Nucl. Phys. **B301**, 157–196 (1988)
4. A. Font, S. Theisen, Introduction to string compactification. Lect. Notes Phys. **668**, 101–181 (2005)
5. Y. Katsuki, Y. Kawamura, T. Kobayashi, N. Ohtsubo, Y. Ono, K. Tanioka, Z(N) orbifold models. Nucl. Phys. **B341**, 611–640 (1990)
6. L.J. Dixon, J.A. Harvey, C. Vafa, E. Witten, Strings on orbifolds (II). Nucl. Phys. **B274**, 285–314 (1986)
7. D. Mumford, in *Tata Lectures on Theta. I (Modern Birkhäuser classics)*. Progress in Mathematics, vol. 28. (Birkhäuser Boston-Basel-Stuttgart, Birkhäuser, 1984)
8. E.T. Whittaker, G.N. Watson, *A Course of Modern Analysis* (Cambridge University, Cambridge, 1927/1996)
9. I. Senda, A. Sugamoto, Orbifold models and modular transformation. Nucl. Phys. **B302**, 291 (1988)
10. J.A. Minahan, One loop amplitudes on orbifolds and the renormalization of coupling constants. Nucl. Phys. **B298**, 36–74 (1988)
11. S.G. Nibbelink, M. Laidlaw, Stringy profiles of gauge field tadpoles near orbifold singularities 1. Heterotic string calculations. JHEP **01**, 004 (2004)

# Non-prime Orbifolds

**9**

Armed with the tools from the partition functions that we considered in the previous chapter, we are now ready to deal with most general toroidal orbifolds. We consider the so-called non-prime orbifold, that is, $T^6/\mathbb{Z}_N$ orbifold with $N$ non-prime or $T^6/(\mathbb{Z}_N \times \mathbb{Z}_M)$.

Unlike prime-order cases, we have non-triviality in the higher twisted sectors. One is from the geometry of fixed points of different orders and the physics of GSO projections. Also, we have more than one choice of lattices, yielding different condition for the Wilson lines. We calculate spectrum of some model, using both projector and conjugacy classes. Finally we consider conditions on Wilson lines and the generalized GSO projectors. The resulting spectrum is rich enough to build realistic models.

## 9.1    The Geometry of Non-prime Orbifold

After carefully examining $T^6/\mathbb{Z}_4$ orbifold with $SU(4) \times SU(4)$ lattice, we consider the standard embedding on it. It shall show how the twisted fields are organized.

### 9.1.1    $T^6/\mathbb{Z}_4$ Orbifold

Novel features in the $T^6/\mathbb{Z}_N$ orbifolds with non-prime $N$, or non-prime orbifold include

- There can be different choices of the twist for the same order $N$.
- The orbifold with the same point group allows for different lattices.
- The fixed points of $\theta^k$-twisted sector $k > 1$ are not invariant under $\theta$ defining the orbifold.
- We may have a fixed torus, whose direction is invariant under $\theta^k$.

© Springer Nature Switzerland AG 2020

K.-S. Choi, J. E. Kim, *Quarks and Leptons From Orbifolded Superstring*, Lecture Notes in Physics 954, https://doi.org/10.1007/978-3-030-54005-0_9

As a prototypical example of non-prime orbifold, we consider $T^6/\mathbb{Z}_4$ orbifolds specified by a twist vector

$$\phi = \frac{1}{4}(2\ 1\ 1). \tag{9.1}$$

It has 16 fixed points in the $\theta$-twisted sector and 16 fixed points in the $\theta^2$-twisted sector not counting those points in the former. This is topological property fixed by the point group of (9.1). Their actual location depends on the choice of the lattice. There are three lattices compatible to the $\mathbb{Z}_4$ action as shown in Table 3.4. Among them, we discuss two lattices $SU(2)^2 \times SO(5)^2$, $SU(4) \times SU(4)$ in detail.

## $SU(2)^2 \times SO(5)^2$ Lattice
First consider the "most orthogonal" Coxeter lattice

$$SU(2) \times SU(2) \times SO(5) \times SO(5),$$

which is depicted in Fig. 9.1, with the fundamental region shaded.

The action of $\theta$ is intuitively understood as $\pi$, $\pi/2$, $\pi/2$ rotations on the two-tori, as the $\phi$ shows. We take the basis $e_i$, $i = 1, \ldots, 6$ satisfying the relation

$$\theta e^1 = -e^1, \quad \theta e^2 = -e^2,$$
$$\theta e^3 = e^4, \quad \theta e^4 = -e^3, \tag{9.2}$$
$$\theta e^5 = e^6, \quad \theta e^6 = -e^5.$$

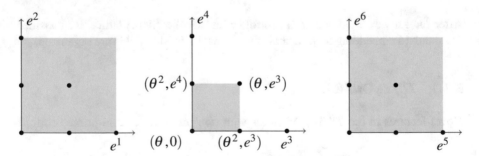

**Fig. 9.1** $T^6/\mathbb{Z}_4$ orbifold with the $SU(2)^2 \times SO(5) \times SO(5)$ lattice. We have relations $(\theta^2, 0) = (\theta, 0)^2$, $(\theta^2, e^3 + e^4) = (\theta, e^3)^2$. Shaded is one possible choice of the fundamental region

This can be expressed by a matrix

$$
\theta = \begin{pmatrix}
-1 & & & & & \\
& -1 & & & & \\
& & -1 & & & \\
& & 1 & & & \\
& & & & -1 & \\
& & & & 1 &
\end{pmatrix}, \tag{9.3}
$$

where the empty entries are zero.

To see the representation of $\theta$ in the $e^i$ basis, we may complexify the basis

$$
\begin{aligned}
e^{\underline{1}} &= e^1 + iU_1 e^2, \\
e^{\underline{2}} &= e^3 + iU_2 e^4, \\
e^{\underline{3}} &= e^5 + iU_3 e^6.
\end{aligned} \tag{9.4}
$$

The SU(2) lattices are essentially one dimensional, so the corresponding complex structure $iU_1$ can be any. The relations (9.2) fix $iU_2 = iU_3 = i$. The above definition fixes the form of the metric

$$
G^{ij} = e^i \cdot e^j = \begin{pmatrix}
A & * & * & * & * & * \\
* & B & * & * & * & * \\
* & * & C & 0 & * & * \\
* & * & 0 & C & * & * \\
* & * & * & * & D & 0 \\
* & * & * & * & 0 & D
\end{pmatrix}, \quad i, j = 1, 2, \ldots, 6,
$$

while starred entries unfixed, as long as the metric is symmetric $G_{ij} = G_{ji}$.

Consider the first twisted sector. We have $\det(1 - \theta) = 16$ fixed points

$$
\begin{aligned}
&(\theta, 0), & &(\theta, e^1), & &(\theta, e^2), & &(\theta, e^1 + e^2), \\
&(\theta, e^3), & &(\theta, e^1 + e^3), & &(\theta, e^2 + e^3), & &(\theta, e^1 + e^2 + e^3), \\
&(\theta, e^5), & &(\theta, e^1 + e^5), & &(\theta, e^2 + e^5), & &(\theta, e^1 + e^2 + e^5), \\
&(\theta, e^3 + e^5), & &(\theta, e^1 + e^3 + e^5), & &(\theta, e^2 + e^3 + e^5), & &(\theta, e^1 + e^2 + e^3 + e^5).
\end{aligned} \tag{9.5}
$$

We see that along the $e_1, e_2$ direction, we have two fixed points and on each of the remaining torus, we have two.

We consider the second twisted sector by $\theta^2$. Since the (12) direction becomes fixed torus, we have $\det(1 - \theta^2) = 0$. Removing this $2 \times 2$ block and letting the resulting matrix $\theta'$, we have $\det(1 - \theta'^2) = 16$ showing that we have sixteen fixed points in the (3456) plane.

The classes (9.5) are equivalently expressed as the classes in the $\theta^2$ twist

$$\left(\theta^2, 0\right), \ \left(\theta^2, e^3 + e^4\right), \ \left(\theta^2, e^5 + e^6\right), \ \left(\theta^2, e^3 + e^4 + e^5 + e^6\right) \qquad (9.6)$$

are invariant under the $\theta$ as well. For example,

$$\left(\theta, e^3\right)^2 = \left(\theta^2, \theta e^3 + e^3\right) = \left(\theta^2, \theta^3 + \theta^4\right). \qquad (9.7)$$

The remaining 12 points are not $\theta$-invariant, but there are pairs always related by $\theta \in \overline{\mathsf{P}}$. For instance,

$$(\theta, 0) \left(\theta^2, e_3\right) (\theta, 0)^{-1} = \left(\theta^2, e^4\right),$$

$$(\theta, e_3) \left(\theta^2, e_4\right) (\theta, e_3)^{-1} = \left(\theta^2, e^3\right). \qquad (9.8)$$

So they belong to the same conjugacy class. We have six conjugacy classes

$$\left[(\theta^2, e^3)\right], \left[(\theta^2, e^5)\right], \left[(\theta^2, e^3 + e^5 + e^6)\right],$$

$$\left[(\theta^2, e^3 + e^4 + e^5)\right], \left[(\theta^2, e^3 + e^5)\right], \left[(\theta^2, e^4 + e^5)\right] \qquad (9.9)$$

There are some degrees of redundancy as in (9.7). The last entry $(\theta^2, e^4 + e^5) \simeq (\theta^2, e^3 + e^6)$ is not redundant to the others. Each conjugacy class forms order two orbit.

In general, each conjugacy defines the centralizer of the space group $\mathsf{S}$ to the generalized point group element $\theta \in \overline{\mathsf{P}}$

$$C(\theta) = \left\{ [(\theta^2, v)] \mid (\theta, u)(\theta^2, v)(\theta, u)^{-1} = (\theta^2, v), \exists u \in \Lambda \right\}. \qquad (9.10)$$

**SU(4) × SU(4) Lattice**

The SU(4) × SU(4) lattice has the representation

$$\theta = \begin{pmatrix} & -1 & & & & \\ 1 & -1 & & & & \\ & 1 & -1 & & & \\ & & & -1 & & \\ & & & 1 & -1 & \\ & & & & 1 & -1 \end{pmatrix} \qquad (9.11)$$

We name the basis vectors $\alpha^i$, since they are identical to the root vectors and we need to distinguish them from those of SU(2) × SO(5) lattice. However they are not normalized so that the metric becomes

$$G^{ij} = \alpha^i \cdot \alpha^j = \begin{pmatrix} 2A & -A & 0 & * & * & * \\ -A & 2A & -A & * & * & * \\ 0 & -A & 2A & * & * & * \\ * & * & * & 2B & -B & 0 \\ * & * & * & -B & 2B & -B \\ * & * & * & 0 & -B & 2B \end{pmatrix}$$

with the starred entries unfixed, as long as $G_{ij} = G_{ji}$. Note that two diagonal $3 \times 3$ block is proportional to the SU(4) Cartan matrix.

Let us focus on the first SU(4) lattice forming $T^3$. The bases are the simple roots $\alpha^1, \alpha^2, \alpha^3$ satisfying the relations

$$\theta\alpha^1 = \alpha^2, \tag{9.12}$$

$$\theta\alpha^2 = \alpha^3, \tag{9.13}$$

$$\theta\alpha^3 = \alpha^0 \equiv -\alpha^1 - \alpha^2 - \alpha^3, \tag{9.14}$$

$$\theta\alpha^0 = \alpha^1, \tag{9.15}$$

showing $\theta^4 = 1$. The basis vectors are not orthogonal. Here $\alpha^0$ in (9.14) is defined from the lowest root. These four roots are schematically displayed in Fig. 9.2. They are all identified on the resulting orbifold.

We have again $\det(1 - \theta) = 16$ invariant fixed points, because each $T^3$ contains 4 fixed points. The first three-torus that we consider has four fixed points

$$0, \quad \frac{1}{4}\left(3\alpha^1 + 2\alpha^2 + \alpha^3\right), \quad \frac{1}{2}\left(\alpha^1 + \alpha^3\right), \quad \frac{1}{4}\left(\alpha^1 + 2\alpha^2 + 3\alpha^3\right).$$

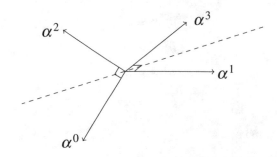

**Fig. 9.2** The three simple roots of SU(4) spanning $T^3$ compatible with the $\mathbb{Z}_4$ twist $\theta$. With the lowest roots $\alpha^0 = -\alpha^1 - \alpha^2 - \alpha^3$ the vectors become more symmetric under the rotation by $\theta$. The designated angles are $\pi/2$ and all the others are $2\pi/3$. Under the twist $\theta^2$, the dashed line becomes invariant and forms a fixed circle

The symmetry becomes clearer if we denote them by conjugacy classes

$$[(\theta, 0)] : \theta 0 = 0, \tag{9.16}$$

$$\left[(\theta, \alpha^1)\right] : \theta \frac{1}{4}\left(3\alpha^1 + 2\alpha^2 + \alpha^3\right) = \frac{1}{4}\left(-\alpha^1 + 2\alpha^2 + \alpha^3\right), \tag{9.17}$$

$$\left[(\theta, \alpha^1 + \alpha^3)\right] : \theta \frac{1}{2}\left(\alpha^1 + \alpha^3\right) = \frac{1}{2}(\alpha^2 + \alpha^0) = -\frac{1}{2}\left(\alpha^1 + \alpha^3\right), \tag{9.18}$$

$$\left[(\theta, \alpha^1 + \alpha^2 + \alpha^3)\right] : \theta \frac{1}{4}\left(\alpha^1 + 2\alpha^2 + 3\alpha^3\right) = \frac{1}{4}\left(-3\alpha^1 - 2\alpha^2 - \alpha^3\right). \tag{9.19}$$

We may also show

$$\left(\theta, \alpha^2\right) \in \left[(\theta, \alpha^1)\right], \quad \left(\theta, \alpha^1 + \alpha^2\right) \in \left[(\theta, \alpha^1 + \alpha^3)\right],$$

which reflect the identification (9.12)–(9.15).

Noting that $\theta$ rotates $\alpha^1 + \alpha^3$ by $\pi$

$$\theta\left(\alpha^1 + \alpha^3\right) = \alpha^2 + \alpha^0 = -\left(\alpha^1 + \alpha^3\right),$$

in fact 0 and $\frac{1}{2}(\alpha^1 + \alpha^3)$ are also invariant under $\mathbb{Z}_2$.

Taking the highest representation, $\frac{1}{2}(\alpha^1 + 2\alpha^2 + \alpha^3)$, a successive subtraction by $\alpha^2, \alpha^1, \alpha^3$ reproduces the fundamental weight of the SU(4), however, two of them are identified by the twist. With equal four root vectors in the second three-torus, we have $4 \times 4 = 16$ fixed points in total.

We have $\theta^2$-twisted sector making use of the following:

$$\theta^2 \alpha^1 = \alpha^3, \tag{9.20}$$

$$\theta^2 \alpha^2 = \alpha^0, \tag{9.21}$$

$$\theta^2 \alpha^3 = \alpha^1, \tag{9.22}$$

$$\theta^2 \alpha^0 = \alpha^2. \tag{9.23}$$

We note that the combination

$$\theta^2 \left(\lambda(\alpha^1 + \alpha^3)\right) = \lambda\left(\alpha^1 + \alpha^3\right), \tag{9.24}$$

is invariant under $\theta^2$, for any $\lambda \in \mathbb{R}$. In the lattice they are redundant so we may limit the range as $0 \leq \lambda < 1$. The space $S : \lambda(\alpha^1 + \alpha^3)$ spans a *fixed circle*. In effect we form $T^3/\mathbb{Z}_2 \simeq T^2/\mathbb{Z}_2 \times S$ orbifold by identifying the opposite points with respect to the axis $S$. For the whole $T^6/\mathbb{Z}_4$ orbifold, $S^2 \simeq T^2$ becomes a fixed torus.

Returning back to the previous $T^3/\mathbb{Z}_2$, we have also four fixed points

$$\left(\theta^2, 0\right) : \theta^2 0 = 0, \tag{9.25}$$

$$\left(\theta^2, \alpha^1 + \alpha^2\right) : \theta^2 \frac{1}{2}\left(\alpha^1 + \alpha^2\right) = \frac{1}{2}\left(\alpha^3 + \alpha^0\right) = -\frac{1}{2}\left(\alpha^1 + \alpha^2\right), \tag{9.26}$$

$$\left(\theta^2, \alpha^2 + \alpha^3\right) : \theta^2 \frac{1}{2}\left(\alpha^2 + \alpha^3\right) = \frac{1}{2}\left(\alpha^0 + \alpha^1\right) = -\frac{1}{2}\left(\alpha^2 + \alpha^3\right), \tag{9.27}$$

$$\left(\theta^2, \alpha^1 + 2\alpha^2 + \alpha^3\right) : \theta^2 \frac{1}{2}\left(\alpha^1 + 2\alpha^2 + \alpha^3\right) = \frac{1}{2}\left(\alpha^3 + 2\alpha^0 + \alpha^1\right)$$

$$= -\frac{1}{2}\left(\alpha^1 + 2\alpha^2 + \alpha^3\right). \tag{9.28}$$

Since $\frac{1}{2}(\alpha^1 + \alpha^3)$ is invariant, no lattice translation is necessary and the corresponding point belongs to the conjugacy class $[(\theta^2, 0)]$. We can also check that $\frac{1}{2}(\alpha^2 - \alpha^1)$ belongs to the conjugacy class $[(\theta^2, \alpha^2 + \alpha^3)]$. We see $(\theta, \alpha^1)^2 = (\theta^2, \theta\alpha^1 + \alpha^1) = (\theta^2, \alpha^2 + \alpha^1)$, so that $(\theta, \alpha^1)$ and $(\theta^2, \alpha^1 + \alpha^2)$ are the same fixed points. Lastly, we have only two points in the fundamental region, while others are outside; they are in the fundamental region of the $T^3/\mathbb{Z}_2$ orbifold, though.

The two $\mathbb{Z}_2$ fixed points are not invariant under $\theta$ but they are exchanged by it

$$\theta \frac{1}{2}\left(\alpha^1 + \alpha^2\right) = \frac{1}{2}\left(\alpha^2 + \alpha^3\right),$$

$$\theta \frac{1}{2}\left(\alpha^2 + \alpha^3\right) = \frac{1}{2}\left(\alpha^3 + \alpha^0\right) = -\frac{1}{2}\left(\alpha^1 + \alpha^2\right), \tag{9.29}$$

possibly accompanied by a lattice translations. There are also unique $\mathbb{Z}_2$ fixed points

$$\frac{1}{2}\left(\alpha^2 + \alpha^3\right), \quad \frac{1}{2}\left(\alpha^1 + 2\alpha^2 + \alpha^3\right) = \frac{1}{2}\left(\alpha^2 - \alpha^0\right), \tag{9.30}$$

which lie in the fundamental region of $\theta$, but become redundant in $\theta^2$. Including the second three-torus, we have $4 \times 4 = 16$ such fixed points. This arises because there is a fixed torus $T^2$ left invariant under the action $\theta^2$, but it has order two twist under $\theta$.

**Fig. 9.3** The relation
between SU(4) and
SU(2) × SO(5) lattice. The
former is primitive and
generates the latter. The three
basis vectors are all
orthogonal

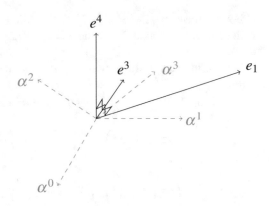

To have the twist eigenstate of (9.1), we introduce a complexified basis

$$e^{\underline{1}} = r_1 \left( \alpha^1 + \alpha^3 + (\beta^1 + \beta^3)t \right),$$

$$e^{\underline{2}} = r_2 \left( \alpha^1 + \alpha^2 - (\alpha^2 + \alpha^3)i \right), \tag{9.31}$$

$$e^{\underline{3}} = r_3 \left( \beta^1 + \beta^2 - (\beta^2 + \beta^3)i \right),$$

where $t$ can be any complex number and $r_1, r_2, r_3$ are real numbers. We can show
that they are eigenstates with eigenvalues $-1, i, -i$, respectively, without need of
lattice translation.

**Relation Between SU(4) and SU(2) × SO(5) Lattices**
The same twist realizes two different lattices SU(4) and SU(2) × SO(5). We study
the relation between them.

Using the SU(4) root vectors $\alpha^1, \alpha^2, \alpha^3$, we can form the SU(2) × SO(5) basis
vectors

$$e_1 = r_1 \left( \alpha^1 + \alpha^3 \right), \quad e^3 = r_3 \left( \alpha^1 + \alpha^2 \right), \quad e^4 = r_3 \left( \alpha^2 + \alpha^3 \right). \tag{9.32}$$

The corresponding vectors are shown in Fig. 9.3. At least they are orthogonal.

$$e_1 \cdot e^3 = e_1 \cdot e^4 = e^3 \cdot e^4 = 0,$$

as desired for the SU(2) × SO(4) lattice. But they cannot be equivalent on torus.

If we want for $e^3, e^4$ to have length 2, we may fix $r_3 = 1$. Then, the coordinate
of the fixed point $\frac{1}{2}(e^3 + e^4) = \frac{1}{2}(\alpha^1 + 2\alpha^2 + \alpha^3)$ is equivalent to $\frac{1}{2}e_1$ up to a lattice
translation by $\alpha^2$. This means that the fixed point for the SU(2) × SO(5) lattice is
not. Although they share the same basis vector, the orbifolding made space different.

To see this further, we note further for $r_3 = 1$

$$\alpha^1 = \frac{1}{2r_1}e_1 + \frac{1}{2}e^3 - \frac{1}{2}e^4,$$

$$\alpha^2 = -\frac{1}{2r_1}e_1 + \frac{1}{2}e^3 + \frac{1}{2}e^4,$$

$$\alpha^3 = \frac{1}{2r_1}e_1 - \frac{1}{2}e^3 + \frac{1}{2}e^4.$$

The translation by $\alpha^i$ in the SU(4) is not in the SU(2) $\times$ SO(5). They are not compatible. They are independent lattices.

## 9.2 Strings on Non-prime $\mathbb{Z}_N$ Orbifolds

Now we come back to physics and consider string theory on $T^6/\mathbb{Z}_N$ orbifold, with non-prime $N$.

### 9.2.1 Eigenstates of Point Group Element

We require the physical state to be invariant under the point group action $\theta \in \overline{\mathsf{P}}$ defining the orbifold. This statement is meaningful only when the state under consideration has definite transformation property under $\theta$. Namely, we require it to be eigenstate of $\theta$ [1].

Consider a state in the $k$-th twisted sector. We calculate the massless spectrum using the formulae (7.37) and (7.38)

$$\left|P + kV\right\rangle_L \otimes \left(\prod \alpha\right)\left|s + k\phi\right\rangle_R \equiv \left|\left(\theta^k, v\right)\right\rangle, \tag{9.33}$$

where we displayed only the spatial information using the conjugacy class. They are further subject to the generalized GSO projection. Under the point group action $\theta^l \in \overline{\mathsf{P}}$, the state acquires the phase

$$\left|(\theta^k, v)\right\rangle \to \left(\Delta_{\theta^k}\right)^l \left|(\theta^k, v)\right\rangle. \tag{9.34}$$

We require that this state must be invariant, $\Delta_{\theta^k} = 1$. However, this only makes sense if the state $|\varphi\rangle$ is an eigenstate under the point group.

Consider a fixed point $f$ in the $\theta^k$-twisted sector

$$\left(\theta^k, v\right), \quad \theta^k f + v = f, \ v \in \Lambda. \tag{9.35}$$

Such fixed point is $\theta$-eigenstate if it is the common fixed point under $\theta^k$ and $\theta$. In this case, there exists $v_0$ such that $(\theta^k, v) = (\theta, v_0)^k$. Then it is invariant under any $\theta$-conjugation.

$$(\theta, u)\left(\theta^k, v\right)(\theta, u)^{-1} = (\theta, u)(\theta, v_0)^k(\theta, u)^{-1} = \left(\theta^k, v\right).$$

In a general a state $k > 1$ is not an eigenstate of $\theta$. If a state is located at a non-fixed point, conjugation by $\theta$

$$(\theta, 0)^l\left(\theta^k, v\right)((\theta, 0)^{-1})^l = \left(\theta^k, \theta^l v\right). \tag{9.36}$$

We generate other fixed point states that are not in the fundamental region of $\theta$ but that of $\theta^k$. Let the order of this conjugation be $L = |[(\theta, v)]|$. This is purely geometric property. Thus it is given by the number of non-invariant fixed points at a given direction $a$

$$L = \left[\chi(\theta)/\chi(\theta, \theta^l)\right]_a. \tag{9.37}$$

If there are more directions we have

$$\text{least common multiplet}\left([\chi(\theta)/\chi(\theta, \theta^l)]_a, [\chi(\theta)/\chi(\theta, \theta^l)]_b\right). \tag{9.38}$$

We can always make an eigenstate of $\theta$ by linearly combining all the conjugates

$$|(\theta^k, v)_l\rangle = \frac{1}{\sqrt{L}}\left(|(\theta^k, v)\rangle + \gamma^{-l}|(\theta^k, \theta v)\rangle + \cdots + \gamma^{-(k-1)l}|(\theta^k, \theta^{k-1}v)\rangle\right), \tag{9.39}$$

where the phase $\gamma$ satisfies

$$\gamma^L = 1. \tag{9.40}$$

Since the total number of states should be invariant after linear combination, we have $L$ possible linear combinations, $l = 1, 2, \ldots, L$. It is the eigenstate of $\theta$, with the eigenvalue $\gamma^l$:

$$\theta|(\theta^k, v)_l\rangle = \frac{1}{\sqrt{L}}\left(|(\theta^k, \theta v)\rangle + \gamma^{-l}|((\theta^k, \theta^2 v)\rangle + \cdots + \gamma^{-(k-1)l}|(\theta^k, v)\rangle\right)$$
$$= \gamma^l|(\theta^k, v)_l\rangle. \tag{9.41}$$

The information on $\gamma$ is contained in the multiplicity factor $\tilde{\chi}(\theta^k, \theta^l)$ in (8.62) counting the number of simultaneous fixed points. See the example below. If $\gamma = 1$, then the corresponding linearly combined states have the same phase as $\theta$-invariant state without linear combination. This should be because $\gamma = 1$ states have $\theta$ eigenvalue 1 and thus invariant. The coefficient of the projector $P_{\theta^2}$ already take into account the multiplicity of these $\theta$-invariant states.

## 9.2.2 The Spectrum of $T^6/\mathbb{Z}_4$ Orbifold Model

Finally we take an example. Consider the $\mathbb{Z}_4$ orbifold (9.1) with the standard embedding [2]

$$V = \frac{1}{4}\left(2\,1\,1\,0^5\right)\left(0^8\right).$$

This also provides a general enough example for non-prime orbifolds. We take $SU(2) \times SO(5)^2$ lattice because we can see the structure of the lattice more intuitively. The six basis vectors $e^i$, $i = 1, \ldots, 6$ satisfy the relations in (9.2).

We obtain the root vectors from the rule $P \cdot V = $ integer in (7.53), so that we identify the gauge group

$$E_6 \times U(1)^2 \times E_8'.$$

In the untwisted sector, we have left-handed matter fields transforming as

$$2(\mathbf{27}, \mathbf{2}) + 2(\mathbf{1}, \mathbf{2}) + (\mathbf{27}, \mathbf{1}) + (\overline{\mathbf{27}}, \mathbf{1}).$$

All the states are neutral under the $E_8'$, so we omit the corresponding information. They all have appropriate right movers making the whole states invariant under $\mathbb{Z}_4$.

**Twisted Spectrum: Operator Method**
In the first twisted sector $k = 1$ we have zero point energy $\tilde{c} = 2f(\frac{1}{2}) + 4f(\frac{1}{4}) + 18f(0) = -11/16$ and the states

$$\prod(\tilde{\alpha}^a_{-n})|P + V\rangle_L$$

with the following highest weight vectors satisfy the mass shell condition

$$(\mathbf{27}, \mathbf{1}) : \tilde{N} = 0, (P + V) = \left(-\tfrac{1}{2}\,\tfrac{1}{4}\,\tfrac{1}{4}\,1\,0^4\right)\left(0^8\right),$$

$$(\mathbf{1}, \mathbf{2}) : \tilde{N} = \tfrac{1}{4}, (P + V) = \left(-\tfrac{1}{2}\,\tfrac{1}{4}\, -\tfrac{3}{4}\,0^5\right)\left(0^8\right),$$

$$(\mathbf{1}, \mathbf{1}) : \tilde{N} = \tfrac{1}{2}, (P + V) = \left(\tfrac{1}{2}\,\tfrac{1}{4}\,\tfrac{1}{4}\,0^5\right)\left(0^8\right).$$

It is sufficient to present for the, since the whole representation is obtained by successive subtractions by simple roots. The oscillator-excited states carry the internal index $a$. For $\tilde{N} = \frac{1}{4}$ excitation, the state carries $a = 2, 3$ giving multiplicity 2. For $\tilde{N} = \frac{1}{2}$, we have four states from the excitations: $\tilde{\alpha}^1_{-1/2}, \tilde{\alpha}^2_{-1/4}\tilde{\alpha}^2_{-1/4}, \tilde{\alpha}^2_{-1/4}\tilde{\alpha}^3_{-1/4}, \tilde{\alpha}^3_{-1/4}\tilde{\alpha}^3_{-1/4}$.

The right mover has the zero point energy $c = 2f(\frac{1}{2}) + 4f(\frac{1}{4}) + 6f(0) = -3/16$ and from the mass shell condition we obtain the states

$$|s + \phi\rangle_R,$$

with

$$(s + \phi) = \left(-\frac{1}{2} \, 0 \, -\frac{1}{4} \, -\frac{1}{4}\right) \text{(R)},$$

$$(s + \phi) = \left(0 \, -\frac{1}{2} \, \frac{1}{4} \, \frac{1}{4}\right) \text{(NS)}.$$

From this we interpreted the state has $-1$ helicity for the untwisted sector states.

Combining the states, we obtain $\Delta_\theta = 1$. We have the projection operator, from (8.59),

$$P_\theta = \frac{1}{4}\left(16 + 16\Delta_\theta + 16\Delta_\theta^2 + 16\Delta_\theta^3\right). \tag{9.42}$$

With $\mathscr{CPT}$ conjugates provided by the third ($= N - k$) twisted sector result in

$$64(\overline{\mathbf{27}}, \mathbf{1}), \quad 32(\mathbf{1}, \mathbf{2}), \quad 16(\mathbf{1}, \mathbf{1}). \tag{9.43}$$

Now consider the $k = 2$ twisted sector, whose effective twist is $2\phi = \frac{1}{2}(2\ 1\ 1)$. Because the first entry is integral, we expect that there is a fixed torus with vanishing $\chi(\theta^2)$. From the mass shell condition, we have

$$|\tilde{N}, P + 2V\rangle_L$$

$$(\mathbf{27}, \mathbf{1}) : \tilde{N} = 0, (P + 2V) = \left(1 \, -\frac{1}{2} \, -\frac{1}{2} \, 0^5\right)\left(0^8\right),$$

$$(\overline{\mathbf{27}}, \mathbf{1}) : \tilde{N} = 0, (P + 2V) = \left(1 \, \frac{1}{2} \, \frac{1}{2} \, 0^5\right)\left(0^8\right),$$

$$(\mathbf{1}, \mathbf{1}) : \tilde{N} = 0, (P + 2V) = \left(-1 \, -\frac{1}{2} \, -\frac{1}{2} \, 0^5\right)\left(0^8\right),$$

$$(\mathbf{1}, \mathbf{1}) : \tilde{N} = 0, (P + 2V) = \left(-1 \, \frac{1}{2} \, \frac{1}{2} \, 0^5\right)\left(0^8\right).$$

Appearance of complex conjugate representations seems natural because now the $k = 2$ twisted sector is the same as the $(N - k) = 2$ twisted sector. The chirality is however determined after combining with the right movers. We obtain ground states

$$|s + 2\phi\rangle_R$$

$$- : (s + 2\phi) = \tfrac{1}{2}(-1\ 1\ 0\ 0),$$

$$+ : (s + 2\phi) = \tfrac{1}{2}(1\ -1\ 0\ 0).$$

They are combined to have multiplicity according to the GSO projector

$$P_{\theta^2} = \frac{1}{4}\left(16 + 4\Delta_{\theta^2} + 16\Delta_{\theta^2}^2 + 4\Delta_{\theta^2}^3\right) \tag{9.44}$$

Applying this, we obtain the following states:

| Left mover | Right mover | $\Delta_{\theta^2}$ | $P_{\theta^2}$ |
|---|---|---|---|
| $(\mathbf{27}, \mathbf{1}), (\mathbf{1}, \mathbf{1})$ | $-$ | 1 | 10 |
| $(\mathbf{27}, \mathbf{1}), (\mathbf{1}, \mathbf{1})$ | $+$ | $-1$ | 6 |
| $(\overline{\mathbf{27}}, \mathbf{1}), (\mathbf{1}, \mathbf{1})$ | $-$ | $-1$ | 6 |
| $(\overline{\mathbf{27}}, \mathbf{1}), (\mathbf{1}, \mathbf{1})$ | $+$ | 1 | 10 |

We have indeed $\Delta_{\theta^2} = 1$ for these states to multiplicity 10 to each $|\mathbf{27}, \mathbf{1}\rangle_L \otimes |-\rangle_R$ and $|\overline{\mathbf{27}}, \mathbf{1}\rangle_L \otimes |+\rangle_R$. Since they are $\mathscr{CPT}$ to each other, there are five Weyl fermions transforming $(\mathbf{27}, \mathbf{1})$ in four dimensions. The same holds true for the pair $|\mathbf{27}, \mathbf{1}\rangle_L \otimes |+\rangle_R$ and $|\overline{\mathbf{27}}, \mathbf{1}\rangle_L \otimes |-\rangle_R$ having multiplicity 6. Thus we have 4D states, all with the left chiralities

$$5(\mathbf{27}, \mathbf{1}) + 5(\mathbf{1}, \mathbf{1}) + 3(\overline{\mathbf{27}}, \mathbf{1}) + 3(\mathbf{1}, \mathbf{1}) \tag{9.45}$$

## Twisted Spectrum: Geometric Method

We can understand the multiplicity of the twisted states reflected in the geometry. In the $k = 1$ twisted sector, the first torus has four fixed points, and each of the second and the third torus has two fixed points. So we have $2 \times 2 \times 4 = 16$ fixed points in total. This is directly translated to the multiplicity of the states in (9.43). All the states are invariant under the projection $\Delta_\theta$, thus survive.

In the $k = 2$ sector, there is non-triviality. To understand the geometric distribution, we decompose (9.44) as

$$P_{\theta^2} = \frac{1}{4}\left(4 + 4\Delta_{\theta^2} + 4\Delta_{\theta^2}^2 + 4\Delta_{\theta^2}^3\right)$$

$$+ \frac{1}{4}\left(6 + 6\Delta_{\theta^2} + 6\Delta_{\theta^2}^2 + 6\Delta_{\theta^2}^3\right) + \frac{1}{4}\left(6 - 6\Delta_{\theta^2} + 6\Delta_{\theta^2}^2 - 6\Delta_{\theta^2}^3\right).$$

$$\tag{9.46}$$

Note that, in each term, all the coefficients are the same up to phases. Their sum

$$4 + 6 + 6 = 16$$

is going to be the number of the fixed points.

Indeed, in the $k = 2$ sector, the two-torus spanned by $e^1$ and $e^2$ is an invariant, "fixed torus" and the remaining four-torus has $4 \times 4 = 16$ fixed points. Among them, only 4 of them

$$| (\theta^2, 0) \rangle, \ | (\theta^2, e^3 + e^4) \rangle, \ | (\theta^2, e^5 + e^6) \rangle, \ | (\theta^2, e^3 + e^4 + e^5 + e^6) \rangle \quad (9.47)$$

are invariant under the $\theta$ as well. This number is reflected in the coefficient 4 in the first term of (9.46). They survive if they acquire no phase $\Delta_{\theta^2} = 1$. This multiplicity explains the contribution in the first line of (9.46). Considering $\mathscr{CPT}$ conjugates in the same $k = N - k = 2$ twisted sector, we have invariant states

$$2(\mathbf{27}, \mathbf{1}) + 2(\mathbf{1}, \mathbf{1}),$$

which have right-handed chirality in four dimensions.

The remaining 12 points are not invariant but can be pairwise combined to make eigenstates under $\theta$ as in (9.39). Each pair is related to one of the conjugacy classes in (9.9). Using (9.37) we find each conjugacy class forms order $L = 2$ orbit and

$$\gamma^2 = 1. \quad (9.48)$$

The expression for the states is

$$| \left( \theta^2, v \right)_l \rangle = \frac{1}{\sqrt{2}} \Big( |(\theta^2, v)\rangle + \gamma |(\theta^2, \theta v)\rangle \Big), \quad (9.49)$$

with $(\theta^2, v)$ can be one of (9.9). We have two possible phase $\gamma = \pm 1$ from (9.48), which are also the eigenvalues of the whole state $|(\theta^2, v)_l\rangle$, $l = 0, 1$. If $\Delta_{\theta^2} = \pm 1$, the $\gamma = \pm 1$ combination survives, respectively. They contribute to the two terms in the second line of (9.46). The overall contribution becomes

$$\frac{1}{4} \left( 12 + 12\Delta_{\theta^2}^2 \right) = 6.$$

Since the order of the conjugacy class is $L = 2$, we always have $\Delta_{\theta^2}^2 = 1$. Taking into account $\mathscr{CPT}$ conjugates, we have three such 4D fields with right-handed chirality for each quantum number

$$3(\mathbf{27}, \mathbf{1}) + 3(\mathbf{1}, \mathbf{1}) + 3(\overline{\mathbf{27}}, \mathbf{1}) + 3(\mathbf{1}, \mathbf{1}).$$

Collecting all the states, we can explain the spectrum (9.45). We can also locate where the twisted fields are distributed.

## 9.3 Strings on $\mathbb{Z}_N \times \mathbb{Z}_M$ Orbifolds

Referring to Table 3.3, there is another way to have $\mathcal{N} = 1$ supersymmetry in four dimensions. Namely, we may act two $\mathcal{N} = 2$ twists in different directions, such that the common intersection becomes $\mathcal{N} = 1$ [3]. For example, we can choose $\phi = \frac{1}{3}(1\ 1\ 0)$ and $\psi = \frac{1}{4}(1\ 0\ 1)$ in Table 3.3. This results in a new type of orbifold of "order" 12, considering all the possible combinations.

We have a number of features in $\mathbb{Z}_N \times \mathbb{Z}_M$ orbifolds.

- We have two independent twists, thus as many projections. This means that a twisting and projecting elements are not identical any more. Local model picture does not work anymore.
- Modular invariance does not completely fix the relative phases. We may have discrete torsion [4].
- Taking into account $\mathcal{CPT}$ conjugates we may fix chirality to be, say, left-handed. In $\mathbb{Z}_N$ orbifold, all the left chiral fields appear in the same twisted sector, not the inversely twisted sector. Then some state appears in a twisted sector and some appears in the inversely twisted sector. For this, we take a convention of displaying only the left-handed fields. Instead we need to show all the twisted sectors including the inversely twisted sectors [5].

### 9.3.1 Combination of Twists

The point group for the $\mathbb{Z}_N \times \mathbb{Z}_M$ orbifold is

$$\mathsf{P} = \{\theta^{j_1}\omega^{j_2},\ j_1 = 0, \ldots, N-1,\ j_2 = 0, \ldots, M-1\} \tag{9.50}$$

so that we have $NM$ twisted sectors. We have defining elements of order $N, M$, respectively, denoted by twist vectors

$$\theta = \mathrm{diag}\left(e^{2\pi i\phi_1}, e^{2\pi i\phi_2}, e^{2\pi i\phi_3}\right), \quad \omega = \mathrm{diag}\left(e^{2\pi i\psi_1}, e^{2\pi i\psi_2}, e^{2\pi i\psi_3}\right). \tag{9.51}$$

Then we choose the lattices. For each two-torus direction, the minimal order is determined by the minimal eigenvalue of $\theta^{j_1}\omega^{j_2}$. It is not necessarily determined by $(j_1, j_2) = (1, 0)$ or $(0, 1)$. For instance, for $\mathbb{Z}_2 \times \mathbb{Z}_3$ with $\phi = \frac{1}{2}(1\ 1\ 0)$, $\psi = \frac{1}{3}(1\ 0\ 1)$, the order of the first torus six is determined by $\phi + 2\psi = (\frac{1}{6}\ \frac{1}{2}\ \frac{1}{3})$. We should take the corresponding lattice $G_2$.

Now we construct string theory on $\mathbb{Z}_N \times \mathbb{Z}_M$ orbifold. For each $j_1, j_2$ of above, we associate the twist with a shift in the group space

$$j_1\phi + j_2\psi \longrightarrow j_1 V + j_2 W. \tag{9.52}$$

The modular invariance condition is basically the same as Eq. (7.50) coming from (7.49) [6],

$$(N\phi)^2 = (NV)^2 \mod 2N,$$

$$(M\psi)^2 = (MW)^2 \mod 2M, \tag{9.53}$$

$$(M\phi) \cdot (M\psi) = (MV) \cdot (MW) \mod 2M.$$

As always, the untwisted sector spectrum is invariant combinations of the left and the right movers. In this case, every state has the transformation property $(\alpha^k, \beta^l)$ under two orbifold actions.

In the presence of Wilson lines, we have further conditions [6]

$$(N_i a_i) \cdot (N_i V) = 0 \mod 2N_i,$$

$$(N_i a_i)^2 = 0 \mod 2N, \tag{9.54}$$

$$(N_{ij} a_i \cdot N_{ij} a_j) = 0 \mod 2N_{ij}, i \neq j,$$

where $N_i$ are the orders of Wilson lines $a_i$ displayed in Table 9.3 and $N_{ij} = \gcd(N_i, N_j)$ are the common orders.

The resulting group is the intersection of the unbroken groups of each action, because two projection conditions (7.53) should be simultaneously satisfied,

$$P \cdot V = \text{integer}, \quad P \cdot W = \text{integer}. \tag{9.55}$$

## 9.3.2  Partition Function and Discrete Torsion

The way we define a theory on $\mathbb{Z}_N \times \mathbb{Z}_M$ orbifold is different from that on $\mathbb{Z}_N$ orbifold. In particular, the projection condition is different.

The partition function is made of the building blocks

$$\mathscr{Z}_{\theta^{j_1}\omega^{j_2}, \theta^{k_1}\omega^{k_2}}(iU, \overline{i}U) = \text{Tr}\left(\theta^{j_1}\omega^{j_2} q^{\tilde{L}_0(\theta^{k_1}\omega^{k_2})+\tilde{c}} \bar{q}^{\tilde{L}_0(\theta^{k_1}\omega^{k_2})+c}\right). \tag{9.56}$$

In the untwisted sector we relate $(1, 1)$ and $(1, \theta)$, $(1, \omega)$ which is $\mathscr{S}$-transformed into $(\theta, 1)$, $(\omega, 1)$. However these are not related to $(\theta, \omega)$, so that the coefficients of the corresponding partition functions are not fixed. We may let them free [3, 4]

$$\mathscr{Z} = \frac{1}{MN} \sum_{j_1, j_2} \sum_{k_1, k_2} \varepsilon\left(\theta^{j_1}\omega^{j_2}, \theta^{k_1}\omega^{k_2}\right) \mathscr{Z}_{\theta^{j_1}\omega^{j_2}, \theta^{k_1}\omega^{k_2}}(\tau, iU) \tag{9.57}$$

but they are subject to consistency conditions from two-loop amplitude [4]

$$\varepsilon(g, hh') = \varepsilon(g, h)\varepsilon(g, h'), \tag{9.58}$$

$$\varepsilon(g, h)\varepsilon(h, g) = 1, \tag{9.59}$$

$$\varepsilon(g, g) = 1. \tag{9.60}$$

Without loss of generality, we may set

$$\varepsilon(\theta, \omega) \equiv \varepsilon, \quad \varepsilon^N = 1. \tag{9.61}$$

and solve them to find

$$\varepsilon\left(\theta^{j_1}\omega^{k_1}, \theta^{j_2}\omega^{k_2}\right) = \varepsilon^{j_1 k_2 - j_2 k_1}. \tag{9.62}$$

We may freely choose discrete torsion $\epsilon = 1, \alpha, \dots, \alpha^{N-1}$. All of the modes give consistent anomaly free vacua.

The twisted sector spectrum is obtained by using the shift vectors (9.52),

$$\frac{1}{4}M_L^2 = \frac{(P + j_1 V + j_2 W)^2}{2} + \tilde{N} + \tilde{c}_{j_1, j_2} = 0,$$

$$\frac{1}{4}M_R^2 = \frac{(s + j_1\phi + j_2\psi)^2}{2} + N + c_{j_1, j_2} = 0, \tag{9.63}$$

$$\frac{1}{2}M^2 = M_L^2 = M_R^2,$$

where the zero point energies are calculated as before by the effective twist $j_1\phi + j_2\psi$. We have generalized GSO projection conditions

$$\exp 2\pi i\Big[(P + j_1 V + j_2 W) \cdot (k_1 V + k_2 W) - (s + j_1\phi + j_2\psi + \rho_R - \rho_L)$$

$$\cdot (k_1\phi + k_2\psi) - \frac{1}{2}((j_1 V + j_2 W) \cdot (k_1 V + k_2 W) - (j_1\phi + j_2\psi) \cdot (k_1\phi + k_2\psi))\Big]. \tag{9.64}$$

Note that the $R$-charge vector becomes $R = s + j_1\phi + j_2\psi + \rho_R - \rho_L$. This should hold true for every $k_1, k_2$, because both of the following GSO projection conditions should be satisfied

$$\left(P + \frac{1}{2}(j_1 V + j_2 W)\right) \cdot V - (\rho_R - \rho_L + \frac{1}{2}(j_1\phi + j_2\psi)) \cdot \phi = 0 \mod 1, \tag{9.65}$$

$$(P + \frac{1}{2}(j_1 V + j_2 W)) \cdot W - (\rho_R - \rho_L + \frac{1}{2}(j_1\phi + j_2\psi)) \cdot \psi = 0 \mod 1. \tag{9.66}$$

Note that we have stronger condition than $T^6/\mathbb{Z}_N$ type orbifold. In particular, local model picture does not hold any more, for which the projection condition should also be written by $j_1 V + j_2 W$ for fixed $j_1$ and $j_2$, but here we have independent $k_1$ and $k_2$. It follows that the change of the shift vector by a lattice vector $\in \Gamma$ does not have the same projection condition. Thus it does give an equivalent model any more. The change of the shift vectors by lattice translations

$$(V, W) \rightarrow (V + \Delta V, W + \Delta W), \quad \Delta V, \Delta W \in \Gamma, \tag{9.67}$$

consistent with the modular invariance conditions

$$
\begin{aligned}
V_i \cdot \Delta V_i &= 0 \mod 1, \quad i = 1, 2, \\
V \cdot \Delta W + \Delta V \cdot \Delta W &= -\Delta V \cdot W \mod 2.
\end{aligned}
\tag{9.68}
$$

Plugging these, the new phase condition reduces to a choice of a discrete torsion [6],

$$e^{2\pi i (j_1 k_2 - j_2 k_1) V \cdot \Delta W} \equiv \varepsilon^{j_1 k_2 - j_2 k_1}. \tag{9.69}$$

### 9.3.3   $\mathbb{Z}_3 \times \mathbb{Z}_3$ Example

Let us illustrate with the simplest $\mathbb{Z}_3 \times \mathbb{Z}_3$ example with the standard embedding. The twists vectors are

$$\phi = \tfrac{1}{3}(1 \; 1 \; 0), \quad \psi = \tfrac{1}{3}(1 \; 0 \; 1)$$

and the shift vectors are

$$V = \tfrac{1}{3} \left(1 \; 1 \; 0^6\right)\left(0^8\right), \quad W = \tfrac{1}{3}\left(1 \; 0 \; 1 \; 0^5\right)\left(0^8\right).$$

The modular invariance condition (9.53) is automatically satisfied. The unbroken gauge group is obtained from (9.55)

$$E_6 \times U(1)^2 \times E_8'$$

In the untwisted sector, each $\alpha^i_{-1}|0\rangle_R$ transforms differently under the action $\theta$ and $\omega$. Thus the untwisted sector is divided into three subsectors, calling $U_{\alpha,1}, U_{1,\alpha}, U_{\alpha^2,\alpha^2}$. Since we have two projections (9.65) and (9.66), each state $|P\rangle_L$ selectively combines to the right mover, depending the value $P \cdot V$ and $P \cdot W$.

We have eight twisted sectors $\theta^{j_1}\omega^{j_2}$, $j_1 = 0, 1, 2$, $j_2 = 0, 1, 2$. States in $T_{j_1, j_2}$ pair up states in $T_{N-j_1, M-j_2}$ by $\mathscr{CPT}$ conjugation. Most of the twisted sectors six dimensional, in the sense that they have one fixed torus, and the other four-torus have 9 fixed points for each. All of them have the same zero point energy $\tilde{c} = -7/9$.

**Table 9.1** All the multiplicities come from the number of fixed points

| Sector | States |
|--------|--------|
| $U_{\alpha,1}$ | **27** |
| $U_{1,\alpha}$ | **27** |
| $U_{\alpha^2,\alpha^2}$ | **27** |
| $T_{1,0}$ | $9 \cdot \mathbf{27} + 9 \cdot \mathbf{27} + 18 \cdot \mathbf{1} + 27 \cdot \mathbf{1} + 36 \cdot \mathbf{1}$ |
| $T_{2,0}$ | $\mathscr{CPT}$ conjugate to $T_{1,0}$ states |
| $T_{0,1}$ | $9 \cdot \mathbf{27} + 9 \cdot \mathbf{27} + 18 \cdot \mathbf{1} + 27 \cdot \mathbf{1} + 36 \cdot \mathbf{1}$ |
| $T_{0,2}$ | $\mathscr{CPT}$ conjugate to $T_{0,1}$ states |
| $T_{1,1}$ | $\mathbf{27} \cdot \mathbf{27}$ |
| $T_{1,2}$ | $9 \cdot \mathbf{27} + 9 \cdot \mathbf{27} + 18 \cdot \mathbf{1} + 27 \cdot \mathbf{1} + 36 \cdot \mathbf{1}$ |
| $T_{2,1}$ | $\mathscr{CPT}$ conjugate to $T_{1,-1}$ states |
| $T_{2,2}$ | $\mathscr{CPT}$ conjugate to $T_{1,1}$ states |

The exceptions are the $T_{1,1}$ and $T_{2,2}$ twisted sectors, which has the effective twist $\phi + \psi = \frac{1}{3}(2\ 1\ 1)$. Since it has the same effective shift as the $\mathbb{Z}_3$ orbifold, we have the same zero point energy. From the mass shell condition $(P + j_1 V + j_2 W)^2/2 + \tilde{N} + \tilde{c} = 0$, we can calculate the massless spectrum. We display the resulting spectrum in Table 9.1.

## 9.4 Wilson Lines on General Orbifolds

Finally, generalizing the discussion on prime orbifolds, we seek the condition for Wilson lines and related GSO projections.

### 9.4.1 Constraints on Wilson Lines

As we saw in Sect. 7.5, discrete Wilson lines are required to be consistent with the space group. Thus the order is determined by the compatibility with the space group.

Recall that in the $\mathbb{Z}_3$ orbifold we are forced to identify Wilson lines in two directions in a SU(3) lattice, e.g. $a_1 = a_2$ modulo a lattice vector, because of lattice symmetry $e^2 = \theta e^1$. From the relation of conjugacy class (3.30), the boundary conditions $(\theta, v_0)$ and $(\theta, v_0 + (1 - \theta)\Lambda)$ describe the equivalent fixed points, thus we set Wilson lines along $v_0$ and $\theta v_0$ same,

$$a_{v_0} \approx a_{\theta v_0}, \tag{9.70}$$

where the sign "$\approx$" means the equivalence up to a lattice translation. In general orbifold, this constraint becomes stronger [1]. Since the Wilson lines are turned along a basis defining lattice, the relations among them are *lattice dependent*, as seen in Table 9.2.

**Table 9.2**  Constraints on Wilson lines for $\mathbb{Z}_N$ orbifold

| P | Lattice | Order | Condition |
|---|---|---|---|
| $\mathbb{Z}_3$ | $SU(3)^3$ | $3a_1 \approx 0, 3a_3 \approx 0, 3a_5 \approx 0$ | $a_1 \approx a_2, a_3 \approx a_4, a_5 \approx a_6$ |
| $\mathbb{Z}_4$ | $SU(2)^2 \times SO(5)^2$ | $2a_1 \approx 0, 2a_2 \approx 0,$ $2a_3 \approx 0, 2a_5 \approx 0$ | $a_3 \approx a_4, a_5 \approx a_6$ |
|  | $SU(2) \times SU(4) \times SO(5)$ | $2a_1 \approx 0, 4a_2 \approx 0, 2a_5 \approx 0$ | $a_2 \approx a_3 \approx a_4, a_5 \approx a_6$ |
|  | $SU(4)^2$ | $4a_1 \approx 0, 4a_4 \approx 0$ | $a_1 \approx a_2 \approx a_3, a_4 \approx a_5 \approx a_6$ |
| $\mathbb{Z}_6$-I | $SU(3) \times G_2^2$ | $3a_1 \approx 0$ | $a_1 \approx a_2$ |
|  | $[SU(3)]^2 \times G_2$ | $3a_1 \approx 0$ | $a_1 \approx a_2 \approx a_3 \approx a_4$ |
| $\mathbb{Z}_6$-II | $SU(2) \times SU(6)$ | $2a_1 \approx 0, 6a_2 \approx 0$ | $a_2 \approx a_3 \approx a_4 \approx a_5 \approx a_6$ |
|  | $SU(3) \times SO(8)$ | $3a_1 \approx 0, 2a_5 \approx 0$ | $a_1 \approx a_2, a_3 \approx a_4 \approx 0, a_5 \approx a_6$ |
|  | $SU(2) \times SU(3) \times SO(7)$ | $2a_1 \approx 0, 3a_2 \approx 0$ | $a_2 \approx a_3, a_4 \approx a_5 \approx a_6 \approx 0$ |
|  | $SU(2)^2 \times SU(3) \times G_2$ | $3a_1 \approx 0, 2a_3 \approx 0, 2a_4 \approx 0$ | $a_1 \approx a_2, a_5 \approx a_6 \approx 0$ |
|  | $SU(2)^2 \times [SU(3)^2]$ | $3a_1 \approx 0, 2a_3 \approx 0, 2a_4 \approx 0$ | $a_1 \approx a_2, a_5 \approx a_6 \approx 0$ |
| $\mathbb{Z}_7$ | $SU(7)$ | $7a_1 \approx 0$ | $a_1 \approx a_2 \approx a_3 \approx a_4 \approx a_5 \approx a_6$ |
| $\mathbb{Z}_8$-I | $SO(9) \times SO(5)^*$ | $2a_1 \approx 0, 2a_6 \approx 0$ | $a_1 \approx a_2 \approx a_3 \approx a_4, a_5 \approx a_6$ |
|  | $[SU(4)^2]$ | $4a_1 \approx 0$ | $a_1 \approx a_2 \approx a_3 \approx a_4 \approx a_5 \approx a_6$ |
| $\mathbb{Z}_8$-II | $SU(2)^2 \times SO(9)$ | $2a_1 \approx 2a_2 \approx 2a_3 \approx 0$ | $a_3 \approx a_4 \approx a_5 \approx a_6$ |
|  | $SU(2) \times SO(10)^*$ | $2a_1 \approx 0, 2a_2 \approx 0$ | $a_2 \approx a_3 \approx a_4 \approx a_5 \approx a_6$ |
| $\mathbb{Z}_{12}$-I | $E_6$ | $3a_1 \approx 0$ | $a_1 \approx a_2 \approx a_3 \approx a_4 \approx a_5 \approx a_6$ |
|  | $SU(3) \times F_4$ | $3a_1 \approx 0$ | $a_1 \approx a_2$ |
| $\mathbb{Z}_{12}$-II | $SU(2)^2 \times F_4$ | $2a_1 \approx 2a_2 \approx 0$ | |

They depend on the choice of the lattices, classified in Table 3.4. Conventions of shift vectors are given in (3.4) and the lattices follow the same orders. The sign "$\approx$" means equivalence up to lattice translation. The order of $F_4$ and $G_2$ Coxeter group is 1, so that we cannot turn on nontrivial Wilson line. On each line, the lattice has the same order of the entries of the twist vectors, except ones with asterisk (*)

We classified the space group in Sect. 3.4. We used the Coxeter lattice as a building block. Each of them has a specific order, we have from (3.57)

$$a_{\alpha^1} \approx a_{\alpha^2} \approx \cdots \approx a_{\alpha^N}. \tag{9.71}$$

A successive translation gives rise to successive gauge transformation of the Wilson lines, we have

$$a_{a\alpha+b\beta} = a a_\alpha + b a_\beta. \tag{9.72}$$

Specific to Coxeter group is the definition of the extended root (3.58)

$$a_{\alpha_N} \approx a_{\sum_i^N v_i \alpha_i} \approx \sum_i^N v_i a_{\alpha_1} \approx a_{\alpha_1} \sum_i^N v_i, \tag{9.73}$$

**Table 9.3** Constraints on Wilson lines for $\mathbb{Z}_N \times \mathbb{Z}_M$ orbifold

| P | Lattice | Order | Condition |
|---|---|---|---|
| $\mathbb{Z}_2 \times \mathbb{Z}_2$ | $SU(2)^6$ | $2a_1 \approx 0, 2a_2 \approx 0, 2a_3 \approx 0,$ $2a_4 \approx 0, 2a_5 \approx 0, 2a_6 \approx 0$ | |
| $\mathbb{Z}_2 \times \mathbb{Z}_4$ | $SO(5) \times SU(2)^2 \times SO(5)$ | $2a_1 \approx 0, 2a_3 \approx 0,$ $2a_4 \approx 0, 2a_5 \approx 0$ | $a_1 \approx a_2, a_5 \approx a_6$ |
| $\mathbb{Z}_3 \times \mathbb{Z}_3$ | $SU(3)^3$ | $3a_1 \approx 0, 3a_3 \approx 0, 3a_5 \approx 0$ | $a_1 \approx a_2, a_3 \approx a_4, a_5 \approx a_6$ |
| $\mathbb{Z}_2 \times \mathbb{Z}_6$-I | $G_2 \times SU(2)^2 \times G_2$ | $2a_3 \approx 0, 2a_4 \approx 0$ | |
| $\mathbb{Z}_2 \times \mathbb{Z}_6$-II | $G_2 \times SU(3) \times G_2$ | $3a_3 \approx 0$ | $a_3 \approx a_4$ |
| $\mathbb{Z}_4 \times \mathbb{Z}_4$ | $SO(5)^3$ | $2a_1 \approx 0, 2a_3 \approx 0, 2a_5 \approx 0$ | $a_1 \approx a_2, a_3 \approx a_4, a_5 \approx a_6$ |
| $\mathbb{Z}_3 \times \mathbb{Z}_6$ | $G_2 \times SU(3) \times G_2$ | $3a_3 \approx 0$ | $a_3 \approx a_4$ |
| $\mathbb{Z}_6 \times \mathbb{Z}_6$ | $G_2^3$ | | |

For conventions, see Table 9.2

where we used (9.71). It determines the order of Wilson lines

$$La_{\alpha_1} \approx \left(1 - \sum_i^N v_i\right) a_{\alpha_1} \approx 0. \tag{9.74}$$

This orders are $N, 2, 4, 3, 1, 1$, respectively, for $SU(N), SO(2n + 1), SO(2n), E_6,$ $F_4, G_2$, which are coincident with the determinant of the corresponding Cartan matrix. Non-Coxeter lattices $[SU(3)^2], [SU(4)^2]$ has order 3, 4, respectively. The result is shown in Table 9.2. If the effective order is 1, we cannot turn on nontrivial Wilson line. We also displayed similar conditions for $\mathbb{Z}_N \times \mathbb{Z}_M$ Wilson lines in Table 9.3.

For example, consider the $T^6/\mathbb{Z}_4$ orbifold defined on the $SU(4) \times SU(4)$ lattice. Considering one of the $SU(4)$ lattice, we have a fixed point

$$\left(\theta, \alpha^1\right). \tag{9.75}$$

Since

$$\left(\theta, \alpha^2\right), \ \left(\theta, \alpha^3\right), \ \left(\theta, \alpha^0\right)$$

can be obtained by conjugation of (9.75), they should have the same Wilson lines up to lattice translation. It follows that

$$a_{\alpha^1} = a_{\alpha^2} = a_{\alpha^3} = a_{\alpha^0} = a_{-\alpha^1-\alpha^2-\alpha^3}. \tag{9.76}$$

The relations tell us that we have

$$a_{\alpha^1} \approx a_{-\alpha^1-\alpha^2-\alpha^3} \approx -3a_{\alpha^1}.$$

The same condition should apply to other directions, thus

$$4a_{\alpha^1} = 4a_{\alpha^2} = 4a_{\alpha^3} \in \Lambda. \tag{9.77}$$

This is easily generalized to SU($N$) and the order of Wilson line $N$.

Now consider the SO(5) lattice with the basis vectors $e^3, e^4$. From the defining relation (9.2), we have

$$a_{e^3} \approx a_{e^4} \approx a_{-e^3} \approx a_{-e^4}. \tag{9.78}$$

In particular the first and the third imply

$$a_{e^3} \approx -a_{e^3},$$

and the same for $a_{e^4}$. Thus we have effective order two

$$2a_{e^3} \approx 2a_{e^4} \approx 0. \tag{9.79}$$

This is generalized to SO($2n + 1$) and the order of Wilson line is 2. Considering all directions, we have

$$2a_1 \approx 0, \quad 2a_2 \approx 0, \quad 2a_3 \approx 2a_4 \in \Lambda, \quad 2a_5 \approx 2a_6 \in \Lambda. \tag{9.80}$$

## 9.4.2 Conjugacy Class

A twisted string is specified by conjugacy class, containing the information about the rotation and translation in the space group. We have seen that the conjugacy class for the element $(\theta^2, e^3)$ is

$$\left(\theta^k, m_3 e^3 + m_4 e^4\right) \left(\theta^2, e_3\right) \left(\theta^k, m_3 e^3 + m_4 e^4\right)^{-1}$$
$$= \left(\theta^2, 2m_3 e^3 + 2m_4 e^4 + \theta^k e^3\right). \tag{9.81}$$

and

$$\left[(\theta^2, e^3)\right] = \left\{\left(\theta^2, (2m_3 + 1)e^3 + 2m_4 e^4\right), \left(\theta^2, 2m_3 e^3 \right.\right.$$
$$\left. +(2m_4 + 1)e^4\right) \mid m_3, m_4 \in \mathbb{Z}\right\}$$
$$= \left\{\left(\theta^2, \pm e^3\right), \left(\theta^2, \pm e^4\right), \left(\theta^2, \pm e^3 \pm 2e^4\right), \left(\theta^2, \pm 2e^3 \pm e^4\right), \dots\right\} \tag{9.82}$$

All the twisted strings by the elements in (9.82) should be regarded as *the same twisted string*. It follows that, for the $[(\theta^2, e^3)]$-twisted strings, the translation by $2e^3$ and $2e^4$ should be neglected. Note the order two relation (9.79). This is consistent with the fact that additional translations by integral multiples of $\pm 2e^3$, $\pm 2e^4$ define the same conjugacy class. Therefore we have effectively order 2 Wilson lines.

### 9.4.3   Generalized GSO Projection

Consider a model with shift vector $V$ and Wilson lines $a_i$. We employ summation convention. In the $(\theta^k, m_i e_i)$-twisted sector, the spectrum satisfies the mass shell conditions (7.94)

$$\frac{1}{4} M_L^2 = \frac{(P + kV + m_i a_i)^2}{2} + \tilde{\mathrm{N}} + \tilde{c}^{(k)},$$

$$\frac{1}{4} M_R^2 = \frac{(s + k\phi)^2}{2} + \mathrm{N} + c^{(k)},$$

$$\frac{1}{2} M^2 = M_L^2 = M_R^2.$$

We may understand all of these in terms of local shifts $V' = kV + m_i a_i$.

The spectrum is also subject to generalized GSO projection

$$P_{(\theta^k, m_i e_i)} = \frac{1}{|C|} \sum_{(\theta^h, l_i e_i) \in C} \tilde{\chi}(\theta^k, m_i e_i; \theta^h, l_i e_i) \Delta_{(\theta^k, m_i e_i; \theta^h, l_i e_i)}, \qquad (9.83)$$

where $|C|$ is the order of the centralizer of $(\theta^k, m_i e_i)$, defined by commuting elements

$$C(\theta^k, m_i e_i) = \left\{ (\theta^h, l_i e_i) \,\middle|\, [(\theta^k, m_i e_i), (\theta^h, l_i e_i)] = 0 \right\}. \qquad (9.84)$$

Its element and $(\theta^k, m_i e_i)$ have the common fixed points. In prime orbifolds, $|C|$ is the order multiplied by the number of fixed points. However in nonprime orbifolds, different elements may belong to the same conjugacy class and hence we have smaller $|C|$.

The phase is obtained by the operator

$$\Delta_{(\theta^k, m_i; \theta^h, l_i)} = \exp 2\pi i \Big[ (P + kV + m_i a_i) \cdot (hV + l_i a_i)$$

$$- (s + k\phi + \rho_R - \rho_L) \cdot (h\phi)$$

$$- \frac{1}{2}(kV + m_i a_i) \cdot (hV + l_i a_i) + \frac{1}{2}(k\phi) \cdot (h\phi) \Big]. \qquad (9.85)$$

Here the dot product is understood if we treat $P$, $V$, $a_i$ as vectors in the current algebra direction and $s$, $\phi$ as ones in the internal space.

Note that the function $\tilde{\chi}(\theta^k, \theta^l)$ in (8.62) counted the number of common fixed points under the two point groups. The degeneracy of the point group is broken because the space group distinguishes different fixed points. We find the number of simultaneous fixed points under the two space groups as

$$\tilde{\chi}(\theta^k, m_i e_i; \theta^h, l_i e_i) = \begin{cases} 1 & (\theta^k, m_i e_i), (\theta^h, l_i e_i) \text{ have a common fixed point,} \\ 0 & \text{otherwise.} \end{cases}$$

$$(9.86)$$

Of course, we consider the orbifolded directions only and do not count the fixed torus directions.

In the absence of Wilson lines $a_i = 0$ for all $i$, the projector reduces to $\Delta_{(\theta^k, m_i; \theta^h, l_i)} \to \Delta_{\theta^k}^h$, as in (8.61). Also, we neglect the translation part in the space group, so $C = \{\theta^h\}$, $|C| = N$. We recover the GSO projector without Wilson lines (8.59).

We can do a similar analysis for $\mathbb{Z}_N \times \mathbb{Z}_M$ type orbifolds [1]. We may modify the projector (9.64) by replacing

$$j_1 V + j_2 W \to j_1 V + j_2 W + m_i a_i \tag{9.87}$$

## Exercise

▶ **Exercise 9.1** Show that the above $T^6/(\mathbb{Z}_3 \times \mathbb{Z}_2)$ orbifold is in fact equivalent to $T^6/\mathbb{Z}_6 - \text{II}$ orbifold by proving the following.

1. Calculate all the effective twists and the shifts.
2. In each twisted sector, show the $\mathbb{Z}_{6-\text{II}}$ projection can be uniquely decomposed into those of $\mathbb{Z}_3$ and $\mathbb{Z}_2$.

## References

1. T. Kobayashi, N. Ohtsubo, Analysis on the Wilson lines of Z(N) orbifold models. Phys. Lett. **B257**, 56–62 (1991)
2. L.E. Ibanez, J. Mas, H.-P. Nilles, F. Quevedo, Heterotic strings in symmetric and asymmetric orbifold backgrounds. Nucl. Phys. **B301**, 157–196 (1988)
3. A. Font, L.E. Ibanez, F. Quevedo, Z(N) X Z(m) orbifolds and discrete torsion. Phys. Lett. **B217**, 272–276 (1989)
4. C. Vafa, Modular invariance and discrete torsion on orbifolds. Nucl. Phys. **B273**, 592–606 (1986)
5. A. Font, L.E. Ibanez, F. Quevedo, A. Sierra, The construction of 'realistic' four-dimensional strings through orbifolds. Nucl. Phys. **B331**, 421–474 (1990)

6. F. Ploger, S. Ramos-Sanchez, M. Ratz, P.K.S. Vaudrevange, Mirage torsion. J. High Energy Phys. **04**, 063 (2007)
7. T. Kobayashi, S. Raby, R.-J. Zhang, Constructing 5-D orbifold grand unified theories from heterotic strings. Phys. Lett. **B593**, 262–270 (2004)

# Interactions on Orbifolds 10

In the effective field theory approach, we have constructed Lagrangian guided by symmetry, for a given spectrum. Since we have the first principle of string theory, we can not only obtain the spectrum but also calculate interaction operator. It turns out that many stringy effects can enhance or suppress the interaction. Stringy selection rules forbid some couplings, and they provide origin of symmetries in the field theory.

In this chapter, we derive low energy theory taking into account stringy effects. First, we calculate the Yukawa interaction, which is of utmost importance in understanding the low energy physics [1–3]. Conformal field theory on the world-sheet provides powerful tool. Then, using the dimensional reduction [4, 5] and the symmetry matching [6, 7], we will construct Kähler potentials and gauge kinetic functions. Most of the stringy interaction is contained in the higher dimensional Lagrangian, for which supersymmetry is strong enough to constrain the interactions. The stringy version of nonrenormalization theorem tells us a lot on the form of the action.

Pedagogical reviews on this topic are Refs. [8, 9]. We will use the coordinates $z = e^{2i(\tau - \sigma)}$ and thus $\partial \equiv \partial_z$.

## 10.1 Conformal Field Theory on Orbifolds

The heterotic orbifold theory also predicts feasible forms for superpotentials, among which we are interested in the Yukawa couplings. In this section, we will take the strategy of calculating string amplitude and compare the supergravity action giving rise to it. There is a very powerful method using conformal field theory [2, 3]. The main discussion will be on the calculation of three-point correlation functions. This low-level method is universally applicable to any string theory, including intersecting brane (see Sect. 17.2). From three point functions, we can extract

© Springer Nature Switzerland AG 2020
K.-S. Choi, J. E. Kim, *Quarks and Leptons From Orbifolded Superstring*,
Lecture Notes in Physics 954, https://doi.org/10.1007/978-3-030-54005-0_10

relative strengths among possible couplings. Then, we will sketch the procedure for obtaining four point functions which will give the absolute normalization.

Knowledge on Kähler potential, whose form we will seek in the next chapter section, is required to have the absolute normalization of physical Yukawa couplings. However, the qualitative result such as relative strengths will not change. This conformal field theory method can also be applied to any other string theory, e.g. intersecting brane models [10].

### 10.1.1  Conformal Field Theory

We use conformal symmetry to facilitate the calculation.

**Vertex Operators**
Consider an interaction diagram for closed strings with external "in" and "out" states, as shown in Fig. 10.1. Its complicated form can be simplified by using the rescaling (6.6), $\delta h_{\alpha\beta} \rightarrow (\delta\Lambda)h_{\alpha\beta}$ with a suitable $\delta\Lambda(\tau, \sigma)$. We can make the diagram into a sphere by shrinking external legs into points. This drastically simplifies the calculation.

Locally, this is viewed as transforming a closed string propagation diagram (a cylinder) into a sphere. It is done by a holomorphic transformation

$$w \rightarrow z = e^{w}, \quad w \equiv \tau + i\sigma. \tag{10.1}$$

We can easily check that the *state* from the infinite past ($\tau \rightarrow -\infty$) corresponds to that on the *point* at the origin. During the transformation, all the information on the state $|\varphi\rangle$ should be kept. Therefore, we are led to have a corresponding local operator $V_\varphi(z, \bar{z})$ at the point. The simplest case is the one carrying no particular quantum number: the tachyon. Since it is an eigenstate of the spacetime translation, it carries a momentum as

$$: e^{ik \cdot X} : \tag{10.2}$$

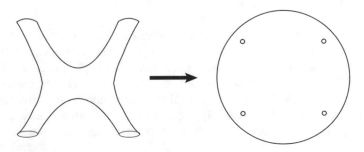

**Fig. 10.1** With a suitable transformation, the $|\text{in}\rangle$ or $|\text{out}\rangle$ state can shrink into a single point, keeping the quantum numbers. Especially, a state becomes an operator, and the mass is converted into the conformal weight

with the understanding of normal ordering.

Now, let us define the conformal weight $h$ as

$$V'(z') = \left(\frac{\partial z'}{\partial z}\right)^h V(z). \tag{10.3}$$

Naively speaking, it shows the scale transformation behavior. From now on, we will concentrate on the left mover only with the argument $z$, since the same can be copied to the right mover also. For example, the fields $\partial_\alpha^n X$ and $: e^{ik \cdot X} :$ have conformal weights $n$ and $-k^2/4$, and so on. Locally, we measure the conformal weight $h$ of an operator $\mathcal{O}$ from the operator product expansion (OPE) of the energy–momentum tensor

$$T(z)\mathcal{O}(w, \bar{w}) \sim \frac{h\mathcal{O}(w, \bar{w})}{(z - w)^2} + \frac{\partial_w \mathcal{O}(w, \bar{w})}{z - w}. \tag{10.4}$$

It turns out that the information on the mass of a given state is converted into the conformal weight. The vertex operator will be inserted into the path integral as $\int d^2 z \sqrt{-h} V_\phi$. Since it should also be conformally invariant, $d^2 z \sqrt{-h}$ has conformal weight $-2$ under the transformation (10.3). Therefore, $V_\phi$ should have the conformal weight of two, $h = 2$. If $V$ is for the tachyon (10.2), we have

$$-\frac{\alpha' k^2}{4} = -\frac{\alpha' M^2}{4} = 2. \tag{10.5}$$

Using the correspondence

$$\begin{aligned}
\partial^n X^\mu &\leftrightarrow \tilde{\alpha}_{-n}^\mu, \\
\bar{\partial}^n X^\mu &\leftrightarrow \alpha_{-n}^\mu,
\end{aligned} \tag{10.6}$$

we may form the vertex operator of graviton. Carrying spin two, we need a second rank tensor index. The vertex operator remains invariant under the worldsheet transformation. Thus, we obtain

$$: \partial_\alpha X_\mu \partial^\alpha X_\nu e^{ik \cdot X} : \tag{10.7}$$

Since each $\partial X$ carries conformal weight 1, the total conformal weight is

$$1 + 1 - \frac{k^2}{4} = 2,$$

and it reduces to $k^2 = 0$. The *graviton should be strictly massless* (Fig. 10.2).

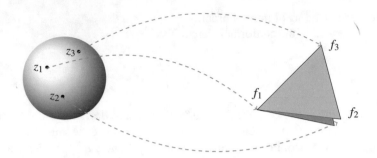

**Fig. 10.2**  The scattering string diagram is a sphere with the twist vertex operators attached at a number of points. These points are mapped to fixed points of the orbifold in the target space

## Yukawa Coupling and Higher Order Generalization

For Yukawa couplings, we need three-point correlation functions. The corresponding tree-level string diagram is the "sphere" with a number of external legs. The external legs of the sphere representing the "in" and the "out" state particles are shrunken at some points, as shown in Fig. 10.1, and the quantum numbers are contained in the vertex operators. Technically, this is not enough to determine overall normalization, so we also need to calculate a related four-point correlation function.

In this section, we employ a covariant approach with the full Lorentz group $SO(1, 9)$, instead of light cone [2, 3]. Thus, we should take into account the conformal ghost $\varphi$. It is known that, from anomaly cancellation for ghost currents, the sum of ghost charges equals $-2$ for the sphere diagram. The renormalizable Yukawa-type interaction, i.e. boson–fermion–fermion coupling ($BFF$), is calculated from a correlation function

$$\left\langle V_{-1} V_{-\frac{1}{2}} V_{-\frac{1}{2}} \right\rangle \equiv \int [DZ] \dots [D\varphi] e^{-S} V_{-1} V_{-\frac{1}{2}} V_{-\frac{1}{2}}, \tag{10.8}$$

where the bracket represents path integration over all the relevant variables, including ghosts, and the subscripts denote their charges. The form of the vertex operator $V$ will be evident shortly. These $BFF$ interactions are in the effective Lagrangian, from which we can deduce the superpotential. Higher order interactions of type $B^n FF$ are given by

$$\mathscr{L} \equiv \left\langle V_{-1} V_{-\frac{1}{2}} V_{-\frac{1}{2}} V_0 \cdots V_0 \right\rangle \tag{10.9}$$

with vertex operators of ghost charge 0. We shall shortly increase the ghost charge by picture-changing operation, so there is no essential difference between $V_{-1}$ and $V_0$.

## 10.1.2  Vertex Operators for Interactions

We proceed to construct pertinent vertex operators. It has one-to-one correspondence with the state and thus encodes all the information about the spectrum.

### Untwisted Fields
Heterotic string has 26 bosonic left movers and supersymmetric 10 right movers. The general vertex operator has the form

$$V_q(z, \bar{z}) = g e^{q\varphi(\bar{z})} e^{i \sum_{M=1}^{26} k_L^M X_L^M(z)} e^{i \sum_{N=1}^{10} k_R^N X_R^N(\bar{z})} e^{is^a H^a(\bar{z})}. \tag{10.10}$$

We conventionally write this as

$$V_q(z, \bar{z}) = g e^{q\varphi(\bar{z})} e^{ik_L^\mu X_L^\mu(z)} e^{ik_R^\mu X_R^\mu(\bar{z})} e^{iP^I X^I(z)} e^{is^a H^a(\bar{z})}. \tag{10.11}$$

We need a ghost field $\varphi(\bar{z})$ in the covariant formalism with the charge $q$ [1–3]. $g$ is the closed string coupling. The factor $e^{ik\cdot X(z,\bar{z})} \equiv e^{ik_L^\mu X_L^\mu(z) + ik_R^\mu X_R^\mu(\bar{z})}$, with only spacetime degrees of freedom, indicates that the vertex operator carries lightlike momentum $k$. The complex field $H^a(\bar{z})$, $a = 0, 1, 2, 3, 4$, is the bosonization of the right-moving worldsheet fermions.

From conformal field theory of ghosts, the NS (spacetime boson) and the R (spacetime fermion) states have $-1$ and $-\frac{1}{2}$ ghost charges, respectively. Therefore for bosonic states,

$$V_{-1}(z, \bar{z}) = g \, e^{is\cdot H} e^{ik\cdot X} e^{-\varphi} e^{iP\cdot F}. \tag{10.12}$$

Thus, it has $s = (\underline{\pm 1\, 0\, 0\, 0\, 0})$, as the vector representation **10** of SO(10). See (6.247). For fermions

$$V_{-\frac{1}{2}}(z, \bar{z}) = g \, e^{is\cdot H} e^{ik\cdot X} e^{-\frac{1}{2}\varphi} e^{iP\cdot F}, \tag{10.13}$$

where $s = ([\frac{1}{2}\, \frac{1}{2}\, \frac{1}{2}\, \frac{1}{2}\, \frac{1}{2}])$ with even numbers of minus signs, which constitute the spinorial representation **16**. The conformal weights for the operator (10.13) are

$$h_{V_{-1}} = \frac{P^2}{2} + \frac{k_L^2}{2} = \frac{2}{2} + 0 = 1, \tag{10.14}$$

$$\bar{h}_{V_{-1}} = \frac{s^2}{2} + \frac{k_R^2}{2} - \frac{1}{2}1(1-2) = \frac{2}{2} + 0 + \frac{1}{2} = 1. \tag{10.15}$$

The conformal weight for the ghost with charge $q$ is

$$h_{q\phi} = -\frac{1}{2}q(q+2). \tag{10.16}$$

To calculate nonrenormalizable interaction (10.9), we should make a zero charge vertex operator. We perform the picture-changing operation [3, 11]

$$V_0(z, \bar{z}) = \lim_{\bar{\omega} \to \bar{z}} e^{\varphi(\bar{\omega})} J_-(\bar{\omega}) V_{-1}(z, \bar{z}), \qquad (10.17)$$

where

$$J_- = \bar{\partial} X^\mu \psi_\mu + \bar{\partial} \bar{Z}^{\bar{a}} \psi_{\bar{a}} + \bar{\partial} Z^a \psi_a \qquad (10.18)$$

is the right-moving worldsheet supercurrent (6.109). Therefore, we have the following zero charge vertex operator:

$$V_0(z, \bar{z}) = g e^{ik \cdot X} \left( i k^\nu \psi_\nu e^{is \cdot H} + \bar{\partial} X^M \right) e^{iP \cdot F}. \qquad (10.19)$$

Here, the $k$-dependent extra term is the connection in the worldsheet sigma model. In the zero momentum limit $k \to 0$, it is simply

$$V_0 = \bar{\partial} X^M,$$

which has the same Lorentz transform property as (10.12). We have apparent discrepancy for Lorentz index in the first and second terms, which we will discuss shortly.

## Twist Fields

We may also have *twisted* states in the correlation function in general. We construct vertex operators for the twisted field. The twisted boundary condition brings about a number of changes. First, the ground state is also twisted, so that we have the corresponding operator. It reminds us of a familiar problem with the fermionic vertex operator. A spin half fermion is characterized by the double-valued function, $\psi(e^{2\pi i} z) = -\psi(z)$. To make a vertex operator, we introduce "spin fields" $S(z)$ and $\tilde{S}(z)$ obeying OPE

$$\psi(z) S(0) \sim z^{-1/2} \tilde{S}(0), \qquad (10.20)$$

where the $\sim$ means that both sides are the same up to irrelevant regular terms. This should be realized by local fields, because the position of singularity is independent of $z$. $S(0)$ and $S(\infty)$ are interpreted as making cuts running between 0 and $\infty$. This $z^{1/2}$ is double-valued function, and for single-valuedness, we introduce the covering space, called *Riemann sheets* [1]. When we transport $z$ once around 0, we cross the branch cut, the coordinate can be thought to be in a different space. This is what we do in the complex analysis when we have a multiple-valued function.

The two-point correlation function now contains spin fields

$$\langle S(\infty)\psi(z)\psi(w)S(0)\rangle = -\frac{1}{2}\frac{1}{z-w}\left(\sqrt{\frac{z}{w}}+\sqrt{\frac{w}{z}}\right). \tag{10.21}$$

Although it has different global structure to the one without spin fields,

$$\langle\psi(z)\psi(w)\rangle = -\frac{1}{2}\frac{1}{z-w}$$

in the $z \to w$ limit, or locally, both behave the same. Surveying conformal weights, one notes that it reproduces the mass shell condition [2,3]. With the spin fields, the theory is not local any more. This is the reason why we introduced GSO projection to make the theory local.

Precisely, the same action is done to bosons by a $\mathbb{Z}_2$ twisting $Z(e^{2\pi i}z) = -Z(z)$. We may introduce a double-valued "twist field" $\sigma(z)$ and an "excited twist field" $\tau(z)$ performing the same action on the bosonic field

$$\partial Z(z)\sigma(0) \sim z^{-1/2}\tau(0). \tag{10.22}$$

This is naturally generalized to $\mathbb{Z}_N$ twist fields $\sigma_k$ and $\tau_k$ (from now on, it is understood that the order $N$ is implicit),

$$\partial Z(z)\sigma_k(0) \sim z^{k/N-1}\tau_k(0). \tag{10.23}$$

Encircling $z$ once around 0, it acquires a phase $e^{2\pi i k/N}$ and we subtract $-1$ to make the power singular.

Here, the reason for acquiring a phase is not due to the spin statistics, but to the property of geometry. For patching to a single-valued space, we introduce a covering space, which is exactly how we made the $\mathbb{Z}_N$ orbifold. The multi-valued function $z^{k/N}$ gives the correct orbifold phase. The twisted ground state (7.9) is now interpreted as

$$|\sigma_k\rangle = \sigma_k(0,0)|0\rangle, \tag{10.24}$$

with the untwisted ground state $|0\rangle$. Therefore, the relation (10.23) can be extracted when we consider the (complexified) mode expansion, (7.5). Consider, for example, the $\partial Z$ part, where only the holomorphic left movers are relevant. By the definition of the ground state, $\tilde{\alpha}^a_{m+k/N}$s with $m \geq 0$ annihilate the ground state (10.24). Therefore, among excited states, the most singular part is

$$\partial Z|\sigma_k^a\rangle = \partial Z\sigma_k^a(0,0)|0\rangle \sim -\frac{i}{2}z^{k/N-1}\tilde{\alpha}^a_{k/N-1}|\sigma_k^a\rangle \equiv z^{k/N-1}\tau_k^a(0,0)|0\rangle, \tag{10.25}$$

where we could also extract the excited twist field $\tau_k$, with the suggestive name. Their conformal weights are

$$h_{\sigma_k} = \tfrac{1}{2}\tfrac{k}{N}(1 - \tfrac{k}{N}), \quad h_{\tau_k} = \tfrac{1}{2}\tfrac{k}{N}(3 - \tfrac{k}{N}).$$  (10.26)

In the similar manner, we obtain

$$\partial Z(z)\sigma_k(0,0) \sim z^{k/N-1}\tau_k(0,0),$$

$$\partial \overline{Z}(z)\sigma_k(0,0) \sim z^{-k/N}\tau_k'(0,0),$$

$$\bar\partial Z(\bar z)\sigma_k(0,0) \sim \bar z^{-k/N}\tilde\tau_k'(0,0),$$

$$\bar\partial \overline{Z}(\bar z)\sigma_k(0,0) \sim \bar z^{k/N-1}\tilde\tau_k(0,0),$$  (10.27)

where the twist fields are in general functions of both coordinates $\omega$ and $\bar\omega$. The excited twisted fields, $\tau, \tau'$, etc., are related with others with slightly different conformal weights. Also, we have similar expressions for $\sigma_{-k}$. Again, to avoid non-locality we need to introduce generalized GSO projection.

The whole vertex operator for a right mover in the $j$th twisted sector has the form

$$V_{-1} = e^{i(P+jV)\cdot F}e^{i(s+j\phi)\cdot H}e^{ik\cdot X}e^{-\varphi}\prod_a \sigma^a_{j\phi^a},$$

with $s$ being vectorial **10** and

$$V_{-\frac{1}{2}} = e^{i(P+jV)\cdot F}e^{i(s+j\phi)\cdot H}e^{ik\cdot X}e^{-\frac{1}{2}\varphi}\prod_a \sigma^a_{j\phi^a},$$

with $s$ being spinorial **16**. Here, $\phi$ is the twist vector, $j\phi = (\tfrac{2j}{3}\ \tfrac{j}{3}\ \tfrac{j}{3}\ 0\ 0)$ for $\mathbb{Z}_3$ example, and $\sigma$ is the twist field. We have calculated $s$ already, e.g. in (7.58–7.60) in the $\mathbb{Z}_3$ example (note that the twisted sector $H$-momentum is not $s$, but $s + k\phi$). Note also that the untwisted sector has various massless solutions of (6.246) as $H$-momentum, depending on what Lorentz components they have. However, in the twisted sector there is essentially a unique solution, because a specification on orbifold determines it. We have tabulated them in Table 10.1.

We can similarly construct higher order twist operators. However, as in the construction of wave function in Sect. 9.2.2, there might be twist fields that are not eigenstates of point group action $\theta \in \overline{P}$ in the higher twisted sectors. If a twisted sector has fixed tori, we can make a physical twist field as an eigenstate of $\theta$ by forming a linear combination

$$\frac{1}{\sqrt{l}}\left(\sigma_{k,f} + \gamma^{-1}\sigma_{k,\theta f} + \gamma^{-2}\sigma_{k,\theta^2 f} + \cdots - \gamma^{l+1}\sigma_{k,\theta^{l-1}f}\right).$$  (10.28)

**Table 10.1** $H$-momenta $(s + k\phi)$ of twisted fields $\theta_k$ in various $\mathbb{Z}_N$ orbifolds. The twisted fields are chosen to give L-handed fermions in four dimensions. The fields are in the NS sector in the $(-1)$-picture. These are the solutions of (7.38) in our convention on twist given in Table 3.4

| Orbifold | $\theta_1$ | $\theta_2$ | $\theta_3$ | $\theta_4$ | $\theta_5$ | $\theta_6$ |
|---|---|---|---|---|---|---|
| $\mathbb{Z}_3$ | $\frac{1}{3}(\text{-}1\ 1\ 1)$ | | | | | |
| $\mathbb{Z}_4$ | $\frac{1}{4}(\text{-}2\ 1\ 1)$ | $\frac{1}{2}(0\ 1\ 1)$ | | | | |
| $\mathbb{Z}_6$-I | $\frac{1}{6}(\text{-}4\ 1\ 1)$ | $\frac{1}{3}(\text{-}1\ 1\ 1)$ | $(0\ 1\ 1)$ | | | |
| $\mathbb{Z}_6$-II | $\frac{1}{6}(\text{-}3\ 2\ 1)$ | $\frac{1}{3}(0\ 2\ 1)$ | $\frac{1}{2}(\text{-}1\ 0\ 1)$ | $\frac{1}{3}(0\ 1\ 2)$ | | |
| $\mathbb{Z}_7$ | $\frac{1}{7}(\text{-}4\ 2\ 1)$ | $\frac{1}{7}(\text{-}1\ 4\ 2)$ | | $\frac{1}{7}(\text{-}2\ 1\ 4)$ | | |
| $\mathbb{Z}_8$-I | $\frac{1}{8}(\text{-}5\ 2\ 1)$ | $\frac{1}{4}(\text{-}1\ 2\ 1)$ | | $\frac{1}{2}(\text{-}1\ 0\ 1)$ | $\frac{1}{8}(\text{-}1\ 2\ 5)$ | |
| $\mathbb{Z}_8$-II | $\frac{1}{8}(\text{-}4\ 3\ 1)$ | $\frac{1}{4}(0\ 3\ 1)$ | $\frac{1}{8}(\text{-}4\ 1\ 3)$ | $\frac{1}{2}(0\ 1\ 1)$ | | $\frac{1}{4}(0\ 1\ 3)$ |
| $\mathbb{Z}_{12}$-I | $\frac{1}{12}(\text{-}7\ 4\ 1)$ | $\frac{1}{6}(\text{-}1\ 4\ 1)$ | $\frac{1}{4}(\text{-}3\ 0\ 1)$ | $\frac{1}{3}(\text{-}1\ 1\ 1)$ | | $\frac{1}{2}(\text{-}1\ 0\ 1)$ |
| $\mathbb{Z}_{12}$-II | $\frac{1}{12}(\text{-}6\ 5\ 1)$ | $\frac{1}{6}(0\ 5\ 1)$ | $\frac{1}{4}(\text{-}2\ 1\ 1)$ | $\frac{1}{3}(0\ 2\ 1)$ | $\frac{1}{12}(\text{-}6\ 1\ 5)$ | $\frac{1}{2}(0\ 1\ 1)$ |
| Orbifold | $\theta_7$ | $\theta_8$ | $\theta_9$ | $\theta_{10}$ | | |
| $\mathbb{Z}_{12}$-I | $\frac{1}{12}(\text{-}1\ 4\ 7)$ | | $\frac{1}{4}(\text{-}1\ 0\ 3)$ | | | |
| $\mathbb{Z}_{12}$-II | | $\frac{1}{3}(0\ 1\ 2)$ | | $\frac{1}{6}(0\ 1\ 5)$ | | |

The same argument applies to the twisted left movers. For example, consider the state $(\bar{\mathbf{3}}, \mathbf{1})$ in the standard embedding of $\mathbb{Z}_3$. The massless state (7.75) is formed with the aid of oscillator $\tilde{\alpha}^i_{-1/3}$; therefore, we expect the fractional oscillator contribution $\partial Z^i/\partial \bar{z}^{1/3}$. Therefore, we have

$$V^i = e^{i(P+V)\cdot F} \partial_{\bar{z}^{1/3}} Z^i e^{ik\cdot X} \sigma_1. \tag{10.29}$$

It carries the spacetime index $i$ and plays an interesting role in understanding geometry. However, $(\mathbf{27}, \mathbf{1})$ has no oscillator

$$V = e^{i(P+V)\cdot F} e^{ik\cdot X} \sigma_1. \tag{10.30}$$

## 10.2 Selection Rules

The Lagrangian and observable should be gauge invariant. From the invariance properties, we deduce the following selection rules arising from the properties of the internal space [1–3,12]. Without calculating couplings explicitly, we can quickly check whether a term with given fields survives or not.

## 10.2.1  Space Group Invariance

We first consider a selection rule from the space group. Let $\theta$ be the twist defining the toroidal orbifold. In view of (3.30), every twisted state in the $\theta^k$ twisted sector is specified by a conjugacy class (3.29)

$$v \in \left\{ \left( \theta^k, \left(1 - \theta^k\right) \left(\theta^l f + u_i\right) \right) \right\}. \tag{10.31}$$

This means, any of $u_i \in \Lambda, l = 0, 1, \ldots, \frac{N}{k} - 1$ gives the same fixed point in the fundamental region (which is usually meant by $f$).

The transformation property of a correlation function is also given by that of the product of the elements given in (10.31). The amplitude is nonzero only when the product is $(1, 0)$. Consider a three-point correlation function. Its space group property is determined by the product

$$\begin{aligned} \left(\theta^{k_1}, v_1\right)\left(\theta^{k_2}, v_2\right)\left(\theta^{k_3}, v_3\right) \\ = \left(\theta^{k_1+k_2+k_3} , v_1 + \theta^{k_1} v_2 + \theta^{k_1+k_2} v_3\right) \tag{10.32} \\ = (1, 0), \end{aligned}$$

where the last equality is the requirement.

We may use the selection rule, for two given twisted fields, to find the space property of the third twisted field.

- Invariance of the rotational part gives the condition for $k_i$

$$k_1 + k_2 + k_3 \equiv 0 \mod N, \tag{10.33}$$

because $\theta$ is of order $N$. This is also called point group selection rule.
- Using the specific form (10.31) (for the moment, take $l = 0$ for each $v_i$), we have

$$\left(1 - \theta^{k_1}\right)(f_1 + u_1) + \theta^{k_1}\left(1 - \theta^{k_2}\right)(f_2 + u_2) + \theta^{k_1+k_2}\left(1 - \theta^{k_3}\right)(f_3 + u_3) = 0. \tag{10.34}$$

It turns out that if this holds for $l = 0$, it also holds for any $l$.

The translation part can be further simplified using $\theta^{k_3} = \theta^{-k_1-k_2}$ from (10.33) as

$$f_1 - f_3 + u_1 - u_3 + \theta^{k_1}(f_2 - f_1 + u_2 - u_1) + \theta^{k_1+k_2}(f_3 - f_2 + u_3 - u_2) = 0. \tag{10.35}$$

This is the condition for forming a triangle $A_1 A_2 A_3$ using the side vectors $\overrightarrow{A_3 A_1} = (f_1 - f_3 + u_1 - u_3)$, $\overrightarrow{A_1 A_2} = \theta^{k_1}(f_2 - f_1 + u_2 - u_1)$, $\overrightarrow{A_2 A_3} = \theta^{k_1+k_2}(f_3 - f_2 +$

**Fig. 10.3** The triangle
surrounded by the three fixed
points
$A_i = f_i + u_i, i = 1, 2, 3$

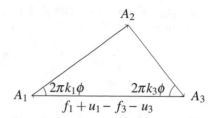

$u_3 - u_2)$, depicted in Fig. 10.3. The angles are $A_i = 2\pi k_i \phi$ with $\phi$ the component of the twist in this direction. The modulo condition in the point group selection alone allows the sum of the angle to be multiple of $\pi$ admitting any polygon. However, the translational part (10.35) restricts it to be the triangle.

A triangle is uniquely formed if we know two angles and one side in between. In this respect, the condition (10.35) seems overconstrained. However, *one* side is defined by difference of *two* fixed points, say $(f_1 + u_1)$ and $(f_2 + u_2)$, fixing the third means the condition for just *one* complex number $f_3 + u_3$. Therefore, a collection of three fixed points does not always make an invariant Yukawa coupling.

Without loss of generality, we can set $u_3 = 0$. Note that letting $k = \gcd(k_1, k_2)$, we can show that $(1 - \theta^k)$ always divides $(1 - \theta^{k_1+k_2})$ and the result is still the entire lattice $\Lambda$. Then, we can solve (10.34)

$$u_1 \in u_2 + (1 - \theta^k)^{-1}(1 - \theta^{k_1+k_2})\Lambda, \tag{10.36}$$

which further restricts $v$.

In prime orbifolds, the space group selection rule completely determines the third fixed point given two fixed points. In non-prime orbifolds, there are number of fields formed by linear combination like (10.28) in the higher twisted sector than first. In fact, the above selection rule is satisfied by a set of fixed points $(\theta^{k_i}, (1-\theta^{k_i})\theta^{l_i}(f_i + u_i))$. If the selection rule holds for one set of $(l_1, l_2, l_3)$, *all the couplings from the same conjugacy classes satisfy the rule.*

Suppose that the coupling satisfy the rule (10.34), or in the special case $(l_1, l_2, l_3) = (0, 0, 0)$. It is enough to consider the space part and show that

$$\left(1 - \theta^{k_1}\right)\theta^{l_1}(f_1 + u_1) + \theta^{k_1}\left(1 - \theta^{k_2}\right)\theta^{l_2}(f_2 + u_2) \tag{10.37}$$

$$+\theta^{k_1+k_2}\left(1 - \theta^{k_3}\right)\theta^{l_3}(f_3 + u_3) = 0,$$

where $l_i = 0, 1, 2, \ldots, N/k_i - 1, i = 1, 2, 3$. Take $k = \gcd(k_1, k_2)$. Having no eigenvalue, we may multiply $(1 - \theta^k)^{-1}$. Using

$$\theta^l \left(1 + \theta + \cdots + \theta^{L-1}\right) = \left(1 + \theta + \cdots + \theta^{L-1}\right), \quad l = 0, 1, \ldots L - 1, \tag{10.38}$$

we may reduce all of this to $(l_1, l_2, l_3) = (0, 0, 0)$. It means that the linear combination (10.28),

$$|f_i\rangle_\gamma = \sum_{l=0}^{N/k_i-1} \gamma^{-l} |\theta^l f_i\rangle,$$

made out of $\theta^l$ has the same transformation property, and hence the linear combination can be really treated as a single field.

An important consequence is that this makes Yukawa couplings *non-diagonal* if we have such nontrivial conjugation. If $k_3$ comes from higher twisted sector $\theta^k$, for given two fields at $f_1$, $f_2$ we may have more than one field allowed by Yukawa couplings.

For a given invariant set of fixed points satisfying (10.35), what happens if we replace one fixed point by lattice translation

$$f_1 + u_1 \to f_1 + u_1 + \lambda_1, \quad f_2 + u_2 \to f_2 + u_2 + \lambda_2, \quad \lambda_1, \lambda_2 \in \Lambda?$$

Then, the condition (10.34) is modified, so that we need to find a new $f_3 + u_3 + \lambda_3$ satisfying

$$\left(1 - \theta^{k_1}\right) \lambda_1 + \left(1 - \theta^{k_2}\right) \lambda_2 = \left(1 - \theta^{k_1+k_2}\right) \lambda_3 = 0.$$

As long as the eigenvalue of $\theta$ is not 1, after dropping the common factor $(1 - \theta^{\gcd(k_1, k_2)})$, we can always find a lattice vector $\lambda_3$. Recall that the inclusion and modification of $u_1$ mean that we allow the fixed point $f_1 + u_1$ not lie in the fundamental region. Thus, we have many selection rules. Shortly, we see that we have as many instanton contributions.

The exception arises that we have the eigenvalue of some power of $\theta$ is 1, when we have fixed torus. Some term in (10.34) vanishes, and we have less constraint. We will see concrete examples shortly.

We may easily generalize the space group selection rule involving $L$-point correlation function, for given twisted fields transforming

$$\left(\theta^{k_i}, \left(1 - \theta^{k_i}\right) (f_i + u_i)\right). \tag{10.39}$$

The rotational part is

$$\sum_{i=1}^{L} k_i \equiv 0 \mod N, \tag{10.40}$$

and the translational part is

$$\sum_{i=1}^{L-1} \theta^{\sum_{j=1}^{i-1} k_i} \left(1 - \theta^{k_i}\right)(f_i + v_i) = 0,$$

$$\text{or } \sum_{i=1}^{L-1} \theta^{\sum_{j=1}^{i-1} k_i} (f_{i+1} - f_i + v_{i+1} - v_i) = 0.$$

(10.41)

### 10.2.2 Lorentz Invariance

String theory provides more fundamental explanation on origin of symmetry and charges. Many of them turn out to be the conservation of momentum. The state containing the vertex operator $e^{ik \cdot X}$ carries the momentum $k$, where $X^M(z, \bar{z})$ is the worldsheet boson carrying the spacetime index $M$.

**Gauge Invariance**

The gauge degrees of freedom are represented by $F^I$ (or $X^I$) whose coefficients $P$s are also conserved. Momentum vectors in the directions of self-dual radii are nothing but the weight vectors of the gauge group. Thus, the gauge invariance is the consequence of the momentum conservation. That is, the gauge invariance is realized as *Lorentz invariance*.

**R-Symmetry in Each Sublattice**

The same is true for the fermionic right movers of the heterotic string. The worldsheet fermions are represented by $H^a(\bar{z})$ by bosonization. The coefficients $(s + k\phi)$ play the role of momentum, previously called the $H$-momentum. We note that this is nothing but a weight vector of the Lorentz group. Thus, the vanishing sum would imply the invariance under the Lorentz group.

In fact, $H$-momentum is not a well-defined notion. We have an explicit expression in terms of the worldsheet fermions

$$e^{is^a \cdot H(\bar{z})} \simeq \psi^a(\bar{z}).$$

It should be noted that here each of

$$s^1 = (1, 0, 0), \quad s^2 = (0, 1, 0), \quad s^3 = (0, 0, 1),$$

is a *vector* (not the $a$th component of a vector $s$). Thus, the operator $e^{is^a \cdot H}$ also behaves like a spacetime Lorentz *vector* with a holomorphic index $a$. We may define $s^{\bar{a}} = -s^a$, so that $e^{is^{\bar{a}} \cdot H}$ behaves antiholomorphic vector. Using these, we can check that $J_-$ in (10.18) is indeed an invariant operator under the spacetime Lorentz symmetry. Thus, the transformation properties of the vertex operators before (for instance (10.12)) and after the picture changing (10.19) should be the same.

This leads us to introduce new conserved charges [12, 13],

$$R^a = s^a + k\phi^a + \rho_L^a - \rho_R^a, \tag{10.42}$$

where $\rho_L^a$ and $\rho_R^{\bar{a}}$ are the numbers of holomorphic and antiholomorphic oscillators, respectively, in the $a$th direction, defined in (7.12). Namely, $R_a$ can be interpreted as the *generalized* number of holomorphic minus antiholomorphic Lorentz indices in the $a$th direction. Note that always the combination (10.42) appears in the spacetime part of the GSO projector (8.61).

This quantity (10.42) is vector, so every component should be independently conserved. We defined the point group action $\theta$ by a simultaneous rotation in the sublattices (sub-two-tori). This means, *in each sublattice has the invariance*. We have the well-defined transformation rules for oscillators and fermions under rotation in the sublattices as given in (7.11) and (7.57). Their vertex operator version is

$$\rho_R^a = -1 : \partial^n Z^a \rightarrow e^{+2\pi i \phi_a} \partial^n Z^a, \tag{10.43}$$

$$\rho_L^a = +1 : \bar{\partial}^n Z^a \rightarrow e^{+2\pi i \phi_a} \bar{\partial}^n Z^a, \tag{10.44}$$

$$\rho_R^a = +1 : \partial^n Z^{\bar{a}} \rightarrow e^{-2\pi i \phi_a} \partial^n Z^{\bar{a}}, \tag{10.45}$$

$$\rho_L^a = -1 : \bar{\partial}^n Z^{\bar{a}} \rightarrow e^{-2\pi i \phi_a} \bar{\partial}^n Z^{\bar{a}}, \tag{10.46}$$

$$e^{i(s+k\phi) \cdot H(\bar{z})} \rightarrow e^{-2\pi i (s+k\phi) \cdot \phi} e^{i(s+k\phi) \cdot H(\bar{z})}, \tag{10.47}$$

where $n$ is an arbitrary integer and there is no summation in each formula. Note that the right movers get rotated in the opposite directions, as in (10.42). Note that all the coefficients in front of $\phi$ is the weight of the Lorentz group, and it is readily checked that the charge (10.42) is invariant combination under these transformations.

Remarkably, the Lorentz symmetry in the internal space can be interpreted as a discrete *R-symmetry* in the sense of conventional supersymmetric field theories. It is because the supercharge, (2.26), can be written as [14]

$$Q = \int \frac{d\bar{z}}{2\pi i} e^{-\frac{1}{2}\varphi(\bar{z})} S(\bar{z}) e^{ir \cdot H(\bar{z})}, \tag{10.48}$$

where $S$ is the spin field (10.20) and hence accompanied by the $R$-vector $r = (\frac{1}{2}, -\frac{1}{2}, -\frac{1}{2})$ in (3.72) and $-\frac{1}{2}$ ghost charge. The spin field $S(\bar{z})$ exchanges a boson and a fermion. From the $H$-momentum transformation (10.47), we also have

$$Q \rightarrow e^{-2\pi i r_a \phi^a} Q, \quad a \text{ not summed.}$$

In the orbifold compactification, the continuous $R$-symmetry is broken into a discrete symmetry

$$SU(4) \simeq SO(6) \to \mathbb{Z}^{N^1} \times \mathbb{Z}^{N^2} \times \mathbb{Z}^{N^3}, \tag{10.49}$$

because of the rotations (10.43)–(10.47). Their orders are determined by the twist vector $\phi$. We define $N^a$ to be the order of $\theta$ along the $a$th direction. So, we have $N^a \phi_a \in \mathbb{Z}$. For example of $\mathbb{Z}_6$-II orbifold specified by $\phi = \frac{1}{6}(3\ 2\ 1)$, we have $(N^1, N^2, N^3) = (2, 3, 6)$.

Since each vertex operator has well-defined transformation under the $R$-symmetry,

$$V \to e^{-2\pi i R_a \phi^a} V, \quad a \text{ not summed} \tag{10.50}$$

and the same for the corresponding low energy field. So, in the low energy field theory, we can mimic this rule by imposing a discrete symmetry. Now, count all the fields as scalars or chiral superfields, so that they are all in the $(-1)$-picture. Then, the superpotential should satisfy the relation

$$\sum R^1 + 2r^1 = 0 \mod N^1,$$

$$\sum R^2 + 2r^2 = 0 \mod N^2, \tag{10.51}$$

$$\sum R^3 + 2r^3 = 0 \mod N^3,$$

where the summation is over the fields forming the product. We have contribution $2r$ because, from (10.9), it is natural to have two fermions and it is done by picture changing involving $r$ as in (7.64).

## 10.3  Three-Point Correlation Function

We calculate the three-point correlation function of the form (10.8) [1–3, 15, 16]. Up to overall normalization, it completely determines Yukawa coupling quantitatively. We will enjoy simplification that is special in the three-point correlation function for the pedagogical reason. Nevertheless, for generalization to higher order orbifold, we take general strategy [2, 17–19].

Consider the classical action

$$S = \frac{1}{2\pi\alpha'} \int d^2z \left( \partial Z_a \bar\partial Z^{\bar a} + \bar\partial Z_a \partial Z^{\bar a} \right). \tag{10.52}$$

In the path integral, the worldsheet field is decomposed into classical and quantum parts

$$Z = Z_{cl} + Z_{qu}, \tag{10.53}$$

and thus the correlation function factorizes as

$$\mathscr{L} = \mathscr{L}_{qu} \cdot \sum_{\{Z_{cl}\}} \exp(-S_{cl}). \tag{10.54}$$

The classical part $Z_{cl}$ satisfies the equation of motion

$$\partial\bar{\partial}Z_{cl} = \partial\bar{\partial}\bar{Z}_{cl} = 0$$

from the classical action given in (6.88). It is going to be described as the local minimum of Euclidian action. It is *instanton* in the sense that it quantifies the difficulty of tunneling between two vacua.

The integration of $X$, $H$ and ghosts leads to the well-known Virasoso–Shapiro amplitude. The nontrivial part involves a correlation between twist fields

$$\mathscr{L} \equiv \left\langle \sigma_{k,f_1}(z_1, \bar{z}_1)\sigma_{l,f_2}(z_2, \bar{z}_2)\sigma_{N-k-l,f_3}(z_3, \bar{z}_3) \right\rangle, \tag{10.55}$$

where the subscripts $k$ and $f_i$ denote the twist $k/N$ and the fixed point, respectively. As shown in Fig. 10.2, we map worldsheet point $z_i$ to the target space $f_i$. This arises for each two-torus, embedded in the complex plane, but the calculation can be independent for each torus, so we suppressed the Lorentz index. The space group selection rule determines, for nonvanishing twists, the relation between fixed points, which is (10.34) in the case at hand, so hereafter we will suppress the fixed point indices $f_i$.

## 10.3.1 The Classical Part

The classical solution satisfies the equation of motion, so that $\partial Z_{cl}, \partial\bar{Z}_{cl}$ are holomorphic and $\bar{\partial}Z_{cl}, \bar{\partial}\bar{Z}_{cl}$ are antiholomorphic. It should reflect the phase change, or *monodromy*. For instance, if a twisted field belonging to $(\theta, v)$ is located at the origin of the worldsheet, the boundary condition

$$Z\left(ze^{2\pi i}, \bar{z}e^{-2\pi i}\right) = e^{2\pi ik/N}Z(z, \bar{z}) + v.$$

It means when we travel by encircling the worldsheet point $z_1$, the worldsheet boson should acquire the phase $e^{2\pi i k/N}$. This behavior at $z_1$, $z_2$, or $z_3$ is as in (10.27), so we have

$$\partial Z_{\rm cl} = a(z - z_1)^{k/N-1}(z - z_2)^{l/N-1}(z - z_3)^{-(k+l)/N},$$
$$\bar\partial Z_{\rm cl} = b(\bar z - \bar z_1)^{-k/N}(\bar z - \bar z_2)^{-l/N}(\bar z - \bar z_3)^{(k+l)/N-1}, \tag{10.56}$$

where we kept the most singular parts. The local properties are encoded in the powers of these ansatz. With SL$(2, \mathbb{C})$, we can fix all the three points and without loss of generality we may send $z_3 \to \infty$. We may consider a case with $k + l > N$ making $\bar\partial Z_{\rm cl}$ divergent. So, we discard this solution by setting $b = 0$.

We may determine the coefficients $a$ by considering the *global property*. Transporting around a contour $C$ encircling a number of fixed points gives rise to a translation $v$ in the target space, if the product of the corresponding space group element is $(1, v)$

$$\Delta_C Z_{\rm cl} = \oint_C dz\partial Z_{\rm cl} + \oint_C d\bar z\bar\partial Z_{\rm cl} = v. \tag{10.57}$$

This holds true if $C$ encircles the multiple points of branch cuts, in which the translational element is given from the product of space group elements. As explained, the twist fields $\sigma_k(z_1)$ and $\sigma_l(z_2)$ introduce a branch cut running between the points $z_1$ and $z_2$, which makes a non-simply connected topology. In this simplest three-point correlation function, we can take a very special contour.

From the definition (10.53), the quantum part satisfies the condition

$$\oint_C dz\partial Z_{\rm qu} + \oint_C d\bar z\bar\partial Z_{\rm qu} = 0. \tag{10.58}$$

That is, the global monodromy from the holomorphic part is the opposite to that of the antiholomorphic part. We will consider the symmetric orbifold, in which the antiholomorphic degrees of freedom is just complex conjugates

$$\bar\partial \overline Z_{\rm cl} = (\partial Z_{\rm cl})^*, \quad \partial \overline Z_{\rm cl} = (\bar\partial Z_{\rm cl})^*. \tag{10.59}$$

Now, consider a closed contour $C$ as depicted in Fig. 10.4 that encircles $z_1$ clockwise, $z_2$ counterclockwise, $z_1$ counterclockwise, and again $z_2$ clockwise. This "Pochhammer loop" $C$ crosses each branch cut inside one and outside once, we have no net monodromy. This requires the so-called *global monodromy* condition.

In the target space, the corresponding transport gives rise to the series of actions (acted from the right)

$$s_2 s_1^{-1} s_2^{-1} s_1$$

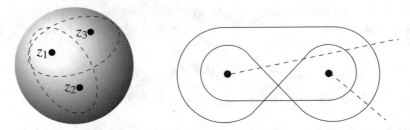

**Fig. 10.4** (Left) There are two independent choices of contours for a three-point function. Due to twisted field, we have branch cut in the target space. Note that each ellipse represents a Pochhammer loop on the right. (Right) Pochhammer loop $C_i$ encircling two branch points $z_i$ and $z_{i+1}$: whichever direction whatever orders the branch cuts have, the nontrivial loop goes in and out, respectively, exactly once

where

$$s_1 = (\omega_1, (1 - \omega_1)(f_1 + v_1)),$$
$$s_2 = (\omega_2, (1 - \omega_2)(f_2 + v_2)).$$

The net effect is the translation because

$$s_2 s_1^{-1} s_2^{-1} s_1 = \left(1, \left(1 - \omega_1^{-1}\right)(1 - \omega_2)(f_2 - f_1 + v_2 - v_1)\right). \tag{10.60}$$

Since $v_2 - v_1$ is again a lattice vector, we may hereafter call it $v$. Recall that $v$ is a complex number for a two dimensional lattice $\Lambda$. The net effect is the translation. With $\partial Z_{\mathrm{cl}}$ given in (10.56) and $\bar{\partial} Z_{\mathrm{cl}} = 0$,

$$\oint_C dz \partial Z_{\mathrm{cl}} = \left(1 - e^{-2\pi i k/N}\right)\left(1 - e^{2\pi i l/N}\right)(f_2 - f_1 + v). \tag{10.61}$$

Here, we use the eigenvalue $\omega_1, \omega_2$ as $e^{2\pi i k/N}, e^{2\pi i l/N}$, respectively. Therefore, the classical solution $Z_{\mathrm{cl}}$ is parameterized by lattice translation $\Lambda$, implied in $v$ in (10.60); the summation of (10.54) is over entire $\Lambda$.

Using the SL(2, $\mathbb{C}$) transformation, we can always fix $z_1 = 0, z_2 = 1, z_3 = \infty$. Because $z_3 \to \infty$, the factor depending on $z_3$ is decoupled. Now, plug the classical solutions (10.56) into the LHS (10.61). We shrink the contour, so that we have only the transport between $z_1$ and $z_2$ (two round trips) and small circles at each $z_1$ and $z_2$ (Exercise). Then, the integral becomes

$$\oint_C dz \partial Z_{\mathrm{cl}} = \left(1 - e^{-2\pi i k/N}\right)\left(1 - e^{2\pi i l/N}\right)$$
$$\times a(-z_3)^{-(k+l)/N} \int_0^1 dz\, z^{k/N-1}(z-1)^{l/N-1}. \tag{10.62}$$

Dropping the common factor, we obtain

$$
\begin{aligned}
f_2 - f_1 + v &= a(-z_3)^{-(k+l)/N} \int_0^1 dz\, z^{k/N-1}(z-1)^{l/N-1} \\
&= a(-z_3)^{-(k+l)/N} \frac{\Gamma\left(\frac{k}{N}\right)\Gamma\left(\frac{l}{N}\right)}{\Gamma\left(\frac{k+l}{N}\right)}.
\end{aligned}
\tag{10.63}
$$

Thus, we can express $a$ in terms of $f_2 - f_1 + v$ to obtain the classical solution (10.56). Inserting this into the action (10.66), we have

$$
\begin{aligned}
S_{cl} &= \frac{1}{2\pi\alpha'} \int d^2 z |\partial Z_{cl}|^2 \\
&= \frac{|a|^2 |z_3|^{-2(k+l)/N}}{2\pi\alpha'} \int d^2 z |z|^{2k/N-2}|z-1|^{2l/N-2}.
\end{aligned}
\tag{10.64}
$$

We use Kawai–Lewellen–Tye (KLT) relation [20]

$$
\int_{\mathbb{C}} d^2 z |z|^{2p}|z-1|^{2q} = -\sin(\pi q) \int_0^1 d\xi\, \xi^p (1-\xi)^q \int_1^\infty d\eta\, \eta^p (1-\eta)^q, \tag{10.65}
$$

showing that a closed string amplitude can be decomposed into doubling of open string amplitude.

We finally obtain

$$
S_{cl} = \frac{1}{2\pi\alpha'} |f_2 - f_1 + v|^2 \frac{\sin\frac{k\pi}{N} \sin\frac{l\pi}{N}}{2\sin\frac{(k+l)\pi}{N}}, \tag{10.66}
$$

where we used $\Gamma(z)\Gamma(1-z) = \pi/\sin(\pi z)$. Plugging it again into (10.54), we can completely determine the size of Yukawa coupling up to the normalization, where the sum is over (10.60).

## Selection Rules and Area Rule

It seems odd that we could completely calculate the action considering only one contour, although there are other independent contours. For instance, taking another loop $C'$ encircling $z_1$ and $z_3$, we should be able to obtain $a$ in terms of $f_3 - f_1 + v'$ from a similar equation to (10.63). This seems to overconstrain the classical solution, but we must have a unique action (10.66). In fact, we have used the above space group selection rules. If we consider an arbitrary twist $k_1 = k, k_2 = l, k_3$ without fixing $k_3$, consideration of two independent contours $C$ and $C'$ should fix

$$
k_3 = -(k+l),
$$

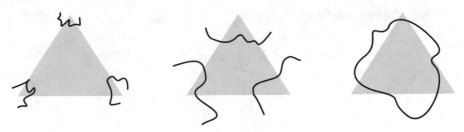

**Fig. 10.5** The area rule summarizes the interaction. The Yukawa coupling is governed by an instanton describing virtual formation of close string from three twisted strings. Its amplitude, or the possibility of forming the instanton, is exponentially suppressed by the area of triangle formed by the corresponding fixed points. The process is possible when the sum of the quantum numbers of the three strings is the same as that of a closed string

proving the point group selection rules. Also, the spatial part gives

$$\frac{|f_2 - f_1 + v|}{\sin \frac{k+l}{N}\pi} = \frac{|f_3 - f_1 + v'|}{\sin \frac{l}{N}\pi} = \frac{|f_3 - f_2 + v''|}{\sin \frac{k}{N}\pi},$$

with appropriate lattice $v'$ and $v''$. Eliminating sine functions, we have

$$\theta^{k+l}(f_3 - f_2 + u_3 - u_2) + \theta^k(f_2 - f_1 + u_2 - u_1) + f_1 - f_3 + u_1 - u_3 = 0, \qquad (10.67)$$

which is nothing but the translational part of the space group selection rule (10.34). Note that the action (10.66) is nothing but the area of an unique triangle, with angles $k\pi/N$ and $l\pi/N$ and a side $|f_2 - f_1 + v|$ between them.

Also, we can take the integration contour $C$ encircle all the vertex operators. There should be neither net rotation nor global monodromy for the whole amplitude. Thus, we have the above relation.

The classical action becomes the area of the triangle formed by the fixed points $f_i + u_i$ divided by coupling $\alpha'$. The path integral is exponentially suppressed. This is instanton effect. Normally, massless twisted strings can be localized at the fixed points, but they can form a virtual closed string as a fluctuation suppressed by $\alpha'$. The difficulty of such formation is proportional to the area of the triangle. The process of semi-classically solving the equation of motion (10.56) and plugging it back to action is in fact the Wentzel–Kramers–Brillouin (WKB) approximation describing instantons. Also, the selection rules state that twisted strings can form a closed string only if the sum of their quantum numbers is the same as that of the closed string (Fig. 10.5).

The quantum part $\mathscr{L}_{\mathrm{qu}}$ is a constant for the three-point function. Because of conformal symmetry, any three points $z_1, z_2, z_3$ can be mapped to $0, 1, \infty$ as above and the correlation function is completely fixed.

## 10.4   Four-Point Correlation Function

We turn to the calculation of four-point correlation function

$$\langle \sigma_{k_1, f_1}(z_1, \bar{z}_1) \sigma_{k_2, f_2}(z_2, \bar{z}_2) \sigma_{k_3, f_3}(z_3, \bar{z}_3) \sigma_{k_4, f_4}(z_4, \bar{z}_4) \rangle. \tag{10.68}$$

This determines the next order couplings to the Yukawa and fixes the normalization of Yukawa couplings. Also, we shall see that it exhibits the modular property clearly [16, 21, 22]. Unlike the previous three-point correlation function, we can calculate the quantum part. We briefly sketch the steps of general strategy and the results [2, 3, 17].

### 10.4.1  The Classical Part

The forms of the classical solutions are fixed by branch structure as in the case of three-point function

$$\partial Z_{cl}(z) = a \omega_k(z),$$
$$\bar{\partial} Z_{cl}(\bar{z}) = b (\omega_{N-k}(z))^*, \tag{10.69}$$

where $k \equiv (k_1, k_2, k_3, k_4)$ and

$$\omega_k(z) \equiv (z - z_1)^{k_1/N-1} (z - z_2)^{k_2/N-1} (z - z_3)^{k_3/N-1} (z - z_4)^{k_4/N-1}. \tag{10.70}$$

Unlike the previous three-point correlation function, both functions in (10.69) contribute finitely, so that both $a$ and $b$ are nonvanishing.

Using the same method as before for the three-point functions, we fix $a$ and $b$ using the global monodromy condition. We need to consider two independent Pochhammer loops, assuming that the correlation function is allowed by the selection rules. We take one $C_1$ encircling $z_1, z_2$ and another $C_2$ encircling $z_2, z_3$

$$\oint_{C_i} dz \partial Z_{cl} + \oint_{C_i} d\bar{z} \bar{\partial} Z_{cl} = \left(1 - e^{-2\pi i k_i / N}\right) \left(1 - e^{2\pi i k_{i+1}/N}\right) (f_{i+1} - f_i + v). \tag{10.71}$$

Again, with SL(2, $\mathbb{C}$) we fix $z_1 = 0$, $z_3 = 1$, $z_4 = \infty$. In the three-point function, we have seen that the factor involving $z_4$ is decoupled and to be cancelled by the same contribution from $a$ and $b$. This happens in the four-point function. We only need to define

$$\tilde{\omega}_k(z) \equiv z^{k_1/N-1} (z - x)^{k_2/N-1} (z - 1)^{k_3/N-1}.$$

Then, we only need to evaluate

$$f_{i+1} - f_i + v = a F_i + b \bar{F}'_i,$$

where we define

$$F_i = \int_{z_i}^{z_{i+1}} dz \tilde{\omega}_k(z), \quad \bar{F}'_i = \int_{\bar{z}_i}^{\bar{z}_{i+1}} d\bar{z} (\tilde{\omega}_{N-k}(z))^*.$$

For the above fixing, these can be expressed in terms of hypergeometric function, summarized in Appendix. We can obtain $a$ and $b$ by solving the equation:

$$\begin{pmatrix} f_2 - f_1 + v_2 - v_1 \\ f_3 - f_2 + v_3 - v_2 \end{pmatrix} = \begin{pmatrix} F_1 & \bar{F}'_1 \\ F_2 & \bar{F}'_2 \end{pmatrix} \begin{pmatrix} a \\ b \end{pmatrix}. \tag{10.72}$$

Plugging the classical solution into the action, we have

$$S = \frac{1}{4\pi\alpha'} \left( |a|^2 I(x) + |b|^2 I'(\bar{x}) \right), \tag{10.73}$$

where

$$\begin{aligned} I(x) &= \int_{\mathbb{C}} d^2 z \, |\tilde{\omega}_k(z)|^2 \\ &= \sin \frac{k_2 \pi}{N} F_2 \bar{F}_1 - \sin \frac{(k_2 + k_3)\pi}{N} F_3 \bar{F}_1 - \sin \frac{(k_1 + k_2)\pi}{N} F_0 \bar{F}_2 \\ &\quad + \sin \frac{k_2 \pi}{N} F_1 \bar{F}_2 \end{aligned} \tag{10.74}$$

can be calculated by using KLT relation again [20]. We can express $F_0$ and $F_3$ in terms of $F_1$ and $F_2$. Thus, we have

$$I(x) = c_{11} |F_1|^2 + c_{12} F_1 \bar{F}_2 + c_{12}^* \bar{F}_1 F_2 + c_{22} |F_2|^2,$$

where

$$c_{11} = \frac{\sin(\pi k_1/N) \sin(\pi(k_2 + k_3)/N)}{\sin(\pi k_4/N)},$$

$$c_{12} = e^{\pi i k_2/N} \sin(\pi k_1/N) \sin(\pi k_3/N), \tag{10.75}$$

$$c_{22} = \frac{\sin(\pi k_3/N) \sin(\pi(k_1 + k_2)/N)}{\sin(\pi k_4/N)}.$$

We obtain a similar expression for $I'(\bar{x}) = \int_{\mathbb{C}} d^2 z \, |\tilde{\omega}_{N-k}(z)|^2$ by replacing $k_i/N \to 1 - k_i/N$ and $x \to \bar{x}$.

**Fig. 10.6** A quadrilateral ABCD. Its area is given by the difference of the areas of the triangles EBC and EAD

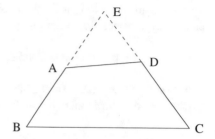

In the path integral, we integrate over $x$. Since the action is positive definite, we have a minimum at

$$\frac{F_2}{F_1} = \frac{f_3 - f_2 + v_3 - v_2}{f_2 - f_1 + v_2 - v_1},$$

making $b = 0$ and the solution becomes purely holomorphic. We have saddle-point approximation

$$S_{\text{cl,min}} = \frac{1}{2\pi\alpha'}\left[\frac{|v_{14}|^2}{2}\frac{\sin(\pi k_1/N)\sin(\pi k_4/N)}{\sin(\pi(k_1+k_4)/N)} - \frac{|v_{32}|^2}{2}\frac{\sin(\pi k_2/N)\sin(\pi k_3/N)}{\sin(\pi(k_2+k_3)/N)}\right].$$
$$(10.76)$$

This is the area of the quadrilateral. As shown in Fig. 10.6, it is understood as the difference between two triangles.

However, the translation part is now restricted to the *intersection of* $v_1$ *and* $v_3$ *regions*. Therefore, we restrict the summation region

$$v \in \left(1 - \theta^k\right)\left(f_2 - f_1 - T_1 + \frac{1 - \theta^l}{1 - \theta^{\gcd(k,l)}}\Lambda\right),$$
$$(10.77)$$

where $T_1$ is, again, a lattice vector satisfying the space group selection rule,

$$\left(1 - \theta^k\right)(f_2 - f_1 - T_1) + \left(1 - \theta^l\right)(f_4 - f_3 - T_2) = 0.$$
$$(10.78)$$

## 10.4.2  The Quantum Part

We can calculate the quantum part $\mathscr{Z}_{\text{qu}}$ using the so-called stress tensor method [2, 3, 17, 23]. The energy–momentum "stress" tensor $T(z)$ has the OPE

$$T(z)\sigma_k(w) \sim \frac{h_{\sigma_k}\sigma_k(w)}{(z-w)^2} + \frac{\partial_w\sigma_k(w)}{z-w},$$
$$(10.79)$$

with the conformal weight for the twisted field $h_{\sigma_k}$ given in (10.26). The idea is to calculate the correlation function by multiplying

$$\sigma_{k_1}(z_1) \text{ on the left and } \sigma_{k_3}(z_3)\sigma_{k_4}(z_4)(z - w) \text{ on the right}$$

to both sides of Eq. (10.79) and taking the expectation value. Replacing $k = k_2$, $w = z_2$, we have a differential equation for the quantum part of (10.68)

$$\partial_{z_2} \log \mathscr{Z}_{\text{qu}} = \lim_{z \to z_2} \left( \frac{z - z_2}{\mathscr{Z}_{\text{qu}}} \langle T(z)\sigma_{k_1}(z_1)\sigma_{k_2}(z_2)\sigma_{k_3}(z_3)\sigma_{k_4}(z_4) \rangle_{\text{qu}} - \frac{h_{\sigma_k}}{z - z_2} \right).$$
(10.80)

We can also know the OPEs for $T$ and $\sigma$s. Since $T(z)$ is the normal ordered version of the free Lagrangian, having the OPE,

$$-\frac{1}{2}\partial_z Z_{\text{qu}}\partial_z \overline{Z}_{\text{qu}} \sim \frac{1}{(z - w)^2} + T(z),$$
(10.81)

it is useful to define a "Green's function,"

$$g(z, w) \equiv \frac{\left\langle -\frac{1}{2}\partial_z Z_{\text{qu}}\partial_w \overline{Z}_{\text{qu}}\sigma_{k_1}(z_1)\sigma_{k_2}(z_2)\sigma_{k_3}(z_3)\sigma_{k_4}(z_4) \right\rangle_{\text{qu}}}{\mathscr{Z}_{\text{qu}}}.$$
(10.82)

Now, the problem is translated to calculating $g(z, w)$, thus $T$, and solving this differential equation. We also know the OPEs between $\partial_z Z$ and $\sigma_k$, etc., from (10.27). This means that we have following limits:

$$\begin{aligned} g(z, w) &\sim (z - z_i)^{k_i/N - 1}, & \text{as } z \to z_i, \quad i = 1, 2, 3, 4, \\ &\sim (w - z_i)^{-k_i/N} & \text{as } w \to z_i, \quad i = 1, 2, 3, 4. \end{aligned}$$
(10.83)

All with the coefficient 1. We can guess that the function having the desired property is proportional to $\omega_k(z)\omega_{N-k}(\omega)$ of (10.70). Thus, we have

$$g(z, w) = \omega_k(z)\omega_{N-k}(w) \left[ \sum_{i<j} a_{ij} \frac{(z - z_i)(z - z_j)\prod_{k \neq i, j}(w - z_k)}{(z - w)^2} + A \right],$$
(10.84)

and a constant $A$ depending on $z_i, \bar{z}_i$. We should also have the behavior as in Eq. (10.81) as we take $z \to w$

$$g(z, w) \sim \frac{1}{(z - w)^2}, \quad \text{as } z \to w,$$
(10.85)

determining

$$\sum_{i<j} a_{ij} = 1. \tag{10.86}$$

To have no residue in $z - z_i$, we also need

$$\sum_{j=1}^{4} a_{ij} = 1 - \frac{k_i}{N}, \tag{10.87}$$

where we define $a_{ji} \equiv a_{ij}$ for $j > i$. Using (10.81), we can calculate the OPE between $T$ and $\sigma$s and hence the differential equation (10.80)

$$\partial_{z_2} \ln \mathscr{Z}_{\mathrm{qu}} = \partial_{z_2} \sum_{i<j} \left[ a_{ij} - \left(1 - \frac{k_i}{N}\right)\left(1 - \frac{k_j}{N}\right) \right] \frac{1}{z_2 - z_j}$$

$$+ \frac{A}{\prod_{i \neq k}(z_2 - z_1)(z_2 - z_3)(z_2 - z_4)}.$$

The undetermined constant $A$ is again determined by the global monodromy condition. Because the classical part in (10.53) took the spatial translation $v_i$ in (10.71), the remaining quantum part has a condition

$$\oint_{C_i} dz \partial Z_{\mathrm{qu}} + \oint_{C_i} d\bar{z} \bar{\partial} Z_{\mathrm{qu}} = 0. \tag{10.88}$$

Note that $\partial Z_{\mathrm{qu}}$ is contained in the function $g$ in (10.82). For $\bar{\partial} Z_{\mathrm{qu}}$, we introduce an auxiliary function containing it

$$h(\bar{z}, w) \equiv \frac{\left\langle -\frac{1}{2}\bar{\partial} Z_{\mathrm{qu}} \partial_w \overline{Z}_{\mathrm{qu}} \sigma_{k_1}(z_1)\sigma_{k_2}(z_2)\sigma_{k_3}(z_3)\sigma_{k_4}(z_4) \right\rangle_{\mathrm{qu}}}{\mathscr{Z}_{\mathrm{qu}}}.$$

We may solve it using the same method to obtain

$$h(\bar{z}, w) = (\omega_{N-k}(z))^* \omega_{N-k}(w) B,$$

in which the constant $B$ should be determined by global monodromy condition. Then, we have an equivalent condition to (10.88)

$$\oint_{C_l} dz g(z, w) + \oint_{C_l} d\bar{z} h(\bar{z}, w) = 0. \tag{10.89}$$

We took $z_4 \to \infty$, we have $\tilde{\omega}$. Also, we have dominant contribution from $(z - z_4)$, and we have only $a_{i4}$ term

$$\omega_{N-k}(w) A \oint_{C_l} \omega_k(z) dz + \omega_{N-k}(w) B \oint_{C_l} (\omega_k(z))^* dz d\bar{z} + \oint_{C_l} \sum_i a_{i4}(z - z_i) \omega_k(z) dz.$$

From two contours $C_1$ and $C_2$, we obtain

$$A = \left( \frac{k_4}{N} - a_{13} \right) x - a_{12} - \left( 1 - \frac{k_2}{N} \right) \frac{k_4}{N} \left[ G_1 \partial_x \overline{G}_2' - G_2 \partial_x \overline{G}_1 \right] / J.$$

Inserting $A$ into Eq. (10.84), we finally have

$$\mathscr{L}_{\text{qu}} = C (\det F)^{-1} (-\bar{x}) x^{-(1-\frac{k_1}{N})(1-\frac{k_2}{N})} (x - 1)^{-(1-\frac{k_2}{N})(1-\frac{k_3}{N})} \bar{x}^{-\frac{k_1}{N} \frac{k_2}{N}} (\bar{x} - 1)^{-\frac{k_2}{N} \frac{k_3}{N}},$$
$$\tag{10.90}$$

with a normalization constant $C$.

## 10.4.3 Factorization and Normalization

Now, we can determine the normalization [2, 3, 15]. When we move $z_2 \to z_4$, or $x \to \infty$, we have the OPE

$$\sigma_{N-k}(z_2) \sigma_{N-l}(z_4) \sim c_{ij} (z_2 - z_4)^{h_{\sigma 2N-k-l} - h_{\sigma N-k} - h_{\sigma N-l}} \sigma_{2N-k-l}(z_4). \tag{10.91}$$

Therefore, the four-point correlation function (10.68) factorizes to the sum of three-point functions. Again, the three-point function is completely determined by standard OPE as

$$\mathscr{L}(x, \bar{x}) \sim |x|^{-h} |z_\infty|^{-h'} \sum_f Y^k_{f_2, f_4, f} \left( Y^k_{f_1, f_3, f} \right)^*, \tag{10.92}$$

where the sum is over the fixed points, and $h, h'$ are some combinations of conformal weights. It is interpreted as the sum of the products of Yukawa couplings, because they are

$$Y^k_{f_2, f_4, f} = \lim_{|x| \to \infty} |x|^{h_{\sigma k}} \langle \sigma_{k, f_2}(x, \bar{x}) \sigma_{l, f_4}(1, 1) \sigma_{-k-l, f}(0, 0) \rangle \tag{10.93}$$

and we can compare it with the known result on the three-point function. Here, $h_{\sigma k}$ is the conformal weight of $\sigma_k$ in (10.14).

Then, we can normalize the correlation function. When we move $z_2$ to $z_1$, we have a similar OPE as (10.91) and the entire correlation function becomes essentially two-point correlation function

$$\sigma_{k+l}(z_1)\sigma_{N-k-l}(z_4) \simeq 1 \cdot |z|^{-h_{k+l}}.$$

So, we can obtain the absolute normalization of (10.90)

$$C = 2\pi\sqrt{\det G_{ij}} \equiv 2\pi A_\Lambda, \qquad (10.94)$$

where $A_\Lambda$ is the area of two-torus.

### 10.4.4 Modular Property

So far, we have considered one complex dimension, suppressing the spatial index. For $T^d/\mathbb{Z}_N$ orbifold, we can multiply the contribution from $d/2$ two-tori. For the $\mathbb{Z}_3$ orbifold, we have only $\theta^1\theta^1\theta^1$ coupling, so that $k = l = 1$. Inserting them into (10.52), we obtain the same coefficients and the resulting contribution to the action along the $a$th direction is

$$S_{cl}^a = \frac{\sqrt{3}}{8\pi\alpha'}\left|f_2^a - f_1^a + u^a\right|^2, \quad a = 1, 2, 3, \qquad (10.95)$$

where $u = (u^1, u^2, u^3)$ is any lattice vector. The total action becomes the exponent, and we have the Yukawa coupling

$$Y_{\theta^1\theta^1\theta^1} = g_s N \sum_{v\in(1-\theta)(f_1-f_2+\Lambda)} \exp\left(-\frac{\sqrt{3}}{8\pi\alpha'}\sum_a\left|f_2^a - f_1^a + u^a\right|^2\right), \qquad (10.96)$$

where the full numerical coefficient is given in (10.99)

$$N = V_\Lambda^{1/2}\frac{3^{3/4}\Gamma^6(2/3)}{8\pi^3\Gamma^3(1/3)}.$$

We may reflect the contribution from the spacetime metric. In the next chapter, we shall see that we may understand the index contraction in (10.52) in terms of metric, viz. $\partial Z_a\bar{\partial}\overline{Z}^a = G_{a\bar{b}}\partial Z^a\bar{\partial}\overline{Z}^b$. Then, we can write the classical action and hence the Yukawa coupling like

$$Y_{\theta^1\theta^1\theta^1} = g_s N \sum_{u\in\mathbb{Z}^6} \exp\left[-2\pi(f_1 - f_2 + u)^\top M(f_1 - f_2 + u)\right], \qquad (10.97)$$

where we defined $M_{ij} = \frac{\sqrt{3}}{16\pi^2\alpha'}G_{ij}$. We perform the sum over $v \in \mathbb{Z}^6$, and the result is the Jacobi theta function

$$Y_{\theta^1\theta^1\theta^1} = g_s N \vartheta \begin{bmatrix} f_1 - f_2 \\ 0 \end{bmatrix} \left(0, \frac{M}{4\pi^2 i}\right). \qquad (10.98)$$

One may expect the target space modular invariance (11.49), and indeed
   multiplying all the two-torus contributions, the final result turns out to be

$$Y^{k,l}_{f_a, f_b, f_c} = g_{YM} \left[\prod_{j=1}^{d/2} 2\pi A_{\Lambda,j} \Gamma_{k_j, l_j}\right]^{1/2} \sum_{v \in \Lambda^d} \exp\left(-\pi v^\dagger \cdot M v\right), \qquad (10.99)$$

where we regard $v = (v_1, \ldots, v_{d/2})$ as a complex vector. The summation range is restricted as in (10.36),

$$v \in \tau_1 - \tau_2 + \frac{1 - \theta^{k+l}}{1 - \theta^{\gcd(k,l)}}\Lambda, \qquad (10.100)$$

and

$$\Gamma_{k,l} = \frac{\Gamma\left(1 - \frac{k}{N}\right)\Gamma\left(1 - \frac{l}{N}\right)\Gamma\left(\frac{k+l}{N}\right)}{\Gamma\left(\frac{k}{N}\right)\Gamma\left(\frac{l}{N}\right)\Gamma\left(1 - \frac{k+l}{N}\right)}. \qquad (10.101)$$

   The universal normalization is given by the string coupling $g_c$, the area of internal manifold, and the order-dependent factor in Eq. (10.99). For $N = 3$, we have $\Gamma_{1,1} \simeq 0.13$, for $N = 6$, $\Gamma_{2,3} \simeq 0.10$ and for $N = 12$, $\Gamma_{2,3} = 0.1$, $\Gamma_{3,4} = 0.12$.
   If we have an invariant plane, we have no sum. A coupling $\theta^1\theta^2\theta^3$ from the $\mathbb{Z}_6$-I orbifold has the form:

$$S^a_{cl} = \frac{1}{4\pi\alpha'} \frac{\sin(2\pi\phi^a)\sin(3\pi\phi^a)}{\sin(\pi\phi^a)}|v^a|^2,$$

where

$$v \in (f_2 - f_3 + \Lambda).$$

We have $\phi = (\frac{2}{6} \frac{1}{6} \frac{1}{6})$. The coefficients are the same $\frac{1}{4\pi\alpha'}\sqrt{3}$ for $a = 1, 2$. Since the $\theta^3$ twist makes the $a = 1$ plane invariant, we have no contribution on this plane. We limit the vector in the second and third planes. We have

$$Y_{\theta^1\theta^2\theta^3} = N \sum_{u \in (f_2 - f_3 + \Lambda)_\perp} \exp\left(-\frac{1}{4\pi\alpha'}\left(\left(u^1\right)^2 + \left(u^2\right)^2\right)\right), \qquad (10.102)$$

where the lattice $(f_2 - f_3 + \Lambda)_\perp$ means that it is limited to the orbifolded directions.

We may generalize the above calculation to higher order coupling cases. Essentially, the holomorphic part is straightforwardly extended. There needs more generalization for the antiholomorphic part. The result is described by multi-variable generalization of hypergeometric function [17, 24–27].

### $\mathscr{CP}$ **Phase**

The above result can be extended, including antisymmetric tensor background. When we include contributions from antisymmetric tensor field as in (11.26) and Wilson lines, it leads to a modification

$$|v|^2 \to \frac{T + A \cdot \bar{A}}{\mathrm{Re}\, U} \left( |v|^2 - 2\,\mathrm{Re}\, v\, \mathrm{Im}\, v\, \mathrm{Im}\, U \right). \tag{10.103}$$

Neglecting Wilson lines $A = 0$ and letting $U$ modulus $U = 1$, it amounts to a simple modification; $|v|^2$ becomes $T|v|^2$ times a number dependent on the volume of torus. It is because $T$ modulus (11.45) unifies the metric and the antisymmetric tensor field which together describes the volume. For example, for the SU(3) lattice it is $\frac{(2\pi)^2}{\sqrt{3}} T|v|^2$. Therefore, we expect that the Yukawa coupling has the target space modular transformation property (11.49) [16, 21, 22]. As explained, the $U$ modulus gets a fixed VEV if the orbifold action is not compatible with the rotation by $\pi$.

## 10.5   Phenomenology of Yukawa Couplings

We deal with concrete examples and see how the above selection rules apply to construct realistic models.

### 10.5.1  Couplings in $\mathbb{Z}_3$ Orbifold

We first study the properties of Yukawa and higher order couplings, starting with a $\mathbb{Z}_3$ orbifold example [28]. We will generalize to general orbifold later [28, 29].

#### Selection Rules

Many couplings are forbidden by the selection rules. Under the point group, untwisted $U$ and twisted fields $T$ transform as

$$U_1, U_2, U_3 : 1, \quad T_1 : \theta,$$

where the subscripts of $U$ denote the Lorentz indices. One immediately sees that, at tree level the coupling for tadpole $T$ and masse $UT, TT$ is not allowed. Thus, from the point group selection rule (10.33), the order of twisted fields should be a multiple of three.

Constraint from the $R$-symmetry invariance along each sublattice is also strong. Consider for simplicity fields without oscillators, in which the $R$-symmetry becomes $H$-momentum found in Table 10.1,

$$U_1 : (-1, 0, 0),$$

$$U_2 : (0, 1, 0),$$

$$U_3 : (0, 0, 1),$$

$$T_1 : \tfrac{1}{3}(-1, 1, 1).$$

For couplings $U_1^k U_2^l U_3^m T^n$, the condition (10.51) applies

$$\left(-k - \tfrac{n}{3}, l + \tfrac{n}{3}, m + \tfrac{n}{3}\right) \equiv (-1, 1, 1) \mod (3, 3, 3). \tag{10.104}$$

We can check that the solutions are only of the following two types:

$$U_1^{3p+1} U_2^{3q+1} U_3^{3r+1} T^{9s}, \quad U_1^{3p} U_2^{3q} U_3^{3r} T^{9s+3}, \tag{10.105}$$

where $p, q, r, s$ are nonnegative integers.

The allowed lowest-order interactions are

$$U_1 U_2 U_3, \quad T_1 T_1 T_1, \tag{10.106}$$

while the couplings of forms $U, UU, UTT, UUT$ are not allowed, either. Therefore, we see that the mass can be given by Higgs mechanism up to by radiative correction, which is suppressed by string unification scale. Thus, we expect a solution of the $\mu$-problem since $UU$ and $TT$ terms for the Higgs masses are forbidden (see Sect. 13.8). Also, there is no order 4 and 5 coupling. The importance of nonrenormalizable operators depends on the singlet VEVs.

The Lorentz invariance rule simply implies the local symmetry under both the gauge group and the spacetime Lorentz group. Also, fields with excitation are constrained. In the standard embedding, $(\mathbf{1}, \mathbf{27})$ contains no derivative, but $(\overline{\mathbf{3}}, \mathbf{1})$ does due to the oscillator mode $\alpha^a_{-1/3}$. Therefore, the $(\overline{\mathbf{3}}, \mathbf{1})^n$ vanishes unless $n$ is a multiple of 3.

Now, let us look into how twisted fields are arranged to form invariant coupling. Recall that we have 27 fixed points and hence as many $T_1$ fields. According to the Lorentz invariance rule, in each sublattice the term should be invariant. In terms of fixed point, the rule translates into the following. Denoting the different fixed points of Fig. 7.3 as

$\bullet : (\theta, 0).$

$\circ : (\theta, e_1) = (\theta, (1 - \theta) f_o)) \sim (\theta, e_2) = (\theta, (1 - \theta) \theta f_o),$

$\times : (\theta, e_1 + e_2) = (\theta, (1 - \theta) f_\times) \sim (\theta, 2e_1).$

The rule for the translational part (10.34) simplifies. The translational part of the space group (10.34) becomes

$$(1 - \theta)(f_1 + u_1) + \theta(1 - \theta)(f_2 + u_2) + \theta^2(1 - \theta)(f_3 + u_3) = 0.$$

The fixed points are invariant by construction $\theta^l f_2 + u_2$ as $f_2 + u_2'$. So, we have equivalently

$$(1 - \theta)(f_1 + u_1) + (1 - \theta)(f_2 + u_2) + (1 - \theta)(f_3 + u_3) = 0.$$

We can drop the common factor $(1 - \theta)$

$$(f_1 + u_1) + (f_2 + u_2) + (f_3 + u_3) = 0. \tag{10.107}$$

So, for a given $(f_1 + u_1)$, $(f_2 + u_2)$ we can uniquely determine $f_3 + u_3$. Also, for the same fixed points if we change $u_1$ to another lattice vector $u_1' = u_1 + \lambda$, still we can always find a new $u_3' = u_3 - \lambda$ satisfying the selection rule.

Solving the constraint, we have the following kinds of combinations only for the couplings:

$$\bullet \circ \times, \quad \bullet \bullet \bullet, \quad \times \times \times, \quad \circ \circ \circ \tag{10.108}$$

and some permutations in each two-torus. In other words, if we choose two fixed points, the third fixed point is determined. Thus, the number of such combinations is $27 \times 27 = 729$, but we will see that the actual possibility reduces.

### Mass Hierarchy

Consider the $\mathbb{Z}_3$ and a coupling of purely untwisted fields $UUU$ or a coupling of twisted sector all sitting at the same fixed points $\theta^1\theta^1\theta^1$. If we use one of them as Higgs field and give a VEV, then the remaining two fields have the degenerate mass. Due to this fact, it is in general hard to explain the quark mass hierarchy with this simple model of $U$ matter only [30].

For the twisted sector fields $\theta^1\theta^1\theta^1$ located at the different fixed points, we have a very fruitful interpretation of the quark mass hierarchy in terms of geometry. The Yukawa coupling for $\theta^1\theta^1\theta^1$ type is totally determined by the given twist and the shape of lattice.

The space group selection rule (10.34) constrains the fixed points to be all at the same point or all at the different points in each sublattice

$$(f_1)_a = (f_2)_a = (f_3)_a, \quad a = 1, 2, 3,$$

or

$$(f_1)_a \neq (f_2)_a \neq (f_3)_a \neq (f_1)_a, \quad a = 1, 2, 3,$$

up to a lattice translation. So, for each $a$ we should satisfy one of the above conditions. There are only eight possible vectors $f_1 - f_2$

$$(0, 0, 0, 0, 0, 0),$$

$$\left(\tfrac{1}{3}, \tfrac{2}{3}, 0, 0, 0, 0\right), \ \left(0, 0, \tfrac{1}{3}, \tfrac{2}{3}, 0, 0\right), \ \left(0, 0, 0, 0, \tfrac{1}{3}, \tfrac{2}{3}\right),$$

$$\left(\tfrac{1}{3}, \tfrac{2}{3}, \tfrac{1}{3}, \tfrac{2}{3}, 0, 0\right), \ \left(\tfrac{1}{3}, \tfrac{2}{3}, 0, 0, \tfrac{1}{3}, \tfrac{2}{3}\right), \ \left(0, 0, \tfrac{1}{3}, \tfrac{2}{3}, \tfrac{1}{3}, \tfrac{2}{3}\right),$$

$$\left(\tfrac{1}{3}, \tfrac{2}{3}, \tfrac{1}{3}, \tfrac{2}{3}, \tfrac{1}{3}, \tfrac{2}{3}\right),$$

(10.109)

depending on how the three fields are distributed. From this, we note that among 729 possible couplings only 14 of them are independent.

The Yukawa coupling is given in (10.97)

$$Y_{\theta^1 \theta^1 \theta^1} = g_s N \sum_{u \in \mathbb{Z}^6} \exp\left[-2\pi (f_1 - f_2 + u)^\top M (f_1 - f_2 + u)\right].$$

For simplicity, we take that each two-torus is orthogonal to others, by setting all the angles to be $\pi/2$ in (3.80), in which case the matrix $M$ can take the simple form

$$M = \begin{pmatrix} T_1 & -\tfrac{1}{2}T_1 & & & & \\ -\tfrac{1}{2}T_1 & T_1 & & & & \\ & & T_3 & -\tfrac{1}{2}T_3 & & \\ & & -\tfrac{1}{2}T_3 & T_3 & & \\ & & & & T_5 & -\tfrac{1}{2}T_5 \\ & & & & -\tfrac{1}{2}T_5 & T_5 \end{pmatrix}$$

(10.110)

with other entries being zero. $T_i$s given in (10.110) measure the length of unit lattice $\mathrm{Im}\, T_i = \frac{\sqrt{3}}{16\pi^2} R_i^2$ associated with $\Lambda$.

We can expand the exponential form (10.97) in powers of $\lambda_i = 3e^{-2\pi T_i/3}$, which explains the exponential suppression along the distance between two fields. In the simplest case, setting all $T_i$s equal to $T$ we have a mass hierarchy $1, \lambda, \lambda^2$, and $\lambda^3$ [31]. As a bottom-up approach, we can try to put some fields on some specific fixed points to explain the observed mass hierarchy. However, from the top-down approach, namely beginning with $E_8 \times E_8$ theory down to the SM, it is very hard just using renormalizable couplings to compromise with phenomenological needs. The Higgs field is located at the fixed point $f_1$ and $M$ is a $6 \times 6$ matrix, taking into account the VEV and couplings. The appearance of $f_1 - f_2$ accounts the fact that the Yukawa couplings are dependent on two fixed points by (10.108).

**Table 10.2** Renormalizable vertices in the $\mathbb{Z}_N$ orbifold

| Orbifold | Pure unt. | Pure twisted | Mixed |
|---|---|---|---|
| $\mathbb{Z}_3$ | $U_1 U_2 U_3$ | $\theta^1 \theta^1 \theta^1$ | |
| $\mathbb{Z}_4$ | $U_1 U_2 U_3$ | $\theta^1 \theta^1 \theta^2$ | $\theta^2 \theta^2 U_3$ |
| $\mathbb{Z}_6$-I | $U_1 U_2 U_3$ | $\theta^1 \theta^2 \theta^3, \theta^2 \theta^2 \theta^2$ | $\theta^3 \theta^3 U_3$ |
| $\mathbb{Z}_6$-II | $U_1 U_2 U_3$ | $\theta^1 \theta^2 \theta^3, \theta^1 \theta^1 \theta^4$ | $\theta^2 \theta^4 U_3, \theta^3 \theta^3 U_2$ |
| $\mathbb{Z}_7$ | $U_1 U_2 U_3$ | $\theta^1 \theta^2 \theta^4$ | |
| $\mathbb{Z}_8$-I | $U_1 U_2 U_3$ | $\theta^1 \theta^2 \theta^5, \theta^2 \theta^2 \theta^4$ | $\theta^4 \theta^4 U_2$ |
| $\mathbb{Z}_8$-II | $U_1 U_2 U_3$ | $\theta^1 \theta^1 \theta^6, \theta^1 \theta^3 \theta^4, \theta^2 \theta^3 \theta^3$ | $\theta^2 \theta^6 U_3, \theta^4 \theta^4 U_3$ |
| $\mathbb{Z}_{12}$-I | $U_1 U_2 U_3$ | $\theta^1 \theta^2 \theta^9, \theta^1 \theta^4 \theta^7, \theta^2 \theta^3 \theta^7$ | $\theta^3 \theta^9 U_2, \theta^6 \theta^6 U_2$ |
| | | $\theta^2 \theta^4 \theta^6, \theta^4 \theta^4 \theta^4$ | |
| $\mathbb{Z}_{12}$-II | $U_1 U_2 U_3$ | $\theta^1 \theta^3 \theta^8, \theta^1 \theta^1 \theta^{10}$ | $\theta^2 \theta^{10} U_3, \theta^4 \theta^8 U_3$ |
| | | $\theta^3 \theta^3 \theta^6, \theta^2 \theta^5 \theta^5$ | $\theta^6 \theta^6 U_3$ |

$\theta^n$ and $U_i$ denote the number of twists and the number of (spacetime Lorentz symmetry) SO(10) vector components, respectively. Some allowed coupling by the space group selection may not be allowed by Lorentz invariance. In the literature, sometimes $\theta^i$ is also written as $T_i$

## 10.5.2 Yukawa Couplings in $\mathbb{Z}_N$ Orbifolds

In Table 10.2, we have listed possible renormalizable couplings arising from general $\mathbb{Z}_N$ orbifolds, satisfying the above selection rules [29, 32]. $\theta^n$ means a twisted field belonging to the $n$th twisted sector fields, and $U_i$ means the untwisted fields with a nonzero spacetime Lorentz SO(10) vector component in the $i$th direction.

In general orbifolds, properties of Yukawa couplings are different from those of the $\mathbb{Z}_3$ case discussed above. For example, in the $\mathbb{Z}_4$ orbifold the space group selection rule does not forbid "mixed" $TTU$-type couplings, as long as both twisted fields belong to the second twisted sector. However, the main difference is that, in the general $\mathbb{Z}_N$ orbifold, the renormalizable couplings would have *off-diagonal* components [29]. We consider two orbifold examples, $\mathbb{Z}_4$ and $\mathbb{Z}_6$-I.

### $\mathbb{Z}_4$ Orbifold

Consider a $\mathbb{Z}_4$ orbifold coupling $\theta^1 \theta^1 \theta^2$. The coupling transforms under the space group

$$\left(\theta, (1-\theta)(f_1 + u_1)\right)\left(\theta, (1-\theta)(f_2 + u_2)\right)\left(\theta^2, \left(1 - \theta^2\right)\theta^l (f_3 + u_3)\right),$$

where we should consider $l = 0, 1$. The translational part requires

$$(1-\theta)(f_1 + u_1) + \theta(1-\theta)(f_2 + u_2) + \theta^2 \left(1 - \theta^2\right)\theta^l (f_3 + u_3) = 0. \quad (10.111)$$

Having no zero eigenvalue, we may multiply $(1 - \theta)^{-1}$ on both sides to have

$$f_1 + u_1 + \theta(f_2 + u_2) + \theta^{2+l}(1 + \theta)(f_3 + u_3) = 0. \tag{10.112}$$

We may put eigenvalue $\theta = \mathrm{diag}(-1, i, i)$. We may write (10.112) componentwise

$$f_1^1 + u_1^1 - f_2^1 - u_2^1 = 0, \tag{10.113}$$

$$f_1^a + u_1^a + if_2^a + iu_2^a - i^l(1 + i)\left(f_3^a + u_3^a\right) = 0, \quad a = 2, 3. \tag{10.114}$$

The condition (10.113) states that the first two fields should sit at the same point in this direction. The third field can be at any place because it sees the first direction untwisted torus. The other components simply state that they add up to a lattice vector along the $a = 1, 2$ directions. In (10.114), we have nontrivial condition. We have symmetric property that holds for any $l = 0, 1$

$$i^l(1 + i) = 1 + i. \tag{10.115}$$

So, we always have two solutions $i^l(f_3^a + u_3^a)$ for any lattice elements $u_1^a, u_2^a$. Therefore, if the selection rule (10.111) is satisfied by any fixed element for a fixed value of $l$, it holds for any $l$. So, we need to check only one of the cases. For instance, two fixed points denoted by $(\theta^2, e_1) = (\theta^2, (1 - \theta^2)\frac{1}{2}e_1)$ and $(\theta^2, e_2) = (\theta^2, (1 - \theta^2)\frac{1}{2}e_2)$ belong to the same conjugacy class, so they indicate the same fixed points. If one of the points satisfies the selection rule, the other must. Also, since there are two different fields satisfying the same selection rules, we may have off-diagonal component in the Yukawa matrix.

In the $\mathbb{Z}_4$ orbifold, we have the coupling

$$Y_{\theta\theta\theta^2} = N \sum_{v \in (f_2 - f_3 + \Lambda)_\perp} \exp\left(-\frac{1}{4\pi}v^\top M v\right)$$

$$= N\vartheta \begin{bmatrix} f_{23} \\ 0 \end{bmatrix}\left[0, \frac{M}{4\pi i}\right], \tag{10.116}$$

with a symmetric matrix

$$M = \begin{pmatrix} R_1^2 & 0 & * & * \\ 0 & R_1^2 & * & * \\ R_1 R_3 \cos\phi_{13} & -R_1 R_3 \cos\phi_{14} & R_3^2 & 0 \\ R_1 R_3 \cos\phi_{14} & R_1 R_3 \cos\phi_{13} & 0 & R_3 \end{pmatrix}.$$

## $\mathbb{Z}_6$-I Orbifold

Not only the pure twisted ones but also the mixed ones are restricted by the Lorentz group selection rules, from the $H$-momentum conservation. For the example of $\mathbb{Z}_6$-I orbifold, although all the following satisfy the point group rule

$$\theta^1\theta^2\theta^3 : (-1, 1, 1),$$

$$\theta^2\theta^2\theta^2 : (-1, 1, 1),$$

$$\theta^1\theta^1\theta^4 : \left(-\tfrac{4}{3}, -\tfrac{1}{6}, -\tfrac{1}{6}\right),$$

the $\theta\theta\theta^4$ coupling is forbidden by the $R$-invariance.

Consider the space group selection rule. We see that the $\theta^2\theta^2\theta^2$ coupling is essentially the same as the $\theta^1\theta^1\theta^1$ coupling in $\mathbb{Z}_3$ orbifold (exercise). It is not related to the fixed torus, $\theta^1\theta^2\theta^3$ does. The coupling transforms under the space group as

$$(\theta, (1-\theta)(f_1+u_1))\left(\theta^2, \left(1-\theta^2\right)\left(\theta^l f_2 + u_2\right)\right)\left(\theta^3, \left(1-\theta^3\right)\left(\theta^m f_3 + u_3\right)\right).$$

$$l = 0, 1, \quad m = 0, 1, 2.$$

The translational part needs

$$(1-\theta)(f_1+u_1) + \theta\left(1-\theta^2\right)\left(\theta^l f_2 + u_2\right) + \theta^3\left(1-\theta^3\right)\left(\theta^m f_3 + u_3\right) = 0.$$

Multiplying $(1-\theta)^{-1}$ and using $\theta(\theta^l f_2 + u_2) = \theta^{l'} f_3 + \theta u_2'$, etc., we have

$$f_1 + (1+\theta)f_2 + \left(1 + \theta + \theta^2\right) f_3 \in \Lambda. \tag{10.117}$$

Again, if one coupling with a particular combination $l, m$ satisfies the relation, couplings with any $l, m$ satisfy the relation.

The coupling $\theta^2\theta^2\theta^2$ is essentially the $\mathbb{Z}_3$ coupling

$$Y_{\theta^2\theta^2\theta^2} = N \sum_{v \in (f_1 - f_2 + \Lambda)_\perp} \exp\left(-2\pi v^\top M v\right)$$

$$= N\vartheta \begin{bmatrix} f_1 - f_2 \\ 0 \end{bmatrix} \left(0, \frac{M}{4\pi i}\right), \tag{10.118}$$

but with different choice of the lattice $SU(3) \times G_2^2$. Thus, we need $R_4 = \sqrt{3}R_3$, etc. We have the moduli, which is symmetric matrix, $M = \frac{\sqrt{3}}{16\pi^2\alpha'}$

$$M = \begin{pmatrix} R_1^2 & * & & & & \\ -\tfrac{1}{2}R_1^2 & R_1^2 & & & & \\ & & R_3^2 & * & * & * \\ & & -\tfrac{3}{2}R_3^2 & 3R_3^2 & * & * \\ & & R_3 R_5 \cos\phi_{35} & \sqrt{3}R_3 R_5 \cos\phi_{45} & R_5^2 & * \\ & & \sqrt{3}R_3 R_5 \cos\phi_{36} & 3R_3 R_5 \cos\phi_{35} & -\tfrac{3}{2}R_5^2 & 3R_5^2 \end{pmatrix},$$

where the empty entries are zeros and $\cos \phi_{45} = -\sqrt{3} \cos \phi_{35} - \cos \phi_{36}$.

We have nontrivial Yukawa coupling formed by $\theta^1 \theta^2 \theta^3$. It is

$$
\begin{aligned}
Y_{\theta \theta^2 \theta^3} &= N \sum_{v \in (f_1 - f_2 + \Lambda)_\perp} \exp\left(-\frac{\sqrt{3}}{4\pi \alpha'} v^\top M v\right) \\
&= N \vartheta \begin{bmatrix} f_1 - f_2 \\ 0 \end{bmatrix} \left(0, \frac{M}{4\pi i}\right).
\end{aligned}
\tag{10.119}
$$

We have moduli

$$
M = \begin{pmatrix}
R_1^2 & & * & * & * \\
-\frac{3}{2} R_1^2 & 3 R_1^2 & & * & * \\
R_1 R_3 \cos \phi_{13} & -R_1 R_3 (3 \cos \phi_{13} + \sqrt{3} \cos \phi_{14}) & R_3^2 & * \\
\sqrt{3} R_1 R_3 \cos \phi_{14} & 3 R_1 R_3 \cos \phi_{14} & -\frac{3}{2} R_1^2 & R_3^2
\end{pmatrix}.
$$

The case study of a complete $\mathbb{Z}_N$ Yukawa coupling was done and tabulated in Ref. [29]. However, we should be careful on the summation range of the lattice vector, restricted by (10.36) [16], as we have seen before.

### 10.5.3  Toward Realistic Yukawa Couplings

**Texture**

One of the important question in the flavor physics is what is the minimal number of nonzero entries. Since Yukawa matrix is basis dependent, some elements and mixing angles can be derived from other elements.

The most famous model for the Yukawa couplings is provided by the Weinberg–Fritzsch ansatz [33, 34]

$$
M = \begin{pmatrix} 0 & A & 0 \\ A & 0 & B \\ 0 & B & C \end{pmatrix}, \quad |A| \ll |B| \ll |C|,
\tag{10.120}
$$

which naturally explains the Cabbibo angle and the quark mass hierarchy

$$
\sin \theta_C \sim \sqrt{m_d / m_s}.
\tag{10.121}
$$

In the following, we try to show that this form is impossible in string models [35].

In the prime order orbifolds, we can prove it easily. We note that the KM mixing can arise from the off-diagonal term(s) of mass matrix in three complex dimensions. It is noted that in the $\mathbb{Z}_3$ and $\mathbb{Z}_7$ orbifolds, i.e. in the prime orbifolds, there do not appear off-diagonal terms among renormalizable couplings. This would imply the mass degeneracy of all the fields arising from the Higgs mechanism, and no $\mathscr{C}\mathscr{P}$

phase. Still in some cases, for example, $\theta^2\theta^2\theta^2$ coupling in the $\mathbb{Z}_6$-I orbifold, also the third fixed point is determined by the space group selection rule. However, in the non-prime orbifold, for instance, the $\theta\theta^2\theta^3$ coupling in the $\mathbb{Z}_6$-I orbifold can have off-diagonal components.

The following explanation is called as "box closing rule" [35]. If we have a subblock filled except one entry, it should be always completely filled

$$\begin{bmatrix} & \times \\ \times & \times \end{bmatrix} \rightarrow \begin{bmatrix} \times & \times \\ \times & \times \end{bmatrix}. \tag{10.122}$$

This is because if the three entries on LHS should satisfy the space group selection rule in the original matrix, and it can be shown that the blank position $(1, 1)$ also satisfies the rule, thus we should fill the blank position. This rule shows that at the renormalizable level, we cannot obtain the matrix of the Weinberg–Fritzsch type (10.120). However, we may use the fact that off-diagonal terms can appear from nonrenormalizable couplings.

Fortunately, we have an alternative version [28, 35] naturally emerging from string theory, and it also explains angles in terms of mass ratios

$$M = \begin{pmatrix} \epsilon & a & b \\ \tilde{a} & A & c \\ \tilde{b} & \tilde{c} & B \end{pmatrix}, \tag{10.123}$$

while those with lowercase letters are very small compared with $|A| \ll |B|$, to explain large masses for the third family members. To take this form, only $A, B$ should emerge from renormalizable operators, while others from nonrenormalizable ones for them to get suppressed. This also explains the Cabbibo angle relation (10.121) and qualitatively gives natural ratios of fermion mass scales.

Non-prime orbifolds allow off-diagonal terms even in the renormalizable couplings. In Table 10.2, we see that there can be more than one coupling.

To conclude, the nonrenormalizable couplings are very crucial ingredients for having realistic Yukawa interactions as shown in [12, 36, 37].

## Constructing Flat Direction

Although renormalizable operators in odd-order orbifolds are diagonal, we can form off-diagonal entries using nonrenormalizable operators. If some mechanism breaks the gauge symmetry and the SM nonsinglet fields develop VEVs, then we can fit nonrenormalizable off-diagonal entries and fit the mass matrix toward realistic couplings [12, 36, 37]. Of course, this idea can also be applied to non-prime orbifolds. This method is very useful in the top-down approach. Accompanying this method, there possibly arises the Fayet–Iliopoulos term generated by the anomalous U(1), breaking several unwanted U(1)s.

## Appendix

## Hypergeometric Functions

The hypergeometric functions are building blocks of four-point correlation functions. They are given as

$$F_1(x) = \int_0^x dz\tilde{\omega}_k(z) = (-1)^{-(k_2+k_3)/N} x^{-1+(k_1+k_2)/N} \tag{10.124}$$

$$\times B\left(\frac{k_1}{N}, \frac{k_2}{N}\right) {}_2F_1\left(\frac{k_1}{N}, 1 - \frac{k_3}{N}; \frac{k_1+k_2}{N}; x\right), \tag{10.125}$$

$$F_2(1-x) = \int_x^1 dz\tilde{\omega}_k(z) = -(-1)^{-k_3/N}(1-x)^{-1+(k_2+k_3)/N} \tag{10.126}$$

$$\times B\left(\frac{k_2}{N}, \frac{k_3}{N}\right) {}_2F_1\left(\frac{k_3}{N}, 1 - \frac{k_1}{N}; \frac{k_2}{N} + \frac{k_3}{N}; 1 - x\right), \tag{10.127}$$

$$\bar{F}_1'(\bar{x}) = \int_0^{\bar{x}} d\bar{z}(\tilde{\omega}_{N-k}(z))^* = F_1(k_i \to N - k_i, x \to \bar{x}), \tag{10.128}$$

$$\bar{F}_2'(1-\bar{x}) = \int_{\bar{x}}^1 d\bar{z}(\tilde{\omega}_{N-k}(z))^* = F_2(k_i \to N - k_i, x \to \bar{x}). \tag{10.129}$$

Some functions are related

$$F_0 = \int_{-\infty}^0 dz\tilde{\omega}_k(z) = e^{\pi i(1-k_4/N)} B\left(\frac{k_4}{N}, \frac{k_1}{N}\right) {}_2F_1\left(\frac{k_4}{N}, 1 - \frac{k_2}{N}; \frac{k_1}{N} + \frac{k_4}{N}; 1 - x\right),$$
$$\tag{10.130}$$

$$F_3 = \int_1^\infty dz\tilde{\omega}_k(z) = B\left(\frac{k_3}{N}, \frac{k_4}{N}\right) {}_2F_1\left(\frac{k_4}{N}, 1 - \frac{k_2}{N}; \frac{k_3+k_4}{N}; x\right), \tag{10.131}$$

and are expressed in terms of hypergeometric function

$$_2F_1(a, b; c; x) = \sum_{n=0}^\infty \frac{(a)_n(b)_n}{(c)_n} \frac{x^n}{n!}, \tag{10.132}$$

with the Pochhammer symbol $(q)_n = \Gamma(q + n)/\Gamma(q)$. We also define

$$F_0(1-x) = \frac{\sin(\pi k_3/N)}{\sin(\pi k_4/N)} \tag{10.133}$$

$$\times e^{i\pi(k_3+k_4)/N}\left[\frac{\sin(\pi(k_2+k_3)/N)}{\sin(\pi k_1/N)} e^{ik_2\pi/N} F_1(x) + F_2(1-x)\right],$$

$$F_3(x) = \frac{\sin(\pi k_1/N)}{\sin(\pi k_4/N)} \tag{10.134}$$

$$\times e^{i\pi(k_2+k_3)/N} \left[ F_1(x) + \frac{\sin(\pi(k_1+k_2)/N)}{\sin(\pi k_1/N)} e^{-ik_2\pi/N} F_2(1-x) \right].$$

## Exercises

▶ **Exercise 10.1** Show (10.62).

▶ **Exercise 10.2** The rule for the translational part (10.34) simplifies. The translational part of the space group (10.34) becomes

$$(1-\theta)(f_1+u_1) + \theta(1-\theta)(f_2+u_2) + \theta^2(1-\theta)(f_3+u_3) = 0.$$

We see that $(\theta, (1-\theta)(f_2+u_2))$ belongs to the same conjugacy class as $(\theta, (1-\theta)(\theta^l f_2+\theta^l u_2))$, $(l=0,1,2)$, indicating the same fixed point $f_2$. So, we can always write $\theta^l f_2$ as $f_2$. So, we have equivalently

$$(1-\theta)(f_1+u_1) + (1-\theta)(f_2+u_2) + (1-\theta)(f_3+u_3) = 0.$$

▶ **Exercise 10.3** Show that the $\theta^2\theta^2\theta^2$ coupling is essentially the same as the $\theta^1\theta^1\theta^1$ coupling in $\mathbb{Z}_3$ orbifold.

## References

1. S. Hamidi, C. Vafa, Interactions on orbifolds. Nucl. Phys. **B279**, 465–513 (1987)
2. L.J. Dixon, D. Friedan, E.J. Martinec, S.H. Shenker, The conformal field theory of orbifolds. Nucl. Phys. **B282**, 13–73 (1987)
3. D. Friedan, E.J. Martinec, S.H. Shenker, Conformal invariance, supersymmetry and string theory. Nucl. Phys. **B271**, 93–165 (1986)
4. E. Witten, Dimensional reduction of superstring models. Phys. Lett. **155B**, 151 (1985)
5. J.P. Derendinger, L.E. Ibanez, H.P. Nilles, On the low-energy limit of superstring theories. Nucl. Phys. **B267**, 365–414 (1986)
6. L.J. Dixon, V. Kaplunovsky, J. Louis, On effective field theories describing (2,2) vacua of the heterotic string. Nucl. Phys. **B329**, 27–82 (1990)
7. C.P. Burgess, A. Font, F. Quevedo, Low-energy effective action for the superstring. Nucl. Phys. **B272**, 661–676 (1986)
8. D. Bailin, A. Love, Orbifold compactifications of string theory. Phys. Rept. **315**, 285–408 (1999)
9. F. Quevedo, Lectures on superstring phenomenology. AIP Conf. Proc. **359**, 202–242 (1996)
10. I.R. Klebanov, E. Witten, Proton decay in intersecting D-brane models. Nucl. Phys. **B664**, 3–20 (2003)
11. D. Friedan, S.H. Shenker, E.J. Martinec, Covariant quantization of superstrings. Phys. Lett. **B160**, 55–61 (1985)

12. A. Font, L.E. Ibanez, H.P. Nilles, F. Quevedo, Degenerate orbifolds. Nucl. Phys. **B307**, 109–129 (1988). [Erratum: Nucl. Phys.B310,764(1988)]
13. T. Kobayashi, S. Raby, R.-J. Zhang, Searching for realistic 4d string models with a Pati-Salam symmetry: orbifold grand unified theories from heterotic string compactification on a Z(6) orbifold. Nucl. Phys. **B704**, 3–55 (2005)
14. T. Kobayashi, S. Raby, R.-J. Zhang, Constructing 5-D orbifold grand unified theories from heterotic strings. Phys. Lett. **B593**, 262–270 (2004)
15. S. Stieberger, D. Jungnickel, J. Lauer, M. Spalinski, Yukawa couplings for bosonic Z(N) orbifolds: their moduli and twisted sector dependence. Mod. Phys. Lett. **A7**, 3059–3070 (1992)
16. J. Erler, D. Jungnickel, M. Spalinski, S. Stieberger, Higher twisted sector couplings of Z(N) orbifolds. Nucl. Phys. B **397**, 379–416 (1993)
17. K.-S. Choi, T. Kobayashi, Higher order couplings from heterotic orbifold theory. Nucl. Phys. **B797**, 295–321 (2008)
18. S. Abel, Cargese lectures: a String phenomenology primer. PoS **CARGESE2007**, 001 (2007)
19. J. J. Atick, L. J. Dixon, P. A. Griffin, D. Nemeschansky, Nucl. Phys. **B298**, 1–35 (1988)
20. H. Kawai, D.C. Lewellen, S.H.H. Tye, A relation between tree amplitudes of closed and open strings. Nucl. Phys. **B269**, 1–23 (1986)
21. M. Spalinski, Duality transformations in twisted Narain compactifications. Nucl. Phys. B **377**, 339–368 (1992)
22. J. Lauer, J. Mas, H.P. Nilles, Twisted sector representations of discrete background symmetries for two-dimensional orbifolds. Nucl. Phys. B **351**, 353–424 (1991)
23. T.T. Burwick, R.K. Kaiser, H.F. Muller, General Yukawa couplings of strings on Z(N) orbifolds. Nucl. Phys. **B355**, 689–711 (1991)
24. E. Emsiz, E.M. Opdam, J.V. Stokman, Periodic integrable systems with delta-potentials. Commun. Math. Phys. **264**, 191–225 (2006)
25. S.A. Abel, A.W. Owen, N point amplitudes in intersecting brane models. Nucl. Phys. **B682**, 183–216 (2004)
26. M. Cvetic, I. Papadimitriou, Conformal field theory couplings for intersecting D-branes on orientifolds. Phys. Rev. **D68**, 046001 (2003). [Erratum: Phys. Rev.D70,029903(2004)]
27. T. Kobayashi, O. Lebedev, Heterotic Yukawa couplings and continuous Wilson lines. Phys. Lett. **B566**, 164–170 (2003)
28. J.A. Casas, C. Munoz, Fermion masses and mixing angles: a test for string vacua. Nucl. Phys. **B332**, 189–208 (1990). [Erratum: Nucl. Phys. **B340**,280 (1990)]
29. J.A. Casas, F. Gomez, C. Munoz, Complete structure of Z(n) Yukawa couplings. Int. J. Mod. Phys. **A8**, 455–506 (1993)
30. H.P. Nilles, M. Olechowski, S. Pokorski, Does a radiative generation of Quark masses provide us with the correct mass matrices?. Phys. Lett. **B248**, 378–386 (1990)
31. S.A. Abel, C. Munoz, Quark and lepton masses and mixing angles from superstring constructions. JHEP **02**, 010 (2003)
32. T. Kobayashi, N. Ohtsubo, Yukawa coupling condition of $Z(N)$ orbifold models. Phys. Lett. B **245**, 441–446 (1990)
33. H. Fritzsch, Weak interaction mixing in the six—Quark theory. Phys. Lett. B **73**, 317–322 (1978)
34. H. Fritzsch, Quark masses and flavor mixing. Nucl. Phys. B **155**, 189–207 (1979)
35. J.A. Casas, F. Gomez, C. Munoz, Fitting the quark and lepton masses in string theories. Phys. Lett. **B292**, 42–54 (1992)
36. L.E. Ibanez, J. Mas, H.-P. Nilles, F. Quevedo, Heterotic strings in symmetric and asymmetric orbifold backgrounds. Nucl. Phys. **B301**, 157–196 (1988)
37. J.E. Kim, Trinification with sin**2 theta(W) = 3/8 and seesaw neutrino mass. Phys. Lett. B **591**, 119–126 (2004)

# Effective Action

We consider four dimensional action reflecting the symmetry of the internal manifold. The observed four spacetime dimensions should be the effective ones, since string theory is defined in ten dimensions and the extra dimensions hide as small compact space.

First, in Sect. 11.1, starting from ten dimensional supergravity coupled to gauge fields, we use dimensional reduction and the symmetry matching, we obtain the four dimensional action [1, 2]. It provides a nice way to construct four dimensional supergravity action reflecting large symmetry related to internal manifold. We can track which physics may originate from higher dimensional theory with unified field contents.

Stringy effect appears in combination with the geometry of the extra dimensions. The geometric parameters of internal geometry become moduli fields, that is, scalar fields in four dimensions and they parameterize flat directions in supersymmetric theory. The symmetry of the internal space, enhanced by string nature, constrains the moduli dependence in the Kähler potential, superpotential and gauge kinetic function. This shall be considered in Sect. 11.2. The fate of such fields has important implication on dynamical compactification and cosmology.

What is special in the orbifold compactification is the presence of the twisted strings. In the previous section, we calculated correlation functions directly with string theory and matched the corresponding Yukawa couplings and its higher order generalizations, reflected to superpotential. Using the same technique, with the help of the target space modular properties, we can also extract information on Kähler potentials in Sect. 11.3 [3, 4].

© Springer Nature Switzerland AG 2020

K.-S. Choi, J. E. Kim, *Quarks and Leptons From Orbifolded Superstring*,
Lecture Notes in Physics 954, https://doi.org/10.1007/978-3-030-54005-0_11

## 11.1   Dimensional Reduction

Dimensional reduction of the ten dimensional supergravity Lagrangian on torus $T^6$ gives $\mathcal{N} = 4$ supergravity Lagrangian in four dimensions [5, 6]. The internal symmetry SO(6) is identical to the $R$-symmetry of the four dimensional supersymmetry.

In the end we need $\mathcal{N} = 1$, so we break the $R$-symmetry or equivalently the internal symmetry by imposing stronger invariance. For instance, it can be the SU(3) holonomy of Calabi–Yau manifold or the $\mathbb{Z}_N$ holonomy of orbifolds [1, 2]. That is, we associate the charged fields with holonomy and keep the invariant Lagrangian under the combined one. In this way, we can obtain the untwisted sector fields or the bulk fields inherited from ten dimensional ones.

### 11.1.1  Dimeansional Reduction

We begin with the ten dimensional $\mathcal{N} = 1$ (16 real supercharges) supergravity coupled with non-Abelian gauge fields. The bosonic Lagrangian is

$$
\frac{\mathcal{L}}{\sqrt{G^{(10)}}} = \frac{1}{2\kappa_{10}^2 e^{2\Phi}} \left[ R^{(10)} + 4\partial_M \Phi \partial^M \Phi - \frac{1}{2} \hat{H}_{MNP} \hat{H}^{MNP} \right]
$$
$$
- \frac{1}{2g_{10}^2 e^{2\Phi}} \operatorname{tr}_v F_{MN} F^{MN},
$$

(11.1)

where $R^{(10)}$ is the Ricci scalar made of 10D metric $G_{MN}^{(10)}$, $G^{(10)} = |\det G^{(10)}|$, and $\kappa_{10}^2$ is the 10D gravitational constant to be fixed shortly. We use the matrix notation for the gauge field as in (5.27). The trace is normalized in units of the vector representation of SO($n$). The ten dimensional supersymmetry is restrictive enough to completely determine the full action [7, 8].

We have overall dilaton coupling $e^{-2\Phi}$ in the "string frame." The field strength of the antisymmetric tensor field $B_{MN}$ is defined as

$$
\hat{H}_{MNP} = \partial_{[M} B_{NP]} - \frac{\kappa_{10}^2}{g_{10}^2} \left( \omega_{MNP} + \omega_{MNP}^{\text{grav}} \right),
$$

with the Chern–Simons form

$$
\omega_{MNP} = \operatorname{tr}_v \left( A_{[M} F_{NP]} - \frac{2}{3} A_{[M} A_N A_{P]} \right),
$$

(11.2)

which is required by supersymmetry. We have Abelian gauge symmetry

$$
\delta A_M = \partial_M \lambda - i[A_M, \lambda], \quad \delta B_{MN} = \frac{\kappa_{10}^2}{g_{10}^2} \operatorname{tr}_v(\lambda \partial_{[M} A_{N]}).
$$

(11.3)

We will later see that, for anomaly cancellation, we have a similar three-form $\omega^{\text{grav}}_{MNP}$ made of Ricci tensor, fixing $\kappa^2_{10}/g^2_{10} = \alpha'/4$ [9]. Also if coupled, from the condition of anomaly cancellation the gauge group is determined as SO(32) or $E_8 \times E_8$ [10]. It is the low-energy limit $\alpha' \to 0$ of the heterotic strings with the same gauge groups. Taking SO(32) gauge group instead of $E_8 \times E_8$, this provides the low-energy action of Type-I and heterotic string theories, with field redefinition

$$G_{IMN} = e^{-\Phi_{\text{h}}} G_{\text{h}MN}, \quad \Phi_{\text{I}} = -\Phi_{\text{h}}, \quad F_{MNP} = \hat{H}_{MNP}, \tag{11.4}$$

where $\hat{F}_{MNP}$ is the field strength for the rank-2 RR tensor field containing also the Chern–Simons terms.

**Field Reduction**

Now we compactify the internal space and obtain the four dimensional theory. We first take the extra six dimensions as a torus $T^6$ and generalize it. We decompose the metric as follows. It is convenient to use the vielbein and its inverse [11, 12]

$$e^A_M = \begin{pmatrix} e^a_\mu & A^n_\mu E^a_n \\ 0 & E^m_\alpha \end{pmatrix}, \quad e^M_A = \begin{pmatrix} e^\mu_a & -e^\nu_a A^m_\nu \\ 0 & E^m_a. \end{pmatrix}. \tag{11.5}$$

This means we have the metric

$$G^{(10)}_{MN} = e^A_M e^B_N \eta_{AB} = \begin{pmatrix} G_{\mu\nu} + A^a_\mu A_{\nu a} & A^n_\mu \\ A_{\nu m} & G_{mn}. \end{pmatrix}, \tag{11.6}$$

with the spacetime metric $G_{\mu\nu} = e^a_\mu e^b_\nu \eta_{ab}$ and the internal metric $G_{mn} = e^a_m e^b_n \delta_{ab}$. We omit in what follows the Kaluza–Klein U(1) fields $A_{\mu n}$ and similar ones from $B_{\mu n}$, because they are not invariant under the holonomy and to be projected out.[1] The only gauge field is those inherited from the ten dimensional fields.

We have contracted Lorentz indices in (11.1) using $G^{(10)}_{MN}$. We assume so-called the cylindrical condition that all the fields have no dependence on the internal direction. However, it should be that they should be consistent with the ten dimensional equation of motion. We have $\sqrt{\det G^{(10)}_{MN}} = \sqrt{\det G_{mn}}\sqrt{\det G_{\mu\nu}}$.

$$R^{(10)} = R + 4\partial_\mu\phi\partial^\mu\phi - \frac{1}{4}G^{mn}G^{pq}\partial_\mu G_{mq}\partial^\mu G_{np}, \tag{11.7}$$

where the 4D Lorentz indices are contracted by the four dimensional metric $G_{\mu\nu}$. Integrating out the internal space, we have the six dimensional volume $V = (2\pi\ell_s)^6\sqrt{\det G_{mn}}$, which provides a suppression factor for the four dimensional

---

[1] They become a triplet under the SU(3)$_{\text{h}}$ holonomy group below and with the above non-Abelian gauge fields, we have enhanced global symmetry [11].

dilaton $\phi$

$$\frac{1}{e^{2\phi}} \equiv \frac{\sqrt{\det G_{mn}}}{e^{2\Phi}}, \tag{11.8}$$

with the string length defined below Eq. (11.1). We have the physical four dimensional gravitational constant

$$\kappa_{10}^2 e^{2\phi} \equiv \kappa^2 = 8\pi G_N, \tag{11.9}$$

where we define $\kappa_{10}^2 \equiv \frac{1}{2}(2\pi)^7 \alpha'^4$ [9].[2]

So, we can measure the 4D gravitational coupling $\kappa$ in the form of the Newton's constant $G_N$, assuming there is a mechanism stabilizing the dilaton.

Likewise we obtain the physical gauge coupling

$$g_{\mathrm{YM}}^2 \equiv g^2 \equiv g_{10}^2 e^{2\phi}. \tag{11.10}$$

And the gauge field is decomposed as[3]

$$\mathrm{tr}_{\mathrm{v}}\, F_{MN} F^{MN} = \mathrm{tr}_{\mathrm{v}}\, F_{\mu\nu} F^{\mu\nu} + 2\, \mathrm{tr}_{\mathrm{v}}\, F_{\mu n} F^{\mu n}, \tag{11.11}$$

where

$$F_{\mu\nu} = D_\mu A_\nu - D_\nu A_\mu \tag{11.12}$$

$$F_{\mu n} = D_\mu A_n. \tag{11.13}$$

Here we assumed that we have only simple unbroken group. If we have more than one such groups, we have as many field strengths having the same form. The normalization and the gauge coupling of each gauge group are the same if we take the trace over the vector representations.

Finally, the three-form field strength is decomposed as

$$\hat{H}_{MNP} \hat{H}^{MNP} = H_{\mu\nu\rho} H^{\mu\nu\rho} + 3 H_{\mu n p} H^{\mu n p} + 3 H_{\mu\nu p} H^{\mu\nu p} + H_{mnp} H^{mnp}, \tag{11.14}$$

---

[2]The combination $\kappa_{10} e^{\Phi}$ is proportional to the tension of D1-brane or D-string. We have SL(2, $\mathbb{Z}$) symmetry exchanging it with the fundamental Type I/II string. This convention introduces no extra factor. The dilaton normalization is fixed by the ratio $e^{\Phi} = T/\tau_{D1}$ where $T$ is the fundamental string tension (6.20) and $\tau_{D1}$ is the D1-string tension. Then we may convert type I string coupling to the heterotic string using the relation (11.4).

[3]We neglect the Kaluza–Klein U(1)'s, otherwise we should convert $F_{MN}$ to those in the tangent space and obtain lower dimensional ones by multiplying $e_\mu^n$ and $E_\alpha^M$.

where

$$H_{\mu\nu\rho} = \partial_{[\mu} B_{\nu\rho]} - \frac{\alpha'}{4} \mathrm{tr_v} \left( A_{[\mu} F_{\nu\rho]} - \frac{2}{3} A_{[\mu} A_\nu A_{\rho]} \right),$$

$$H_{\mu np} = \partial_\mu B_{np} - \frac{\alpha'}{4} \mathrm{tr_v} \left( (D_\mu A_n) A_p - A_n D_\mu A_p \right), \qquad (11.15)$$

$$H_{mnp} = -\frac{\alpha'}{6} \mathrm{tr_v} A_m [A_n, A_p],$$

and we assume that every field has no dependence on the internal dimensions.

## SU(3) and $\mathbb{Z}_3$ Invariant Action

Regarding the internal index $m$ as that of the holonomy SO(6), we may keep invariant states. This gives $\mathcal{N} = 4$ supergravity action. We are interested in the effective action from orbifold construction (and Calabi–Yau that we meet later) that realizes $\mathcal{N} = 1$ supersymmetry. To this end, we first consider SU(3)$_h$ holonomy group. We embed it to one of the E$_8$ gauge groups, by identifying SU(3)$_h$ Ricci tensor with SU(3) field strength. Then, the unbroken gauge group is the commutant to the embedded group, E$_6$.

We may rearrange the gauge bosons of the E$_8$ in the representations of SU(3) $\times$ E$_6$ :

$$\mathbf{248} \rightarrow (\mathbf{8, 1}) + (\mathbf{1, 78}) + (\mathbf{3, 27}) + (\bar{\mathbf{3}}, \overline{\mathbf{27}}).$$

The zero modes of the "off-diagonal" components become charged scalars in four dimensions. They are going to describe the matter sfermions. They are complex representations, so we complexify the gauge field

$$B_1 = \frac{1}{\sqrt{2}} (A_4 + i A_5), \quad B_2 = \frac{1}{\sqrt{2}} (A_6 + i A_7), \quad B_3 = \frac{1}{\sqrt{2}} (A_8 + i A_9).$$

The corresponding generators $T_x^a$ are also obtained by linearly combining the E$_8$ generators, where $a, x$ are, respectively, the SU(3) and E$_6$ indices. We could use the same index as the spacetime index because of the embedding. We will later see that if the internal manifold is *Kähler,* in Sect. 15.1.2, holomorphic and antiholomorphic indices do not mix.

The charged scalar field $Q^x$ carries the

$$B_a = \sum_x T_x^a Q^x, \quad \text{SU(3) holonomy, no summation over } a. \qquad (11.16)$$

There is no summation over $a$ if the SU(3) is broken. The $Q_a^x$ field carries the SU(3) index $a$ and E$_6$ index $x$. We have

$$\mathrm{Tr}\, T_x^a T_y^b T_z^c = \epsilon^{abc} d_{xyz}, \qquad (11.17)$$

where $d_{xyz}$ is totally symmetric invariant of $E_6$. If we start instead with the SO(32) gauge group, then we have no nonvanishing tensor corresponding to $d_{xyz}$ for any subgroup. This means that we cannot have down-type Yukawa coupling from perturbative string.

We identify the SU(3) holonomy group with the SU(3) gauge subgroup of $E_8$. The four dimensional gauge fields are untouched but we have only unbroken gauge group SU(3) $\times E_6 \times E_8$. Thus $Q_a$ transform as $(\mathbf{3}, \mathbf{27})$, sfermions unifying the sqarks, sleptons, and Higgses. Some of the field strengths change as

$$H_{\mu a \bar{b}} = \partial_\mu B_{a\bar{b}} - \frac{\alpha'}{2} i \, \mathrm{tr}_v \left( (D_\mu Q_a) Q_{\bar{b}}^* - Q_a D_\mu Q_{\bar{b}}^* \right),$$

$$H_{abc} = \epsilon^{abc} d_{xyz} Q_a^x Q_b^y Q_c^z,$$

$$F_{ab} = \mathrm{tr}[Q_a, Q_b],$$                                            (11.18)

$$F_{a\bar{b}} = \mathrm{tr}[Q_a, Q_{\bar{b}}].$$

The last two gives the scalar potential. $H_{abc}$ and $F_{ab}$ contribute to the superpotential and $F_{a\bar{b}}$ contributes to D-term potential.

The Einstein frame $g_{\mu\nu} \equiv e^{-2\phi} G_{\mu\nu}$ has the desired normalization for the Einstein–Hilbert term. Since the Ricci scalar contains the second derivative of the metric, it introduces extra kinetic term for the dilaton $R \rightarrow R - 6\partial_\mu \phi \partial^\mu \phi$ to make the overall sign desirable. The resulting four dimensional Lagrangian becomes [1, 13, 14]

$$
\frac{\mathscr{L}}{\sqrt{\det g}} = \frac{1}{2\kappa^2} \left[ R - 2\partial_\mu \phi \partial^\mu \phi - \frac{1}{2} e^{-4\phi} H_{\mu\nu\rho} H^{\mu\nu\rho} \right.
$$

$$
- \frac{1}{2} G^{a\bar{b}} G^{c\bar{d}} \left( \partial_\mu G_{a\bar{d}} \partial^\mu G_{\bar{b}c} + \partial_\mu \left( B_{a\bar{d}} - i \frac{\alpha'}{2} \, \mathrm{tr}_v \, Q_a Q_{\bar{d}}^* \right) \right.
$$

$$
\left. \left. \times \, \partial^\mu \left( B_{\bar{b}c} - i \frac{\alpha'}{2} \, \mathrm{tr}_v \, Q_{\bar{b}}^* Q_c \right) \right) \right]
$$                                            (11.19)

$$
- \frac{1}{4g^2} \left[ \mathrm{tr} \, F_{\mu\nu} F^{\mu\nu} + \mathrm{tr} \, D_\mu Q^* D^\mu Q + V(Q) \right].
$$

Here, we changed the normalization for the trace to that of fundamental representaion $\mathrm{tr}_v \, T^a T^b = 2 \, \mathrm{tr} \, T^a T^b$. $V(Q)$ is the scalar potential of the charged scalars $Q$ that we obtain later.

In the orbifold compactification, we need some modification. First, the holonomy group is $\mathbb{Z}_3$ and it is the center of SU(3) proportional to identity, thus we recover the gauge group SU(3) $\times E_6$. Since the $Q$ transform as the full $(\mathbf{3}, \mathbf{27})$, now the decomposition

$$B_a = \sum_{x,a} T_x^a Q_a^x, \quad \mathbb{Z}_3 \text{ holonomy, summation over } a,$$                                            (11.20)

now includes the summation over $a$. We have the same field contents and they become the untwisted sector field. We also have twisted sector fields, which should also contribute both Kähler and superpotential. We will calculate these contribution later in this chapter.

## 11.2 General Backgrounds

In the construction of heterotic string, we obtained 10 dimensional theory by compactifying 26 dimensional left movers. To do this, we formally assumed that we have both left and right movers in 26 dimensions, or setting $d = 16$ in Sect. 6.3.2, and eliminated the latter by assigning the constraint of "no right movers" (6.214). This means, we may compactify any bosonic string theory, symmetric and asymmetric, using the same techniques. In principle, we can compactify more dimensions (or even all the dimensions except our four) to obtain various gauge groups. In this section, we consider such general compactification. We also introduce general background field including the metric and antisymmetric tensor.

### Modular Invariance

For consistent theory, we required modular invariance. Even with general background, we see here that we have essentially the same condition of even and self-dual lattice. For this, we study the lattice generated by the dimensionless momenta.

For constructing vertex operators with momenta, as in (10.2), we define dimensionful momenta

$$k_{Li} \equiv \sqrt{\frac{\alpha}{2}} p_{Li}, \quad k_{Ri} \equiv \sqrt{\frac{\alpha}{2}} p_{Ri}. \tag{11.21}$$

Consider OPE of two vertex operators of bosonic strings

$$: e^{ik_L \cdot X_L + ik_R \cdot X_R} : (z) : e^{ik'_L \cdot X_L + ik'_R \cdot X_R} : (0)$$
$$\sim z^{p_L \cdot p'_L} \bar{z}^{p_R \cdot p'_R} : e^{i(k_L + k'_L) \cdot X_L + i(k_R + k'_R) \cdot X_R} : (0). \tag{11.22}$$

Here and from now on, the dot product means the product using the metric tensor.

Encircling one state once around the other state, the RHS acquires the phase $e^{2\pi i(p_L \cdot p'_L - p_R \cdot p'_R)}$ [15, 16]. We should not observe such phase in local theory, so we require

$$p \circ p' \equiv p_L \cdot p'_L - p_R \cdot p'_R \in \mathbb{Z}. \tag{11.23}$$

We need an integral lattice. In the original construction of heterotic string, we took 16 extra dimensions and eliminated the right movers. In $d < 10$ noncompact dimensions, we have other contributions from right movers also carrying momenta.

The lattice now is Lorentzian. We still need even and self-dual lattice. Typical one is $\Gamma_{k,k}$ with $k \in \mathbb{Z}$. Thus we may have in general modular invariant lattice if we compactify $d$ further dimensions, the left and right movers in the same way. The resulting space has the signature $(26 - d, 10 - d)$.

## 11.2.1 Moduli Space

We may find a modular invariant lattice satisfying the above condition. Consider a representative lattice $\Gamma_0$. By the Lorentz transformation $\Lambda \in O(26 - d, 10 - d, \mathbb{R})$, we can span all the possible lattice $\Lambda\Gamma_0$ (meaning acting $\Lambda$ on all the points in the lattice $\Gamma_0$). This is an over-counting, however, because from the Hamiltonian (the mass shell condition (6.224)) rotations in the left and right movers $\Lambda' \in O(26 - d, \mathbb{R}) \times O(10 - d, \mathbb{R})$ are independent symmetries so that these symmetries lead to equivalent ones. Modding out by such equivalent classes, we have a group space of symmetries of moduli space,

$$\frac{O(26 - d, 10 - d, \mathbb{R})}{O(26 - d, \mathbb{R}) \times O(10 - d, \mathbb{R})}. \tag{11.24}$$

In fact, we have another equivalent class that leaves $\Gamma_0$ invariant, by shuffling the lattice vectors. It is denoted conventionally as

$$O(26 - d, 10 - d, \mathbb{Z}). \tag{11.25}$$

The resulting moduli space is obtained from (11.24) by a further modding by (11.25). This symmetry contains (11.49), generalzing the axionic symmetry and $T$-duality.

We can understand what it means, by taking the $d = 4$ example. From the left movers, we expect the group of rank $26 - d = 22$. From the right movers, one obtains $U(1)^6$ which is not enhanced any more due to $p_R^2 = 0$ for the right movers. The number of parameters is counted as

$$22 \cdot 6 = \frac{(22 + 6)(22 + 6 - 1)}{2} - \frac{22 \cdot 21}{2} - \frac{6 \cdot 5}{2}.$$

It indicates that the fields in this space can be identified as the Kaluza–Kelin modes of graviton, antisymmetric tensor and gauge fields: $\frac{1}{2}6(6 + 1)$ from $G_{ij}$, $\frac{1}{2}6(6 - 1)$ from $B_{ij}$, and $6 \cdot 16$ from $A_i^I$ corresponding to the Wilson lines. Vacuum expectation values (VEV) of moduli such as $\langle G_{ij} \rangle$ describe the geometry. They are called *moduli field* and determine geometry of the compact space. They span $22 \cdot 6$ dimensional *moduli space*. A certain phase of low-energy theory corresponds to some special point in the moduli space. In the construction of heterotic string, a special point of the critical radius was chosen to enhance the gauge symmetry.

Target space metric of the low-energy action takes into account of the above symmetry as a footprint.

## 11.2.2 Narain Compactification

So far we have assumed the background as orthonormal Minkowski spacetime without antisymmetric tensor field. We may consider a non-orthonormal internal space described by metric $G_{ij}$ as in (3.9) and the background $B_{ij}$ [15, 16]. It can be implemented by plugging them into the worldsheet action,

$$S = \frac{1}{2\pi\alpha'} \int d\sigma d\tau \left( G_{ij}\partial_\alpha X^i \partial^\alpha X^j + \epsilon^{\alpha\beta} B_{ij}\partial_\alpha X^i \partial_\beta X^j \right), \tag{11.26}$$

with the contraction by worldsheet metric understood as in Chap. 6. Since the momentum and the winding are properties of zero modes, we do not consider the oscillators at this moment

$$X^i = 2L^i\sigma + 2q^i(\tau),$$

recalling that we had $q^i(\tau) = 2\alpha' p^i \tau$ in (6.190). For more general momenta, so we describe the center-of-mass motion by general function $q^i(\tau)$. As done in (6.206), we have quantized winding number

$$L^i = n^i R, \quad n_i \in \mathbb{Z}. \tag{11.27}$$

Indeed, if there is no such winding, $X^i$ is single-valued and the second term in (11.26) is a total derivative $\partial_\alpha(\epsilon^{\alpha\beta} B_{ij} X^i \partial_\beta X^j)$ [16]. Plugging them into (11.26), we have

$$S = \frac{1}{\pi\alpha'} \int d\tau \left( G_{ij}\dot{q}^i \dot{q}^j - G_{ij}n^i n^j R^2 + 2B_{ij}\dot{q}^i n^j R \right).$$

The canonical momentum is

$$\pi_i = \frac{\partial \mathscr{L}}{\partial \dot{q}^i} = \frac{1}{\pi\alpha'} \left( 2G_{ij}\dot{q}^j + 2B_{ij}n^j R \right), \tag{11.28}$$

where in the compact dimension this momentum should be quantized as in (6.208),

$$\alpha' p_i = \alpha' \int_0^\pi \pi_i d\sigma = \frac{m_i}{R}, \quad m_i \in \mathbb{Z}. \tag{11.29}$$

Inverting the metric gives

$$\dot{q}^j = \alpha' \frac{m^j}{R} - 2B^j{}_k n^k R,$$

where we can raise and lower spacetimes with the metric, for example, $m^j = G^{np} m_p$. Thus enables us to generalize the dimensionless momenta (6.193)

$$
\sqrt{\alpha'} p_R^i \equiv \dot{q}^i - L^i = \alpha' \frac{m^i}{R} - (B^{ij} + G^{ij}) n_j R,
$$

$$
\sqrt{\alpha'} p_L^i \equiv \dot{q}^i + L^i = \alpha' \frac{m^i}{R} - (B^{ij} - G^{ij}) n_j R.
$$

(11.30)

The mass formulae (6.219) become

$$
\frac{1}{2} \alpha' M_L^2 = \frac{1}{2} G_{ij} p_L^i p_L^j + \tilde{N} - 1,
$$

$$
\frac{1}{2} \alpha' M_R^2 = \frac{1}{2} G_{ij} p_R^i p_R^j + N - 1,
$$

(11.31)

and the level matching condition (6.225) becomes

$$
N - \tilde{N} = G_{ij}(p_L^i p_L^j - p_R^i p_R^j) = n_i m^i = 0. \tag{11.32}
$$

To sum up, when we quantize involving the metric and the antisymmetric tensor, we may shift the momentum as in (11.30) and do all the inner product using the internal metric.

We can do the similar thing for the Wilson line $A_i^I$ to have the above result, by adding the term [15, 16]

$$
\epsilon^{\alpha\beta} A_i^I \partial_\alpha X^i \partial_\beta X^I. \tag{11.33}
$$

Again, we impose a constraint for the absence of the right mover (6.213). This affects the quantization and using Dirac bracket we can find the following result [16] become

$$
p_R^i = \alpha' \frac{m^i}{R} - (B^{ij} + G^{ij}) n_j R - \alpha' \frac{m^I}{R} A_I^i - \frac{1}{2} \alpha' A_I^i A_I^j n_j, \tag{11.34}
$$

$$
p_L^i = \alpha' m^i R - (B^{ij} - G^{ij}) n_j R - \alpha' \frac{m^I}{R} A_I^i - \frac{1}{2} \alpha' A_I^i A_I^j n_j, \tag{11.35}
$$

$$
P^I = 2p^I = 2\alpha' m^i R + A_j^I n^j, . \tag{11.36}
$$

$$
* * *
$$

Knowing these expressions, we can explicitly construct the even and self-dual lattice of signature $(26 - d, 10 - d)$ containing the background gauge fields [17]. Take a reference lattice $\Gamma_0$ spanned by the following basis vectors:

$$k^i = (0, \tfrac{1}{2}e^{\vee i}; \tfrac{1}{2}e^{\vee i}),$$

$$\bar{k}^i = (a_i^J, -b_{ji}e^{\vee j} - \tfrac{1}{4}a_j^K a_i^K e^{\vee j} + e_i; -b_{ji}e^{\vee j} - \tfrac{1}{4}a_j^K a_i^K e^{\vee j} - e_i), \quad (11.37)$$

$$\ell_\alpha = (f_\alpha^I, -\tfrac{1}{2}f_\alpha^K a_i^K e^{\vee i}; -\tfrac{1}{2}f_\alpha^K a_i^K e^{\vee i}).$$

They have the same product as in (11.23) as the one without the background gauge fields $A_m^I = B_{mn} = 0$, although they are

$$W(A) \equiv \exp\left[\frac{1}{2}\begin{pmatrix} 0 & A_n^I & -A_{n'}^I \\ -A_m^J & -B_{mn} & B_{mn'} \\ -A_{m'}^J & -B_{m'n} & B_{m'n'} \end{pmatrix}\right]$$

$$= \begin{pmatrix} \delta_J^I & \tfrac{1}{2}A_n^I & -\tfrac{1}{2}A_n^I \\ -\tfrac{1}{2}A_m^J & \delta_{mn} - \tfrac{1}{2}B_{mn} - \tfrac{1}{8}A_m^K A_n^K & \tfrac{1}{2}B_{mn'} + \tfrac{1}{8}A_m^K A_{n'}^K \\ -\tfrac{1}{2}A_{m'}^J & -\tfrac{1}{2}B_{m'n} - \tfrac{1}{8}A_{m'}^k A_n^K & \delta_{m'n'} + \tfrac{1}{2}_{m'n'} + \tfrac{1}{8}A_{m'}^K A_{n'}^K \end{pmatrix}.$$

$$(11.38)$$

Finally, we parameterize the background metric $G_{mn}$, which parametrize the angles and the radii. This is done to the unit vectors

$$e_i \rightarrow e_i' = e^{-\alpha_{mn}}e_i \quad (11.39)$$

which is represented as boosts

$$\exp\begin{pmatrix} 0 & 0 & 0 \\ 0 & 0 & \alpha_{mn'} \\ 0 & \alpha_{m'n} & 0 \end{pmatrix} = \begin{pmatrix} \delta_J^I & 0 & 0 \\ 0 & \cosh\alpha_{mn} & \sinh\alpha_{mn'} \\ 0 & \sinh\alpha_{m'n} & \cosh\alpha_{m'n'} \end{pmatrix}. \quad (11.40)$$

The $T$-duality transformation is contained here. For each complex plane, a judicious choice of $\alpha_{mn}$ takes the radius $|e_i| = R$ to $|e_i'| = \sqrt{\alpha'}/R$. This can be generalized to larger duality symmetry. For special values of $\alpha_{mn}, B_{mn}, A_m^I$ the lattice $\Gamma_0$ is invariant although the basis vectors are shuffled. This is what we called $O(26 - d, 10 - d, \mathbb{Z})$ in (11.25).

## 11.2.3 Moduli Fields

The moduli fields discussed above play an important role in low-energy physics and cosmology. Before discussing their dynamics in Sect. 11.1, we briefly inspect some properties that can be deduced from the above symmetries.

We consider again the example of the two-torus $T^2$ with coordinates $y^1$, $y^2$. Rewriting the action (11.26) using holomorphic coordinates, we have

$$
S = \frac{1}{2\pi\alpha'} \int d^2z \left[ G_{11} \partial Y^1 \bar{\partial} Y^1 + G_{22} \partial Y^2 \bar{\partial} Y^2 + G_{12} \left( \partial Y^1 \bar{\partial} Y^2 + \bar{\partial} Y^1 \partial Y^2 \right) \right.
$$
$$
\left. - i B_{12} \left( \partial Y^1 \bar{\partial} Y^2 - \bar{\partial} Y^1 \partial Y^2 \right) \right].
$$
$$(11.41)$$

The metric tensor $G_{ij}$ for the internal geometry becomes a number of scalars in four dimensions. By the supersymmetry, it forms a chiral multiplet; thus should be a part of complex scalar. We collect the target space bosons in terms of a complex field

$$
iU = \frac{R^2}{G_{11}} \left( G_{12} + i\sqrt{\det G} \right)
$$
$$
= R^2 \frac{|e_2|}{|e_1|} e^{i\phi_{12}}.
$$
$$(11.42)$$

Here $\phi_{12}$ is the angle between the two basis vectors $e_1$ and $e_2$ and we have $G_{12} = e_1 \cdot e_2$. It provides a natural basis for the complexification

$$
Z = Y^1 + iUY^2,
$$
$$(11.43)$$

so that $U$ is the complex structure introduced in (3.5). Thus we have

$$
S = \frac{1}{2\pi\alpha'} \int d^2z \left( T\partial Z \bar{\partial} Z^* + T^* \bar{\partial} Z^* \partial Z \right).
$$
$$(11.44)$$

Here the field responsible for the volume pairs up with the $B$-field

$$
iT \equiv \frac{R^2}{\alpha'} (B_{12} + i\sqrt{\det G})
$$
$$
\equiv b_{12} + i\frac{A}{4\pi^2 \alpha'},
$$
$$(11.45)$$

which is called the Kähler modulus, where $R$ and $A$ are the radius and the volume of torus and the determinant is over the two dimensional metric $G$. Remarkably, in string dynamics, the antisymmetric tensor field $B$ is also responsible for the volume. In view of discussion below (11.26), a winding string should be able to measure the size of the torus. Also note that it has a discrete (due to (11.28)) shift symmetry, the axionic symmetry

$$
b_{ij} \to b_{ij} + 1,
$$
$$(11.46)$$

known to be present to the all orders of perturbation expansion.

**Target Space Modular Group**

We can rewrite the internal metric as

$$ds^2 = \frac{\text{Re}\,T}{R^2\,\text{Re}\,U}\left|dY^1 + iU\,dY^2\right|^2 = \frac{\text{Re}\,T}{R^2\,\text{Re}\,U}dZd\overline{Z}, \tag{11.47}$$

in the similar fashion as (6.71). We expect that these moduli acquire VEV's dynamically to spontaneously generate the geometry.

In the previous example, there are modular symmetries

$$iT \to \frac{aiT + b}{ciT + d}, \quad iU \to \frac{a'iU + b'}{c'iU + d'} \tag{11.48}$$

with integer parameters satisfying $ad - bc = a'd' - b'c' = 1$. They form a modular group

$$\text{PSL}(2,\,\mathbb{Z}) \times \text{PSL}(2,\,\mathbb{Z}) \ltimes \mathbb{Z}_2^2. \tag{11.49}$$

The two PSL(2,$\mathbb{Z}$)s act in the same way to modular transformation on the torus parameter $\tau$ but this time act on the target space.

They unify the axionic symmetry and $T$-duality. For example, the symmetries $iT \to iT + 1$ are the axionic symmetry (11.46). $iT \to -1/(iT)$ correspond to simultaneous $T$-duality along $y^1$ and $y^2$ directions. We have similar interpretation for those acting on $U$. The last two $\mathbb{Z}_2$'s in (11.49) are known as the special case of "mirror symmetry" and correspond to exchanging $T \leftrightarrow U$ or $T \leftrightarrow U^*$ and orbifold action.

## 11.2.4  Duality Between Two Heterotic String Theories

We have *T-duality* between SO(32) and $E_8 \times E_8$ heterotic string theories. Compactifying the SO(32) theory on a circle and turn an Wilson line along it

$$a_9 = \tfrac{1}{2}(1^8\,0^8).$$

This breaks the gauge group to SO(16) × SO(16). This group is also obtainable from $E_8 \times E_8$ theory on a circle with a Wilson line

$$a_9' = \tfrac{1}{2}(2\,0^7)(2\,0^7).$$

Now we focus on the neutral states under $SO(16) \times SO(16)$, or $P^I = 0$. We have massless states when $n^9 \equiv 2n$ is an even number. Plugging back into (11.34) and (11.35) and also Wilson lines above we obtain the momentum is

$$
\begin{aligned}
p^9_{L,R} &= \frac{m + 2n}{R} \mp \frac{2nR}{\alpha'}, \\
p'^9_{L,R} &= \frac{m' + 2n'}{R} \mp \frac{2n'R'}{\alpha'},
\end{aligned}
\tag{11.50}
$$

where $m^9 \equiv m$ and restored dimensionful parameter $R$ and $\alpha'$. The primed quantity is of $E_8 \times E_8$. Exchanging $R \leftrightarrow R'/4$ and $(m + 2n, n) \leftrightarrow (n', m' + 2n')$, so that $(p^9_L, p^9_R) \leftrightarrow (p'^9_L, -p^9_R)$. We have the same spectrum. Note that now the radius does not have the critical value of $R = \sqrt{\alpha'}$.

The duality should hold for the full spectrum, because their partition functions are the same as in (6.243). It is interesting to see that we may think of a unified theory with a current algebra containing both $SO(32)$ and $E_8 \times E_8$ [17].

$$
W(A) = \exp\left[ \frac{1}{2} \begin{pmatrix} 0 & A_9 & -A_9 \\ -A_9^\top & 0 & 0 \\ -A_9^\top & 0 & 0 \end{pmatrix} \right]
\tag{11.51}
$$

by $W(A_9')$ and $RW(A)R$, we may relate between them.

By this chain of dualities, all the superstrings in ten dimensions with 16 supersymmetries are related.

## 11.3  Supersymmetric Action and Twisted Fields

As discussed in Sect. 2.3, the $\mathcal{N} = 1$ supergravity Lagrangian in four dimensions is obtained from the Kähler potential, the superpotential and the gauge kinetic function. We rearrange the above Lagrangian accordingly. This clearly shows the modular properties, which is strong enough to accommodate the twisted fields and obtain higher order corrections. We consider effective action now taking into account moduli field of the full orbifold.

### 11.3.1 Kähler Potential

Let us go back to the above dimensional reduction. If we introduce the complex field of the holomorphic coupling

$$
S = e^{-2\phi} + ia,
\tag{11.52}
$$

(we will identify the complex part $a$ in Sect. 11.3.3) and the Kähler moduli

$$T_{a\bar{b}} = g_{a\bar{b}} + i\left(B_{a\bar{b}} - i\frac{\alpha'}{2}\,\mathrm{tr}_\mathrm{v}(Q_a Q_b^*)\right),\tag{11.53}$$

all the kinetic terms are compactly written as

$$2\kappa^2\frac{\mathscr{L}_{\mathrm{kin}}}{\sqrt{g}} = R - \frac{\partial_\mu S \partial^\mu S^*}{2(\mathrm{Re}\,S)^2} - \frac{\partial_\mu T_{a\bar{d}}\partial^\mu T_{c\bar{b}}^*}{2\,\mathrm{Re}\,T_{a\bar{b}}\,\mathrm{Re}\,T_{c\bar{d}}}.\tag{11.54}$$

Noice the formal similarity of the two fields and the corresponding kinetic terms. We have a model-independent axion $a$ and model dependent axions $B_{a\bar{b}}$, depending on details of compactification. Although it is natural to complexify these, the kinetic term for the $S$ field contains $e^{-4\phi}$, we should have only this part in the Kähler potential. The kinetic terms in (11.54) are reproduced by the Kähler potential

$$\kappa^2 K = -\log(S + S^*) - \log\det\left(T_{a\bar{b}} + T_{a\bar{b}}^* - \alpha'\,\mathrm{tr}_\mathrm{v}\,Q_a Q_b^*\right).\tag{11.55}$$

In general, the Kähler potential can be Taylor-expanded in terms of matter fields $Q_\alpha$

$$\begin{aligned}\kappa^2 K =&\,\kappa^2 K_{\mathrm{geom}}(\Sigma, \Sigma^*) + Z_{\alpha\bar{\beta}}(\Sigma, \Sigma^*)Q_\alpha Q_\beta^*\\&+ Z_{\alpha\beta}(M, M^*)(Q_\alpha Q_\beta + Q_\alpha^* Q_\beta^*) + \dots,\end{aligned}\tag{11.56}$$

where $\Sigma$ denotes the general moduli like $S, T, U$ and the ellipsis contains higher order terms in $Q_\alpha$. The flavor index $\alpha$ is not necessarily related to the spatial index $a$ in general. Except the first term, we have no dependence on the holomorphic coupling $S$. We shall also show $Z_{\alpha\beta} = 0$.

Further loop expansion gives

$$K_{\mathrm{geom}} = -\log(S + S^*) + \hat{K}(M, M^*) + \frac{V^{(1)}(M, M^*)}{8\pi^2(S + S^*)} + \frac{V^{(2)}(M, M^*)}{64\pi^4(S + S^*)} + \dots,\tag{11.57}$$

where now the moduli $M$ involves only $T, U$, not $S$, and

$$Z_{\bar{\alpha}\beta} = Z_{\bar{\alpha}\beta}^{(0)}(M, M^*) + \frac{Z_{\bar{\alpha}\beta}^{(1)}(M, M^*)}{8\pi^2(S + S^*)} + \frac{Z_{\bar{\alpha}\beta}^{(2)}(M, M^*)}{64\pi^4(S + S^*)} + \dots.\tag{11.58}$$

The dependence of $S + S^*$ is due to the axionic symmetry (11.97) that we see shortly.

The explicit calculation is possible for the standard-embeddings of Calabi–Yau manifold and orbifolds [3]. A consequence is that, after the canonical normalization, every matter interaction involves also the moduli fields. Their leading order contribution give the coupling strengths, and higher order terms gives interactions.

## The Geometrical Part from Orbifolds

We calculate the Kähler potential for the moduli fields [13, 18–21]. For toroidal compactification, we have the following generic form:

$$\kappa^2 \hat{K} = -\log \det(T_{a\bar{b}} + T_{a\bar{b}}^*) - \sum_{m=1}^{h_{2,1}} \log(U_m + U_m^*), \quad a, b = 1, 2, 3, \qquad (11.59)$$

where $h_{2,1}$ is for the moment understood as the number of unfixed complex structures. We shall define the numbers $h_{2,1}$ and $h_{1,1}$ in Chap. 15.

There are three cases for six dimensional toroidal orbifolds $T^6/\mathbb{Z}_N$, $T^6/(\mathbb{Z}_N \times \mathbb{Z}_M)$, depending on the factorization of the internal dimensions and the fixing of the complex structure. The first is the $\mathbb{Z}_3$ orbifold, for which all the complex structure moduli are fixed thus $h_{2,1} = 0$. There are $h_{1,1} = 9$ unfixed Kähler moduli, as shown in (10.110) and defined in (11.53). The moduli space (11.24) becomes more special

$$\mathscr{M} = \left[ \frac{SU(3, 3)}{SU(3) \times SU(3) \times U(1)} \right]_T. \qquad (11.60)$$

The Kähler potential (11.59) with $h_{2,1} = 0$ reflects this symmetry.

In the second case, all the Kähler moduli vanish except diagonal ones $T_a \equiv T_{a\bar{a}}$, having $h_{1,1} = 3$. For $\mathbb{Z}_7$, $\mathbb{Z}_8$, $\mathbb{Z}_{12}$, etc., all the complex structure is fixed $h_{2,1} = 0$ since all the Coxeter basis are related. For $\mathbb{Z}_{6-I}$, $\mathbb{Z}_8$, we have one (in our convention the first) fixed torus so $h_{2,1} = 1$. In this case we may also label the complex structure $U_1$. For $\mathbb{Z}_2 \times \mathbb{Z}_2$ we have three fixed tori so that $h_{2,1} = 3$. In this case we may label the complex structure as $U_a$, $a = 1, 2, 3$. They have the same target space symmetry

$$\mathscr{M} = \left[ \frac{SU(1, 1)}{U(1)} \right]_T^3 \times \left[ \frac{SU(1, 1)}{U(1)} \right]_U^{h_{2,1}},$$

$$\log \det(T_{a\bar{b}} + T_{a\bar{b}}^*) = -\sum_{a=1}^{3} \log(T_a + T_a^*). \qquad (11.61)$$

Still $m$ denotes the complexified torus. Depending on the structure of orbifold, $m = 1$ may label 2nd or 3rd torus. The classical symmetry $SU(1, 1)/U(1)$ is isomorphic to $SL(2, \mathbb{R})$.

The last case is $\mathbb{Z}_{6-II}$ and $\mathbb{Z}_4$, where we have one fixed torus while in the unfixed direction we have unfixed angles, so that we have $h_{1,1} = 5$. The complex structure for $\mathbb{Z}_{6-I}$ is completely fixed $h_{2,1} = 0$ but we have unfixed for $\mathbb{Z}_4$ so that $h_{2,1} = 1$

$$\mathscr{M} = \left[ \frac{SU(1, 1)}{U(1)} \times \frac{SU(2, 2)}{SU(2) \times SU(2) \times U(1)} \right]_T \times \left[ \frac{SU(1, 1)}{U(1)} \right]_U^{h_{2,1}},$$

$$\log \det(T_{a\bar{b}} + T_{a\bar{b}}^*) = \log(T_1 + T_1^*)$$

$$- \log \det(T_{a\bar{b}} + T_{a\bar{b}}^*) \quad a, b = 2, 3. \qquad (11.62)$$

## Matter Dependence

Now we consider the matter fields that we collectively write $Q_\alpha$. The Kähler potential

$$\kappa^2 K_{\text{matter}} \equiv Z_{\alpha\bar\beta} Q_\alpha Q_\beta^* \tag{11.63}$$

provides the metric of the target space. For convenience we consider the case (11.61) where we have diagonal Kälher moduli $T_a \equiv T_{a\bar a}$.

Dimensional reduction (11.55) has shown that for the untwisted field, we have

$$Z_{\alpha\bar\beta} = \delta_{\alpha\beta} \prod_{a=1}^{3} (T_a + T_a^*)^{-1} \prod_{m=1}^{h_{2,1}} (U_m + U_m^*)^{-1}. \tag{11.64}$$

We had $h_{2,1} = 0$ We know that the kinetic term should be invariant under the modular group.

This is valid when the $Q_\alpha$ are untwisted matter fields because it is obtained from dimensional reduction of gaugino. We also need such transformation for the twisted fields. The symmetry of the target space is strong enough to constrain the form of Kähler potential. We may assume that the Kähler potential for the twisted matter fields has a similar form as (11.64),

$$Z_{\alpha\bar\beta} = \delta_{\alpha\beta} \prod_{a=1}^{3} (T_a + T_a^*)^{n_\alpha^a} \prod_{m=1}^{h_{2,1}} (U_m + U_m^*)^{l_\alpha^m}. \tag{11.65}$$

Thus $Q_\alpha$ transforms under the modular transform (11.48) as

$$Q_\alpha \to \Upsilon_\alpha^\beta Q_\beta \prod_{a=1}^{3} (i c_a T_a + d_a)^{n_\alpha^a} \prod_{m=1}^{h_{2,1}} (i c_m' U_m + d_m')^{l_\alpha^m}, \tag{11.66}$$

where $\Upsilon$ is the unitary matrix having dependence on the moduli. This transform makes the total Kähler potential (11.63) and the gauge kinetic term invariant. We say that $Q_\alpha$ is a modular form of weights $n_\alpha^a$ and $l_\alpha^m$ with $a = 1, \ldots, 3, m = 1, \ldots, h_{2,1}$.

Untwisted fields are always associated with a two-torus, so it carries an internal index like $Q_b$, $b = 1, 2, 3$. Their modular weights are deduced from the dimensional reduction result

$$n_b^a = -\delta_b^a. \tag{11.67}$$

Recall that the superscript $a$ is associated to the moduli $T_a$. If there survives $U$ moduli along the $b$-th directions, the complexification of $Q_b$ field along those directions gives

$$l_b^a = -\delta_b^a. \tag{11.68}$$

## Calculating Metter Kähler Potential

We will calculate the modular weights for the twisted fields. Interestingly, we may relate them with the generalized $H$-momenta or $R$-charges. Recall that the above modular weights basically counts the oscillator numbers and the $R$-charges are generalization of the $H$-momenta including the oscillators.

$$Z_{a\bar{b}} = \prod (T_a + T_a^*)^{R_a - 1}. \tag{11.69}$$

where

$$n_\alpha^a = R_a - 1. \tag{11.70}$$

Note that for the untwisted sector field, we have $R_a = 1$ if it has the spacetime index $a$.

Following [22], we calculate the matter Kähler potential (11.65) by considering scattering amplitude between two matter and two moduli fields. Using the Kähler potential (11.56), the field theory calculation gives

$$\mathscr{A}_{\mathrm{QFT}}^{(0)}(M_{\bar{a}}, Q_{*\bar{a}}, Q_\beta, M_b) = \kappa^2 \frac{su}{t} Z_{\bar{\alpha}\beta} G_{\bar{a}b} + s \frac{Z_{\bar{\alpha}\beta}}{\partial \overline{M}^a \partial M_b} - s \frac{\partial Z_{\bar{\alpha}\gamma}}{\partial \overline{M}^a} \left( Z^{-1} \right)^{\gamma\bar{\delta}} \frac{\partial Z_{\bar{\delta}\beta}}{\partial \overline{M}^a},$$
$$\tag{11.71}$$

where $s = (k_1 + k_2)^2$, $t = (k_2 + k_3)^2$, $u = (k_1 + k_3)^2$ are Mandelstam variables and $G_{\bar{a}b} = \kappa^{-2} \partial_{\bar{a}} \partial_b K_{\mathrm{geom}}$.

This amplitude can also be obtained from the string calculation

$$\mathscr{A}_{\mathrm{string}}^{(0)}(M^{*\bar{a}}, Q_{*\bar{a}}, Q_\beta, M^a) = g_s^2 \int d^2 z \langle Q_\alpha | T V_{M^{*\bar{a}}}(w) V_{M^a}(z) | Q_\beta \rangle. \tag{11.72}$$

Here the vertex operators for the moduli fields are given as

$$V_{T^b} = e^{ik_\mu X^\mu} \partial \overline{X}^b \left( \bar{\partial} X^b + \frac{1}{2} k_\mu \psi^\mu \psi^b \right),$$

$$V_{T^{\bar{a}}} = e^{ik_\mu X^\mu} \partial X^a \left( \bar{\partial} \overline{X}^a + \frac{1}{2} k_\mu \psi^\mu \overline{\psi}^a \right),$$

$$V_{U^b} = e^{ik_\mu X^\mu} \partial X^b \left( \bar{\partial} X^b + \frac{1}{2} k_\mu \psi^\mu \psi^b \right), \tag{11.73}$$

$$V_{U^{\bar{a}}} = e^{ik_\mu X^\mu} \partial \overline{X}^a \left( \bar{\partial} \overline{X}^a + \frac{1}{2} k_\mu \psi^\mu \overline{\psi}^a \right).$$

Firstly, the index structure simplifies. Since the moduli come from the untwisted sector, the only space group quantum numbers come from the matter fields. As we see in Fig. 11.1, there is a Lorentz frame where a single matter field flows and is

**Fig. 11.1** Scattering of two moduli and matter fields yielding the matter Kähler potential

scattered by the moduli fields. Thus we have $\alpha = \beta$. Thus all the quantum number of the moduli fields should be conserved including $a = b$.

Consider then the left movers first. The vertex operators for the moduli fields are given in (10.19). The OPE for the momentum part is nothing but the Virasoro–Shapiro amplitude $w^{-s/8}z^{-u/8}(z-w)^{-t/8}$. The oscillator part gives

$$\langle Q_\alpha|T\partial X^a(w)\partial\overline{X}^a(z)|Q_\alpha\rangle_L$$

$$= w^{k/N}z^{1-k/N}\left[\frac{(k/N)w + (1-k/N)z}{(z-w)^2} + \frac{1-k/N}{z}N_\alpha^a + \frac{k/N}{w}\overline{N}_\alpha^a\right],$$
(11.74)

where the state $Q_\alpha$ is excited by $N_\alpha^a$ and $\overline{N}_\alpha^a$ oscillators as in (7.12).

For the right movers, we have the bosonic and the fermionic contributions

$$\langle Q_\alpha|TV_{M^{*\bar{a}}}(w)V_{M^a}(z)|Q_\beta\rangle_R = \bar{w}^{-s/8}\bar{z}^{-u/8}(\bar{z}-\bar{w})^{-t/8}$$

$$\times \langle Q_\alpha|T\bar{\partial}\overline{X}^a(w)\bar{\partial}X^a(z) + \frac{t}{8}T\psi^\mu(\bar{w})\psi^\mu(\bar{z})|Q_\beta\rangle_R.$$
(11.75)

The bosonic part is the complex conjugation of that of the left movers (11.74). The fermionic contribution is similarly obtained as

$$\langle Q_\alpha|T\psi^\mu(\bar{w})\psi^\mu(\bar{z})|Q_\beta\rangle_R = \frac{1}{\bar{w}-\bar{z}}\left(\frac{\bar{w}}{\bar{z}}\right)^{k/N},$$
(11.76)

where we should replace the power $k/N$ with 1 when $R = k/N$ for all the twisted sectors and 0 for untwisted sector states with $l_I \neq i$ and 1 for untwisted sector states with $l_I = a$.

Plugging all the results into (11.72), we obtain

$$\mathcal{A}_{\text{string}}^{(0)}(T^{\bar{a}}, Q_{\alpha*}, Q_\beta, T^b) = \frac{1}{8}g_s^2\delta_{ab}\delta_{IJ}\left(\frac{su}{t} - sn_I^a + s\mathcal{O}(s,t,u)\right),$$
(11.77)

with

$$R_I^a = \begin{cases} k/N + \rho_R^a - \rho_L^a, & \text{for twisted states,} \\ \delta_a^b, & \text{for untwisted states.} \end{cases} \tag{11.78}$$

Note that this is exactly the $R$-charge introduced in (10.42).

For the calculation of the complex structure $U^i$, we modify some of the above. The right mover is identical. So we replace $\partial X^a$, $N_I^a$, $k^a/N$ with $\partial \overline{X}^a$, $\rho_L^a - \rho_R^a$, respectively. We have

$$\mathscr{A}_{\text{string}}^{(0)}(U^{\bar{a}}, Q_{\alpha *}, Q_J, U^b) = \frac{1}{8} g_s^2 \delta_{ab} \delta_{IJ} \left( \frac{su}{t} - s\overline{n}_I^a + s\mathscr{O}(s,t,u) \right), \tag{11.79}$$

with the above $R^a$ is replaced witt $\overline{R}^a$.

Using (11.65), we can calculate the corresponding string diagram

$$\begin{aligned} \mathscr{A}_{\text{string}}^{(0)}(M^{\bar{a}}, Q_{\alpha *}, Q_\beta, M^b) &= \kappa^2 \delta_{\bar{a}b} \delta_{\bar{\alpha}\beta} \left( \frac{su}{t} - sn_\beta^a + s\mathscr{O}(s,t,u) \right), \\ &\to \kappa^2 G_{\bar{a}b} Z_{\bar{\alpha}\beta} \left( \frac{su}{t} - sn_\beta^a + s\mathscr{O}(s,t,u) \right). \end{aligned} \tag{11.80}$$

Here $n_\beta^a$ are the modular weights in (11.66) for the moduli fields $T_a$. For the moduli $U_m$, we may replace $n_\beta^a$ with $l_\beta^m$. In the last line, we have changed from the canonical normalization of the $Q_\beta$ to same normalization as the moduli fields. The amplitude agrees except the terms proportional to $s$, so we equate them

$$\begin{aligned} \left( Z^{-1} \partial_{\bar{a}} \partial_b Z - Z^{-1} \partial_{\bar{a}} Z Z^{-1} \partial_b Z \right)_\beta^\alpha &= \partial_{\bar{a}} (Z^{-1} \partial_b Z)_\beta^\alpha \\ &= -\kappa^2 G_{\bar{a}b} \delta_\beta^\alpha n_\beta^a \\ &= -\frac{\delta_{\bar{a}b} \delta_\beta^\alpha n_\beta^b}{(M_b + M_b^*)^2}. \end{aligned} \tag{11.81}$$

Solving this for $T$ and $U$ moduli, we have

$$Z_{\bar{\alpha}\beta} = \sum_\gamma F_{\bar{\alpha}\gamma}^* F_{\beta\gamma} \prod_a (T_a + T_a^*)^{n_\alpha^a} \prod_m (U_m + U_m^*)^{l_\alpha^a}. \tag{11.82}$$

Here $F_{\beta\gamma}$ are some non-degenerate matrix whose components are holomorphic functions of the moduli. They parameterize the choice of the reference coordinate system, so we may rotate as $F_{\beta\gamma} = \delta_{\beta\gamma}$.

We cannot have $Z_{\alpha\beta}$ in the Kähler potential because scattering between holomorphic fields gives purely holomorphic amplitude. It only contributes to the superpotential we consider next.

## 11.3.2 Superpotential

The superpotential is a holomorphic function of superfields. In view of the matter superfields $Q_\alpha$ (using the same notation as the scalar fields) we may expand the superpotential as

$$W = \mu_{\alpha\beta}(T, U)Q_\alpha Q_\beta + \chi_{\alpha\beta\gamma}(T, U)Q_\alpha Q_\beta Q_\gamma + \ldots, \tag{11.83}$$

where the ellipsis denote higher order couplings of $Q_\alpha$ and we collectively denoted the gauge group indices by $\alpha, \beta, \gamma$. The coefficients are holomorphic functions of the geometric moduli. We have no dependence on $S$ because the gauge coupling is $1/(S + S^*)$.

For instance, the scalar potential $V(Q)$ of the above dimensionally reduced Lagrangian (11.19)

$$V = e^{\kappa^2 K}\left[Z^{\alpha\bar\beta}(D_\alpha W)(D_{\bar\beta}\overline{W}) - 3\kappa^2|W|^2\right] + \frac{g^2}{8}\sum_\alpha\left(K_\alpha q_\alpha^A Q_\alpha + \bar{Q}_{\bar\alpha} q_{\bar\alpha}^A K_{\bar\alpha}\right)^2, \tag{11.84}$$

can be compactly written by the superpotential

$$W = \epsilon^{abc}d_{xyz}Q_a^x Q_b^y Q_c^z, \tag{11.85}$$

where the coefficients are given in (11.17). In particular, $H_{abc}, F_{ab}, F_{\bar{a}b}^2$ give the $W, D_\alpha W, K_\alpha q Q_\alpha$ terms, respectively. Here $Z^{\alpha\bar\beta}$ is the inverse of the Kähler metric (11.65). This inheritance again works only for the untwisted fields. For general (un)twisted fields, we may calculate the $N$-point coupling as done in the Chap. 10. We may supersymmetrize the result to obtain the superpotential.

Consider again the $PSL(2, \mathbb{Z})$ modular symmetries on each $a$th complex plane. Since $K \to K + \log|c_a T_a + d_a|^2$ should be Kähler and $G = K + \log|W|^2$ is invariant, we have

$$W \to W(ic_a T_a + d_a)^{-1}, \quad a = 1, 2, 3, \tag{11.86}$$

up to a phases. The overall superpotential $W$ is a modular form of weight $-3$ for the factorizable tori case. We have a similar relation $W \rightarrow W(ic_m U_{+} d_m)^{-1}, m = 1, \ldots, h_{2,1}$, for the complex structures. Therefore the Yukawa coupling should transform as

$$\chi_{IJK}(T) \rightarrow \chi_{IJK}(T) \prod_{\alpha} \prod_{a=1}^{3} (ic_a T_a + d_a)^{-1-n_\alpha^a} \prod_{m=1}^{h_{2,1}} (ic_m U_m + d_m)^{-1-l_\alpha^a}. \tag{11.87}$$

Non-renormalization theorem restrains the coefficient from perturbative correction. We cannot make a holomorphic polynomial in $T$ field in (11.55) obeying the axionic symmetry, the superpotential cannot receive perturbative corrections to all orders of sigma model expansion. Because it does not depend on $S$ modulus which appears in the string coupling as Re $S = 1/g_s$, there is no correction from string perturbation. This is the stringy version of the non-renormalization theorem [23]. There is no self-coupling of moduli $S, T, U$; thus they remain in flat directions, which explains the name "*moduli fields*." A non-perturbative correction of the gauge coupling to the holomorphic coefficient $\chi_{\alpha\beta\gamma}(T)$ has a form $\sim e^{-aS}$, which may preserve the axionic symmetry. We have seen that they are exponentially suppressed following the area rule, still obeying modular symmetry (11.49).

We can take into account the mirror symmetry $T \leftrightarrow U$ if there is a $U$ field present.

### 11.3.3 Gauge Kinetic Function

We have seen that the four dimensional gauge coupling is given by the real part of the $S$ modulus (11.52)

$$e^{-2\phi} = g^{-2} = \text{Re } S, \tag{11.88}$$

where $g$ is the four dimensional coupling defined in (11.10) at the string scale $M_s^2$. For toroidal compactification, this is the coupling of the gauge group inherited from the 10D theory. If we break it using geometry of background gauge fields, at tree level, all the four dimensional gauge groups $\{G_A\}$ arise from the breaking of the 10D gauge group and all of their couplings are unified

$$\frac{1}{g_A^2} = k_A \text{ Re } S. \tag{11.89}$$

Here $k_A$ are the level of the Kac–Moody algebra that we see in Sect. 12.6. We show that this is generalized to holomorphic gauge coupling whose imaginary part is the axion. We have only one-loop correction due to non-renormalization theorem

$$f_A = k_A S + \frac{1}{16\pi^2} f_A^{(1)}(M). \tag{11.90}$$

**Antisymmetric Tensor and Axion**
In four dimensions, the rank-two antisymmetric tensor field has one real degree of freedom. It is dual to a pseudoscalar

$$H_{\mu\nu\rho} = \frac{1}{6}\epsilon^{\mu\nu\rho\sigma} e^{4\phi} \partial^\sigma a. \tag{11.91}$$

Note that this includes the metric implicitly $\epsilon^{\mu\nu\rho\sigma} = \sqrt{g}\epsilon_{\mu\nu\rho\sigma}$. The pseudoscalar $a$ is called the model-independent axion since it does not depend on the details of the compactification. Equivalently, we may employ $a$ as the Lagrangian multiplier for the Bianchi identity (11.133) [24]

$$-\frac{1}{4\kappa^2}\int d^4x \sqrt{g} e^{-4\phi} H_{\mu\nu\rho} H^{\mu\nu\rho}$$
$$+\frac{1}{4\kappa^2}\int d^4x\, a\epsilon^{\mu\nu\rho\sigma}\left[\partial_{[\mu} H_{\nu\rho\sigma]} + \frac{\alpha'}{4}\left(\mathrm{tr_v}\, F_{\mu\nu}F_{\rho\sigma} - \mathrm{tr_v}\, R_{\mu\nu}R_{\rho\sigma}\right)\right], \tag{11.92}$$

where the normalization for the second line is chosen to make the interaction dimensionless. We may check that the equation of motion for $H_{\mu\nu\rho}$ becomes the dual transformation (11.91). Integrating out $H$, we obtain

$$\frac{1}{4\kappa^2}\int d^4x \sqrt{g} e^{4\phi}(\partial_\mu a)^2 - \frac{1}{4g^2}\int d^4x \sqrt{g}\, \mathrm{tr_v}\, F_{\mu\nu}F^{\mu\nu}$$
$$+\frac{1}{16\pi^2}\int a\,(\mathrm{tr_v}\, F \wedge F - \mathrm{tr_v}\, R \wedge R)\,. \tag{11.93}$$

By canonical normalization, we have the axion decay constant

$$F_{\mathrm{MI}} = \frac{g^2}{192\pi^{5/2}\kappa}. \tag{11.94}$$

We have supersymmetric description

$$\int d^4x\, d^2\theta\, S \mathcal{W}^\alpha \mathcal{W}_\alpha + \mathrm{h.c.}. \tag{11.95}$$

We have a universal gauge kinetic function [22]

$$f_{AB} = k_A S \delta_{AB}, \tag{11.96}$$

Since it is diagonal, we define $f_A = f_{AA}$. We see also there is symmetry

$$a \to a + 1, \quad \text{or} \quad iS \to iS + 1. \tag{11.97}$$

**One-Loop Correction**

The gauge kinetic function (11.96) is model independent, because all the gauge groups come from that of heterotic string. The traditional non-renormalization theorem applies and there is only one-loop correction, because of the holomorphy of $f_{AB}$. The running gauge coupling at the scale $\mu$ from $M_s$ is

$$\frac{16\pi^2}{g_A^2(\mu)} = k_A \frac{16\pi^2}{g^2} + b_A \log \frac{M_s^2}{\mu^2} + \Delta_A, \tag{11.98}$$

where $\Delta_A$ is threshold correction. The universal coupling $g$ also receives correction $\Delta_A^{\text{univ}}$ depending on the moduli

$$\frac{1}{g^2} = \text{Re } S + \frac{1}{16\pi^2} \Delta^{\text{univ}}. \tag{11.99}$$

If a transformation mixes the charged fields, we have Konishi anomaly

$$f_A \to f_A - \frac{1}{4\pi^2} \sum_a \alpha_A^a \log(ic_a T_a + d_a) - \frac{1}{4\pi^2} \sum_m \alpha_A^m \log(ic_m U_m + d_m), \tag{11.100}$$

where we have the so-called modular anomaly coefficients

$$\alpha_A^a = \sum_{\mathbf{R}_\alpha} l(\mathbf{R}_\alpha) \left(1 - 2n_\alpha^a\right) - C(G_A),$$

$$\alpha_A^m = \sum_{\mathbf{R}_\alpha} l(\mathbf{R}_\alpha) \left(1 - 2l_\alpha^m\right) - C(G_A), \tag{11.101}$$

and $n_\alpha^a, l_\alpha^m$ are the modular weights (11.70). What is the form of gauge kinetic function giving rise to this? A holomorphic function transforms in this way is the Dedekind eta function (6.84), thus we find [22],

$$f_A = k_A S - \sum_{a=1}^{3} \frac{\alpha_A^a}{4\pi^2} \log \eta(iT_a) - \sum_{m=1}^{h_{2,1}} \frac{\alpha_A^m}{4\pi^2} \log \eta(iU_m) + p_A, \tag{11.102}$$

where $p_A$ are modular invariant holomorphic functions of the moduli.

The one-loop effective gauge coupling, dependent on moduli field, in SUSY [25] and SUGRA [26] is [22]

$$\frac{16\pi^2}{g_A^2} = 16\pi^2 \text{Re } f_A + b_A \log \frac{M_s^2}{\mu^2}$$

$$+ c_A K + 2C(G_A) \log g_A^{-2} \left(\mu^2\right) - \sum_{\mathbf{R}_\alpha} l(\mathbf{R}_\alpha) \log \det Z_{\mathbf{R}_\alpha}^{\text{eff}} \left(\mu^2\right), \tag{11.103}$$

where $b_A = \sum n_{\mathbf{R}_\alpha} l(\mathbf{R}_\alpha) - 3C(G_A), c_A = \sum n_{\mathbf{R}_\alpha} l(\mathbf{R}_\alpha) - C(G_A), l(\mathbf{R}_\alpha)$ is the index of the representation of $\mathbf{R}_\alpha$ introduced in (2.16) and $C(G_A)$ is the quadratic Casimir of the adjoint of $G_A$. The total correction is

$$
\begin{aligned}
\Delta_A + k_A \Delta^{\text{univ}} = &-\sum_a \alpha_A^a \log |\eta(iT_a)|^4 (\text{Re } T_a) \\
&-\sum_m \alpha_A^m \log |\eta(iU_m)|^4 (\text{Re } U_m) + \text{Re } p_A.
\end{aligned}
\tag{11.104}
$$

The last term $\text{Re } p_A$ is in general dependent on the moduli, but for factorizable orbifold, it is constant [22].

## Threshold Correction

For estimating gauge coupling constants [27], we assume that the masses of the fields taking part in the running are the same as the unification scale $M_{\text{U}}$. By the decoupling theorem, we can neglect massive fields heavier than the running scale. However, around the unification scale the effects of mass differences become sizable because the masses themselves are of order of the unification scale $M_{\text{U}}$. We should take into account this effect which is known as the *threshold correction* [28,29].

We calculate the stringy threshold correction $\Delta_a$ by string one-loop amplitude we considered in Chap. 8, Eq. (6.75),

$$
\Delta_a = \int \frac{d^2\tau}{\tau_2} [\mathscr{B}_a(\tau, \bar{\tau}) - b_a],
\tag{11.105}
$$

integrated over the $\text{PSL}(2, \mathbb{Z})$ fundamental region (6.74). The modular invariant amplitude $\mathscr{B}_a$ is just a partition function (8.6) weighted by the squared charge generator $Q_a^2$,

$$
\mathscr{B}_a(\tau, \bar{\tau}) = \sum_{h \in \bar{\mathsf{P}}} \left[ \frac{1}{N} \sum_{g \in \bar{\mathsf{P}}} \text{Tr}(Q_a^2 \, g \, q^{\tilde{L}_0(h)} \bar{q}^{L_0(h)}) \right].
\tag{11.106}
$$

It is analogous to the field theory case where $b_a$ is given by the trace of the squared charge generators over the charged fields leading to $-\frac{11}{3} \text{Tr } Q_a^2$. In the field theory limit, note that $\tau_2 \to \infty$, and $\mathscr{B}_a(\tau, \bar{\tau})$ reduces to $b_a$, to match the low-energy running (11.98).

In the orbifold theory, we can calculate $\Delta_a$ explicitly. The important result obtained in Refs. [29, 30] is that the only nontrivial threshold correction emerges *in the twisted sector having fixed tori* or twisted sectors with $\mathcal{N} = 2$ local supersymmetry. The moduli describe the geometry of the torus, but if the twisted sector fields are localized at the fixed points, so that they cannot see the torus geometry. However, if a twisted sector contains an invariant torus, it can see the geometry of the torus and contributes the threshold correction. The moduli can freely move in this direction, thus has the corresponding coordinate dependence. Then it would be that the untwisted fields should contribute the threshold correction, but their contribution is cancelled by strong ($\mathcal{N} = 4$ in 4D) SUSY. As a corollary, the $\mathbb{Z}_3$ orbifold having no fixed torus receives no threshold correction except for a trivial constant correction of order 5%.

Thus we need to consider twisted sectors having fixed tori. For instance, consider the second twisted sector of the $\mathbb{Z}_4$ orbifold of Sect. 9.2.2. Here, we have a twist $2\phi = \frac{1}{2}(2\ 1\ 1) \simeq \frac{1}{2}(0\ 1\ 1)$. The first two-torus remain untouched. This twisted sector has effective order 2 and the twist $2\pi$ forms $\mathbb{Z}_2$ point group. We call $D_a$ the little subgroup of $\mathsf{P}$ leaving the $a$th complex plane invariant. For the above $\mathbb{Z}_4$, we have $D_1 = \mathbb{Z}_2$, $D_2 = D_3 = 1$. For the $\mathbb{Z}_6-\mathrm{II}$ orbifold, the twist vector is $(\frac{1}{2}\ \frac{1}{3}\ \frac{1}{6})$. The little groups are $D_1 = \mathbb{Z}_3$, $D_2 = \mathbb{Z}_2$, $D_3 = 1$. For cases where this factorization is not possible, see [31].

The result is simply,

$$\Delta_a = - \sum_{\text{moduli}} \sum_i b_a^i \frac{N_i}{N} \log\left[|\eta(T_i)|^4 \operatorname{Re} T_i\right] + \log\left[|\eta(U_i)|^4 \operatorname{Re} U_i\right] + c_a,$$

$$(11.107)$$

where $c_a$ is a constant term, $N'$ is the order of the subsector action (2 for the second twisted sector of the above $\mathbb{Z}_4$ example), and $\eta$ is the Dedekind eta function (6.84), a regular customer in the modular form. All the moduli fields have the same dependence [29].

The most important issue is whether this threshold correction is large enough to fill the gap between grand unification scale and string scale (or the Planck scale). In a large $T$ limit, from the asymptotic behavior of eta function, it behaves as

$$\Delta_a \sim \sum_i b_a^i \frac{N'}{N} \frac{\pi(T + T^*)}{6}.$$

$$(11.108)$$

However, it seems that the threshold correction alone is not enough to make the three couplings meet at the string scale. But, we note that in field theoretic calculation of the running we must take into account the above string threshold corrections which are present near the string scale (Fig. 11.2).

### The Final Form

We may separate the contribution from the universal and non-universal pieces

$$\alpha_A^i = b_A^{\mathcal{N}=2}(i) \frac{N'}{N} + k_A \delta_A^{\mathrm{GS}}.$$

$$(11.109)$$

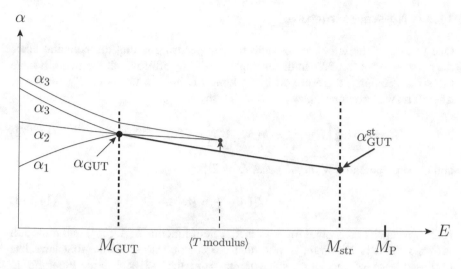

**Fig. 11.2** A view on the threshold correction

This makes the gauge kinetic function non-invariant. Plugging this to (11.101), we may obtain $\delta_A^{GS}$. The final Kähler potential has the form [20]

$$
\kappa^2 K^{(1)} = -\log\left( S + S^* + \frac{k_A \delta_A^{GS}}{4\pi^2}\Big[ \sum_{a=1}^{3} \log|\eta(iT_A)|^4 \operatorname{Re} T_a \right.
$$

$$
\left. + \sum_{m=1}^{h_{2,1}} \log|\eta(iU_A)|^4 \operatorname{Re} U_a \Big]\right). \tag{11.110}
$$

Consider the $\mathbb{Z}_3$ orbifold with diagonal moduli $T_a$ for simplicity. We have no $\mathcal{N} = 2$ plane thus no contribution $b_A^{\mathcal{N}=2}$. From (11.101), we see that the coefficient is the same for $a = 0$. For the standard embedding, using the spectrum obtained in Sect. 7.4, we obtain

$$
\alpha_{E_6}^a = \alpha_{SU(3)}^a = \alpha_{E_8}^a = -30, \tag{11.111}
$$

verifying the universality of the anomalous U(1). For other embedding we may verify $\delta_A^{GS}$ is universal but we may have different values of $\alpha_A^a$. Fpr tje $\mathbb{Z}_4$ orbifold, the twist vector is $(\frac{1}{2}\ \frac{1}{4}\ \frac{1}{4})$. The only little group is $D_1 = \mathbb{Z}_2$. We have

$$
\alpha_A^3 = k_A \delta_{GS}^1 + \frac{1}{2} b_A^{\mathcal{N}=2}, \qquad \alpha_A^{2,3} = k_A \delta_{GS}^{2,3}. 
$$

We have $\delta_{GS}^1 = 0$. For the $\mathbb{Z}_6-\text{II}$ orbifold, we have $\delta_{GS}^1 = 0$ but $\delta_{GS}^{2,3} \neq 0$.

## 11.3.4 No-scale Structure

One feature of the effective potential from superstring is that the potential takes the no scale form [32, 33]. In the example of $T^6/\mathbb{Z}_3$ orbifold, all the two-tori have the same geometry, therefore all the volume moduli are the same $T_{a\bar{b}} = T\delta_{ab}$. Therefore, we have the Kähler potential (11.55) of a form

$$K = -3\log\left[T + T^* - h(A^a, A^{\bar{a}})\right] - \tilde{K}(S^n, S^{n*}).  \quad (11.112)$$

and the superpotential $W$, *not depending on $T$*,

$$W = W_1(A^a) + W_2(S^n).  \quad (11.113)$$

Here, we consider a general form for matter field function $h(A^a, A^{\bar{a}})$ and $S$ moduli $\tilde{K}(S^n, S^{n*})$. It is important to note that the *coefficient* $-3$ of the first term has a consequence of cancelling some terms from the $-3|W|^2$ term. Plugging $K$ into (2.41), it has the form [32, 33]

$$V = \frac{1}{3(T + T^* - h)^3}\left(N^{-1}\right)^i_{\ j}\left(\frac{\partial W}{\partial A_i}\right)\left(\frac{\partial W^*}{\partial A^{*j}}\right)$$
$$+ (G^{-1})^n_{\ m}(D_n W)(D_m W)^*,  \quad (11.114)$$

where

$$N^i_{\ j} = \frac{\partial^2 h}{\partial A_i \partial A^{*j}}.$$

Here, $G^n_{\ m}$ and $N^a_{\ b}$ are positive definite. Even if we include the $D$-term scalar potential, (the last term in (2.41)), the scalar potential (11.114) is positive definite and has the minimum at zero. This is similar to the global supersymmetry case, but the minimum *does not* correspond to the supersymmetry preserving one, since in view of Eq. (2.44)

$$\delta_\epsilon \Psi^i \sim D^i W = -\frac{3}{|T + T^* - h|}\frac{\partial h}{\partial A_i}W  \quad (11.115)$$

need not vanish. It is better to have the zero minimum of the scalar potential with broken supersymmetry. This helps fitting with the almost zero cosmological constant [34].

Without $S$, the supersymmetry breaking scale $M_{\text{SUSY}}$ (2.46) and the gravitino mass $m_{3/2}$ (2.47) are not determined at tree level, which is then generated by radiative corrections. Because of this absence of scale parameter, it is called *no scale supergravity*. This scale invariance property is the characteristic feature in the string interactions [1]. However, the supersymmetry transformation of the $S$ field relates gravitino mass to the supersymmetry breaking scale.

This no-scale structure is tracked to the ten dimensional action, which in turn originates from the scale invariance structure of string [1].

## 11.4 Shift Vector

We have learned that the projection associated with orbifold symmetry imposes a boundary conditions on the fields and thus breaks some symmetries.

Suppose that we have $SO(2n)$ gauge field $A_M$ on which the following projector is acting:

$$P = \text{diag}\left(e^{2\pi i V_1 H^1}, \ldots, e^{2\pi i V_n H^n}\right), \tag{11.116}$$

where $\{H^I\}$ is the Cartan subalgebra of the $SO(2n)$. Beside of these, the remaining generators of a Lie Algebra can make linear combinations and becomes ladder operators $E^\alpha$ satisfying

$$[H^I, E^\alpha] = \alpha^I E^\alpha. \tag{11.117}$$

Here $\alpha$ are vectors having the components $(\alpha^I)$ are called roots of the $SO(2n)$. We have unbroken generators, this generator satisfies the (5.35), or

$$P E^\alpha P^{-1} = e^{2\pi i [V^I H^I, E^\alpha]} E^\alpha = e^{2\pi i V^I \alpha^I} E^\alpha, \tag{11.118}$$

where we used Baker–Campbell–Hausdorff formula. Thus the generators $E^\alpha$ of the unbroken gauge groups satisfy the relation

$$\alpha \cdot V = \text{integer}, \tag{11.119}$$

where the dot product is the summation in (11.118). Indeed, the field theory with extra dimensions provides a low-energy limit of string theory. Historically the doublet–triplet splitting from orbifold, discussed in Sect. 2.3.3, was observed long time ago in string orbifolds in this way.

## 11.5 Anomaly Cancellation

The field theory limit of heterotic string is ten dimensional supergravity coupled with non-Abelian gauge fields. Here we show that gauge and gravitational anomalies can be cancelled by the rank 16 gauge groups $E_8 \times E_8$ or $SO(32)$ [35]. The mixed anomaly can be cancelled with the participation of the antisymmetric gauge field $B_{MN}$ by the Green–Schwarz mechanism [10]. This motivated to the discovery of heterotic string theories based on the same gauge group [36], which takes two corners of the moduli space in Fig. 2.10.

### 11.5.1 Anomaly Polynomial

Anomaly is expressed as a failure of classical current conservation. The non-conserved source is the nonvanishing change $G(\lambda)$ of the effective action under the gauge transformation $\lambda$. It satisfies Wess–Zumino consistency condition [37]

$$\delta_{\lambda_1} G(\lambda_2) - \delta_{\lambda_2} G(\lambda_1) = G([\lambda_1, \lambda_2]). \tag{11.120}$$

This can be solved by the gauge-invariant polynomial in the field strength. For gauge anomaly it is field strength $F_{\mu\nu}$ and for the gravitational one it is Riemann tensor $R_{\mu\nu}$, regarding the internal Lorentz index as gauge index of SO($D$) in $D$-dimension, which are all contracted with the generator of the corresponding algebra. If the polynomial $I_D$ satisfies the conditions

$$I_{D+2} = dI_{D+1} \tag{11.121}$$

$$\delta I_{D+1} = dI_D^1, \tag{11.122}$$

we can solve the condition (11.120). It will be useful to use $\hat{I} = i(2\pi)^{D/2} I$.

As an example, we consider ten dimensional anomaly [35].

- For the $n$ Majorana–Weyl fermions, we have

$$\hat{I}_8(F, R) = -\frac{\operatorname{Tr} F^6}{1440} + \frac{\operatorname{Tr} F^4 \operatorname{tr} R^2}{2304} - \frac{\operatorname{Tr} F^2 \operatorname{tr} R^4}{23040} - \frac{\operatorname{Tr} F^2 \left(\operatorname{tr} R^2\right)^2}{18432}$$

$$+ \frac{n \operatorname{tr} R^6}{725760} + \frac{n \operatorname{tr} R^4 \operatorname{tr} R^2}{552960} + \frac{n \left(\operatorname{tr} R^2\right)^3}{1327104}. \tag{11.123}$$

- For the Majorana–Weyl **56**, we have

$$\hat{I}_{56}(R) = -\frac{495 \operatorname{tr} R^6}{725760} + \frac{225 \operatorname{tr} R^4 \operatorname{tr} R^2}{552960} - \frac{63(\operatorname{tr} R)^3}{1327104}. \tag{11.124}$$

- For the rank-two self-dual tensor, we have

$$\hat{I}_{SD}(R) = \frac{992 \operatorname{tr} R^6}{725760} - \frac{448 \operatorname{tr} R^4 \operatorname{tr} R^2}{552960} + \frac{128 \left(\operatorname{tr} R^2\right)^3}{1327104}. \tag{11.125}$$

For heterotic string, we have one gravitino, $n = 1$ dilatino and $n = \dim g$ gauginos, all of which are Majorana–Weyl. The sum of anomaly polynomial is

$$\hat{I} = \hat{I}_{56}(R) - \hat{I}_8(R) + \hat{I}_8(F, R)$$

$$= \frac{1}{1440}\left(-\operatorname{Tr} F^6 + \frac{1}{48}\operatorname{Tr} F^2 \operatorname{Tr} F^4 - \frac{1}{14400}(\operatorname{Tr} F^2)^3\right)$$

$$+ (\dim g - 496)\left(\frac{\operatorname{tr} R^6}{725760} + \frac{\operatorname{tr} R^4 \operatorname{tr} R^2}{552960} + \frac{(\operatorname{tr} R^2)^3}{1327104}\right) \qquad (11.126)$$

$$+ \frac{Y_4 X_8}{768},$$

where

$$Y_4 = \operatorname{tr} R^2 - \frac{1}{30}\operatorname{Tr} F^2 \qquad (11.127)$$

$$X_8 = \operatorname{tr} R^4 + \frac{(\operatorname{tr} R^2)^2}{4} - \frac{\operatorname{Tr} F^2 \operatorname{tr} R^2}{30} + \frac{\operatorname{Tr} F^4}{3} - \frac{(\operatorname{Tr} F^2)^2}{900}. \qquad (11.128)$$

The second line of (11.126) contains $F$ only and yield pure gauge anomaly. Likewise the third line contains $R$ only and gives pure gravitational anomaly contribution. It suggests that a gauge group of dimension 496 may cancel the anomaly [10]. There is a unique simple group SO(32) whose dimension is 496. For SO($n$), we have

$$\operatorname{Tr} T^2 = (n - 2)\operatorname{tr_n} T^2,$$

$$\operatorname{Tr} T^4 = (n - 8)\operatorname{tr_n} T^4 + 3(\operatorname{tr_n} T^2)^2, \qquad (11.129)$$

$$\operatorname{Tr} T^6 = (n - 32)\operatorname{tr_n} T^6 + 15\operatorname{tr_n} T^2 \operatorname{tr_n} T^4,$$

where $\mathbf{n}$ is the vector representation of SO($n$). Also, if we have $n = 32$, the second line of (11.126) for the pure gauge anomaly, vanishes. It comes from SO(32) heterotic and type I strings.

We also have semisimple group $E_8 \times E_8$, with dimension $248 + 248 = 496$. It satisfies

$$\operatorname{Tr} T^4 = \frac{\operatorname{Tr} T^2}{100},$$

$$\operatorname{Tr} T^6 = \frac{(\operatorname{Tr} T^2)^3}{7200}. \qquad (11.130)$$

Thus it makes the second and the third lines vanish. It arises from $E_8 \times E_8$ heterotic string.

It can be shown that there is no other non-Abelian gauge groups satisfying the above. An Abelian group U(1) can have trivial anomaly cancellation if we have appropriate number. However, there is no consistent top-down construction from string theory.

The only remaining one is the last term in (11.126). It is factorized and this property is important in every dimension. We have a counterterm in the Lagrangian [10]

$$B \wedge X_8, \tag{11.131}$$

where $B$ is the antisymmetric tensor field that is always accompanied by the metric tensor.

The Green–Schwarz mechanism cancels the anomaly. It satisfies the Wess–Zumino consistency condition and the total anomaly polynomial vanishes. Under the gauge transformation, we have

$$\delta A = d\lambda$$

$$\delta \omega = d\Theta \tag{11.132}$$

$$\delta B = \text{Tr}(\lambda dA) - \frac{1}{30}\text{tr}(\Theta d\omega).$$

We have the Bianchi identity for the antisymmetric tensor field,

$$\frac{2}{\alpha'}dH = d^2 B + \text{tr } R^2 - \frac{1}{30}\text{Tr } F^2 = Y_4, \tag{11.133}$$

where $d^2 B = 0$. Here we used the fact that the trace over vector representation of $SO(n)$ is converted to that of the adjoint representation of $E_8$,

$$\text{tr}_\text{v} T^2 = \frac{1}{30}\text{Tr } T^2.$$

Thus the counterterm (11.131) cancels the anomaly polynomial in the last term of Eq. (11.126).

## 11.5.2 Elliptic Genus

If the holomorphic partition function has modular invariance, we have anomaly-free low-energy field theory. A heterotic string in $8m + 2$ dimensions with the self-dual lattice in the bosonic sector is anomaly free. The heterotic string is the special case $m = 2$. The anomaly polynomial is generated by *elliptic genus* [Schellekens and Warner].

The absence of anomaly can be shown in another way, which is more appropriate to orbifold [38–40]. We tweak the partition function to elliptic genus, we may extract information on anomaly, we do not consider the full spectrum. We just consider the partition function for the left mover, which we require holomorphic and modular invariant. We twist the partition function and define elliptic genus

$$\mathscr{Z}_{T^2}(\tau) = \mathrm{Tr}_R(-1)^F q^{\tilde{L}_0 + \tilde{c}} \bar{q}^{L_0 + c} y^J \prod_a x_a^{K_a}, \tag{11.134}$$

where the trace is over the R sector, $q$ is the modular parameter (6.78), and $K_a$ are the global charges. Because of the worldsheet fermion number $F$ of the right movers spacetime supersymmetric pairs cancel and giving the left-moving zero modes. It is

$$\mathscr{Z}_{T^2}(\tau) = \frac{1}{16\pi^4} \hat{A}(R) \mathscr{Z}^{E_8}_{(kV, lV)}(\tau) \mathscr{Z}^{E_8}_{(0,0)}(\tau) \mathscr{Z}^X_{(\theta^k, \theta^l)}(\tau). \tag{11.135}$$

The Dirac genus (15.69) counts the "the number of fixed points" under the twisting (8.32). The next factor $P(q, F)$ is the generating function for Chern characters.

If we take all $\phi_a = 0$, we have the partition function for the untwisted heterotic string, multiplied by the Dirac genus. For nonzero $\phi_a$ we *formally* have the partition function for the untwisted sector. It is understood that it is the partition function with a general background. The trick is we replace

$$\phi_a \to R.$$

The anomaly generating function is expanded in even powers of $q$. The field theory contribution is contained in the constant term of $q$.

It is invariant under $\mathscr{T}$. Under $\mathscr{S}$, we have

$$A(q(-1/\tau), \tau^{-1}F, \tau^{-1}R) = \tau^{-4m} \exp\left[\frac{i}{4\pi\tau}\left(-\frac{1}{8\pi^2}\mathrm{tr}\,R^2 + \frac{1}{8\pi^2 C_A}\mathrm{Tr}\,F^2\right)\right]. \tag{11.136}$$

This means that the $A$ is modular function of weight two. Since all the theta function involved here is holomorphic for $\mathrm{Im}\,\tau > 0$, If the Bianchi identity holds, the only poles of $C(q)$ occur at $q = 0$. Since there is no modular form of weight two, the constant term in $q$ is zero. This means that there is no anomaly.

Chiral fields in even spacetime dimensions can lead to anomalies. They are spinors and antisymmetric tensor fields that is self-dual or anti-self-dual under Hodge duality.

We have experienced that anomaly cancellation is a hint of larger symmetry: for example, chiral anomaly cancellation of the SM is automatically implemented when we consider GUTs such as SU(5), SO(10), etc. Although the resulting low-energy theory is chiral, the whole combination resulting from the fields of the GUT groups cancel the anomaly.

In the field theoretic orbifold, there are arbitrariness: the number of dimensions, the gauge group, and so on. We cannot determine which fields should live in the bulk or at the fixed points, either. The only consistency condition is the anomaly cancellation.

Construction from string theory is more restrictive and predictive. We always obtain anomaly-free theory if we obtain it by compactifying string theory, satisfying *global consistency condition*. In the orbifold compactification it is modular invariance of the holomorphic partition function [41]. It regularizes the divergence of one-loop diagram in a gauge-invariant way. In the field theory limit we have no divergent one-loop diagram, from where we may have potential anomaly.

Once we parameterize the symmetry breaking using a single vector $V$ in (11.118) determining gauge symmetry breaking, it also determines matter spectrum in the bulk and the fixed points without arbitrariness.

# References

1. E. Witten, Dimensional reduction of superstring models. Phys. Lett. **155B**, 151 (1985)
2. J.P. Derendinger, L.E. Ibanez, H.P. Nilles, On the low-energy limit of superstring theories. Nucl. Phys. **B267**, 365–414 (1986)
3. L.J. Dixon, V. Kaplunovsky, J. Louis, On effective field theories describing (2,2) vacua of the heterotic string. Nucl. Phys. **B329**, 27–82 (1990)
4. C.P. Burgess, A. Font, F. Quevedo, Low-energy effective action for the superstring. Nucl. Phys. **B272**, 661–676 (1986)
5. A.H. Chamseddine, N=4 supergravity coupled to N=4 matter. Nucl. Phys. **B185**, 403 (1981)
6. E. Bergshoeff, M. de Roo, B. de Wit, P. van Nieuwenhuizen, Ten-dimensional Maxwell-Einstein supergravity, its currents, and the issue of its auxiliary fields. Nucl. Phys. **B195**, 97–136 (1982)
7. G.F. Chapline, N.S. Manton, Unification of Yang-Mills theory and supergravity in ten-dimensions. Phys. Lett. **120B**, 105–109 (1983). [105(1982)]
8. M. Dine, R. Rohm, N. Seiberg, E. Witten, Gluino condensation in superstring models. Phys. Lett. **156B**, 55–60 (1985)
9. J. Polchinski, *String Theory. Vol. 2: Superstring Theory and Beyond.* Cambridge Monographs on Mathematical Physics (Cambridge University Press, Cambridge, 2007)
10. M.B. Green, J.H. Schwarz, Anomaly cancellation in supersymmetric D=10 gauge theory and superstring theory. Phys. Lett. **149B**, 117–122 (1984)
11. J. Maharana, J.H. Schwarz, Noncompact symmetries in string theory. Nucl. Phys. **B390**, 3–32 (1993)
12. Y. Tanii, *Introduction to Supergravity.* Springer Briefs in Mathematical Physics, vol. 1 (Springer, Tokyo, 2014)
13. S. Ferrara, C. Kounnas, M. Porrati, General dimensional reduction of ten-dimensional supergravity and superstring. Phys. Lett. **B181**, 263 (1986)
14. A. Sen, An introduction to nonperturbative string theory, in *Duality and Supersymmetric Theories. Proceedings, Easter School, Newton Institute, Euroconference, Cambridge, April 7–18, 1997*, pp. 297–413 (1998)
15. K.S. Narain, New heterotic string theories in uncompactified dimensions < 10. Phys. Lett. **169B**, 41–46 (1986)
16. K.S. Narain, M.H. Sarmadi, E. Witten, A note on toroidal compactification of heterotic string theory. Nucl. Phys. **B279**, 369–379 (1987)

17. P.H. Ginsparg, Comment on toroidal compactification of heterotic superstrings. Phys. Rev. D **35**, 648 (1987)

18. M. Cvetic, J. Louis, B.A. Ovrut, A string calculation of the Kahler potentials for moduli of Z(N) orbifolds. Phys. Lett. **B206**, 227–233 (1988)

19. L.E. Ibanez, D. Lust, Duality anomaly cancellation, minimal string unification and the effective low-energy Lagrangian of 4-D strings. Nucl. Phys. **B382**, 305–361 (1992)

20. J.-P. Derendinger, S. Ferrara, C. Kounnas, F. Zwirner, All loop gauge couplings from anomaly cancellation in string effective theories. Phys. Lett. **B271**, 307–313 (1991)

21. G.L. Cardoso, B.A. Ovrut, A Green-Schwarz mechanism for D = 4, N=1 supergravity anomalies. Nucl. Phys. **B369**, 351–372 (1992)

22. V. Kaplunovsky, J. Louis, On Gauge couplings in string theory. Nucl. Phys. **B444**, 191–244 (1995)

23. M. Dine, N. Seiberg, Nonrenormalization theorems in superstring theory. Phys. Rev. Lett. **57**, 2625 (1986)

24. P. Svrcek, E. Witten, Axions in string theory. J. High Energy Phys. **6**, 051 (2006)

25. M.A. Shifman, A.I. Vainshtein, Solution of the anomaly puzzle in SUSY gauge theories and the Wilson operator expansion. Nucl. Phys. **B277**, 456 (1986). [Sov. Phys. JETP64,428(1986); Zh. Eksp. Teor. Fiz.91,723(1986)]

26. V. Kaplunovsky, J. Louis, Field dependent gauge couplings in locally supersymmetric effective quantum field theories. Nucl. Phys. **B422**, 57–124 (1994)

27. A.J. Buras, J.R. Ellis, M.K. Gaillard, D.V. Nanopoulos, Aspects of the grand unification of strong, weak and electromagnetic interactions. Nucl. Phys. **B135**, 66–92 (1978)

28. V.S. Kaplunovsky, Mass scales of the string unification. Phys. Rev. Lett. **55**, 1036 (1985)

29. L.J. Dixon, V. Kaplunovsky, J. Louis, Moduli dependence of string loop corrections to gauge coupling constants. Nucl. Phys. **B355**, 649–688 (1991)

30. V.S. Kaplunovsky, One loop threshold effects in string unification (1992)

31. D. Bailin, A. Love, W.A. Sabra, S. Thomas, String loop threshold corrections for Z(N) Coxeter orbifolds. Mod. Phys. Lett. A **9**, 67–80 (1994)

32. J.R. Ellis, C. Kounnas, D.V. Nanopoulos, Phenomenological SU(1,1) supergravity. Nucl. Phys. B **241**, 406–428 (1984)

33. A.B. Lahanas, D.V. Nanopoulos, The road to no scale supergravity. Phys. Rep. **145**, 1 (1987)

34. S. Weinberg, The cosmological constant problem. Rev. Mod. Phys. **61**, 1–23 (1989)

35. L. Alvarez-Gaume, E. Witten, Gravitational anomalies. Nucl. Phys. **B234**, 269 (1984). [269 (1983)]

36. D.J. Gross, J.A. Harvey, E.J. Martinec, R. Rohm, The heterotic string. Phys. Rev. Lett. **54**, 502–505 (1985)

37. J. Wess, B. Zumino, Consequences of anomalous Ward identities. Phys. Lett. **37B**, 95–97 (1971)

38. A.N. Schellekens, N.P. Warner, Anomalies and modular invariance in string theory. Phys. Lett. **B177**, 317–323 (1986)

39. A.N. Schellekens, N.P. Warner, Anomaly cancellation and self-dual lattices. Phys. Lett. **B181**, 339–343 (1986)

40. A.N. Schellekens, N.P. Warner, Anomalies, characters and strings. Nucl. Phys. **B287**, 317 (1987)

41. K.-S. Choi, S.-J. Rey, Elliptic genus, anomaly cancellation and heterotic M-theory (2017)

# Algebraic Structure

# 12

In gauge theory, we understand matter and gauge fields by transformation under Lie algebras. We have seen that string theory naturally realizes such algebra. In heterotic string, momentum-winding quantum numbers provide representation of a state. Considering massive states and twisted states, this algebra is generalized it to the affine Lie algebra. It is generalized to twisted Lie algebra that describes twisted sector spectrum, which is a special feature in orbifold theory. Thus, it is crucial to understand the structure of the algebra.

We break gauge symmetry of string theory by modding out the symmetry of a given algebra, which forms automorphism. The shift vectors parameterize such automorphism and hence breaking of the symmetry. If we have more than one automorphism actions, in general we cannot express the information using the shift vectors and they can reduce the rank. We can understand the patterns of symmetry breaking so that the classification become possible. This general method is also used to construct asymmetric orbifolds.

## 12.1 Lie Algebra

We briefly review basic facts on Lie algebra, following Ref. [1].

### 12.1.1 Lie Algebra

A Lie algebra $\mathfrak{g}$ is defined by generators $T^a$ satisfying the commutation relation

$$[T^a, T^b] = i f^{abc} T^c, \tag{12.1}$$

with the structure constants $f^{abc}$ specifying the algebra. The number $d$ of the involved generators is called dimension of the algebra $\mathfrak{g}$, which is also denoted

© Springer Nature Switzerland AG 2020
K.-S. Choi, J. E. Kim, *Quarks and Leptons From Orbifolded Superstring*,
Lecture Notes in Physics 954, https://doi.org/10.1007/978-3-030-54005-0_12

as dim $\mathsf{g}$. The following Cartan–Weyl basis is useful. There is a set of commuting generators $\{H^i\}$, called Cartan subalgebra (CSA)

$$[H^I, H^J] = 0, \quad I, J = 1, \ldots, r. \tag{12.2}$$

Here $r$, the rank, is an invariant of the algebra, independent of the choice of the CSA. After separating CSA, the remaining generators $E^P$ satisfy the relation

$$[H^I, E^P] = P^I E^P, \quad I = 1, \ldots, r. \tag{12.3}$$

That is, every generator is an eigenstate of the adjoint operator by CSA elements

$$\mathrm{ad}_A \, B = [A, B]. \tag{12.4}$$

The vector $P = (P^I)$ defines *root vectors,* or simply *roots.* It is convenient to consider the space of roots.

Root vectors are generated by linear combinations of *simple roots* $\alpha^i, i = 1, \ldots, r$. We display the simple roots of some Lie algebras in the orthogonal representation in Table 12.1. Note that each $\alpha^i$ is a vector of rank $r$ with suppressed index in the sense of (12.3) and we may define an inner product. The structure of the Lie algebra in (12.1) is equivalently contained in the *Cartan matrix*

$$A^{ij} \equiv \frac{2\alpha^i \cdot \alpha^j}{\alpha^j \cdot \alpha^j} = 2\frac{|\alpha^i|}{|\alpha^j|} \cos\theta_{ij}, \tag{12.5}$$

where $\theta_{ij}$ is the angle between the two root vectors $\alpha^i$ and $\alpha^j$. Note that the definition is asymmetric. We may define dual root

$$\alpha^\vee = \frac{2\alpha}{\alpha \cdot \alpha} \tag{12.6}$$

and make it more systemic $A^{ij} = \alpha^i \cdot \alpha^{j\vee}$.

A state transforms under the Lie algebra. For each $E^P$ in (12.3), we always have $E^{-P}$. We can normalize as

$$[E^P, E^{-P}] = P \cdot H \equiv H^P. \tag{12.7}$$

They form a set of ladder operators. The Cartan matrix tells us the relative steps of ladder operations. To deal with this, it is useful to define the fundamental weights $\Lambda_j$

$$\alpha^{i\vee} \cdot \Lambda_j \equiv \delta^i_j. \tag{12.8}$$

The fundamental weights of some algebras are shown in Table 12.1. From (12.8), the extended root is always $\alpha^0 \equiv -\Lambda_1$.

**Table 12.1** Simple roots of Lie algebras in the orthogonal representations

| Algebra | Simple roots | Fundamental weights |
|---|---|---|
| $A_{n-1}$ $SU(n)$ | $\alpha^1 = (1, -1, 0, 0, \ldots, 0, 0)$ $\alpha^2 = (0, 1, -1, 0, \ldots, 0, 0)$ $\alpha^3 = (0, 0, 1, -1, \ldots, 0, 0)$ $\vdots$ $\alpha^n = (0, 0, 0, 0, \ldots, 1, -1)$ | $\Lambda_1 = \frac{1}{n}(n-1, -1, -1, \ldots, -1, -1)$ $\Lambda_2 = \frac{1}{n}(n-2, n-2, -2, \ldots, -2, -2)$ $\Lambda_3 = \frac{1}{n}(n-3, n-3, n-3, \ldots, -3, -3)$ $\vdots$ $\Lambda_n = \frac{1}{n}(1, 1, 1, \ldots, 1, n-1)$ |
| $D_n$ $SO(2n)$ | $\alpha^1 = (1, -1, 0, 0, \ldots, 0, 0)$ $\alpha^2 = (0, 1, -1, 0, \ldots, 0, 0)$ $\alpha^3 = (0, 0, 1, -1, \ldots, 0, 0)$ $\vdots$ $\alpha^{n-1} = (0, 0, 0, 0, \ldots, 1, -1)$ $\alpha^n = (0, 0, 0, 0, \ldots, 1, 1)$ | $\Lambda_1 = (1, 0, 0, 0, \ldots, 0, 0)$ $\Lambda_2 = (1, 1, 0, 0, \ldots, 0, 0)$ $\Lambda_3 = (1, 1, 1, 0, \ldots, 0, 0)$ $\vdots$ $\Lambda_{n-1} = (+, +, +, \ldots, +, -)$ $\Lambda_n = (+, +, +, \ldots, +, +)$ |
| $E_8$ | $\alpha^1 = (0, 1, -1, 0, 0, 0, 0, 0)$ $\alpha^2 = (0, 0, 1, -1, 0, 0, 0, 0)$ $\alpha^3 = (0, 0, 0, 1, -1, 0, 0, 0)$ $\alpha^4 = (0, 0, 0, 0, 1, -1, 0, 0)$ $\alpha^5 = (0, 0, 0, 0, 0, 1, -1, 0)$ $\alpha^6 = (0, 0, 0, 0, 0, 0, 1, -1)$ $\alpha^7 = (+, -, -, -, -, -, -, +, )$ $\alpha^8 = (0, 0, 0, 0, 0, 0, 1, 1)$ | $\Lambda_1 = (1, 1, 0, 0, 0, 0, 0, 0)$ $\Lambda_2 = (2, 1, 1, 0, 0, 0, 0, 0)$ $\Lambda_3 = (3, 1, 1, 1, 0, 0, 0, 0)$ $\Lambda_4 = (4, 1, 1, 1, 1, 0, 0, 0)$ $\Lambda_5 = (5, 1, 1, 1, 1, 1, 0, 0)$ $\Lambda_6 = (\frac{7}{2}, +, +, +, +, +, +, -)$ $\Lambda_7 = (2, 0, 0, 0, 0, 0, 0, 0)$ $\Lambda_8 = (\frac{5}{2}, +, +, +, +, +, +, +)$ |

$+$ and $-$ denote in the spinor forms filling 8 entries, respectively, $\frac{1}{2}$ and $-\frac{1}{2}$

The fundamental weights provide another complete basis $\{\Lambda_i\}$ which is called the *Dynkin basis*. A vector in the Dynkin basis will be denoted by a square bracket [ ]. We have highest weight states, and the ladder operator takes one state into another. Every highest weight representation has integral nonnegative entries in the Dynkin basis. Thus, we can associate the nodes of the Dynkin diagram with tensoral representations. For instance, $\Lambda_1, \Lambda_n$ of $SU(n)$ corresponds to the fundamental weights $\mathbf{n}, \bar{\mathbf{n}}$, respectively.

From (12.8), we see that the inverse Cartan matrix

$$A_{ij}^{-1} = \Lambda_i \cdot \Lambda_j \tag{12.9}$$

plays the role of a metric tensor. Multiplying $\alpha^j$ and contracting the $j$ index, we have

$$\Lambda_i = A_{ij}^{-1} \alpha^j. \tag{12.10}$$

It provides the components of the simple roots for fixed $i$ in Dynkin basis, which is convenient basis for obtaining irreducible representations from a highest weight vector.

**Fig. 12.1** Dynkin diagrams of finite dimensional simple Lie algebras. Numbers indicate the order of the corresponding simple roots. Solid nodes correspond to short simple roots

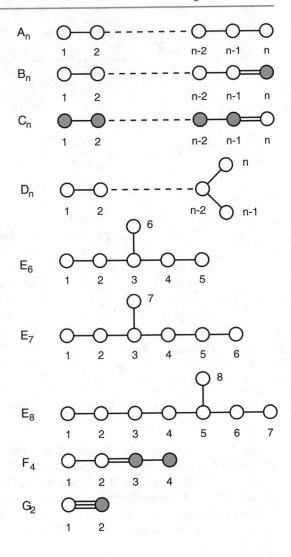

It is useful to picture the above information in *Dynkin diagrams*. It is set of nodes representing the roots and they are connected by one or more lines. The number of lines connecting $i$th and $j$th nodes is the minus of the Cartan matrix. All the possible ladder operations are classified, and we have shown all the possible Dynkin diagrams in Fig. 12.1 and extended diagrams in Fig. 12.2 with the Coxeter labels below the nodes.

Because the Cartan matrices are asymmetric in general, we have also asymmetric Dynkin diagrams. To have the above cosines (12.5), some roots have different lengths. In Fig. 12.1, solid nodes correspond to the short nodes. Thus, a single line represents 120° between the simple roots connected to the line. In $A$, $D$, $E$ type algebras, all the roots have the same length and all the connecting lines are single

**Fig. 12.2** Extended Dynkin diagrams of simple Lie algebras. They are also Dynkin diagrams of the so-called untwisted affine Lie algebras [1]. The numbers inside the nodes are dual Coxeter numbers which are mostly the same as Coxeter numbers, but some are different, which are expressed in italicized numbers. The numbers outside are the ordering, and the extended root is the zeroth

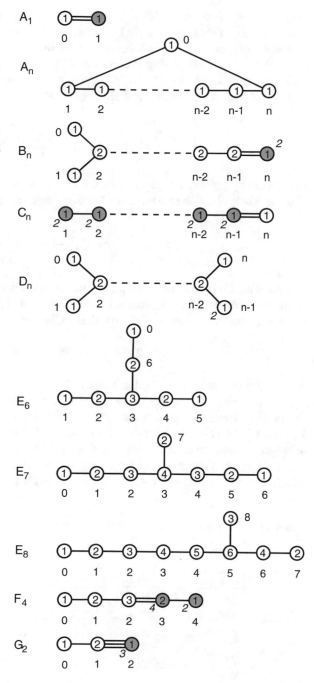

lines. Such algebras are called *simply laced*. For these, we can fix $\alpha \cdot \alpha = 2$, so that they are self-dual $\alpha^\vee = \alpha$ and we can forget about the dual roots. The angles 135 and 150° between simple roots are represented by the double and triple lines, respectively. The ratio of simple roots connected to the double (triple) line is $\sqrt{2}$ ($\sqrt{3}$) among which the smaller simple root is represented by a solid node.

For each algebra, there is a unique, highest root

$$\theta \equiv \sum_{i=1}^{r} a_i \alpha^i, \qquad \frac{2}{\theta \cdot \theta}\theta \equiv \sum_{i=1}^{r} a_i^\vee \alpha^{i\vee}, \tag{12.11}$$

determined by positive integers $a_i$ called the *Coxeter labels*. We have displayed the Coxeter labels below the nodes of the Dynkin diagrams in Fig. 12.2. The sums

$$g \equiv 1 + \sum_{i=1}^{r} a_i, \qquad g^\vee \equiv 1 + \sum_{i=1}^{r} a_i^\vee \tag{12.12}$$

are called the *Coxeter number* and *dual Coxeter number*, respectively. We can define similarly the dual Coxeter number using the dual Coxeter labels. Especially, the dual Coxeter number coincides with the quadratic Casimir

$$2g^\vee \delta^{ab} = f^{acd} f^{bcd}. \tag{12.13}$$

These information are displayed in Table 12.2.

This highest root provides an extended root $\alpha^0 \equiv -\theta$. Its inner products with the original simple roots give the extended Dynkin diagram shown in Fig. 12.2. In view of (12.12), it is natural to define $a_0 = 1$. We can easily find a maximal subalgebra using the extended Dynkin diagram. The simple roots and the extended root are linearly dependent, and projecting out one of them gives a maximal regular subalgebra of the original algebra $E_8$. However, it is known that there are five

**Table 12.2**  Some characteristic number of Lie algebras $g$

| $g$ | $\dim g$ | $g^\vee$ | $g$ | $I$ | Casimir dims. |
|---|---|---|---|---|---|
| $A_{n-1}$ | $n^2 - 1$ | $n$ | $n$ | $n$ | $2, 3, \ldots, n$ |
| $B_n$ | $n(2n+1)$ | $2n-1$ | $2n$ | $2$ | $2, 4, \ldots, 2n$ |
| $C_n$ | $n(2n+1)$ | $n+1$ | $2n$ | $2$ | $2, 4, \ldots, 2n$ |
| $D_n$ | $n(2n-1)$ | $2n-2$ | $2n-2$ | $4$ | $2, 4, \ldots, 2n-2, n$ |
| $E_6$ | $78$ | $12$ | $12$ | $3$ | $2, 5, 6, 8, 9, 12$ |
| $E_7$ | $133$ | $18$ | $18$ | $2$ | $2, 6, 8, 10, 12, 14, 18$ |
| $E_8$ | $248$ | $30$ | $30$ | $1$ | $2, 8, 12, 14, 18, 20, 24, 30$ |
| $F_4$ | $52$ | $9$ | $12$ | $1$ | $2, 6, 8, 12$ |
| $G_2$ | $14$ | $4$ | $6$ | $1$ | $2, 6$ |

$g^\vee$ and $g$ are (dual) Coxeter numbers. $I$ is the index or the determinant of the Cartan matrix

exceptions to this rule on "maximality": when one projects out the third root of $F_4$, the third root of $E_7$, and the third or the fifth or the sixth root of $E_8$ [1,2].

## 12.1.2 Affine Lie Algebra

We have seen that the group generators, Eqs. (6.233) and (6.234), are represented by vertex operators [3]. The operator product expansion between two currents is given as

$$j^a(z) j^b(0) \sim \frac{k\delta^{ab}}{z^2} + \frac{if^{ab}{}_c}{z} j^c(0). \qquad (12.14)$$

By $j^a$ is triangularly decomposed to $H^I$, $E^\alpha$, and $E^{-\alpha}$. The current can be expanded as

$$j^a(z) = \sum_{n=-\infty}^{\infty} \frac{j_n^a}{z^{n+1}} \qquad (12.15)$$

to give

$$[j_m^a, j_n^b] = if^{ab}{}_c j_{m+n}^c + m\delta_{m+n,0}\delta^{ab} K, \qquad (12.16)$$

where $a = 1, 2, \ldots, d$ with the dimension of the adjoint representation $d$. This extended algebra is called the *affine Lie algebra* or the *Kac–Moody* algebra. The zero mode of this algebra $m = n = 0$ reduces to the simple Lie algebra.

We have introduced two additional generators. One is the *central element K*, commuting with all the generators

$$[K, j_n^a] = 0, \qquad (12.17)$$

whose eigenvalue $k$ is called the *level* of the algebra. The other is the *grade D*, whose eigenvalue is the Kaluza–Klein mode number, and satisfies the following commutation relations

$$[D, j_n^a] = nj_n^a, \quad [D, K] = 0. \qquad (12.18)$$

Without the central extension, we just have replicas of the simple Lie algebra. In our formulation, the relative normalization is fixed as $k = 1$ if we identify $z^{-1}$ coefficient as the structure constant $f^{abc}$ of Lie algebra with normalization (12.12).

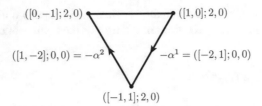

**Fig. 12.3** **3** of SU(3) in the $k = 2$ level. The highest weight is ([1 0]; 2, 0). We obtain all the weights by applying $-\alpha^1 = ([-2\ 1];\ 0,\ 0)$ to obtain ([−1 1]; 2, 0) and then applying $-\alpha^2 = ([1\ -2]; 0, 0)$ to obtain ([0 − 1]; 2, 0)

With the central extension, now we have another ladder operator $T_n$ with $n \neq 0$, raising and lowering the grade number $n$. In addition, the elements of the Cartan subalgebra are not mutually commuting

$$[H_n^I, H_{-n}^J] = \delta^{IJ} n K . \tag{12.19}$$

The Cartan subalgebra elements also raise and lower the eigenvalues of $D$ by $nk$.

Since the generators $(H_0; K, D)$ are mutually commuting, we can consider their simultaneous eigenvector $(\lambda; k, n)$. As usual, we define the inner product as

$$(\lambda; k, n) \cdot (\lambda'; k', n') = \lambda \cdot \lambda' + kn' + k'n. \tag{12.20}$$

With this definition, still the gauge generators belong to the $k = 0$ level. For the simple SU(3) case, it is illustrated in Fig. 12.3.

### 12.1.3 Twisted Algebra

The twisting (7.2) makes the algebra also twisted

$$j^a(\sigma + \pi) = \omega j^a(\sigma) = e^{2\pi i \eta^a} j^a(\sigma) \tag{12.21}$$

with $N\eta^a = \mathbb{Z}$. This is an automorphism, so that it preserves the commutation relation (12.47). For inner automorphism, we will see shortly that we may always write this automorphism using the shift vector, e.g., $\eta^I = V^I$ for the first twisted sector, while other components of $\eta^a$ vanishing. We may generalize it to any twisted sectors and those in the presence of Wilson lines. Hence, we have the twisted affine Lie algebra [1, 4]

$$[j_{m+\eta^a}^a, j_{n+\eta^b}^b] = i f^{abc} j_{m+n+\eta^a+\eta^b}^c + (m + \eta^a) \delta_{m+n+\eta^a+\eta^b,0} \delta^{ab} K. \tag{12.22}$$

Having Cartan subalgebra as the maximal set of commuting generators of the modes with $n = 0$, we have a tower of generators with the modes with $n \neq 0$ as well. Then, we shift the grade

$$\tilde{D} = D - \eta \cdot H_0 \tag{12.23}$$

for these modes to have

$$\tilde{H}_n^I = H_n^I + \eta^I \delta_{n,0} K. \tag{12.24}$$

The level operator $K$ commutes with all the others and hence is not changed. With the rest of the generators, we make ladder operators as

$$\tilde{E}_n^\alpha = E_{n+\eta \cdot \alpha}^\alpha. \tag{12.25}$$

The newly defined (tilded) operators satisfy the same commutation relations (12.16). That means, the twisted algebra is isomorphic to the untwisted algebra. The adjoint representations are not affected by $\eta$. Thus, the gauge group is that of the common untwisted sector. However, the weights are shifted by $\eta$.

## 12.2 Matter Fields

We study properties of matter fields. Their charges are understood by representation theory. Other physical properties such as mass and spin are described by conformal field theory.

### 12.2.1 Highest Weight Representations

The matter spectrum can be understood in terms of the highest weight representation. That is, a multiplet state $|\Lambda\rangle$ is annihilated by all the raising operators

$$E_0^\alpha |\Lambda\rangle = 0, \quad \text{for all } \alpha > 0, \tag{12.26}$$

$$H_n^I |\Lambda\rangle = 0, \quad \text{for all } n > 0. \tag{12.27}$$

The last equation is also the requirement for the highest weight along the direction of $D$. For example, for the untwisted matter $(\mathbf{3}, \mathbf{27})$, the highest weight vector is represented as $[10][100000]$ in the $SU(3) \times E_6$ Dynkin basis. It is annihilated by all the simple roots, among which the nontrivial conditions are

$$E^{\alpha^1} |[1\ 0][1\ 0^5]\rangle = 0 \tag{12.28}$$

$$E^{\alpha^3} |[1\ 0][1\ 0^5]\rangle = 0 \tag{12.29}$$

where we used the original name of $E_8$ simple roots. Then, by successive sub-tractions of simple roots, according to Cartan matrices, we list all the states in the representation. In fact, this can be obtained by the branching rule, since they come from the adjoint **248** of $E_8$. So, in principle we have the untwisted sector representations by reading off the branching tables [3].

Can we do the same for the twisted sector fields? That should be, because they transform under the untwisted gauge group. Looking at the (12.24) we see that the CSA of grade zero ($n = 0$) is shifted by the twist $\eta^I$, which we identify with the local shift vector $kV + m_i a_i$. It follows that their eigenvalues, the weight vectors $P$, are shifted by this. They form representations

$$|P + kV + m_i a_i\rangle, \tag{12.30}$$

where the state is in the $k$th twisted sector and we have Wilson lines $m_i a_i$. For example, the vectors forming $(\mathbf{3}, \mathbf{1})$ in the twisted sector are charged under $SU(3) \times E_6$ and obtained from

$$(-\tfrac{1}{3} \tfrac{1}{3} - \tfrac{2}{3} \, 0^5) \xrightarrow[-\alpha^1]{} (-\tfrac{1}{3} \, -\tfrac{2}{3} \tfrac{1}{3} \, 0^5) \xrightarrow[-\alpha^0]{} (\tfrac{2}{3} \tfrac{1}{3} \tfrac{1}{3} \, 0^5), \tag{12.31}$$

where $\alpha^0$ is now one of the simple roots of the $SU(3)$ subgroup. The weight $(0^8)$ *without* oscillator $\alpha^I_{-1}$ belongs to the $E_8$ *lattice* but is not a root. This implies that this representation does not come from branching of an adjoint like **248**. In obtaining this representation, we just resorted to the twisted mass shell condition (7.73), and it seems that we have another representation something like **248**, whose broken representation gives such state as $(\mathbf{3}, \mathbf{1})$. We will see in the next section that this is easily understood when we use the algebra in a twisted form.

## 12.2.2 Integrability and No-adjoint Theorem

It is sufficient to consider the highest weight vector only, of a form

$$\Lambda = \sum_{i=1}^{r} t_i \Lambda_i, \tag{12.32}$$

with nonnegative integers $t_i$. In the "ket" state notation, it is $|\hat{\Lambda}\rangle \equiv |\Lambda; k, n\rangle$ of (12.19). Looking at the eigenvalue in the $(\alpha^0; 0, 1)$ direction, we have

$$\hat{H}^{(\alpha^0;\, 0,1)} |\hat{\Lambda}\rangle = \left(\alpha^0 \cdot \Lambda + k\right) |\Lambda\rangle, \tag{12.33}$$

where $\alpha^0$ is the extended root we discussed before.

In analogy with the simple harmonic oscillator algebra, one can see that the eigenvalues of $\hat{H}$ are nonnegative integers. In the simple Lie algebra, note that the eigenvalues of $\Lambda \cdot \theta$ are already integers. Therefore, the eigenvalue for level $k$ is a nonnegative integer, which translates into the so-called integrability condition

$$k \geq \sum_{i=1}^{r} a_i t_i \geq 0 \,,$$

or, using the extended root,

$$k = t_0 + \sum_{i=1}^{r} a_i t_i \,, \tag{12.34}$$

where $t_0$ is a nonnegative integer. It is noted that for the level one $(k = 1)$ algebra only a few can satisfy this condition because $a_i \geq 1$. As shown in Fig. 12.1, for the SU($n$) algebra the Coxeter label corresponding to every fundamental weight[1] is always 1; thus, every antisymmetric representation $\Lambda_i$, with the dimension $\binom{n}{i}$, is possible. However, for the other groups $a_i = 1$ is possible only for the outer most nodes of the Dynkin diagram. For $E_6$, for example, $\Lambda_1$ and $\Lambda_6$ have $a_i = 1$. In other words, **27** and $\overline{\mathbf{27}}$ can satisfy the $k = 1$ condition.

A corollary of this observation is the "no-adjoint theorem" that the adjoint representation, needed for breaking the SU(5), SO(10), and $E_6$ GUTs, cannot satisfy this condition. Look at Dynkin diagrams of Fig. 12.1. The highest weight vector for the adjoint representation of SU($n$) is $\Lambda_1 + \Lambda_{n-1}$, and hence it has the sum of the Coxeter level greater than 1: $a_1 + a_{n-1} = 2 > 1$. For SO($n$), the adjoint representation $\mathbf{n(n-1)/2}$ is $\Lambda_2$, and the level is $a_2 = 2$. For the other groups, the adjoint representations have the Coxeter label greater than 1.

When $k > 1$, this constraint is relaxed; however, there exists some upper limit for dimension of a highest weight. We will come back to this point later.

### 12.2.3 Mass and Conformal Weight

The Virasoro operators for group degree of freedom are constructed by the Sugawara method [5]. Consider a biproduct of the current

$$: jj(z_1) := \lim_{z_2 \to z_1} \left( j^a(z_1) j^a(z_2) - \frac{kd}{z_{12}^2} \right). \tag{12.35}$$

---

[1] Completely antisymmetric representation in this case.

We may show the following OPE using (12.14) [6]

$$: jj(z_1) : j^c(z_3) \sim (k + g^\vee)\theta^2 \left( \frac{j^c(z_3)}{z_{13}^2} + \frac{\partial j^c(z_3)}{z_{13}} \right), \qquad (12.36)$$

which is nothing but the OPE between $T(z_1)$ and $j^c(z_3)$ up to normalization. Here, $\theta$ is the highest root. Thus, we have $(k + g^\vee)\theta^2 T(z) \equiv\ : jj(z)\ :$. Mode expansion (6.40) gives

$$\tilde{L}_n = \frac{1}{\theta^2(k + g^\vee)} \sum_{m \in \mathbb{Z}} \sum_{a=1}^{d} : j^a_{m+n} j^a_{-m} :, \qquad (12.37)$$

where $k$ is the level defined below Eq. (12.17) and $g^\vee$ is the dual Coxeter number in (12.12). This normalization gives the same Virasoro algebra as (6.44) with the central charge $c = kd/(k + g^\vee)$. So, we have the minimal nonnegative eigenvalue of Hamiltonian operator $\tilde{L}_0$ as

$$[\tilde{L}_m, j^a_{-n}] = n j^a_{m-n}. \qquad (12.38)$$

It means that the eigenvalue of $L_0$ is $n$, or the eigenvalue $D$.

Consider again the highest weight $|\Lambda\rangle$ defined by (12.26) and (12.27). Being normal ordered $\tilde{L}_0$ is proportional to $\sum_a^d T_0^a T_0^a$, which is the Casimir operator $C_2(G)$,

$$C_2(G) = (\Lambda + 2\rho) \cdot \Lambda, \qquad (12.39)$$

where $\rho \equiv \sum \Lambda_i$ [1]. Since $\Lambda$ is given by Eq. (12.32), Eq. (12.37) becomes

$$\tilde{L}_0|\Lambda\rangle = \frac{(\Lambda + 2\rho) \cdot \Lambda}{2(k + g^\vee)}|\Lambda\rangle = \frac{1}{2(k + g^\vee)} \sum_i (t_i + 2)\Lambda_i \cdot \sum_j t_j \Lambda_j |\Lambda\rangle.$$

Thus, the eigenvalue is given by

$$h_\Lambda = \frac{1}{2(k + g^\vee)} \sum_{i,j=1}^{r} (t_i + 2)t_j A_{ij}^{-1}, \qquad (12.40)$$

with $A_{ij}^{-1}$ in (12.9). Consider a special case where a state carries a single fundamental weight $\Lambda = \Lambda_i$ without oscillator at level $k = 1$. In this case, using (12.12) we obtain

$$h_{\Lambda_i} = \frac{1}{2}\Lambda_i^2 = \frac{1}{2}A_{ii}^{-1} \quad \text{(no summation of } i\text{)}. \qquad (12.41)$$

Considering $k$th twisted sector, the mass shell condition (12.33) for a state is translated into

$$\frac{1}{4}M_L^2 = \frac{(P+kV)^2}{2} + \tilde{N} + \tilde{c} = h_\Lambda + \tilde{c} = 0. \qquad (12.42)$$

In other words, from this condition we can find the highest weight state having the above $h_\Lambda$. This is the only mass condition of all the possible combinations of $h_\Lambda$ satisfying this condition is the highest weight spectrum. The conformal weight $h$ of $|\Lambda\rangle$ appears as

$$\tilde{L}_0|\Lambda\rangle = (h_\Lambda + \tilde{c})|\Lambda\rangle = [\tfrac{1}{2}(P+kV)^2 + \tilde{c}]|\Lambda\rangle. \qquad (12.43)$$

The problem of finding states satisfying the mass shell condition (12.33) is converted to the finding $h_\Lambda$ satisfying it [7]. Although it looks like a nontrivial task, only a few $\Lambda$ can satisfy the above condition, because of the integrability (12.34).

We are usually in a situation where a gauge group $g$ (like $E_8 \times E_8$ or SO(32)) is broken down to $\oplus h$. For each simple group $h$, the inverse Cartan matrix $(A_{ij}^h)^{-1}$ is completely determined as in Table 12.3, independent of the basis. If a state has the conformal weight of each simple algebra as $h_{\Lambda^h}$ in (12.41), since the conformal weight is additive, we may replace the $h_\Lambda$ in (12.41) as the sum over the whole algebra

$$h_\Lambda = \sum h_{\Lambda^h}. \qquad (12.44)$$

If it satisfies the mass shell condition (12.42), then its highest weight is given by the sum of the fundamental weights

$$P + kV = \sum \Lambda_i^h = \sum (A^h)_{ij}^{-1}\alpha^j, \qquad (12.45)$$

where $A^h$ is the Cartan matrix for the subgroup $h$, but now $\alpha^j$ is a simple root of $g$.

For example, we can understand the $(\mathbf{3}, \mathbf{1})$ representation of (12.31) in the twisted sector, in terms of the highest weight representation. From Eq. (12.42), we should have $h_\Lambda = -\tilde{c} = \frac{2}{3}$. Since the oscillator $\alpha_{-1/3}^i$ has conformal weight $\frac{1}{3}$ (from $\tilde{N} = \frac{1}{3}$), we expect that $\frac{1}{3}$ comes from the $\mathbf{3}$ of SU(3) with as many multiplicity. Indeed, by reading the inverse Cartan matrix $\frac{1}{2}(A^{A_2})_{11}^{-1} = \frac{1}{3}$ presented in Table 12.3, we obtain the corresponding highest weight representation (here $P + V$)

$$P + V = (A^{A_2})_{1j}^{-1}\alpha^j = (-\tfrac{1}{3}\ \tfrac{1}{3}\ -\tfrac{2}{3}\ 0^5)(0^8), \qquad (12.46)$$

where $\alpha^j$ are the $E_8$ simple roots. This result agrees with the one obtained in Sect. 7.4.2.

**Table 12.3** The Cartan and the inverse Cartan matrices of Lie algebras

| Algebra | Cartan matrix $A^{ij}$ | Inverse Cartan matrix $A_{ij}^{-1}$ |
|---|---|---|
| $A_{n-1}$ | $\begin{pmatrix} 2 & -1 & 0 & \cdots & 0 & 0 \\ -1 & 2 & -1 & \cdots & 0 & 0 \\ 0 & -1 & 2 & -1 & \cdots & 0 \\ & & & \ddots & & \\ 0 & 0 & 0 \cdots & -1 & 2 & -1 \\ 0 & 0 & 0 \cdots & 0 & -1 & 2 \end{pmatrix}$ | $\dfrac{1}{n}\begin{pmatrix} n-1 & n-2 & n-3 & \cdots & 2 & 1 \\ n-2 & 2n-4 & 2n-6 & \cdots & 4 & 2 \\ n-3 & 2n-3 & 3n-3 & \cdots & 6 & 3 \\ & & & \ddots & & \\ 2 & 4 & 6 & \cdots & 2n-4 & n-2 \\ 1 & 2 & 3 & \cdots & n-2 & n-1 \end{pmatrix}$ |
| $D_n$ | $\begin{pmatrix} 2 & -1 & 0 & \cdots & 0 & 0 & 0 \\ -1 & 2 & -1 & \cdots & 0 & 0 & 0 \\ 0 & -1 & 2 & \cdots & 0 & 0 & 0 \\ & & & \ddots & & \\ 0 & 0 & 0 & \cdots & 2 & -1 & -1 \\ 0 & 0 & 0 & \cdots & -1 & 2 & 0 \\ 0 & 0 & 0 & \cdots & -1 & 0 & 2 \end{pmatrix}$ | $\dfrac{1}{2}\begin{pmatrix} 2 & 2 & 2 & \cdots & 2 & 1 & 1 \\ 2 & 4 & 4 & \cdots & 4 & 2 & 2 \\ 2 & 4 & 6 & \cdots & 6 & 3 & 3 \\ & & & \ddots & & \\ 2 & 4 & 6 & \cdots & 2n-4 & n-2 & n-2 \\ 1 & 2 & 3 & \cdots & n-2 & \frac{n}{2} & \frac{n-2}{2} \\ 1 & 2 & 3 & \cdots & n-2 & \frac{n-2}{2} & \frac{n}{2} \end{pmatrix}$ |
| $E_6$ | $\begin{pmatrix} 2 & -1 & 0 & 0 & 0 & 0 \\ -1 & 2 & -1 & 0 & 0 & 0 \\ 0 & -1 & 2 & -1 & 0 & -1 \\ 0 & 0 & -1 & 2 & -1 & 0 \\ 0 & 0 & 0 & -1 & 2 & 0 \\ 0 & 0 & -1 & 0 & 0 & 2 \end{pmatrix}$ | $\dfrac{1}{3}\begin{pmatrix} 4 & 5 & 6 & 4 & 2 & 3 \\ 5 & 10 & 12 & 8 & 4 & 6 \\ 6 & 12 & 18 & 12 & 6 & 9 \\ 4 & 8 & 12 & 10 & 5 & 6 \\ 2 & 4 & 6 & 5 & 4 & 3 \\ 3 & 6 & 9 & 6 & 3 & 6 \end{pmatrix}$ |
| $E_8$ | $\begin{pmatrix} 2 & -1 & 0 & 0 & 0 & 0 & 0 & 0 \\ -1 & 2 & -1 & 0 & 0 & 0 & 0 & 0 \\ 0 & -1 & 2 & -1 & 0 & 0 & 0 & 0 \\ 0 & 0 & -1 & 2 & -1 & 0 & 0 & 0 \\ 0 & 0 & 0 & -1 & 2 & -1 & 0 & -1 \\ 0 & 0 & 0 & 0 & -1 & 2 & -1 & 0 \\ 0 & 0 & 0 & 0 & 0 & -1 & 2 & 0 \\ 0 & 0 & 0 & 0 & -1 & 0 & 0 & 2 \end{pmatrix}$ | $\begin{pmatrix} 2 & 3 & 4 & 5 & 6 & 4 & 2 & 3 \\ 3 & 6 & 8 & 10 & 12 & 8 & 4 & 6 \\ 4 & 8 & 12 & 15 & 18 & 12 & 6 & 9 \\ 5 & 10 & 15 & 20 & 24 & 16 & 8 & 12 \\ 6 & 12 & 18 & 24 & 30 & 20 & 10 & 15 \\ 4 & 8 & 12 & 16 & 20 & 14 & 7 & 10 \\ 2 & 4 & 6 & 8 & 10 & 7 & 4 & 5 \\ 3 & 6 & 9 & 12 & 15 & 10 & 5 & 8 \end{pmatrix}$ |

## 12.3   Automorphism

The actions we performed in the previous chapters are characterized by a set of shift vectors $V, a_1, a_2, \ldots$. These actions are understood as automorphisms, which we will discuss here.

### 12.3.1 Shift Vector

An automorphism $\omega$ of an algebra $\mathsf{g}$ is an isomorphism from $\mathsf{g}$ to itself (hence the prefix auto). In other words, it preserves the commutation

$$\omega([x, y]) = [\omega(x), \omega(y)] \tag{12.47}$$

for $x, y \in \mathsf{g}$, and the mapping is a one-to-one correspondence. Roughly speaking, it is a permutation among the roots. An *order* $N$ of an automorphism is defined to be the minimum integer such that $\omega^N$ becomes identity. We are interested in a finite automorphism, where $N$ is a finite number. If an automorphism is generated by the algebraic elements of $\mathfrak{g}$ itself, by adjoint operation, then the automorphism is called the *inner automorphism*; otherwise, it is *outer*.

By linear combinations with complex coefficients, we can always decompose $\mathfrak{g}$ into eigenstates of the automorphism $\omega$

$$\mathfrak{g} = \bigoplus_{k=0}^{N-1} \mathfrak{g}_k, \tag{12.48}$$

where each $\mathfrak{g}_k$ is a subalgebra with eigenvalue $e^{2\pi i k/N}$,

$$\omega(x) = e^{2\pi i k/N} x, \quad x \in \mathfrak{g}_k. \tag{12.49}$$

In general, we can make a state into a definite eigenstate $\mathfrak{g}_k$ by forming a linear combination

$$x_k = \frac{1}{N} \sum_{j=0}^{N-1} e^{-2\pi i k j/N} \omega^j(x), \tag{12.50}$$

for which we can show $\omega(x_k) = e^{2\pi i k/N} x_k$. With this form, we will encounter another way of breaking the gauge group.

This provides a way to obtain matter representations from an adjoint of the unified group by modding out by automorphism $\omega$. The representation gauge fields come from the invariant subgroup $\mathfrak{g}_0$. Matter representations come from eigenstates $\mathfrak{g}_k$.

There is an important theorem that *a finite inner automorphism* is always represented by the *shift vector* (12.21). The key idea is that, for the invariant algebra under $\omega$, i.e. $\mathfrak{g}_0$ of $\mathfrak{g}$, we can *always find the elements of the Cartan subalgebra of $\mathfrak{g}$, invariant under $\omega$*. A sketch of the proof is as follows. They are some special linear combinations of elements of $\mathfrak{g}$, such that they commute each other and form a maximal Abelian subalgebra of $\mathfrak{g}_{(0)}$. It implies that $\omega(E^\alpha)$ is still eigenstate with respect to $H^I$, since

$$[H^I, \omega(E^\alpha)] = [\omega(H^I), \omega(E^\alpha)] = \omega([H^I, E^\alpha]) = \alpha^I \omega(E^\alpha). \tag{12.51}$$

Because there is only one $E^\alpha$ corresponding to the root $\alpha$, $\omega(E^\alpha)$ should be proportional to $E^\alpha$ up to a phase. Evidently, this $\omega$ is mapped by $\mathrm{ad}_{H^I} \cdot \omega$ where $\mathrm{ad}_A B$ is defined in (12.4) and the phase is determined by a linearly combined CSA. Thus, we define the shift vector $V$, introduced in (7.50), as

$$\omega = \exp\left(2\pi i \sum_{I=1}^{r} V^I \, \mathrm{ad}_{H^I}\right), \tag{12.52}$$

where each $NV^I$ is an integer.

## Example

We consider automorphism of SU(3) and how it affects the algebra. For this, it is natural to complexify the Lie algebra by defining the generators using the Gell-Mann matrices (5.32) as

$$
\begin{aligned}
H_\pm &\equiv \lambda_3 \pm i\lambda_8, \\
E^{\pm\rho_1} &\equiv \lambda_1 \pm i\lambda_2, \\
E^{\pm\rho_2} &\equiv \lambda_4 \pm i\lambda_5, \\
E^{\pm\rho_3} &\equiv \lambda_6 \pm i\lambda_7,
\end{aligned}
\tag{12.53}
$$

Using the commutation relation (12.3),

$$
[H_\pm, E^{\pm\rho_i}] = \pm\rho_i E^{\pm\rho_i}, \quad i = 1, 2, 3,
\tag{12.54}
$$

where the signs are correlated here and in what follows. This means that we have roots

$$
\rho_1 = \sqrt{2}, \quad \rho_2 = \frac{1}{\sqrt{2}} + \frac{3}{\sqrt{6}}i, \quad \rho_3 = \frac{1}{\sqrt{2}} - \frac{3}{\sqrt{6}}i,
\tag{12.55}
$$

where we can take two of them as simple roots and the rest being the negative sum of the two. We took the normalization $\rho^2 = 2$. The fundamental representation **3** is a set of states $|d_i\rangle$ with weights

$$
d_1 = \frac{1}{\sqrt{2}} + \frac{1}{\sqrt{6}}i, \quad d_2 = -\frac{2}{\sqrt{6}}i, \quad d_3 = -\frac{1}{\sqrt{2}} + \frac{1}{\sqrt{6}}i.
\tag{12.56}
$$

We verify that the roots are differences of weights.

Consider inner automorphism generated by

$$
T^a \to wT^a, \quad w \equiv e^{2\pi i/3}.
\tag{12.57}
$$

It is an order 3 action. This implies, from (12.54),

$$
[e^{\pm 2\pi i/3}H_\pm, e^{\pm 2\pi i/3}E^{\pm\rho_i}] = \pm(e^{\pm 2\pi i/3}\rho_i)e^{\pm 2\pi i/3}E^{\pm\rho_i}.
\tag{12.58}
$$

All the weights are also rotated as

$$
w\rho_1 = \rho_2, \quad w\rho_2 = \rho_3, \quad w\rho_3 = \rho_1,
\tag{12.59}
$$

$$
wd_1 = d_2, \quad wd_2 = d_3, \quad wd_3 = d_1.
\tag{12.60}
$$

Since the Cartan subalgebra is not invariant as in (12.58), we are tempted to say that the rank is reduced. However, we have invariant linear combinations of the roots

$$H'_\pm \equiv E^{\pm\rho_1} + E^{\pm\rho_2} + E^{\pm\rho_3}, \tag{12.61}$$

where the signs are correlated. These provide another CSA of a rank two algebra

$$SU(2) \times U(1). \tag{12.62}$$

The other CSA elements are projected out and the resulting gauge group becomes smaller.

We have linear combinations of these other CSA elements form a state transforming line $\alpha = e^{2\pi i/3}$ and $\alpha^2$

$$\begin{aligned} E'_\pm &\equiv E^{\pm\rho_1} + \alpha^2 E^{\pm\rho_2} + \alpha E^{\pm\rho_3}, \\ E''_\pm &\equiv E^{\pm\rho_1} + \alpha E^{\pm\rho_2} + \alpha^2 E^{\pm\rho_3}. \end{aligned} \tag{12.63}$$

We may verify that each of $E'_\pm$ and $E''_\pm$ form a doublet under each SU(2) in (12.62). There is another doublet formed by $\{H_+, H_-\}$.

For the weights, we have invariant combination forming the generators

$$|d_1\rangle + |d_2\rangle + |d_3\rangle, \tag{12.64}$$

$$\{|d_1\rangle + \alpha^2|d_2\rangle + \alpha|d_3\rangle, |d_1\rangle + \alpha|d_2\rangle + \alpha^2|d_3\rangle\}. \tag{12.65}$$

The first and the second, respectively, form a singlet and a doublet under the SU(2).

## 12.3.2 Weyl Group

Regarding automorphism as shuffling the roots, obvious actions are reflections. Consider a reflection $\sigma_\alpha$ of a root $P$ with respect to the plane perpendicular to a root $\alpha$,

$$P \to \sigma_\alpha P = P - 2\frac{\alpha \cdot P}{\alpha^2}\alpha = P - (\alpha^\vee \cdot P)\alpha. \tag{12.66}$$

This is readily extended to weight space thus to shift vectors. Being a reflection, we have $\sigma_\alpha = \sigma_{-\alpha} = \sigma_\alpha^{-1}$. We see that these reflections form a group, called *Weyl group*. Like the root system, every Weyl reflection is decomposed into product of *fundamental reflections* $\sigma_{\alpha^i}$ with respect to simple roots $\alpha^i$, but not every element of Weyl group is a reflection. Also, each Weyl reflection acts as an automorphism since it just permutes the roots.

A Weyl reflection in orthogonal bases will provide a good example. With Weyl reflections, we can understand many operations on the weight lattices and also on the shift vectors. For instance, Weyl reflections of $\mathscr{P}$ with respect to the roots of the type $(1, -1, 0^6)$,

$$
\begin{aligned}
P = (P_1, P_2, \ldots, P_8) &\to \\
&= (P_1, P_2, \ldots, P_8) - [(1, -1, 0^6) \cdot (P_1, P_2, \ldots, P_8)](1, -1, 0^6) \quad (12.67) \\
&= (P_2, P_1, \ldots, P_8),
\end{aligned}
$$

correspond to exchanging two components. This statement can be generalized. With respect to the roots of the type $(1, 1, 0^6)$, the Weyl reflections give

$$
(P_1, P_2, \ldots, P_8) \to (-P_2, -P_1, \ldots, P_8). \tag{12.68}
$$

Then, if we apply both actions together, we have the result that two elements just obtain minus signs.

What will be the case when we apply Weyl reflections on the half-integral elements? In general, it results in a quite complicated action. Fortunately, it is known that three such actions are redundant to two actions up to integral reflections. This fact is crucially used when we check an equivalence between two shift vectors. Later, we will see that the Weyl reflection of affine Lie algebra accompanies a translation. Thus, we confirm that all the symmetries are generated by Weyl reflections.

Weyl refection is inner automorphism, because it is generated by generator themselves $\sigma_\alpha = \exp[i\pi(E^\alpha + E^{-\alpha})/2]$, so that

$$
\sigma_\alpha H^I \sigma_\alpha^{-1} = H^I - \alpha^I \alpha \cdot H. \tag{12.69}
$$

Since the Weyl group is a group of reflections, we can always rearrange the roots to have bases of simple roots from an initially given set of simple roots. The only remaining ambiguity is rearranging simple roots within themselves. This cannot be arbitrary but should be in the way that preserves the relations between the simple roots. That is, it is a symmetry that leaves the Cartan matrix or the Dynkin diagram invariant. Therefore, we have

$$
\text{Automorphism of } \mathbf{g} = (\text{Symmetries of Dynkin diagram}) \ltimes (\text{Weyl group}). \tag{12.70}
$$

For example, the simple $SU(n)$ Dynkin diagram possesses a $\mathbb{Z}_2$ symmetry along the vertical axis, which corresponds to a complex conjugation. We also have complex conjugation in the $SO(2n)$ case, which corresponds to a $\mathbb{Z}_2$ reflection along the long axis of the Dynkin diagram. Some extended diagram has an enhanced symmetry. For example, the extended $E_6$ has the order 3 permutation symmetry $S_3$, and the extended $SU(n)$ Dynkin diagram is a regular $n$-gon, so we have $D_n$ symmetry. Since the $E_8$ Dynkin diagram has no symmetry, the only automorphism is the Weyl reflections.

There is a special element, called the *Coxeter element*, defined by the product of all the fundamental reflections

$$w = \prod \sigma_{\alpha^i}, \tag{12.71}$$

where $\alpha^i$ are the simple roots. Although its form is also dependent on the lattice choice, it has a definite order, which turns out to be the Coxeter number $g$ given in (12.12), i.e. $w^g = 1$ [1]. For the SU($n$), we have $g = n$ and can show that the coxeter element is $2\pi i/n$ rotation.

## 12.4   General Action on Group Lattice

We consider a point group action in the current algebra space $T_{G,L}^{16}$. Like shift vector, this point group also mods out the lattice to break the gauge group. In general, such actions do not commute, so some of the Cartan generators are also projected out.

### 12.4.1   Point Group Embedding

In Chaps. 5 and 6, we have broken a group by associating the space action $\theta$ with a translation $V$ in the group space, as in (7.27),

$$(\theta, v)x \longrightarrow (1, V)X = X + V.$$

The most general possibility is that with an automorphism on the group space $\Theta$ [8],

$$(\theta, v)x \longrightarrow (\Theta, V)X = \Theta X + V. \tag{12.72}$$

As with the shift vector (7.53), the unbroken gauge group is obtained by the invariance condition under this action. We have matter fields that are also given definite transformation properties under this unbroken group.

Therefore, we can form a point action $\Theta$ on the group lattice, by a suitable product of Weyl reflections. In fact, then the form and, in particular, the order of $\Theta$ are dependent on the choice of lattice.

For the adjoint representation, every element $x$ has a corresponding state labeled by a root vector $|x\rangle$. For the heterotic string, all the states begin from roots, and thus the above action is applicable. The action is defined by a bracket relation $y|x\rangle = |[y, x]\rangle$. The string states can be treated in this way. In this case, the group sum $+$ is a linear superposition of the states, which is clearly understood in terms of vertex operators (6.234) and (6.233).

For an action $w$, the invariant states with the invariant right movers become gauge bosons. The non-invariant states with suitable right movers form matter representations.

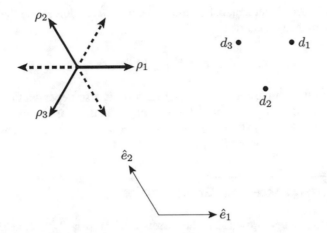

**Fig. 12.4**  SU(3) roots and weights of **3**

## $\mathbb{Z}_3$ **Example: $E_8 \times E_8$ Through $SU(3)^8$**

We associate the order three point group of the $T^6/\mathbb{Z}_3$ orbifold with an automorphism of the $E_8 \times E_8$ [8]. Since the only group with an order three Coxeter element is SU(3), we consider a multiple embedding made by a product of (12.71). This is useful, because the elements of $E_8$ can be understood by those of $SU(3)^4$. Here, the **248** branches into four adjoints,

$$(\mathbf{8, 1, 1, 1}), \ (\mathbf{1, 8, 1, 1}), \ (\mathbf{1, 1, 8, 1}), \ (\mathbf{1, 1, 1, 8})$$

which constitute 32 states, including Cartan subalgebra, and eight kind of multi-fundamentals in the form

$$(\mathbf{3, 3, 3, 1}), (\mathbf{\bar{3}, 3, 1, 3}), (\mathbf{3, 1, \bar{3}, 3}), (\mathbf{1, \bar{3}, 3, 3}), \text{(complex conjugates)}, \qquad (12.73)$$

which are 216 states. All the SU(3)s are equivalent, seen by a suitable complex conjugation $\mathbf{3} \leftrightarrow \mathbf{\bar{3}}$ (exercise).

Analogously to the spacetime twisting, it is convenient to use complexified coordinates. Each element can be represented as a vector in the weight space. For example, the roots $(\mathbf{8, 1, 1, 1})$ are represented by CSA $\alpha^I_{-1}|0\rangle_L$, $I = 1, 2$, and root vectors $|\pm\rho_i, 0, 0, 0\rangle_L$, $i = 1, 2, 3$, with $\rho_i$ in (12.55). It satisfies $\rho^2 = 2$. Weights of $(\mathbf{3, 3, 3, 1})$ are the eight dimensional vectors of the form $|d_a, d_b, d_c, 0\rangle_L$ with each $d$ being the SU(3) weight in (12.56). Those corresponding to $\mathbf{\bar{3}}$ have the opposite sign. Their lengths are chosen such that $d_a^2 + d_b^2 + d_c^2 = 2$. They are shown in Fig. 12.4. It is another way of representing *the same* $E_8$, resulting to the same lattice $\Gamma_8$ in a different basis.

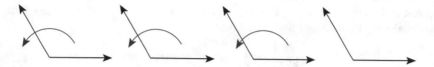

**Fig. 12.5** $E_8$ is decomposed into four SU(3)s. We associate $\mathbb{Z}_3$ action $\theta$ with the rotation on the first three SU(3) lattices by $2\pi/3$. It is realized by products $\Theta = w_1 w_2 w_3$ of the SU(3) Coxeter elements $w_a$ in the $a$th torus

The modular invariance condition again restricts the possible automorphisms. Essentially, it comes from (7.49), by the way of guaranteeing massless states. Now, a crucial difference is that we have a point group action $\Theta$ on the group space; thus, states got twisted by $\Theta$ cannot have a momentum contribution to mass (but possibly has a fractional oscillator number). To guarantee a massless state, the total number of action $w$ on the eight SU(3) lattices should be a multiple of three [8]. These possibilities are listed in Table 12.7.

**Gauge Group**

We act the Coxeter element on the *first three SU(3)s* of eight SU(3)s of $E_8 \times E_8$

$$\Theta = w_1 w_2 w_3,$$

where $w_a, a = 1, 2, 3$, is the SU(3) Coxeter element in the $a$th torus. We draw the situation for the first $E_8$ in Fig. 12.5.

As we have seen in quantization of string on orbifolds in Sect. 7.2, oscillator states having definite transformation under $\Theta$ are

$$\alpha_{-1}^A |0\rangle_L \equiv (\alpha_{-1}^{2A-1} + i\alpha_{-1}^{2A})|0\rangle_L, \quad A = 1, 2, 3, \tag{12.74}$$

$$\alpha_{-1}^{\bar{A}} |0\rangle_L \equiv (\alpha_{-1}^{2A-1} - i\alpha_{-1}^{2A})|0\rangle_L, \quad \bar{A} = \bar{1}, \bar{2}, \bar{3}, \tag{12.75}$$

$$\alpha_{-1}^I |0\rangle_L, \quad I = 7, 8. \tag{12.76}$$

The states in the first, the second, and the third lines, respectively, transform as $\alpha \equiv e^{2\pi i/3}, \alpha^2, 1$. Like in the previous example, although the first Cartan generators are projected out, the rank is not reduced, but there are new states invariant under $\Theta$ among the momentum-winding states. They are

$$| \pm \rho_1^A \rangle_L + | \pm \rho_2^A \rangle_L + | \pm \rho_3^A \rangle_L \tag{12.77}$$

for each SU(3) $A = 1, 2, 3$ with the roots $\pm \rho_i$'s in (12.55). So, we have as many surviving states, forming CSA, so that the rank is not reduced. Their vertex operators are

$$V \simeq\, : e^{i\rho_1^A Z^1} + e^{i\rho_2^A Z^1} + e^{i\rho_3^A Z^1} : \tag{12.78}$$

up to coefficients, where $Z^A = X^{2A-1} + iX^{2A}$, $A = 1, 2, 3$. Note that they form a $\mathfrak{g}_0$ subalgebra containing invariant states

$$\mathfrak{g}_0 = \{|P\rangle + |\Theta P\rangle + |\Theta^2 P\rangle\}. \tag{12.79}$$

Consider then the charged generators. First, the states $(\bar{3}, 3, 1, 3)$ are represented by weight vectors $(-d_a, d_b, 0, d_d)$. Each $d_a, d_b, d_d$ can be any of $d_i$s in (12.56). Under the action $\Theta$, the first two entries $d_a, d_b$ are changed. We make an invariant state by forming a linear combination of the form (12.79)

$$|(-d_a, d_b, 0, d_d)\rangle_L + |(-wd_a, wd_b, 0, d_d)\rangle_L + |(-w^2 d_a, w^2 d_b, 0, d_d)\rangle_L, \tag{12.80}$$

using (12.60), with an arbitrary $d_d$. So, we have $3 \cdot 3 \cdot 3/3 = 9$ such choices. We have also 9 complex conjugates.

Similarly, the representation $(3, 3, 3, 1)$ has 9 invariant combinations

$$|(d_a, d_b, d_c, 0)\rangle_L + |(wd_a, wd_b, wd_c, 0)\rangle_L + |(w^2 d_a, w^2 d_b, w^2 d_c, 0)\rangle_L, \tag{12.81}$$

and as many complex conjugates.

Collecting all the states in (12.73) along with the Cartan generators, we have $8 + 78$ states which form the adjoint representation of

$$SU(3) \times E_6.$$

**Untwisted Matter**

Now, consider matter states in the untwisted sector. In $\mathbb{Z}_3$ orbifold, there are only ones belong to grade one subalgebra $\mathfrak{g}_1$ in (12.50)

$$\mathfrak{g}_1 = \{|S\rangle = |P\rangle + \alpha^2 |\Theta P\rangle + \alpha |\Theta^2 P\rangle\}. \tag{12.82}$$

They transform like

$$\Theta|S\rangle = |\Theta S\rangle = \alpha|S\rangle.$$

For $(\bar{3}, 3, 1, 3)$, we can make a $\mathfrak{g}_1$ state

$$\begin{aligned}
|(-d_1, d_2, 0, d_d)\rangle &+ \alpha^2 |\Theta(-d_2, d_3, 0, d_d)\rangle + \alpha|\Theta^2(-d_3, d_1, 0, d_d)\rangle \\
&= |(-d_1, d_2, 0, d_d)\rangle + \alpha^2 |(-d_3, d_1, 0, d_d)\rangle + \alpha|(-d_2, d_3, 0, d_d)\rangle.
\end{aligned} \tag{12.83}$$

We obtain 9 combinations, whose vertex operators are, as deduced from (12.78)

$$V \simeq\, :e^{i(-d_1, d_2, 0, d_d)\cdot Z} + \alpha^2 e^{i(-d_3, d_1, 0, d_d)\cdot Z} + \alpha e^{i(-d_2, d_3, 0, d_d)\cdot Z}:.$$

We have as many states in the representation $(\mathbf{3}, \mathbf{1}, \overline{\mathbf{3}}, \mathbf{3})$, $(\mathbf{1}, \overline{\mathbf{3}}, \mathbf{3}, \mathbf{3})$. Also, we have nine combinations from $(\mathbf{3}, \mathbf{3}, \mathbf{3}, \mathbf{1})$. We also obtain similar states from the complex conjugations.

Similarly, we have the following states:

$$|(\pm\rho_1, 0, 0, 0)\rangle + \alpha^2|(\pm\rho_2, 0, 0, 0)\rangle + \alpha|(\pm\rho_3, 0, 0, 0)\rangle, \qquad (12.84)$$

where the signs and permutations are correlated. We have 6 such states. With three oscillator states (12.74), we have 81 states in total, forming $(\mathbf{3}, \mathbf{27})$.

**Twisted Sector**

We considered a twist in the current algebra direction (12.72),

$$X^I(\pi) = (\Theta X(0))^I + V^I, \qquad (12.85)$$

where we allowed for a translation by lattice vectors of $E_8$. This looks like a space group action; however, with only the left movers the action is not well-defined. We can introduce auxiliary right movers, make the lattice symmetric and project out the right movers again. Then, we have a well-defined meaning for the orbifolding action. In this way, we may define the fixed points arising from (12.85). After removing the right movers, the number of fixed points reduces to its square root

$$n_f = \sqrt{9} = 3. \qquad (12.86)$$

Alternatively, we may just read off this factor from the partition function as given in (8.59). In the spacetime, we also have a similar action as (12.85) and the number of fixed points is 27.

Since we twist gauge degrees of freedom as well, the zero point energy is $\tilde{c} = 12f(0) + 4f(\frac{2}{3}) + 8f(\frac{1}{3}) = -\frac{1}{3}$. The mass shell condition becomes

$$\tfrac{1}{4}M_L^2 = \tfrac{1}{2}P^2 + \tilde{N} - \tfrac{1}{3}. \qquad (12.87)$$

Note that the twisted string in the current direction cannot have a nontrivial momentum. The only nonvanishing components in $P^I$ are for $I = 7, 8$ and we have fractional oscillators in the $A = 1, 2, 3, \bar{1}, \bar{2}, \bar{3}$ directions. The resulting spectrum is summarized in Table 12.4.

**Table 12.4** Twisted sector states for the "standard" $\mathbb{Z}_3$ orbifold by automorphism

| State | Group spc. mult. | Spacetime mult. | Representation |
|---|---|---|---|
| $|0, 0, 0, d_d\rangle$ | 3 | 27 | Part of $27(\mathbf{1}, \mathbf{27})$ |
| $\alpha^A_{-1/3}|0\rangle$ | 3 | 27 | Part of $27(\mathbf{1}, \mathbf{27})$ |
| $\alpha^a_{-1/3}|0\rangle$ | 3 | $27 \cdot 3$ | $27 \cdot 3(\overline{\mathbf{3}}, \mathbf{1})$ |

Only the left movers are shown. $d_d$ can be any six SU(3) weights

The oscillator contributions for $P^2 = 0$ and $\tilde{N} = \frac{1}{3}$ make up the state of a form $\alpha_{-1/3}^{M_+}|0\rangle_L$, where $M_+$ is a holomorphic index. If $M_+$ becomes a spacetime holomorphic index $M_+ = a$, the corresponding state transforms as **3** under the holonomy group, so we have three copies of the state at each spacetime fixed point. It is important to note that there are three fixed points in the group space, providing the representation $(\overline{\mathbf{3}}, \mathbf{1})$.

The remaining oscillators in the group space $\alpha_{-1/3}^{A}|0\rangle$ with $M_+ = A = 1, 2, 3$ give nine states because of the fixed points. Considering the state with $P^2 = 2/3$ and $\tilde{N} = 0$, we have the possibility of $P = (0, 0, 0, \pm d_d)$. These six states with the multiplicity 3 in the current algebra fixed points make 18 states out of $(\mathbf{1}, \mathbf{27})$. With the above nine states, these make the complete $(\mathbf{1}, \mathbf{27})$ at each spacetime fixed point. As before, we obtain a $\mathscr{CPT}$ conjugate states in the second twisted sector.

## Total Spectrum

We combine the above left-moving states with the right movers and perform the generalized GSO projection. The model has gauge group SU(3) × E₆. We have three $(\mathbf{3}, \mathbf{27})$s in the untwisted sector and $27[3(\overline{\mathbf{3}}, \mathbf{1}) + (\mathbf{1}, \mathbf{27})]$, which is anomaly free. We see that this model is identical to that with the shift vector with the standard embedding, considered in Sect. 7.4. We see that, if we take the Cartan subalgebra using (12.76) and (12.77), the rest of the states provides the eigenstates under this CSA and can parameterize the symmetry breaking with the shift vector used in Sect. 7.4.

## 12.4.2 Reducing the Rank by Orbifolding

We have seen that, solely the automorphism embedding $\Theta$ in (12.85) in the group space gives a spectrum that is equivalently described by a shift vector. When we have more than one such actions, generally they do not commute. Then, if the Cartan subalgebra is not invariant under all such actions, the rank is reduced [8]. This is the case when we include the Wilson line of the type $(1, v) \to (1, a)$, so that

$$(\Theta, 0)(1, a) \neq (1, a)(\Theta, 0). \tag{12.88}$$

This means $(\Theta, \Theta a) \neq (\Theta, a)$. In other words, this happens when the rotated Wilson line is not equal to the original: $\Theta a \neq a$. In this case, this Wilson line is not subject to any order $N$ condition, that is, the strength of Wilson line $\langle a \rangle$ is free. For this reason, we call it *continuous Wilson line*. Note that in a specific decomposition like (12.73), the possible Wilson line shift $a_i$ is provided by weight vectors of the unbroken group. In this case, it is the translation part in Eq. (12.85). When a vector $a$ is invariant under a rotation $\Theta$, they can be treated as independent shift vectors, and even the action $\Theta$ can be converted into an equivalent shift vector $V$.

As always, the modular invariance condition for the additional Wilson lines (7.99) should be checked. We note that only the part of $a$ invariant under $\Theta$ fulfills this condition, whereas the non-invariant part does not.

The rank reduction is important in the process of obtaining the SM. In addition to the above mechanism of rank reduction by the non-commuting point rotation and the Wilson line shift, there are simpler field theoretic methods. For example, we can use the Higgs mechanism via VEVs of some massless fields in the spectrum and also the Fayet–Iliopoulos term generated by the anomalous U(1).

## 12.5 Asymmetric Orbifold

So far, the twisting (7.2) acted equally on the left and right movers. We may relax this condition and have a different twisting on each mover. We may take asymmetric orbifold

$$O = T^{16}_{\text{group},L}/\mathscr{V} \times T^6_L/\bar{\mathsf{P}}_1 \times T^6_R/\bar{\mathsf{P}}_2. \tag{12.89}$$

We decomposed the six dimensional torus into two pieces, although the geometry of torus is not decomposable. It only means that the string has independent symmetries on the left and the right movers.

|        | $x^0 \dots x^3$ | $x^4 \dots x^9$      | $x^{10} \dots x^{25}$ |
|--------|-----------------|----------------------|-----------------------|
| Left   |                 | $(\theta_L, \phi_L)$ | $(\Theta, V)$         |
| Right  |                 | $(\theta_R, \phi_R)$ |                       |

We can even think of an action $\theta_L \oplus \Theta$ acting on the whole left movers. This is an *asymmetric orbifold* [9].

### 12.5.1 Extending Group Lattice

We take six dimensional radius to the critical one $R = 1/\sqrt{2}$ to enhance the rank of group to $16 + 6 = 22$. We can choose any group which has rank 22. Let us fix such extra lattice coming from the rank 6 part to that of $SU(3)^3$. It corresponds to fixing $G_{ij}$ as

$$\langle G_{ij} \rangle = \frac{1}{4} \begin{pmatrix} 2 & -1 & & & & \\ -1 & 2 & & & & \\ & & 2 & -1 & & \\ & & -1 & 2 & & \\ & & & & 2 & -1 \\ & & & & -1 & 2 \end{pmatrix}, \tag{12.90}$$

where we understand that the empty entries are zero. In view of Eqs. (11.34, and 11.35), we want to simplify the mass shell condition. The choice for $B_{ij}$ is

$$\langle B_{ij} \rangle = \frac{1}{4} \begin{pmatrix} 0 & -1 & & & & & \\ +1 & 0 & & & & & \\ & & 0 & -1 & & & \\ & & +1 & 0 & & & \\ & & & & 0 & -1 & \\ & & & & +1 & 0 \end{pmatrix} \tag{12.91}$$

We can always choose $p_{Ri} = 0$ by assigning relations between $m_i$ and $n_i$. This is the constraint that in the original construction of heterotic string we have no right mover. The mass shell condition is

$$\frac{M_L^2}{4} = \frac{P^2}{2} + \frac{p_L^2}{2} + \tilde{N} - 1 = 0,$$

$$\frac{M_R^2}{4} = \frac{p_R^2}{2} + N - \frac{1}{2} = 0. \tag{12.92}$$

As always, $P$ is the momentum in the group space, and $p_L$ is the momentum in the compact dimension. Besides the solution $P^2 = 2$, $p_L^2 = 0$, we have another class of solutions with $P^2 = 0$, $p_L^2 = 2$, which has the form $p_L = (\rho, 0, 0)$ with the SU(3) roots $\rho$. They form the roots of the $SU(3)^3$. For the right mover, we do not have invariant state

$$b^i_{-\frac{1}{2}} |0\rangle_R$$

where signs are correlated. We have the same situation as in the previous subsection and obtain the same kinds of twisted states. We have an enhanced gauge group

$$E_8 \times E_8 \times SU(3)^3 \times U(1)^6$$

with rank $22 + 6$, where rank 6 comes from right movers whose generators take the form of Eq. (12.77),

## Example

Consider asymmetric orbifolds such that left movers are untwisted and right movers are twisted by $\mathbb{Z}_3$. Then, under the $\mathbb{Z}_3$ action $\theta$, let the lattice get twisted as

$$|P, p_L, p_R\rangle \rightarrow \exp(2\pi i P \cdot V)|P, p_L, \theta p_R\rangle. \tag{12.93}$$

In Eq. (12.93), we have explicitly shown the spacetime momenta $p_L$, $p_R$, because here they are treated equally with $P$. Therefore in the untwisted sector, the invariant L-moving states with $\alpha^\mu_{-1}|0\rangle$ and $|P\rangle$ with $P \cdot V$ = integer make up the adjoint representation. It breaks $E_8$ down to $E_6 \times SU(3)$. Including the original $SU(3)^3$ adjoint $\alpha^i_{-1}|0\rangle$ and $|p_L, p_R\rangle$ with $p_L^2 = 2$, $p_R^2 = 0$, $\mathcal{N} = 1$, we have

$$E_6 \times E_8 \times SU(3)^4 \times U(1)^6$$

with the same matter representation as in the standard $\mathbb{Z}_3$ case.

In the twisted sector, the mass shell condition becomes

$$\frac{M_L^2}{4} = \frac{(P+V)^2}{2} + \frac{p_L^2}{2} + \tilde{N} - 1 = 0, \qquad \frac{M_R^2}{4} = \frac{p_R^2}{2} + N - \frac{2}{3} = 0. \qquad (12.94)$$

Note that we have no twist on the left mover, thus no shift on the zero point energy. While $P$ can become the original vectors in the $E_8 \times E_8$ lattice, the $p_L$ can assume the SU(3) roots $\rho$ and weights $d$ with $d^2 = 2/3$ in particular. The relevant spectrum is listed in Table 12.5. The right mover has the same spectrum as in the standard $\mathbb{Z}_3$ case, since the twisting and hence the zero point energy are the same.

## 12.5.2 Symmetrizing Lattice

The asymmetric orbifold suffers from some ambiguities such as in defining the notion of fixed point. Here, we give up the geometric meaning since, for example, we cannot imagine fixed points present only on the left mover. The cure comes from introducing a mirror lattice to make the theory *symmetric and Euclidian*; then, after a consistent calculation we project out the right movers. This was the idea from which the heterotic string was first formulated.

Let us begin with the original, even and self-dual lattice $\Gamma^{p,q}$ (think of $(p, q) = (22, 6)$). To treat all the coordinates equally, consider a Euclidianized lattice denoted by $\tilde{\Gamma}^{p,q}$, by treating the R-handed part in the same way as the L-handed part. Then, it is not self-dual any more. Nevertheless, for vector $(p_1, p_2) \in \tilde{\Gamma}^{p,q}$, vector

**Table 12.5** Spectrum of the asymmetric $\mathbb{Z}_3$ orbifold acted on the right mover only

| State | Representation |
|---|---|
| *Untwisted sector* | |
| $\tilde{N} = 0, n^2 = 0, P^2 = 2$ | $3(\mathbf{27, 3; 1, 1, 1})$ |
| *Twisted sector* | |
| $\tilde{N} = 0, p_L^2 = 0, (P+V)^2 = 2$ | $(\overline{\mathbf{27}}, \mathbf{3; 1, 1, 1})$ |
| $\tilde{N} = 0, p_L^2 = \frac{2}{3}, (P+V)^2 = \frac{4}{3}$ | $(\mathbf{27, 1; \underline{3, 1, 1}}) + (\mathbf{27, 1; \overline{3}, 1, 1})$ |
| $\tilde{N} = 0, p_L^2 = \frac{4}{3}, (P+V)^2 = \frac{2}{3}$ | $(\mathbf{1, 3; \underline{3, 3, 1}}) + (\mathbf{1, \overline{3}; \overline{3, 3}, 1}) + (\mathbf{1, \overline{3}; \overline{3}, 3, 1})$ |

The underline denotes permutations. The twisted sector has multiplicity 1

$(p_1, -p_2)$ belongs to the dual lattice since it comes from the Euclidianization. We will consider windings on $\frac{1}{2}\tilde{\Gamma}^{p,q}$ and momenta on its dual lattice. They have the form for momenta $p = 2(k_1, -k_2)$ and windings $w = \frac{1}{2}(k_3, k_4)$ with $(k_1, k_2)$ and $(k_3, k_4)$ belonging to $\tilde{\Gamma}^{p,q}$.

Now, let us construct a symmetrized lattice $\Gamma^{p+q;p+q}$, whose element consists of momenta (11.34,11.35),

$$(p_L; p_R) = (\tfrac{1}{2}p - Bw + Gw; \tfrac{1}{2}p - Bw - Gw), \tag{12.95}$$

where we renamed the quantized momenta $m$ and windings $n$ as $p$ and $w$, respectively, and for brevity we suppressed the vector and matrix indices. It is convenient to consider the $\frac{1}{2}p$ part only in $p_L$ in scalar products. So, we choose $B_{ij}$ such that,

$$e \cdot Be = e \cdot Ge, \qquad \text{mod } 2, \tag{12.96}$$

as we did just above. $G_{ij}$ is already fixed by orbifolding. The unit lattice of $\tilde{\Gamma}^{p,q}$ is $e_i$. $G_{ij}$ has the Lorentzian signature, having $+$ sign for $p$ entries (thus $Bw = Gw$) and $-$ for $q$ Entries (thus $Bw = -Gw$). Then, vectors $(p_L; p_R)$ span a Euclidian lattice $\Gamma^{p+q;p+q}$ with elements of the form

$$(k_3, k_2 - k_4; k_3 - k_1, -k_4). \tag{12.97}$$

They are generated by vectors having the form

$$(k_1, 0; 0, -k_2), \quad (0, k_2; -k_1, 0). \tag{12.98}$$

In other words, we have a well-defined projection to select just one of them to make the asymmetric lattice

$$\Gamma_{p,q} \ni (k_1, k_2) \equiv (k_1, 0; 0, -k_2). \tag{12.99}$$

For the oscillators, we use only the first $p$ left-moving and the last $q$ right-moving ones.

## Number of Fixed Points

Since $\tilde{\Gamma}$ is a Euclidian lattice, we can define the number of fixed points without ambiguity [9]. Let $I$ denote an invariant sublattice of $\tilde{\Gamma}$ by the operation $\theta$, and let $N$ be the sublattice orthogonal to $I$. Every fixed point $x_f$ satisfy $(1 - \theta)x_f = v$ by virtue of Eq. (3.22). Also, for every $w \in I$ we have $(1 - \Theta)w = 0$. It follows that

$v \cdot w = 0$, which means that the fixed points lie in $N$. We find the number of fixed points as

$$n_{\mathrm{f}}^{\mathrm{symm}} = \frac{N}{(1 - \theta)\tilde{\bar{\Gamma}}}, \tag{12.100}$$

in view of Eq. (3.30).

Since we have the mirror twin of the original asymmetric lattice, we may reasonably define the number of fixed points as [9]

$$n_{\mathrm{f}}^{\mathrm{asymm}} = \sqrt{n_{\mathrm{f}}^{\mathrm{symm}}}. \tag{12.101}$$

In the group lattice, we have the same result just by replacing $\theta$ with $\Theta$. This reasoning can be easily generalized to the twisted sector.

Although this has a direct geometric interpretation, it is hard to apply in the practical sense. Rather we use a simple formula

$$n_{\mathrm{f}}^{\mathrm{asymm}} = \sqrt{\frac{\det(1 - \theta)}{|I^*/I|}}. \tag{12.102}$$

Here, $\det(1 - \theta)$ is the number of fixed points of invariant lattice $I$ (assuming it is symmetric lattice), given by Lefschetz fixed point theorem (3.82). $I^*$ is dual lattice to $I$, generated by fundamental weights. $|I^*/I|$ is called index of $I$ in $I^*$, and it is inverse of the volume of unit lattice $I$.

They are the same because $|N^*/(1-\theta)N^*| = \det(1-\theta)$ and $(1-\theta)\Gamma^{p,q} = (1-\theta)N^*$ hold [9]. Its interpretation is simple as follows. By modular transformation $\mathscr{S}$, untwisted sector goes to twisted sector, which has a number of fixed points we are counting. There appears an extra factor $|I^*/I|$ as a volume difference factor of the partition function. From (12.10), we observe that it is given by

$$|I^*/I| = \det A, \tag{12.103}$$

where $A$ is Cartan matrix for the root system generating $I$. When $I$ is semi-simple and/or contains Abelian group, $\det A$ is simply the product of simple subgroup determinants.

In the example discussed in Sect. 12.4.1, we have rotated three SU(3) subgroup of visible $E_8$, counting SU(3) fixed points for three lattices, we have $\det(1-\theta) = 27$. Now, we check invariant lattice $I$. There remains one SU(3) lattice in this $E_8$ and thus $\det A_{\mathrm{SU}(3)} = 3$. The hidden sector $E_8$ is untouched, so there is no fixed points, and $\det A_{E_8} = 1$ as the famous self-dual lattice. Therefore, we have

$$n_{\mathrm{f}}^{\mathrm{asymm}} = \sqrt{\frac{27}{3} \cdot \frac{1}{1}} = 3.$$

We can verify the number of fixed points in Table 12.6. Note that $\det A_{E_6} = 3$.

**Table 12.6** Possible $\mathbb{Z}_3$ shift vectors for each $E_8$ sector, in the Dynkin basis $[s_i | s_0]$ and in the Cartan–Weyl basis $3V$

| Case | $[s_i | s_0]$ | $3V$ | Group | Number of $w$ |
|---|---|---|---|---|
| 0 | $[00000000|0]$ | $(0^8)$ | $E_8$ | 0 |
| 1 | $[01000000|0]$ | $(2\,1\,1\,0^5)$ | $SU(3) \times E_6$ | 3 |
| 2 | $[00000001|0]$ | $(\frac{3}{2}\,\frac{1}{2}^7)$ | $SU(9)$ | 4 |
| 3 | $[10000000|1]$ | $(1\,1\,0^6)$ | $E_7 \times U(1)$ | 2 |
| 4 | $[00000010|1]$ | $(2\,0^7)$ | $SO(14) \times U(1)$ | 1 |

The number of SU(3)s under the Coxeter action $w$ given in Eq. (12.71) is also listed. The convention on fundamental weights is presented in Table 12.1

### Equivalence to Symmetric Orbifold

In the literature, the asymmetric orbifold is used not only for those with the asymmetric orbifolding action but also for those with the radius of the compact internal space taking the critical value $R = \sqrt{\alpha'}$, as introduced in the beginning of this subsection. With the critical radius, we have the extra gauge group $SU(3)^3$. Of course, this critical radius is not necessary for asymmetric orbifolds.

However then, there is *no way of distinguishing* whether this is the compactified internal space or the original gauge group space. It is just a matter of assigning coordinates. Note that we have used the Coxeter action for breaking the gauge group, e.g. breaking SU(3) in Sect. 12.4.1. We can use the same trick for breaking $SU(3)_{st}$ arising from spacetime in asymmetric orbifold. Indeed, if we do exchange one of SU(3)s from group space and $SU(3)_{st}$, we have a situation that there appears a symmetric orbifold with three Coxeter elements with $SU(3)^3$ in the group space. Also, the number of fixed points and the mass shell condition is the same, as it should be.

## 12.6  Group Structure

An apt reader might have noticed that the *massless spectrum from untwisted sector*, resulting from breaking by shifts, obeys the *branching rule* of Lie algebra. We will observe here its group theoretical origin. In particular, if there is an Abelian algebra in the unbroken algebra, there are some subtleties, which is cured only by understanding the group structure.

Moreover, we are interested in the massive states also. Interestingly, they still form the representations of a certain algebra, which we will discuss here. In usual case, we neglected them, because they have mass of order $1/\alpha'$ so that decoupled in the low energy limit. Nevertheless, these massive states are relevant for the following reasons.

In the twisted sector, we have observed that the spectrum does not belong to the untwisted sector spectrum, since it is in a different Hilbert space. For example, there appears a state like $P = (1^4\,0^4)(0^8)$ not belonging to $E_8 \times E_8$ roots since $P^2 \neq 2$.

It belongs to the root *lattice*. These states appear to be mixed with massive states because the twisting (7.2) and the shift $P \rightarrow P+V$ give massless states, like (7.73). We will see that this originates from just isomorphic algebra in $P + V$ basis [7].

In this section, we will also show that there cannot be an adjoint matter representation from the *level one* algebra. Thus, the higher-level algebra must be used if the adjoint Higgs of SU(5), SO(10), or E$_6$ GUTs are used for gauge symmetry breaking. It can be obtained by projecting out some components of massive states with $P^2 > 2$ to $P^2 = 2$ states [9–11].

There are good references for the topics discussed in this section [1, 13].

## 12.6.1 Classification of the Gauge Group

The information on group breaking is entirely contained in the shift vector $V$. By representing $V$ in the Dynkin basis, we can track the group theoretical origin. Armed with it we can have a deeper understanding of its structure and handily classify all the possible breaking.

**Shift Vector**
The shift vector of the $\mathbb{Z}_3$ example given in Sect. 7.4 is

$$V = \frac{1}{3}(\Lambda_2)(0), \tag{12.104}$$

where

$$\Lambda_2 = (2\ 1\ 1\ 0^5), \tag{12.105}$$

which can be read from Table 12.6. Then, by definition of the fundamental weights, the condition (7.53) for unbroken roots is satisfied for every (simple) root except $\alpha^2$. The remaining root vectors make up the root system of the unbroken algebra SU(3)×E$_6$ (Fig. 12.6). This is Cartan's general procedure of obtaining the maximal subalgebra and explains why the untwisted sector spectrum obeys the branching rule. We can generalize this method as follows.

Denoting the shift vector in terms of the fundamental weights in the Dynkin basis,

$$V = \frac{1}{N} \sum_{i=1}^{r} s_i \Lambda_i = \frac{1}{N}[s_i], \tag{12.106}$$

we can always satisfy the following condition [14]:

$$N = s_0 + \sum_{i=1}^{r} a_i s_i, \qquad s_0, s_i \in \mathbb{Z}_{\geq 0} \tag{12.107}$$

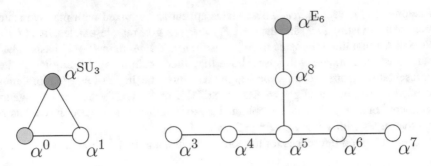

**Fig. 12.6** With a further breaking with a Wilson line, we just apply the same rule to the subgroups

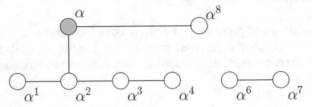

**Fig. 12.7** Some shift vectors (here $V = \frac{1}{3}\Lambda_5$) are redundant to the "standard form" ($V = \frac{1}{3}\Lambda_2$)

with *nonnegative integers* $s_0, s_i$. Surveying the condition (7.53), one observes that it is sufficient to consider only simple root vectors since any root vector is made of successive addition of simple roots. Now, consider a product

$$V \cdot \alpha^i = \frac{1}{N}\sum_{j=1}^{r} s_j(\alpha^i \cdot \Lambda_j) = \frac{1}{N}s_i. \qquad (12.108)$$

If $s_i = 0$, then the condition (7.53) is satisfied. If not, the corresponding $i$th root is not a root of the unbroken group. The unbroken group is obtained from the extended Dynkin diagram: remove this $i$th circle from the extended Dynkin diagram. However, the Cartan generators are invariant, so that the rank is preserved as in the original group.

Since all the numbers are nonnegative in the condition (12.107), only a finite number of set of integers $[s_i]$ can satisfy it. The same applies to the other $E_8'$. For the $\mathbb{Z}_3$ orbifold, there are therefore only five cases shown in Table 12.6.

What happens for the shift vector not satisfying (12.107), for example $V = \frac{1}{3}(\Lambda_5)(0)$? $\Lambda_5 = (5\ 1\ 1\ 1\ 1\ 1\ 0\ 0)$ is given in Table 12.1. Then, it is redundant, and it can be explicitly checked that it yields again SU(3)×$E_6$ [14]. The $\alpha$ seems to be projected out; however, there is another "simple root" $\alpha = \alpha^3 + 2\alpha^4 + 3\alpha^5 + 2\alpha^6 + \alpha^7 + \alpha^8$ connected to $\alpha^8$ and $\alpha^2$. With a number of Weyl reflections, this shift becomes $V = \frac{1}{3}(\Lambda_2)(0)$, satisfying (12.107). This situation is depicted in Fig. 12.7.

**Table 12.7** The only modular invariant combinations without Wilson lines, breaking $E_8 \times E_8$

| Combination | Group | No. of fixed points |
|---|---|---|
| $0 \times 0$ | $E_8 \times E_8'$ | 1 |
| $1 \times 0$ | $SU(3) \times E_6 \times E_8'$ | 3 |
| $1 \times 1$ | $SU(3) \times E_6 \times SU(3)' \times E_6'$ | 9 |
| $2 \times 3$ | $SU(9) \times E_7' \times U(1)'$ | 9 |
| $3 \times 4$ | $E_7 \times U(1) \times SO(14)' \times U(1)'$ | 1 |

The numbers in the first column are from Table 12.6. We used prime to discriminate two $E_8$'s, which can be exchanged. The number of fixed points in the group space emerges when we mod by the point group action $\Theta$, not by the shift vector

So far, we have just analyzed the pattern of symmetry breaking in terms of group theory. Combining with another $E_8'$, the modular invariance condition (7.50) restricts possible combinations. The only possible cases are those presented in Table 12.7.

**With Wilson Lines**

We can generalize the above Dynkin diagram technique in the presence of Wilson lines. We have more shift vectors providing additional projection conditions (9.85). One may expect that the unbroken group is the common intersection of the groups obtained by the shift vector $V$ and Wilson lines $a_i$, which is not true however. The unbroken group is the common intersection of the fixed point gauge groups which are given by $V + \sum m_i a_i$.

At every stage, we begin with the unbroken Dynkin diagram and apply essentially the same rules as explained in [14]. Here, we just summarize the following results:

1. Break the group according the rule above (12.107).
2. For a further breaking, apply the same rule to each subgroup, now with the dual Coxeter labels $a_i'$ of the subalgebra

$$N = s_0' + \sum_{i=1}^{r} a_i' s_i' . \tag{12.109}$$

This will give a possible shift vector corresponding to the Wilson line.
3. However, we cannot satisfy the previous condition with these new $s_i'$ at this stage. We relax the rule of step 1, by

$$N = s_0' + \sum_{i=1}^{r} a_i s_i', \qquad \text{mod } N. \tag{12.110}$$

4. Iterate the procedure.

## 12.6.2 Abelian Charge

Recall that under the shift (12.106) the Cartan generators $\{H^I\}$ remain invariant, even when their roots are prevented by the condition (12.106). This means that in the unbroken group, their linear combination $\sum_I q_i^I H^I$ plays the role of the $U(1)_i$ generator and the rank is preserved. The corresponding charge generator that projects the state vector (12.30) to give the $U(1)$ charge is

$$q_\alpha \cdot (P + kV + m_i a_i). \tag{12.111}$$

We will fix the normalization later. All the Abelian generators are orthogonal to shift vector

$$q_\alpha \cdot V = 0, \quad q_\alpha \cdot a_i = 0, \quad \text{for all } \alpha, i, \tag{12.112}$$

and each other

$$q_\alpha \cdot q_\beta = 0, \quad \text{for all } \alpha, \beta. \tag{12.113}$$

For instance, the SU(5) model considered in Sect. 7.5.4 with shift vectors

$$V = (-\tfrac{2}{3} \tfrac{1}{3} \tfrac{1}{3} \tfrac{1}{3} \tfrac{1}{3} \, 0\,0\,0)(-\tfrac{2}{3}\,0\,0\,0\,0\,0\,0\,0)',$$

$$a_1 = (0\,0\,0\,0\,0\,0\,0\,\tfrac{2}{3})(0\,\tfrac{1}{3}\,\tfrac{1}{3}\,0\,0\,0\,0\,0)',$$

leaves four U(1) groups with the generators

$$(1\,1\,1\,1\,1\,0\,0\,0)(0\,0\,0\,0\,0\,0\,0\,0), \tag{12.114}$$

$$(0\,0\,0\,0\,0\,0\,0\,1)(0\,0\,0\,0\,0\,0\,0\,0), \tag{12.115}$$

$$(0\,0\,0\,0\,0\,0\,0\,0)(1\,0\,0\,0\,0\,0\,0\,0), \tag{12.116}$$

$$(0\,0\,0\,0\,0\,0\,0\,0)(0\,1\,1\,0\,0\,0\,0\,0). \tag{12.117}$$

We can easily understand that they are all proportional to the component of shift vectors. The exception is (12.114), which in fact reflects the symmetry of the unbroken SU(5).

We can easily generalize that the Abelian generators are proportional to the fundamental weights used (corresponding to $s_i \neq 0$) in the shift vector (12.106),

$$q_i \propto \Lambda_i \tag{12.118}$$

if the extended root of the original algebra is projected out $s_0 \neq 0$. This is true because $q_i$ should be orthogonal to the rest of the (simple) roots; otherwise, this vector would be the root vector of the corresponding non-Abelian group. We use the same index $i$ since we have one-to-one correspondence between $\Lambda_i$ and U(1) subgroups. If the extended root survives $s_0 = 0$, we can always find the following Abelian generators $q$ (as many as the number of Abelian groups in the fixed point algebra). By making linear combinations between the fundamental weights used in the shift vector (12.106), allowing the negative coefficient we have

$$q \propto \sum s_i' \Lambda_i, \quad s_i' \in \mathbb{Z} \tag{12.119}$$

satisfying

$$q \cdot \theta = 0,$$

for it should be orthogonal to the extended root $-\theta$ of the original algebra.

The normalization of $q_i$ is related to the level $k$ and determined by normalization of the current $j^a(z)$ [15]. The corresponding vertex operator in this direction is $q_i \cdot \partial_z X$ and has a different coefficient from (12.14). From (12.14), by fixing normalization of $f^{abc}$, as in (12.13), the relative normalization of the $z^{-2}$ term should be $k = q_i^2$ in this direction. For Abelian groups, the structure constants vanish, and the normalization has to be fixed in another way. However, at the compactification scale of an orbifold, this U(1) generator is embedded in $E_8 \times E_8$ and thus has a definite normalization

$$q_i^2 = k \tag{12.120}$$

to 1, as discussed before. The conformal weight for a state is

$$h_{Q_i} = \tfrac{1}{2} Q_i{}^2 = \tfrac{1}{2} (q_i \cdot P)^2. \tag{12.121}$$

Comparing to the similar relation (12.41), we can determine the U(1) charged piece of vector $P$. Interestingly, it is also proportional to $q_i$: The other parts of $P$ are fundamental weights of the unbroken non-Abelian group, which should not be charged under this U(1),

$$q_i \cdot P = q_i \cdot r, \quad r \propto \Lambda_i \propto q_i. \tag{12.122}$$

This means that we can decompose the shift vector into completely disconnected parts. The resulting state vector is

$$P = \sum A_{ij}^{-1} \alpha^{j\vee} + r. \tag{12.123}$$

The normalization of $q$ is fixed by (12.121). In general, states may be charged under more than one U(1)s: then, the vector is simply the addition of each U(1) part.

There are potential anomalous U(1)s. Since they are embedded in the original group SO(32) or $E_8 \times E_8$, they can be rearranged to one U(1). This is cancelled by the Green–Schwarz mechanism, if the U(1) charges of the whole spectrum satisfy a specific "universality" condition. It also fixes normalization [15–17], and our normalization gives the correct answer. Using this [18, 19], if we have at least one anomalous U(1), the GS mechanism fixes the normalization in four dimensional theory, regardless of the origin of group breaking, which in this case is orbifolding.

This is the only way to find Abelian generator systematically. The information is extracted, from which Cartan subalgebras became Abelian generators after symmetry breaking.

## Example
Consider the $T^6/\mathbb{Z}_3$ example with the shift vector

$$V = \tfrac{1}{3}(2\ 0^7)(1\ 1\ 0^6) = \tfrac{1}{3}(\Lambda_7)(\Lambda_1).$$

We can check that the modular invariance condition is satisfied and the resulting gauge group is SO(14)×U(1)×$E_7$×U(1). The two U(1) generators are $q_7 = \tfrac{1}{2}(\Lambda_7; 0)$ and $q_1' = \tfrac{1}{\sqrt{2}}(0; \Lambda_1)$ by the normalization (12.120). Note that this gives the correct normalization [17] for the Green–Schwarz mechanism. In view of the branching rule, in the untwisted sector we obtain

$$3(\mathbf{14})(\mathbf{1}) + 3(\mathbf{64})(\mathbf{1}) + 3(\mathbf{56})(\mathbf{1}) + 3(\mathbf{1})(\mathbf{1}). \tag{12.124}$$

In the twisted sector, the zero point energy is still $\tilde{c} = -\tfrac{2}{3}$. The SO(14) vector with $h_{\mathbf{14}} = \tfrac{1}{2}$ alone cannot be massless but should have other components to fulfill the mass shell condition. The missing mass is provided by other vectors $r_7$ and $r_1'$ charged under U(1)s.

The corresponding highest weight vector has the form

$$P + V = \sum_j (A^{SO(14)})_{1j}^{-1} \alpha^j + r_7 + r_1'. \tag{12.125}$$

The first term is $\Lambda_1$ of SO(14). The $r_7$ and $r_1'$ are also proportional to $(\Lambda_7)(0)$ and $(0)(\Lambda_1)$, respectively. They are completely fixed by the condition

$$h_Q = \frac{1}{2}(q_7 \cdot r_7)^2 + \frac{1}{2}(q_1' \cdot r_1')^2 = \frac{1}{6}, \tag{12.126}$$

and the generalized GSO projection condition. The resulting vector is

$$P + V = (0\ 1\ 0^6)(0^8) + (-\tfrac{1}{3})(1\ 0^7)(0^8) + \tfrac{1}{3}(0^8)(1\ 1\ 0^6), \tag{12.127}$$

and charged as $(\mathbf{14})(\mathbf{1})$. The Lorentz $\mathbf{3}$ of SU(3) (by $\alpha^i_{-1/3}$) can contribute $h = \tfrac{1}{3}$, and it provides another charged state, $(\mathbf{1})(\mathbf{1})$. In addition, there is a state which is a singlet under the whole non-Abelian group $(\mathbf{1})(\mathbf{1})$. They all have multiplicity $\chi = 27$.

## 12.6.3 Complete Spectrum of SO(32) String

The group structure of SO(32) is particularly simple. It is because there are roots with integral elements only. Hence, the Weyl reflection, which is the only meaningful automorphism, rearranges their positions and signs only, by (12.67) and (12.68). As a consequence, the shift vectors are classified only by the number of elements. Any vectorial shift can be brought to the following form:

$$V = \frac{1}{2N}(0^{n_0}, \ldots, N^{n_N}), \quad \text{with} \quad \sum_{k=0}^{N} n_k = 16, \tag{12.128}$$

and lead to the symmetry breaking pattern

$$SO(32) \to SO(2n_0) \times U(n_1) \times \ldots \times U(n_{N-1}) \times SO(2n_N). \tag{12.129}$$

We employ a subscript to make the distinction between the various factors, for example, $U(n_k) = U(1)_k \times SU(n_k)$ and $U(1)_0 = SO(2n_0)$ when $n_0 = 1$. Of course, these shift vectors should satisfy conditions on the modular invariance and the evenness of the sum.

Looking at the mass shell condition, we conclude that only a few kinds of representations can appear. The result is tabulated in Table 12.8. All U(n) representations are totally antisymmetric $k$-form tensor representations of the vector

**Table 12.8** The twisted matter of SO(32) orbifold models: the $k$-form representations $[\mathbf{n}]_k^{\pm}$ of U(n) and the vector $\mathbf{2n}$ and spinor $\mathbf{2}_{\pm}^{n-1}$ representations of SO(2n)

| $\tilde{V}$ | Group | Repr. | Weights | Prop. |
|---|---|---|---|---|
| $\tilde{V} = 0$ | $SO(2n)$ | $[\mathbf{2n}]^k$ | $(\pm 1^k, 0^{n-k})$ | $k = 0, 1$ |
| $0 < |\tilde{V}| < \tfrac{1}{2}$ | $U(n)$ | $[\mathbf{n}]_k^{\alpha}$ | $\alpha(1^k, 0^{n-k})$ | $\alpha = \pm$ $k \geq 0$ |
| $\tilde{V} = \tfrac{1}{2}$ | $SO(2n)$ | $\mathbf{2}_{\alpha}^{n-1}$ | $(-\tfrac{1}{2}^k, \tfrac{1}{2}^{n-k}) - (\tfrac{1}{2}^n)$ | $\alpha = (-)^k$ |

For the three cases, $\tfrac{1}{2}(P + \tilde{V})^2$ are $\tfrac{k}{2}$, $k\left(\tfrac{1}{2} + \alpha\tilde{V}\right) + \tfrac{1}{2}n\tilde{V}^2$, and $\tfrac{n}{8}$, respectively

**n** or its complex conjugate $\bar{\mathbf{n}}$, denoted by $[\mathbf{n}]_k^+$ and $[\mathbf{n}]_k^-$, respectively. (In particular, $[\mathbf{n}]_0^\pm = \mathbf{1}$, $[\mathbf{n}]_1^+ = \mathbf{n}$, $[\mathbf{n}]_1^- = \bar{\mathbf{n}}$, and $[\mathbf{n}]_k = [\bar{\mathbf{n}}]_{n-k}$.) The representations of SO(2n) that arise are the fundamental representation $[\mathbf{2n}]^k$ or the spinor representation $\mathbf{2}_\alpha^{n-1}$ of $\alpha = \pm$ chirality. The index $k = 0, 1$ is used to simultaneously treat the fundamental and the singlet representation. In the SO(32) heterotic string case, we can explicitly prove the no-adjoint theorem by showing that only the weights, whose entries in the Dynkin basis does not exceed level $k$, can appear.

Using these representations, we can identify the irreducible *twisted states*. The vectorial weights give rise to representations of the form

$$\mathbf{R} = \left( [\mathbf{2n_0}]^{k_0}, [\mathbf{n_1}]_{k_1}^{\alpha_1}, \ldots, [\mathbf{n_{m-1}}]_{k_{m-1}}^{\alpha_{m-1}}, \mathbf{2}_{\alpha_m}^{n_m-1} \right), \tag{12.130}$$

where $\alpha_a = \pm$, $k_0 = 0, 1$ and $k_a \geq 0$. The mass contribution of this state reads

$$\frac{M_L^2}{4} = \frac{k_0}{2} + \sum_{a=1}^{m-1} k_a \left( \frac{1}{2} + \alpha_a \tilde{v}_{pa} \right) + \frac{1}{2} (\tilde{v}_p)^2. \tag{12.131}$$

The GSO projection on the vectorial weights require that

$$\frac{1 - \alpha_m}{4} + \frac{1}{2} \sum_{a=0}^{m-1} k_a \in 2\mathbb{Z}. \tag{12.132}$$

So, all we need to do is to solve these linear equations. As a result, we obtain general spectrum, including twisted sector, which is presented in [20].

### 12.6.4 Higher-Level Algebra

In the ordinary setup of string embedding, only the level 1 algebra is allowed, because the whole normalization is fixed by (12.13). The relative normalization of $z^{-2}$ term in (12.14) should be $k = 1$ [6]. To have an adjoint representation based on SU(5), SO(10), or $E_6$, evading the no-adjoint theorem (12.34), we should have higher-level algebra $k \geq 2$.

Although the normalization is fixed from the beginning, we can modify the theory by embedding the group into a larger group and changing the GSO projection to have a different normalization [10]. We take one heuristic example of level 2 SU(2)$_2$ algebra from two SU(2)$_1$'s, where the subscript denotes the level of algebra [21]. Let us call them $SU(2)^A \times SU(2)^B$ and form root space orthogonal to each other. Simple root vectors $\alpha^A$ and $\alpha^B$ of both groups have length squared

**Fig. 12.8** The level 2 algebra SU(2)$_2$ is obtained by projecting to symmetrical currents from two level 1 groups [SU(2)$_1$]$^2$

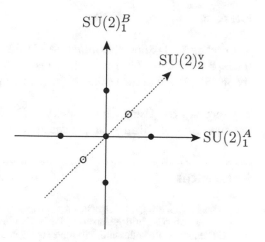

two as before. Projecting the root vectors of each group by keeping symmetrical combinations

$$j^v(z) = j^A(z) + j^B(z),  \tag{12.133}$$

we have another SU(2)$_2$. Since they come from the same group, so that $j^v$ have twice the length, thus they lead to a new normalization $k = 2$. It is easily viewed in the root space depicted in Fig. 12.8. The •'s indicate the original roots of each SU(2)s. After applying some suitable GSO projection (dashed line), we have new SU(2) roots o having length squared 1.

This shows a typical *regular diagonal embedding*, $g_k \subset g_1^k$. In this notation, the example in the preceding paragraph is SU(2)$_2$ $\subset$SU(2)$_1^2$. However, there is another kind of embedding such as SU(2)$_4$ $\subset$SU(3)$_1$ or SU(2)$_{28}$ $\subset$G$_2$ [1].

Beginning with E$_8$ × E$_8$ and SO(32), a limited number of semi-simple groups can be obtained. In other words, an embedding into an algebra with rank $r > 16$ is impossible, when we require level 3 algebra (E$_6$)$_3$ diagonally embedded in (E$_6$)$^3$. Most embeddings in these semi-simple groups are introduced in [1]. A systematic analysis is done in [10], and there are some models in such higher-level algebra realizing SO(10) [12] and E$_6$ [10]. Also, there is a very powerful model-independent criterion on SO(10) GUT from higher-level algebra [10]. A more detailed study revealed more strong constraint: although **126** of SO(10) is a tensor representation and thus is not ruled out by the no-go theorem, there is no way to obtain it [10]. This representation **126** is needed to allow see-saw neutrino masses purely from SO(10) representations and *R*-parity as well.

## Exercises

▶ **Exercise 12.1**  Show that all the SU(3) subgroups in $E_8$ are equivalent. That is, the representation (12.73) has manifest permutation symmetry of three entries, if we fix one SU(3) and make a suitable complex conjugation.

▶ **Exercise 12.2**  Using the symmetry of Dynkin diagrams (12.1) and (12.2), find the outer automorphism of simple and affine Lie algebras.

## References

1. J. Fuchs, C. Schweigert, *Symmetries, Lie Algebras and Representations: A Graduate Course for Physicists* (Cambridge University Press, Cambridge, 2003)
2. M. Golubitsky, B. Rothschild, Primitive subalgebras of exceptional lie algebras. Pac. J. Math. **39**(2), 371–393 (1971)
3. I.B. Frenkel, V.G. Kac, Basic representations of affine Lie algebras and dual resonance models. Invent. Math. **62**, 23–66 (1980)
4. L.J. Dixon, J.A. Harvey, C. Vafa, E. Witten, Strings on orbifolds. 2. Nucl. Phys. **B274**, 285–314 (1986)
5. H. Sugawara, A Field theory of currents. Phys. Rev. **170**, 1659–1662 (1968)
6. J. Polchinski, in *String Theory. Vol. 2: Superstring Theory and Beyond*. Cambridge Monographs on Mathematical Physics (Cambridge University Press, Cambridge, 12 2007)
7. K.-S. Choi, Spectrum of heterotic string on orbifold. Nucl. Phys. B **708**, 194–214 (2005)
8. L.E. Ibanez, H.P. Nilles, F. Quevedo, Reducing the rank of the gauge group in orbifold compactifications of the heterotic string. Phys. Lett. B **192**, 332–338 (1987)
9. K. S. Narain, M. H. Sarmadi, C. Vafa, Asymmetric orbifolds. Nucl. Phys. **B288**, 551–577 (1987)
10. K.R. Dienes, J. March-Russell, Realizing higher level gauge symmetries in string theory: new embeddings for string GUTs. Nucl. Phys. B **479**, 113–172 (1996)
11. K.R. Dienes, New constraints on SO(10) model building from string theory. Nucl. Phys. B **488**, 141–158 (1997)
12. Z. Kakushadze, S.H. Henry Tye, Three family SO(10) grand unification in string theory. Phys. Rev. Lett. **77**, 2612–2615 (1996)
13. P. Goddard, D.I. Olive, Kac–Moody and Virasoro algebras in relation to quantum physics. Int. J. Mod. Phys. **A1**, 303 (1986)
14. K.-S. Choi, K. Hwang, J.E. Kim, Dynkin diagram strategy for orbifolding with Wilson lines. Nucl. Phys. **B662**, 476–490 (2003)
15. A. Font, L.E. Ibanez, F. Quevedo, A. Sierra, The construction of 'realistic' four-dimensional strings through orbifolds. Nucl. Phys. **B331**, 421–474 (1990)
16. J.A. Casas, E.K. Katehou, C. Munoz, U(1) Charges in orbifolds: anomaly cancellation and phenomenological consequences. Nucl. Phys. **B317**, 171–186 (1989)
17. T. Kobayashi, H. Nakano, 'Anomalous' U(1) symmetry in orbifold string models. Nucl. Phys. **B496**, 103–131 (1997)
18. P.H. Ginsparg, Gauge and gravitational couplings in four-dimensional string theories. Phys. Lett. **B197**, 139–143 (1987)

19. L.E. Ibanez, Computing the weak mixing angle from anomaly cancellation. Phys. Lett. B **303**, 55–62 (1993)
20. K.-S. Choi, S.G. Nibbelink, M. Trapletti, Heterotic SO(32) model building in four dimensions. J. High Energy Phys. **12**, 063 (2004)
21. A. Font, L.E. Ibanez, F. Quevedo, Higher level Kac–Moody string models and their phenomenological implications. Nucl. Phys. B **345**, 389–430 (1990)

# Orbifold Phenomenology **13**

We are now ready to use the toolkits we have studied and build concrete models. After describing the construction procedure, we review some generic phenomenological features from string constructed model.

## 13.1 Model Building

Usual orbifold construction takes the following procedure.

1. *Choice of orbifold and shift vector:* We choose an orbifold and take a shift vector. The orbifold geometry is determined by a space group, as summarized in Table 3.4. It determines the order of shift vector associated with the point group. The space group determines possible configurations of Wilson lines and hence the distribution of matter fields. We choose a shift vector guided by Grand Unification, since with only the shift vector, we cannot have the Standard Model (SM) gauge group. Shift vectors are classified in Ref. [1], however, we may simply apply the Dynkin diagram strategy discussed in Sect. 12.6.1.
2. *Choice of Wilson lines:* We take Wilson lines associated with the translational elements of the space group to break the gauge group further. At this stage, we completely fix the low-energy gauge group. We may have the SM at the string scale. Alternatively, we may obtain a GUT. A set of Wilson lines may be equivalent and redundant.
3. *Spectum check:* We quantize strings on orbifold and find the spectrum and check anomaly cancellation. We may check whether we obtain desirable spectrum of quarks and leptons, as discussed below. We need also Higgs multiplets. We may obtain the universal anomalous U(1). If we obtained the SM group, we need to make sure that the hypercharge is not anomalous.
4. *Vacuum configuration:* Usually we have bunch of singlets of the SM or GUT group, which make the superpotential very convoluted. Applying the selection

© Springer Nature Switzerland AG 2020
K.-S. Choi, J. E. Kim, *Quarks and Leptons From Orbifolded Superstring*,
Lecture Notes in Physics 954, https://doi.org/10.1007/978-3-030-54005-0_13

rules discussed in Sect. 10.2 and generating superpotential terms, we need to assign singlet VEVs to have effective normalizable potentials. This breaks unwanted Abelian and non-Abelian gauge symmetries and generates Yukawa couplings. First, we apply the selection rules to generate possible superpotentials. Then we check SUSY consistency $F$ and $D$ flat conditions.

These are all the data to construct realistic models, which is strikingly simple. Especially the localized fields at the orbifold fixed points are not arbitrary but completely fixed as a consequence of modular invariance. It gives us the flavor symmetry reflecting the geometrical distributions. Detailed examples will be discussed in the next chapter.

Finally we compare the obtained model with experimental data. Here is an incomplete list which are needed from superstring:

- *Three families of chiral fermions and Higgs*: The string constructed model must allow three families of quarks of leptons. Also we have observed Higgs scalar, which must also be obtained.
- *Hypercharge quantization*: If the fifteen chiral fields are put in a GUT without exotic particles, the gauge coupling constants measured at $M_Z$ evolve to meet at the GUT point with the GUT scale weak mixing angle of $\sin^2 \theta_W^0 \simeq \frac{3}{8}$. This kind of GUT has an appropriate hypercharge quantization. Some GUTs allow automatic hypercharge quantization.
- *Gauge coupling unification*: The Georgi–Quinn–Weinberg mechanism suggests that the running gauge couplings are unified at a scale close to the Planck scale. We may have the Standard Model gauge group or an intermediate GUT groups.
- *Flavor problem*: Three families in GUTs from superstring must lead to correct mass and/or mixing angle relations.
- *$\mu$-problem*: In addition to the scale problem of $\mu$ in the minimal supersymmetric standard model(MSSM), there is the problem on removing the color-triplet partners of Higgs doublets $H_u$ and $H_d$.
- *Proton longevity problem*: Generic models beyond the standard model like SSMs or GUTs allow rapid nucleon decays. The proton longevity problem in SUSY models was noted right after SUSY was applied to the gauge hierarchy problem. Because scalar partners of colored quarks exist, they can trigger proton decay by a dimension-3 term, $u^c d^c d^c$, in the effective superpotential $W$. This was resolved by introducing an R-parity [2]. Even if this dimension-3 term is absent, a dimension-4 term[1] such as $qqq\ell$ in the effective superpotential is problematic [3].

If we arrive at a consistent model, we may also learn what the constructed model predicts.

---

[1] Below, we will call this "dimension-5" term in the Lagrangian $\mathscr{L}$, or simply "dimension-5 term."

## 13.2 String Unification

### 13.2.1 Gauge Coupling Unification

String theory provides unification relations without intermediate field-theoretical GUT. It is due to the unified group of the string theory.

In ten dimension, we have a simple relation between string $g_{10}$ and gravitational $\kappa_{10}$ constants [4]

$$g_{10}^2 = \frac{4\kappa_{10}^2}{\alpha'}. \tag{13.1}$$

Here, the inverse string tension $\alpha' \sim M_s^{-2}$ measures the string size and can be considered as the mass of the lightest massive string excitation [5]. It is fixed by supersymmetry and extra term responsible for Green–Schwarz mechanism and also by calculating three point correlation functions of gravitons and gauge bosons. This relation still holds when we reduce the space dimensions by compactifying the internal space, as in (11.8),

$$k_A g_A^2 = g^2 = \frac{4\kappa^2}{\alpha'}, \tag{13.2}$$

where $k_A$ is the level of affine Lie algebra for a gauge group labelled by $A$.

Phenomenologically, the unification scale is around $(2–3) \times 10^{16}$ GeV [6–8]. The standard folklore has been that it is better to be realized around the string scale $M_s$. Although they are very close, it still differ by a factor of O(10), assuming order one $g_{10}$. Many ideas are suggested to remedy this discrepancy, including threshold correction, shown in Eq. (11.107). Also it is noted that the Type II construction can pull down the string scale, as we will see in Chap. 17.

The relation (13.2) is the basis for claiming unification of gauge couplings *without* any intermediate GUT. For instance, if we obtain an SU(3) × SU(2) × U(1) model with the gauge couplings $g_3, g_2, g_1$, respectively, at the scale $M_s$, we have the relation

$$k_3 g_3^2 = k_2 g_2^2 = k_1 g_1^2 = g^2 = \frac{32\pi G_N}{\alpha'}, \tag{13.3}$$

where $k_3, k_2, k_1$ are the levels of SU(3), SU(2), and U(1) algebras, respectively, at the compactification scale. Although the level is defined for non-Abelian gauge group, we always obtain a U(1) as a subgroup of the unified group predicted by string theory. So we may define its level as in (12.120). Also we usually take different normalization for the hypercharge

$$C g' \equiv g_1$$

as in (2.18).

Unless we need an adjoint Higgs scaler, discussed in Sect. 12.6.4, we have always $k = 1$. Therefore, the relation

$$g_1 = g_2 = g_3 = g, \tag{13.4}$$

reduces to the conventional one [9]. Here $G_N$ is the Newton's constant obtained in (11.9).

Note that, determination of gauge couplings at the compactification scale is not so trivial because there is a small difference between the compactification scale and the Planck scale.

## 13.2.2 Standard-Like Models

Even though one obtains factor groups such as those in the SM, there is a possibility that the couplings are unified at the string scale, as in (13.4). Thus, it is not a bad idea to obtain a (supersymmetric) SM directly via some orbifold compactification of string theory.

Soon after orbifold compactification was introduced, the first model building attempt was to try $SU(3) \times SU(2) \times U(1)_Y$ from $T^6/\mathbb{Z}_3$ orbifold [10, 11]. This class of models is known as *standard-like models*. For instance, consider a model with three Wilson lines [14],

$$V = \frac{1}{3}(1\ 1\ 1\ 1\ 2\ 0\ 1\ 1)(1\ 1\ 0\ 0\ 0\ 1\ 1\ 2),$$

$$a_1 = \frac{1}{3}(1\ 1\ 1\ 2\ 1\ 0\ 0\ 0)(0\ 0\ 0\ 0\ 0\ 2\ 0\ 0),$$

$$\tag{13.5}$$

$$a_3 = \frac{1}{3}(0\ 0\ 0\ 0\ 0\ 0\ 0\ 2)(1\ 1\ 1\ 2\ 0\ 0\ 0\ 1),$$

$$a_5 = \frac{1}{3}(0\ 0\ 0\ 0\ 0\ 0\ 0\ 2)(0\ 1\ 1\ 0\ 0\ 0\ 0\ 0).$$

It gives the gauge group $SU(3) \times SU(2) \times U(1)^{13}$.

All orbifold models toward this objective used the following type of shift vector and Wilson line(s) [10, 12–15],

$$V = \frac{1}{3}(1\ 1\ 1\ 1\ 2\cdots)(\cdots)$$

$$\tag{13.6}$$

$$a_1 = \frac{1}{3}(1\ 1\ 1\ 2\ 1\cdots)(\cdots),$$

where $\cdots$ are chosen to satisfy the modular invariance conditions discussed in Chap. 7 but not to enhance $SU(3)$ and $SU(2)$ to some larger groups. A kind of skewing results in the SU(5) group space, as implied by the first five entries of Eq. (13.6), along with many extra particles beyond those of the SM. Even though we ambitiously began without the need for the adjoint representation of some GUT group, we may have reached the limitation of the standard-like models.

The standard-like models require the following features toward a supersymmetric standard model.

- It must already contain the SM gauge group SU(3)×SU(2)×U(1)$_Y$, i.e. there is no need for breaking a unified group by an adjoint representation to obtain the SM.
- Fifteen chiral fermions with the correct electroweak hypercharges are contained in the spectrum.
- Three families are in the spectrum.
- There must exist at least a (pair of) Higgs doublet(s) $\mathbf{H}_1$ (and $\mathbf{H}_2$ for the case of supersymmetry) for the electroweak symmetry breaking.

Supersymmetry may be added to this list. Because of the three families requirement, $\mathbb{Z}_3$ orbifold models with two Wilson lines are helpful from the outset. In some orbifold models, it is possible to exclude extra colored scalars [10, 11], which achieves the doublet-triplet splitting. But a standard-like model has some flaws since it is not quite a supersymmetric standard model yet for the following reasons.

- It contains too many extra U(1)s which have to be broken.
- There are too many extra chiral fields with exotic electroweak hypercharges.
- There are too many Higgs doublets (a minimum of six) in $\mathbb{Z}_3$ orbifold models with two Wilson lines. It is hoped that with three Wilson lines, there is a possibility to obtain just two Higgs doublets in the twisted sector; however, such a model has not been found yet.

Later, higher order orbifolds are used to construct models having the SM or the GUT gauge groups. There it is easier to see the unification structure. It is because the shift vector alone can break large part of the E$_8$. For instance, in the $\mathbb{Z}_6$ orbifold models, using the Dynkin diagram strategy in Sect. 12.6.1, we may easily obtain traditional unification groups along the E$_n$ chain as

$$E_6 \times U(1)^2 : V_1 = \frac{1}{6}(\Lambda_1 + \Lambda_2),$$

$$E_6 \times SU(2) \times U(1) : V_2 = \frac{1}{6}\Lambda_2,$$

$$SO(10) \times SU(2) \times U(1)^2 : V_3 = \frac{1}{6}(\Lambda_2 + \Lambda_7),$$

where the fundamental weights of $E_8$ in Table 12.1 are used.[2] The common intersection of the $V_1$ and $V_3$ is the exact $SO(10)$. We may associate one with the point group and the other with the Wilson line. With the help of the shift vectors

$$SU(5) \times SU(2)^2 : V_4 = \frac{1}{6}(\Lambda_1 + \Lambda_6),$$

$$SO(14) \times U(1) : V_5 = \frac{1}{6}(\Lambda_6 + \Lambda_7),$$

we can easily obtain the $SU(5)$ and the SM gauge group.

## 13.3  U(1) Charges

$U(1)$ gauge group plays an interesting role. Its normalization is key to understanding the unification. Also there is a unique feature of anomalous $U(1)$. This is related to axion and strong CP problem.

### 13.3.1 Hypercharge

To distinguish the SM chiral fields among the plethora of fields, one has to identify the $U(1)$ quantum numbers. Many $U(1)$s arise from the left-over Cartan sub-algebras, and all the $U(1)$ charges are determined at the compactification scale. Among them, identifying the hypercharge $U(1)_Y$ in the SM is of utmost importance. The method of obtaining $U(1)$ generators is described in Sect. 12.6.2. The essential feature is that they are proportional to the linear combinations of shift vectors and can be made to be orthogonal to each other. The OPE fixes the normalization of hypercharges to be, Eq. (12.120),

$$q_i^2 = k = 1. \tag{13.7}$$

As an example, consider the Wilson lines of the standard-like model (13.6), which contains the gauge group $SU(3) \times SU(2) \times U(1)$ and a number of other $U(1)$s. In this model, the untwisted sector has the following weights for quark doublets:

$$P_u = (\underline{1\ 0\ 0}\ 1\ 0\ 0\ 0\ 0)(0^8)',$$

$$P_d = (\underline{1\ 0\ 0\ 0}\ -1\ 0\ 0\ 0)(0^8)'. \tag{13.8}$$

---

[2] We may also have other shift vectors yielding an $SO(10)$, but they do not follow the $E_n$ unification chain.

We find that the following basis vectors gives the charges of the quark doublets

$$q_1 = \frac{1}{\sqrt{3}}(1\,1\,1\,0\,0\,0\,0\,0)(0^8)' \tag{13.9}$$

$$q_2 = \frac{1}{\sqrt{2}}(0\,0\,0\,1\,-1\,0\,0\,0)(0^8)' \tag{13.10}$$

which are properly normalized as in (13.7). We identify the hypercharge generator $q_Y$ as

$$q_Y = -\frac{1}{\sqrt{3}}q_1 + \frac{1}{\sqrt{2}}q_2. \tag{13.11}$$

One verifies that the corresponding generator is nothing but $Y$ of Eq. (2.18). The up and down quark representations appear in the right places as in (2.12). Indeed, it gives rise to the desired hypercharges

$$\begin{aligned} u : q_Y \cdot P_u &= \tfrac{1}{6} \\ d : q_Y \cdot P_d &= \tfrac{1}{6}. \end{aligned} \tag{13.12}$$

It seems that we should look for the hypercharge with a clever insight. However, the regular pattern inherits a structure of the unified group SU(5). One is easily convinced that the above weight vectors of quark doublet (13.8) is a part of representation **10** of SU(5)

$$(1\,-1\,0\,0\,0\,0^3)(0^8). \tag{13.13}$$

Indeed, without Wilson line vector $a$ in (13.6), the unbroken group contains SU(5). Without such a unified group, we cannot expect the desired charge pattern for quarks and leptons. However, since the problem arises if some fields are outside the complete multiplet such as **5** or **10** of SU(5), the hypercharge quantization is not the same as that of the SU(5) GUT. Not only this kind of model cannot give the desired hypercharge, but also gives rise to "exotic" particles having strange charges.

## 13.3.2 Weak Mixing Angle

The U(1) charge normalization is an important issue. The sizes of the gauge couplings become the same if we compare $Cg_Y$, $g_2$, $g_3$, with $C = \sqrt{5/3}$. In fact, this value of $C$ is predicted if we break the SU(5) unification group and identify the hypercharge as in (2.18). This holds true if we embed the Standard Model group in larger E-series unification group, like SO(10), $E_6$, $E_8$. This normalization can be expressed in terms of the weak mixing angle, whose value at the GUT scale is given in Eq. (2.19) [9],

$$\sin^2 \theta_W^0 = \frac{1}{1 + C^2}. \tag{13.14}$$

The discussion of the weak mixing angle is somehow intricate, since the spectrum of heterotic string is described by the affine Lie algebra with diverse twisted sectors. In standard-like models, we have no simple picture of conventional grand unification. For example, the convenient form (2.20) cannot be used here.

The untwisted sector is described by zero modes which is simple Lie algebra, and the spectrum pattern is rather simple. So, if there is no matter fields from the twisted sector, then the particle spectrum is essentially given by the original $\mathbf{248 + 248'}$ of $E_8 \times E_8$. In this case, the U(1) charges are determined at the string scale just by the branching rule since the non-Abelian groups and the U(1) groups come from the same $E_8 \times E_8$. Nonzero roots of $E_8 \times E_8$ are defined to have $(\text{length})^2 = 2$. The U(1)s come from the center and basically they are defined by the sixteen independent $Q$s of the previous subsection or by any sixteen independent combinations.

After determination of hypercharge by surveying the spectrum,

$$q_Y = Cq_{\mathrm{u}} = \frac{1}{\sqrt{2}} \sum_A c_A q_A, \tag{13.15}$$

where $q_{\mathrm{u}}$ is the unified generator having the normalization (12.120) and the factor $1/\sqrt{2}$ comes from the normalization convention $l(\text{fund}) = \frac{1}{2}$. Here, $c_{\mathrm{SU}(2)} = \frac{1}{2}$ and $c_{\mathrm{SU}(3)} = -\frac{1}{3}$. Since all the $q_i$s are orthonormal $C^2 = l^{-1} \sum_A c_A^2$, and from (13.14) we have

$$\sin^2 \theta_W^0 = \frac{1}{1 + 2 \sum_A c_A^2}. \tag{13.16}$$

For example, consider the case where the SU(5) space is spanned by the first five entries of (13.5). Then, the electroweak hypercharge is given only by two U(1)s (13.11),

$$C^2 = 2(\tfrac{1}{2} + \tfrac{1}{3}) = \tfrac{5}{3}, \quad \sin^2 \theta_W^0 = \tfrac{3}{8}. \tag{13.17}$$

Of course, this result is due to the fact $q_Y$ in (13.11) has the desirable structure. The desirable U(1) charges for unbroken subgroup and spectrum will be obtained if it belongs to the chain of the E type unification group SU(5)$\subset$E$_8$. It amounts to the unification assumption(here the GUT group as E$_8$) and the bare weak mixing angle is related to the untwisted sector spectrum by (2.20).

After the U(1) normalization from the untwisted sector is known, one looks for the locations of the SM fields. The gauge group and charge assignments are not changed even if there are fields in the twisted sector. *The problem is that whether a set of given fields belongs to a GUT multiplet belonging to a unified group.*

To see more how it works, let us consider the above model (13.5). In this example, three quark doublets are in the untwisted sector, yet the lepton doublets are in the twisted sectors. If all the SM fields were in the untwisted sector, then one could be sure that $\sin^2 \theta_W^0 = \frac{3}{8}$. However, the model (13.5) has the possibility of $\sin^2 \theta_W^0 \neq \frac{3}{8}$. To show it explicitly, one has to identify these SM fields and express their electroweak hypercharges in terms of the above normalized (length $= \sqrt{2}$) U(1) charges. We obtain

$$q_Y = -\frac{32}{75}\sqrt{\frac{3}{2}}q_1 + \frac{13}{50}q_2 - \frac{19}{25}\frac{1}{\sqrt{2}}q_4 + \frac{23}{25}\frac{1}{\sqrt{2}}q_7 + \frac{52}{25}\frac{1}{\sqrt{2}}q_8 + \frac{21}{25}\frac{1}{\sqrt{2}}q_{12} \qquad (13.18)$$

leading to $C^2 = \frac{5353}{375}$. Thus, we obtain $\sin^2 \theta_W^0 = \frac{375}{5728} \sim 0.0655$. Another example is the model discussed in Ref. [15] where $q_Y = \frac{1}{3}q_1 - \frac{1}{2}q_2 + q_4$. By the same method we employed above, this model gives $C^2 = \frac{11}{3}$, and hence, $\sin^2 \theta_W^0 = \frac{3}{14}$.

In general, most standard-like models have a small $\sin^2 \theta_W^0$. From the preceding discussion, it is obvious that $\sin^2 \theta_W^0 = \frac{3}{8}$ requires $C^2 = \frac{5}{3}$ which is possible if $q_Y$ is given with the form seen in (13.11), with $q_1$ and $q_2$ only. If extra pieces appear beyond (13.11), then $C^2 > \frac{5}{3}$ and $\sin^2 \theta_W^0$ is smaller than $\frac{3}{8}$. But this statement is not valid if the coefficients of $q_1$ and $q_2$ are more complicated than those given in Eq. (13.11). A simple flipped SU(5) model may be given with a hypercharge $Y = -\frac{1}{3}q_1 + \frac{1}{2}q_2 + \sum_{i \geq 3} x_i q_i$ with at least one nonvanishing $x_i$ to have $e^+$ in an SU(5) singlet representation. In this case, $\sin^2 \theta_W^0 < \frac{3}{8}$, which usually happens with extra charged singlets. On the other hand, it will be extremely lucky if nonvanishing $x_i$s conspire to give $\sin^2 \theta_W^0 = \frac{3}{8}$ with more complicated coefficients in front of $q_1$ and $q_2$.

Note that we cannot use the simple formula (2.20) in general. To recapitulate how it is obtained, the basic reason stems from the unification assumption where the ratio $g^2/g'^2$ is traded for $\mathrm{Tr}\, Y^2/\mathrm{Tr}\, \tilde{Y}^2$. This formula is very powerful only when we know the spectrum of the model under the unification assumption. Of course, in GUTs such as SU(5), only the knowledge of the fundamental representation is needed and hence, this formula is already very powerful. But in standard-like models the formula is not necessarily helpful.

### 13.3.3 Anomalous U(1)

The rank-two antisymmetric tensor field $B_{MN}$, required by ten dimensional supergravity, plays an important role in anomaly cancellation in ten dimensions. After compactifying 10D down to 4D, as seen in Chap. 11, we obtain a number of pseudoscalars.

In four dimensions, the component $B_{\mu\nu}$, $\mu$, $\nu = 0, \ldots, 3$, is equivalent to and is dualized to a pseudoscalar, as in (11.91),

$$F_1 \partial^\mu a_{MI} = \frac{1}{96\pi^2} \epsilon^{\mu\nu\rho\sigma} H_{\nu\rho\sigma}. \tag{13.19}$$

This $a_{MI}$ is the *model-independent* (MI) axion. In this case, the MI axion survives as a physical degree for which the axion constant $F_a$ is around $10^{16}$ GeV [16].

We have the Green–Schwarz counterterm, as in (11.131). After dimensional reduction, under the background geometry and gauge field, we may have four dimensional effective action [17–19]

$$\int B \wedge X_8 \to \int d^4 x \epsilon^{M_1 M_2 \cdots M_{10}} B_{M_1 M_2} F_{M_3 M_4} \int_K d^6 x \langle F_{M_5 M_6} F_{M_7 M_8} F_{M_9 M_{10}} \rangle$$

$$\to \int d^4 x \epsilon^{\mu\nu\rho\sigma} B_{\mu\nu} (\partial_\rho A_\sigma - \partial_\sigma A_\rho) M,$$

$$\tag{13.20}$$

where $M$ is the above expectation value up to a numerical factor. Thus $A_\sigma$ transfers one derivative of $F_{\rho\sigma}$ to $B_{\mu\nu}$ in Fig. 13.1a. We will calculate this background contribution in Chap. 15. The corresponding Feynman diagram is shown in Fig. 13.1a.

In general we have many U(1) gauge groups. In practice we found most of them anomalous. However, under such transformation, the antisymmetric tensor $B_{\mu\nu}$ transforms as

$$\delta B = \frac{1}{30} \operatorname{Tr}(\lambda d A) - \operatorname{tr}(\Theta d\omega). \tag{13.21}$$

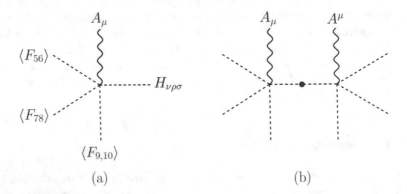

(a)                                                         (b)

**Fig. 13.1** The Green–Schwarz mechanism gives rise to gauge boson mass via Stückelberg mechanism, making the symmetry global

This is accompanied by the gauge transformation for the YM and graviton. In 4D, the ten dimensional group is broken, and the transformation (13.21) must be that of a U(1) gauge boson. Thus it cancels the anomaly of the form

$$\hat{I}_6 = \left(-\frac{1}{30} \operatorname{Tr} F^2 + \operatorname{tr} R^2\right) \wedge F. \tag{13.22}$$

Integration by part, and by dual transformation (13.19), we obtain the coupling $MA_\mu \partial^\mu a_{\mathrm{MI}}$. This means, we obtain the mass term for the gauge boson

$$\frac{1}{2}M^2(\partial_\mu a_{\mathrm{MI}} + A_\mu)^2. \tag{13.23}$$

The gauge boson acquires mass by eating the model-independent axion and both of them are removed at low energy. Note that this Lagrangian is different from the covariant derivative of a scalar, used in Higgs mechanism. This symmetry breaking is called Stückelberg mechanism. The resulting U(1) is called the *(pseudo)-anomalous* U(1). As we have seen, there is no actual anomaly. We are left with global symmetry. The corresponding diagram is Fig. 13.1b. This is consistent with the fact that the gauge transformation has no effect

$$a_{\mathrm{MI}} \to a_{\mathrm{MI}} + \lambda. \tag{13.24}$$

We may modify the Kähler potential reflecting this [20]

$$K = -\log(S + S^* - \delta^{\mathrm{GS}} V_X), \tag{13.25}$$

where $V_X$ is the vector superfield for the U(1)$_X$. This is supersymmetric version of the above Stückelberg mechanism: the chiral superfield $S$ is eaten by the vector superfield $V_X$. The gauge transformation

$$V_X \to V_X + \frac{1}{2}i(\Lambda - \Lambda^\dagger) \tag{13.26}$$

is undone by the transformation

$$S \to S + \frac{1}{2}i\delta^{\mathrm{GS}}\Lambda. \tag{13.27}$$

We can calculate [18]

$$\delta^{\mathrm{GS}} = \frac{1}{192\pi^2\sqrt{k_A}}\operatorname{tr} q_A. \tag{13.28}$$

It generates a Fayet–Ilipoulos term in the Lagrangian, so that the resulting $D$-term potential is

$$\frac{1}{S + S^*} \left| \delta^{GS} \frac{1}{(S + S^*)^2} + \sum q_\alpha^A |q_\alpha|^2 \right|^2. \tag{13.29}$$

The effect is the same for every U(1) gauge group. There is only one anomalous U(1) in the following sense, as Eq. (13.19) shows only one such pseudoscalar field. In the orbifold compactification, in general we get many U(1)s and more than one U(1)s can have nonvanishing anomalies. We may always redefine their charges by linear combinations without changing the normalization: $q'_A = \alpha q_A - \beta q_B$, $q'_B = \beta q_A + \alpha q_B$ with $\alpha^2 + \beta^2 = 1$. All the anomalous charges $q_A$ for each U(1)$_A$ obey the following relations:

$$\frac{1}{k_A} \text{tr}_A \, l(\mathbf{R}) q_A = \frac{1}{3} \text{tr} \, q_A^3 = \frac{1}{24} \text{tr} \, q_A = 8\pi^2 \delta^{GS}, \tag{13.30}$$

where the trace $\text{tr}_A$ is over fundamental representation of $a$ and $l(\mathbf{R})$ is the index of the representation $\mathbf{R}$, defined in (2.16). This means, by linear combination [21] we should have universal relations

$$\frac{1}{3} \text{tr} \, q_A^3 = \text{tr} \, q_A q_B^2. \tag{13.31}$$

The hypercharge of the SM stays as gauge symmetry so it should not be anomalous. From the above universality, the hypercharge generator $q_Y$ should have no component along such anomalous U(1) generated by $q_A$, so we demand them to be orthogonal

$$q_Y \cdot q_A = 0.$$

## 13.4 Three Families

We briefly discuss the potential origin of three families. String theory enables us to understand the number of families and their flavor structure in terms of the internal geometry.

### 13.4.1 The Number of Fixed Points

Untwisted sector fields carry spacetime index, thus are labelled as $U_1$, $U_2$, and $U_3$. The choice of orbifold relates some of them. For instance, $\mathbb{Z}_3$ orbifold always gives three repeated matter fields. However, we have a problem in quark masses that we discuss shortly.

In $\mathbb{Z}_3$ orbifold models, the chiral fermions arise naturally in multiples of three if we use two Wilson lines, because we have equal field contents in three set of fixed points. Other orbifolds having such order three substructure have the same feature. It was argued long ago that three families might arise from $\mathbb{Z}_6 = \mathbb{Z}_3 \times \mathbb{Z}_2$ orbifolds [12]. Also, a $\mathbb{Z}_{12-I}$ orbifold model, we may have the same Wilson lines in three twist sectors. If Wilson lines are to not break such degeneracy, we need a condition

$$P \cdot a = \text{same for all the } \mathbb{Z}_3 \text{ subsectors.} \tag{13.32}$$

We observe that if three families of the quark doublets $(\mathbf{3}, \mathbf{2})$ appear in the untwisted sector and its hypercharge is correctly given, then the resulting spectrum will lead to three generations. It is because global consistency condition, which is stronger than anomaly cancellation, restricts the spectrum to have as many quark singlets. Indeed, this was shown in cases with three Wilson line models of Refs. [14, 22].

There is a potential problem, however, on the mass spectrum with all the families from the same (un)twisted sector [11, 23]. If the quark singlets also appear from the untwisted sector, then we have undesirable relation $m_c = m_t$ [24], not overcoming the flavor problem listed in the beginning of this chapter. It is better for the quark doublets to appear from twisted sectors. Usually it is difficult to break the degeneracy. Therefore, for fitting the observed flavor data, it is necessary for the model to forbid a degeneracy.

If all three families appear from different twisted sectors in three Wilson line models, it will be of great interest because any degeneracy among families can be avoided. For example, it was pointed out in Eq. (10.97) that the Yukawa couplings can be exponentially suppressed, $e^{-l_{ij}}$ where $l_{ij}$ is proportional to the distance between the fixed points $f_i$ and $f_j$ housing two respective chiral fermions. This can be a geometrical understanding of the quark mass hierarchy [23].[3]

There is another interesting option. If we have two generations in twisted sectors and one generation in the untwisted sector, we may understand the nature of the third generation [27–29]. If we have a $\mathbb{Z}_2$ symmetry subgroup of the point group, then it is natural to have a doublet.

Since the untwisted sector matter fields are branched from gauginos of ten dimensional non-Abelian gauge field, we may have natural unification between the top quark Yukawa coupling and the unified gauge coupling [30].

### 13.4.2 Number of Internal Dimensions

We may also relate the number of internal dimensions with the number of families [23]

$$(\text{the number of families}) = (\text{the number of internal dimensions})/2 = 3.$$

---

[3]In intersecting brane models also, the geometrical interpretation can be considered [25, 26].

Each untwisted matter field carries the Lorentz index for the internal space, so the number of complex internal dimensions is related to the multiplicity. Also, the $\mathbb{Z}_2 \times \mathbb{Z}_2$ orbifold has three twisted sectors, as shown relating the number of complex internal dimensions to the multiplicity. The point group is generated by two $\mathbb{Z}_2$ with the twists $(\frac{1}{2}, \frac{1}{2}, 0)$ and $(0, \frac{1}{2}, \frac{1}{2})$. We have in fact three twist sectors, because we also have a point group element $(\frac{1}{2}, 0, \frac{1}{2})$, that is the combination of the two. Each twisted sector has the same geometrical structure, so we naturally have multiplicity three. Models along this line are [28, 29, 31].

This idea has been put forward by stringy flipped SU(5) model with so-called the Antoniadis–Ellis–Hagelin–Nanopoulos (NAHE) set and its follow-ups. They are standard-like models where the fermionic construction of 4D string was employed [32–36], which will be briefly reviewed in Sect. 17.1. As noted there, we have a set $\mathbb{Z}_2 \times \mathbb{Z}_2$ boundary conditions, parameterized by the twist vectors $b_1, b_2$ and we also have the combined twisted sector $b_3 = b_1 + b_2$ [37]. Although it is difficult to visualize the symmetry geometrically, the construction results in three twisted sectors with the same structures [38]. Of course, to have more realistic model, we need several more boundary conditions that does not ruin the $\mathbb{Z}_2 \times \mathbb{Z}_2$ symmetry. It must be supplied with other conditions like (13.32).

### 13.4.3 Family Symmetry

In Eq. (13.6), if we do not introduce $a_1^I$, then we obtain a GUT group SU(5). If there is non-adjoint Higgs field mimicking $a_1^I$ for breaking SU(5), then that will do the job of obtaining the factor SU(3)×SU(2). In the SU(5) space, it introduces 4 and 5 indices together, and hence a fundamental representation[4] with 4 and 5 indices will do the job. A GUT not needing an adjoint representation for the GUT breaking to the SM is called anti-SU($N$) [41]. Anti-SU(5) [42] is usually called "flipped SU(5)" [43, 44].

A similar situation is encountered in the family unification GUT with skewing of the SO($4n + 2$) group space. Since it is intuitive to understand why these kinds of skewing lead inevitably to queerly charged particles, let us consider the spinor representation **64** of SO(14) [39]. When SO(14) breaks down to SU(5), the branching rule of **64** is $2 \cdot (\mathbf{10} + \bar{\mathbf{5}} + \mathbf{1} + \overline{\mathbf{10}} + \mathbf{5} + \bar{\mathbf{1}})$, i.e. there are two families of SU(5) but with a vectorlike form, so no chiral family results at low energy. In terms of the SM quantum numbers, these vectorlike representations are

$$\begin{pmatrix} v_e \\ e \end{pmatrix}_L, \quad \begin{pmatrix} u \\ d \end{pmatrix}_L, \quad \begin{pmatrix} v_\mu \\ \mu \end{pmatrix}_L, \quad \begin{pmatrix} c \\ s \end{pmatrix}_L,$$

$$\begin{pmatrix} v_e' \\ e' \end{pmatrix}_R, \quad \begin{pmatrix} u' \\ d' \end{pmatrix}_R, \quad \begin{pmatrix} v_\mu' \\ \mu' \end{pmatrix}_R, \quad \begin{pmatrix} c' \\ s' \end{pmatrix}_R,$$

$$(13.33)$$

---

[4]An extended definition of fundamental can be completely antisymmetric representation, $\Phi^{[\alpha\beta\cdots]}$ where $\alpha, \beta, \cdots$ are SU(N) indices [39, 40].

where we have not shown 32 SU(2)-singlet charge-conjugated fields with the same chiralities($L$ for the charge-conjugated unprimed singlets and $R$ for the primed ones). The aforementioned skewing raises the electromagnetic charges by one unit for the right-handed $e'$ family and lowers them by one unit for the right-handed $\mu'$ family, which is allowed in the SO(14) model. These shifts give [39,40],

$$
\begin{pmatrix} v_e \\ e \end{pmatrix}_L , \quad \begin{pmatrix} u \\ d \end{pmatrix}_L , \quad \begin{pmatrix} v_\mu \\ \mu \end{pmatrix}_L , \quad \begin{pmatrix} c \\ s \end{pmatrix}_L ,
$$

$$
\begin{pmatrix} \tau^+ \\ \bar{v}_\tau \end{pmatrix}_R , \quad \begin{pmatrix} q_{5/3} \\ t \end{pmatrix}_R , \quad \begin{pmatrix} E^- \\ E^{--} \end{pmatrix}_R , \quad \begin{pmatrix} b \\ q_{-4/3} \end{pmatrix}_R .
$$

(13.34)

The model (13.34) gives three standard lepton families and two standard quark families, with the rest being queer states. In particular, $t$ and $b$, even though their electromagnetic charges coincide with those in the SM, do not belong to the same doublet, which was proven wrong by the decay modes of $b$ in models with $t$ and $b$ in different doublets [45]. Also, there appear queerly charged particles, $E^{--}$, $q_{5/3}$, and $q_{-4/3}$.

As in the above skewed model, in most standard-like models from string compactification, the appearance of queer particles is ubiquitous, which is one of the reasons these models are *not called standard but called standard-like*.

## 13.5  Discrete Symmetries

### 13.5.1  Global and Discrete Symmetries

There is a belief that there does not exist global symmetries in string theory. It is not conserved by Planck scale suppressed operators from gravitational interactions, including the black hole formation and evaporation and wormhole effects [46]. Because a black hole has "no-hair" it cannot carry a global charge. Therefore, the global charge of a field thrown into a black hole is not conserved. Most of all, the remarkable prediction from string theory is that there is no global symmetry except that corresponding to the model-independent axion [47]. Any global symmetry introduced in the worldsheet is lifted to a gauge symmetry. It is because in the target space, this symmetry is coordinate ($X$) dependent and becomes the local symmetry as discussed in Chap. 17.

In contrast to global symmetries, discrete "gauge" symmetries are possible in string theory as a subgroup of continuous symmetries of string theory. Such discrete charge survives well below the Planck scale. In supersymmetry, the $R$ parity is $(-1)^{3B-L}$, and is sometimes traced back to a part of gauge symmetry such as Pati–Salam group or trinification or SO(10). If so, then it would be a good quantum number, too.

There are "fundamental" discrete symmetries in quantum field theory. The parity $\mathscr{P}$ is the inversion of spatial direction. It exchanges helicities of fermions. From discussions below Eq. (1.6), $\mathscr{P}$ is not a good symmetry, and the resulting theory is called a *chiral theory*. As can be deduced from the definition, Eq. (4.22) and below, the charge conjugation operation $\mathscr{C}$ conjugates the gauge charges. Still $\mathscr{C}$ is not a good symmetry in the SM. Also, time reversal $\mathscr{T}$ is not a good symmetry, but there is the renowned $\mathscr{C}\mathscr{P}\mathscr{T}$ theorem that any local, Lorentz invariant quantum field theory preserves the product $\mathscr{C}\mathscr{P}\mathscr{T}$.

The discrete symmetries $\mathscr{C}$, $\mathscr{P}$, and $\mathscr{T}$ can be a part of *gauge symmetry*, or Lorentz symmetry in higher dimensional theory [48]. It is possible if the spacetime dimensions are $8k + 1$, $8k + 2$, and $8k + 3$ so that Majorana fermions and invariant gauge groups such as $E_8$ or $SO(4n)$ are allowed. These naturally arise in the string theory framework. In terms of weights $P$, charge conjugation operation $\mathscr{C}$ amounts to changing $P^I \to -P^I$. Consequently, in the bosonic description, it has the form

$$X^I \to -X^I, \quad I = 1, 2, \ldots, 16. \tag{13.35}$$

This is the property of spacelike (extended version of four dimensional) $\mathscr{P}$ operation

$$\begin{aligned} X^M &\to X^M, \quad M = 0; 4, 6, 8, \\ X^M &\to -X^M, \quad M = 1, 2, 3; 5, 7, 9. \end{aligned} \tag{13.36}$$

Note that the 4D part($M = 0, 1, 2, 3$) is the familiar parity operation. The above $\mathscr{C}$ and $\mathscr{P}$ operations are the elements of proper Lorenz transformation $SO(1,9) \times SO(16)$, thus they are gauge symmetries. This means that the symmetries are *exact* and survives even if nonperturbative and gravitational interactions are considered. Therefore, its breaking (for example, the $\mathscr{C}\mathscr{P}$ violation) must be achieved spontaneously if introduced at the 10D heterotic string level. We cannot do a similar reasoning for $\mathscr{T}$, but at least perturbatively it is a good symmetry in string theory.

Besides $\mathscr{P}$ being four dimensional inversion in $M = 1, 2, 3$ directions, we can freely choose the inversions in extra dimensions, as long as the transformation is proper. However, taking fermions into account, it turns out that the choice of (13.36) is unique, otherwise it does not commutes with the GSO projection. In terms of complexified coordinates, it corresponds to complex conjugation.

## 13.5.2 R-Parity

Baryon ($B$) and lepton ($L$) numbers, being global, are broken. The degree of breaking depends on compactification models. We start with $B$ and $L$ conserving dimension-4 operators, which has led to the R-party conservation. In standard-like models from string, a vacuum with R-parity was explicitly shown to exist first in

**Fig. 13.2** A diagram for
$\Delta B \neq 0$ without R parity.
The cubic couplings in this
diagram break R-parity

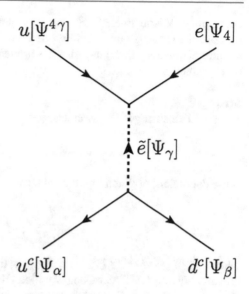

$u[\Psi^{4\gamma}]$          $e[\Psi_4]$

$\tilde{e}[\Psi_\gamma]$

$u^c[\Psi_\alpha]$          $d^c[\Psi_\beta]$

[49]. The standard R-parity or $\mathbb{Z}_{2R}$, however, have been known to be dangerous for the proton longevity due to the dimension-5 operators [3, 50]. Without R parity, a dangerous dimension-5 operator appears as shown in Fig. 13.2 [51].

Without R-parity, forbidding dimension-5 B vioalting operators involves considering all standard-like model singlets which can obtain GUT scale VEVs in principle. Therefore, it will be economic if the model contains some kind of R-parity. Dimension-5 $B$ violating operators and the $\mu$ term are required to be suppressed but dimension-5 $L$ violating Weinberg operator needs to be allowed [52]. The relevant dimension-5 proton decay and neutrino mass operators are

$$W^{\Delta B} \equiv qqq\ell, \tag{13.37}$$

$$W^{\nu \, \text{mass}} \equiv \ell\ell H_u H_u, \tag{13.38}$$

where the subscripts $q$, $\ell$, and $H_u$ denote matter field doublets and up-type Higgs doublet, respectively.

Separately from these, the $\mu$-problem must be resolved [53, 54].

### 13.5.3 $\mathscr{CP}$ Violation

Whatever four dimensional $\mathscr{CP}$ is given from a string theory, its observed violation can be realized as the choice of vacuum in the process of compactification. The symmetry (13.36) is not preserved in the presence of background field as in (10.103), such as antisymmetric tensor field $B$ or $T$ modulus. Generically, it can be block-diagonalized and one can easily verify that it is not preserved under (13.36). Showing its dependence on Wilson line $A$ is nontrivial. It turned out [55] that, by continuous Wilson lines, $\mathscr{C}\mathscr{P}$ is not violated in the standard embedding case, because the space group selection rule associates the gauge group action to make the interaction term invariant. Also it is argued that discrete Wilson lines can violate that symmetry.

**Strong $\mathscr{C}\mathscr{P}$**

The effective strong $\mathscr{C}\mathscr{P}$ parameter

$$\bar{\theta} = \theta_{\mathrm{QCD}} + \arg \det M_{\mathrm{quark}} \tag{13.39}$$

is the coefficient of gluon topological term

$$\frac{\bar{\theta}}{32\pi^2} F^a_{\mu\nu} \tilde{F}^{a\,\mu\nu}, \tag{13.40}$$

where $\tilde{F}^{a\,\mu\nu} = \frac{1}{2}\epsilon^{\mu\nu\rho\sigma} F^a_{\rho\sigma}$, and $\theta_{\mathrm{QCD}}$ is the parameter introduced at the scale, presumably at a GUT scale, below which QCD becomes an exact confining gauge symmetry. $M_{\mathrm{quark}}$ is the quark mass matrix(including heavy quarks) needed for describing the electroweak $\mathscr{C}\mathscr{P}$ violation. The observed absence of neutron electric dipole moment [56] gives a bound on $\bar{\theta}$

$$|\bar{\theta}| < 10^{-10} \tag{13.41}$$

which is the basis for questioning the strong $\mathscr{C}\mathscr{P}$ problem, "Why is the observed value of $\bar{\theta}$ so small?" [57] The true indication of the strong $\mathscr{C}\mathscr{P}$ solution can be accessed only after a realistic MSSM from superstring is found.

Here at field theory level, we list the following three types [58] for solutions of the strong $\mathscr{C}\mathscr{P}$ problem, which need some kind of symmetry. Indeed, all of these are implementable in superstring.

- **The axion solution:** It has been already commented briefly in Sect. 13.3.3 and will be discussed more in Sect. 13.6.3. The so-called invisible axion solution needs a very light axion [59] derivable from the Peccei–Quinn global symmetry [60] at an intermediate scale.
- **The massless up-quark solution:** The massless up-quark solution can be addressed in orbifold compactification, just by observing $\det . M_{\mathrm{quark}} = 0$ from the spectrum obtained by orbifold compactification [24]. The problem is whether this leads to a phenomenologically viable mass matrix or not [61–63]. At the 4D field theory level, this solution also has an axial global U(1) symmetry for the phase shift of the massless up-quark, which is unbroken however.
- **The set-$\theta$-zero solution:** The set-$\theta$-zero solution sets $\theta_{QCD} = 0$ at the tree level, presumably by a symmetry, and requires that loop corrections to $\theta_{QCD}$ are sufficiently small. As we have done in the orbifold compactification, discrete symmetries can be introduced in string theory. So in principle, the Nelson–Barr type solution with the discrete $\mathscr{C}\mathscr{P}$ symmetry is realizable in string theory.

Among the set-$\theta$-zero solutions, the Nelson–Barr type solutions [64,65] attracted some attention recently. The Nelson–Barr type solution needs heavy vectorlike quarks. At this heavy scale, $\mathscr{CP}$ is spontaneously broken and the source for the weak $\mathscr{CP}$ violation is introduced this high scale. Below this spontaneous $\mathscr{CP}$ violation scale which is above the electroweak scale, the Yukawa couplings become complex due to the introduction of $\mathscr{CP}$ violation. The massless up-quark solution makes the vacuum angle unphysical. The axion solution chooses $\bar{\theta} = 0$, via the axion potential. Since the weak $\mathscr{CP}$ violation seems to prefer the Kobayashi–Maskawa type weak $\mathscr{CP}$ violation, then the Nelson–Barr type, massless up-quark and the axion solutions are the allowable ones. But, the spontaneous $\mathscr{CP}$ violation at the electroweak scale is not favored.

The set-$\theta$-zero solution needs a discrete $\mathscr{CP}$ symmetry so that the $\theta_{\text{QCD}}$ of QCD is vanishing at the bare Lagrangian level. Assuming a $\mathscr{CP}$ invariant Lagrangian, the discrete $\mathscr{CP}$ symmetry chooses the weak $\mathscr{CP}$ violation as a spontaneously broken one. Here, the spontaneous $\mathscr{CP}$ violation is not at the string scale but at a much smaller scale. Then, $\theta_{\text{QFD}}$ is calculable below the scale of the spontaneous $\mathscr{CP}$ violation, and is required in order to have the limit (13.41). Both the massless up quark solution and the axion solution need a global symmetry $U(1)_{\text{PQ}}$ where the divergence of the corresponding current, which is proportional to (13.40) is nonvanishing.

Another method to gain a strong $\mathscr{CP}$ solution in string models is an accidental PQ symmetry [66]. There can be discrete symmetries in string models which may allow a PQ symmetry up to some level, for example, up to dimension 9 operators in the superpotential. Then the $\bar{\theta}$ parameter is within the limit $10^{-10}$. This model is similar to the Nelson–Barr type in that discrete symmetries are used. String constructions of these types of solutions may contain both features.

## Electroweak $\mathscr{CP}$

Electroweak $\mathscr{CP}$ violation is implemented by complex phases in the Yukawa couplings, which are not invariant under $\mathscr{CP}$. The CKM type complex Yukawa couplings can introduce the needed electroweak $\mathscr{CP}$. In general texture, there is a basis independent $\mathscr{CP}$ measure, called the Jarlskog invariant [67,68]

$$J \propto \text{Im} \det[Y^u Y^{u\dagger}, Y^d Y^{d\dagger}], \tag{13.42}$$

or more succinctly just by looking at the CKM matrix $V$ for $\det V = 1$ [69],

$$J = |\text{Im} \, V_{31} V_{22} V_{13}|. \tag{13.43}$$

String models are constructed in 10D($8k + 2 = 10$) and hence they provide the possibility that the elecroweak $\mathscr{CP}$ can be a discrete gauge symmetry [70] where the covering continuous symmetry is the 10D Lorentz symmetry times the 10D gauge symmetry. Or it can appear as a subgroup of continuous symmetries of string theory. Whatever four dimensional $\mathscr{CP}$ is given from a 10D theory, the observed electroweak $\mathscr{CP}$ violation can be realized as the choice of vacuum in the process of compactification of 10D string models. So the phenomenological consideration of the elecroweak $\mathscr{C}\mathscr{P}$ violation reduces to the study of Yukawa couplings and the flavor problem from string models. The above KM type complex Yukawa couplings can introduce the needed electroweak $\mathscr{C}\mathscr{P}$.

The Nelson–Barr type introduction of electroweak $\mathscr{C}\mathscr{P}$ violation is possible with this discrete gauge symmetry spontaneously broken below $10^{10}$ GeV so that it is consistent with the $\mathscr{C}\mathscr{P}$ violation in the kaon system [48]. But one drawback is that one introduces another small mass parameter here.

## 13.6 "Invisible" Axion from String

### 13.6.1 't Hooft Mechanism

The 't Hooft mechanism states, "If a gauge symmetry and a global symmetry are broken by one complex scalar field through the Higgs mechanism, then the gauge symmetry is broken and a global symmetry remains unbroken" [71, 72]. It is obvious that the gauge symmetry is broken because the gauge boson obtains mass by the Higgs mechanism. The original continuous symmetry $U(1)_{\text{gauge}} \times U(1)_{\text{global}}$ introduces two continuous parameters $\alpha(x)$ and $\beta$ for tht symmetry transformations on the fields,

$$\phi \to e^{i\alpha(x)Q_{\text{gauge}} + i\beta Q_{\text{global}}} \phi, \tag{13.44}$$

where $Q_{\text{gauge}}$ and $Q_{\text{global}}$ are the generators of the transformations. Redefining the local parameter as $\alpha'(x) = \alpha(x) + \beta$, the generator for the surviving global symmetry is

$$Q'_{\text{global}} = Q_{\text{global}} - Q_{\text{gauge}}. \tag{13.45}$$

If there is no gauge symmetry to be broken by a VEV of the scalar VEV, then a global symmetry is broken by the VEV. In the compactification of 10D string models, there appear many $U(1)$ gauge symmetries as we will show in the next chapter. If compactification introduces $U(1)^n$ gauge symmetries at the GUT scale, then we need at least $n$ independent scalar VEVs before considering breakdown of a global symmetry. This is fundamental making a global symmetry survive below the GUT scale. In this regard, note that if compactification introduces an anomalous $U(1)$ gauge symmetry discussed in this section, then the gauge symmetry $U(1)_{\text{anom}}$ is broken. Was there a global symmetry in string compactification? The needed continuous parameter direction in string compactification is $a_{\text{anom}}$, which serves as the longitudinal degree of the anomalous gauge boson. What happened to the original global direction $a_{\text{anom}}$? It survives as a global symmetry by the 't Hooft mechanism, first used in [14] for an "invisible" axion from string.

In the literature, the Fayet–Iliopoulos terms (FI-term) for $U(1)_{\text{anom}}$ have been discussed extensively. In the hierarchical scheme, the VEVs of scalars are assumed to be at the GUT scale, somewhat smaller than the string scale. Even if one adds the FI-term for U(1)$_{anom}$, $|\phi^* Q^a \phi - \xi|^2$ with $\xi \ll M_{string}^2$, it breaks a gauge symmetry U(1)$_{anom}$ and a global symmetry, not just the global symmetry alone. So, the 't Hooft mechanism works here also. Then, one can consider the global symmetry U(1)$_{anom}$, surviving down from string compactification.

If U(1)$_{anom}$ survives as a global symmetry by the 't Hooft mechanism, the surviving global charge generator is given in Eq. (13.45). The matter fields in string compactification do not carry a charge corresponding to $H_{\mu\nu\rho}$. Thus, the surviving global charge or the PQ charge is negative of the original U(1)$_{anom}$ charge which belonged to E$_8$ × E$_8$. The PQ charge example is given, for example, in [73].

### 13.6.2 Domain Wall Number of "Invisible" Axion

So, it is definite to list the PQ quantum numbers for the "invisible" axion from string. Cosmological effects of "invisible" axion are to the cold dark matter energy density and domain walls in the Universe. If the "invisible" axion is not tuned to satisfy these cosmological constraints, then the model does not work for our Universe and must be discarded.

It has been noted that the domain wall number $N_{DW}$ in axion models must be effectively 1 [75]. There are two mechanisms to fulfil $N_{DW}$ effectively 1 [76, 77]. The strategy with the Goldstone boson direction was discussed in [78], for which the global U(1)$_{anom}$ in string compactification is repeated in Fig. 13.3a. Since $N_{DW} = 1$ in the MI-axion direction [79] ($\alpha_1$ in Fig. 13.3a), the red dash arrow direction identifies all vacua. In Fig. 13.3b, it is re-drawn on the familiar torus. The red arrows show that $\alpha_2$ shifts by one unit for one unit shift of $\alpha_1$. In this case, all the vacua are identified and we obtain effectively $N_{DW} = 1$. In Fig. 13.3b, the green lines show that $\alpha_2$ shifts by two units for one unit shift of $\alpha_1$. If $N_2$ is even, then we obtain $N_{DW} = 2$ since only halves of $N_2$ are identified by green lines.

To find out $N_{DW}$ from compactification, we consider all the VEVs of scalar fields needed for generating the SM gauge group. In this case, we consider the maximum number, dividing all the U(1)$_{anom}$ quantum numbers of the scalar fields, as $N_{DW}$.

### 13.6.3 String Perspective

In Sect. 13.3.3, the anomaly cancellation by the Green–Schwarz mechanism has been discussed. The QCD axion from the PQ symmetry has its root on the global anomaly with QCD gauge bosons U(1)$_{PQ}$-SU(3)$_c$-SU(3)$_c$. Therefore, the potential for the QCD axion derives from the anomalous term.

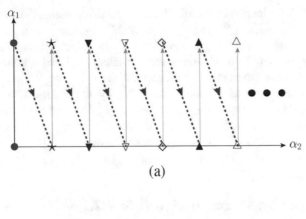

(a)

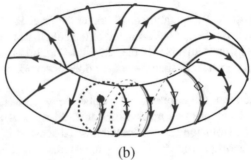

(b)

**Fig. 13.3** The MI-axion example of [74]: (**a**) the standard torus identification, and (**b**) identification by the winding direction in the torus

We have the model-independent axion, by taking the dual of the second rank antisymmetric tensor in the transverse direction $B_{\mu\nu}$, as (Eq. (13.19)). Taking the divergence, we obtain

$$\partial^2 a_{MI} = \frac{1}{32\pi^2 F_1}\left(\mathrm{tr}R_{\mu\nu}\tilde{R}^{\mu\nu} - \frac{1}{30}\mathrm{Tr}F_{\mu\nu}\tilde{F}^{\mu\nu}\right), \qquad (13.46)$$

which shows that $a_{MI}$ is an axion with the decay constant $F_1$. $F_1$ has been estimated as [16],

$$F_1 \simeq \frac{\tilde{g}^2}{192\pi^{5/2}}M_P \sim 1.5 \times 10^{15}\,\mathrm{GeV}, \qquad (13.47)$$

which is too large compared to the allowed axion window, $10^{10}$–$10^{12}$ GeV [57].

In (11.55), we have seen that the imaginary part of $T$ modulus have the axionic shift symmetry and hence more additional pseudoscalars can appear depending on the compactification scheme. They are four dimensional scalars and zero modes of $B_{ij}$. In generic Calabi–Yau compactification, the second Hodge number counts these massless modes as in (15.32)

$$h_{0,2} = h_{2,0} = 0 \, , \quad h_{1,1} \geq 1, \qquad (13.48)$$

where the inequality is valid for all Kähler manifold due to the Kähler two from. Then, the coupling (13.40) is possible due to the Green–Schwarz term. These pseudoscalars are called the *model-dependent axion* [79]. For $h_{0,2} = h_{2,0} = 0$, and $h_{1,1} = 1$, there exists one more axion $a_{MD}$, and the axion couplings in a Calabi–Yau compactification are calculated as [80]

$$\frac{1}{32\pi^2 F_1} a_{MI}(F\tilde{F} + F'\tilde{F}') + \frac{1}{32\pi^2 F_2} a_{MD}(F\tilde{F} - F'\tilde{F}'), \qquad (13.49)$$

where $F_2$ is the decay constant of the model-dependent axion. It is again at the compactification scale. Axions in M-theory have been discussed in Ref. [81].

Nevertheless, it has been argued that even if the model-dependent axions are introduced, the worldsheet instantons would make them massive by generating a superpotential violating the shift symmetry of the model-dependent axions [47, 82].

The axion solution seems to have a problem because there is no global symmetry except that corresponding to the model-independent axion $H_{\mu\nu\rho} \sim \partial_\mu B_{\nu\rho}$ [47, 82]. The problem with the model-independent axion is that its decay constant is at a GUT scale [16]. But, as discussed in the 't Hooft mechanism, one global symmetry U(1)$_\Gamma$ can survive down to low energy in models with the anomalous U(1)$_X$ [14, 22]. This U(1)$_X$ can be used also for the massless up-quark solution.

Sometimes, accidental global symmetries can appear from string compactification, leading to light pseudoscalars. In the literature, these have been discussed extensively under the name of axion-like particles (ALPs) [83, 84].

## 13.7 Phenomenology on Electroweak $\mathscr{C}\mathscr{P}$

The main ingredients for Yukawa couplings from string compactification are string selection rules and their moduli dependence. Therefore, it is crucial to know how such moduli are stabilized to have nontrivial complex phases.

Although highly model dependent, many properties are known for renormalizable interactions having only the $T$ modulus dependence, thanks to the target space modular properties (11.49). In Sect. 10.5.3 we discussed the structure of Yukawa couplings and will summarize it again in Sect. 16.3. In the simplest $\mathbb{Z}_3$ orbifold, at the renormalizable level the mass matrix is always diagonal by the selection rules, therefore we cannot expect nontrivial phases. By the standard procedure, a number of phases can be absorbed by redefining fermion fields

$$Y_u \to U^\dagger Y^u V^u, \quad Y_d \to U^\dagger Y^d V^d, \qquad (13.50)$$

where matrices $U$, $V^{u,d}$ are unitary matrices. From Table 10.2, we observe that Yukawa couplings have a similar structure for prime orbifolds and for some non-prime orbifolds. In non-prime orbifolds, nontrivial $\mathscr{CP}$ phase can emerge at the renormalizable level [85]. In the simplest case, where the dependence is only from $T$ moduli, there is always vanishing $\mathscr{CP}$ phase when $\operatorname{Im} T = \pm\frac{1}{2}$ as a consequence axionic symmetry in (11.55). In general, we expect that the boundary of fundamental region has such properties, if the Jarlskog parameter is invariant under the moduli transformation, but the proof is tricky [85]. This interval includes fixed point at $T = e^{i\pi/6}$ of PSL(2,$\mathbb{Z}$) action (11.49) which can be likely the extremum of the potential of the $T$ modulus. An analysis was done [86] by assuming empirical form for a superpotential having the desired modular properties.

Of course, the possibility of nonrenormalizable interactions are always open, whose strength can still be sizable. In general, a scalar can acquire a VEV with a nontrivial phase. It is also related to supersymmetry breaking and the above discussion could not take into account supersymmetry breaking effects. There is a scenario proposing that some charged scalars acquire VEVs and $F$ terms have complex phases from the FI-term (13.29) generated by the GS mechanism [87].

The ultimate goal for electroweak phenomenology from string is to obtain the CKM matrix shown in Eq. (2.11) from discrete symmetries allowed in the compactification.

▶ **Exercise 13.1** Show that the index of fundamental representation of SU($N$) is $\frac{1}{2}$.

## 13.8    The $\mu$-Problem

As discussed in Chap. 2, we introduce two Higgs doublets, $H_u$ ($Y = +\frac{1}{2}$) and $H_d$ ($Y = -\frac{1}{2}$), in supersymmetric standard models. Therefore, we can introduce a dimension 2 term $-\mu H_u H_d$ in the superpotential. For the electroweak symmetry breaking at the TeV scale, $\mu$ needs to be as small as the electroweak scale. The $\mu$-problem [53] is, "Why is $\mu$ so small even though the SUSY $\mu$ term is allowed at the GUT scale?" In the standard-like models from superstring, there are two issues related to the $\mu$-problem. How come only one pair of Higgs doublets, out of several Higgs doublets from string compactification, survives down to the electroweak scale? It has been argued that some kind of symmetry is needed for this [88]. One such symmetry is the PQ symmetry such that the PQ charge of $H_u H_d$ is not vanishing. With a SM singlet(s) $S_i$, if the PQ charge of $S_i S_j$ is the opposite that of $H_u H_d$, then a nonrenormalizable superpotential $\frac{1}{M_P} H_u H_d S_i S_j$ can generate the needed electroweak scale $\mu$ by the intermediate scale VEV of $S_i S_j$ which is the so-called Kim–Nilles mechanism [53]. So, in string compactification we attempt to answer the above two questions.

One can also obtain the electroweak scale $\mu$ from the Kähler potential [54]. Indeed, such a scenario was shown to be realizable from $\mathbb{Z}_3$ orbifold [89]. With $\mathbb{Z}_3$ orbifolds, the bare mass term $H_u H_d$ is not allowed because $UU$ (with untwisted sectors fields $U$) and $TT$ (with twisted sectors fields $T$) are not allowed as discussed in Chap. 8.[5] But, among the $UUU$ and $TTT$ type couplings, one $U$ or $T$ can develop a GUT scale VEV, $\langle U \rangle UU$, and $\langle T \rangle TT$, in which case $\mu$ is of order the GUT scale. Assuming that the Planck scale Higgsino mass term is forbidden in the superpotential, then the $\mu$ term is generated if the Kähler potential mixes Higgs doublets with neutral scalars which acquire the GUT scale VEVs [89], which is a Giudice–Masiero mechanism.

Another attempt for the electroweak scale $\mu$ term is by strong dynamics at the intermediate scale, as suggested in [22].

# References

1. Y. Katsuki, Y. Kawamura, T. Kobayashi, N. Ohtsubo, Y. Ono, K. Tanioka, *Tables of Z(N) Orbifold Models*, vol. 5 (1989)
2. L.J. Hall, J.D. Lykken, S. Weinberg, Supergravity as the messenger of supersymmetry breaking. Phys. Rev. **D27**, 2359–2378 (1983)
3. N. Sakai, T. Yanagida, Proton decay in a class of supersymmetric grand unified models. Nucl. Phys. **B197**, 533 (1982)
4. P.H. Ginsparg, Gauge and gravitational couplings in four-dimensional string theories. Phys. Lett. **B197**, 139–143 (1987)
5. V.S. Kaplunovsky, mass scales of the string unification. Phys. Rev. Lett. **55**, 1036 (1985)
6. U. Amaldi, W. de Boer, H. Furstenau, Comparison of grand unified theories with electroweak and strong coupling constants measured at LEP. Phys. Lett. **B260**, 447–455 (1991)
7. P. Langacker, M.-X. Luo, Implications of precision electroweak experiments for $M_t$, $\rho_0$, $\sin^2 \theta_W$ and grand unification. Phys. Rev. **D44**, 817–822 (1991)
8. C. Giunti, C.W. Kim, U.W. Lee, Running coupling constants and grand unification models. Mod. Phys. Lett. **A6**, 1745–1755 (1991)
9. H. Georgi, Helen R. Quinn, S. Weinberg, Hierarchy of interactions in unified gauge theories. Phys. Rev. Lett. **33**, 451–454 (1974)
10. L.E. Ibanez, J.E. Kim, H.P. Nilles, F. Quevedo, Orbifold compactifications with three families of $SU(3) \times SU(2) \times U(1)^n$. Phys. Lett. **B191**, 282–286 (1987)
11. J.A. Casas, C. Munoz, Three generation $SU(3) \times SU(2) \times U(1)_Y$ models from orbifolds. Phys. Lett. **B214**, 63–69 (1988)
12. L.E. Ibanez, J.Mas, H.-P. Nilles, F. Quevedo, Heterotic strings in symmetric and asymmetric orbifold backgrounds. Nucl. Phys. **B301**, 157–196 (1988)
13. J.A. Casas, E.K. Katehou, C. Munoz, U(1) charges in orbifolds: anomaly cancellation and phenomenological consequences. Nucl. Phys. **B317**, 171–186 (1989)
14. J.E. Kim, The strong CP problem in orbifold compactifications and an $SU(3) \times SU(2) \times U(1)^n$ model. Phys. Lett. **B207**, 434–440 (1988)
15. A. Font, L.E. Ibanez, F. Quevedo, A. Sierra, The construction of 'realistic' four-dimensional strings through orbifolds. Nucl. Phys. **B331**, 421–474 (1990)
16. K. Choi, J.E. Kim, Harmful axions in superstring models. Phys. Lett. **B154**, 393 (1985). [Erratum: Phys. Lett. **B156**, 452 (1985)]
17. E. Witten, Some properties of O(32) superstrings. Phys. Lett. **B149**, 351–356 (1984)
18. J.J. Atick, L.J. Dixon, A. Sen, String calculation of Fayet–Iliopoulos D-terms in arbitrary supersymmetric compactifications. Nucl. Phys. **B292**, 109–149 (1987)

---

[5]In $\mathbb{Z}_3$ orbifolds, it was shown in [90].

19. M. Dine, I. Ichinose, N. Seiberg, F terms and D-terms in string theory. Nucl. Phys. **B293**, 253–265 (1987)
20. M. Dine, N. Seiberg, X.G. Wen, E. Witten, Nonperturbative effects on the string world sheet. 2. Nucl. Phys. **B289**, 319–363 (1987)
21. T. Kobayashi, H. Nakano, 'Anomalous' U(1) symmetry in orbifold string models. Nucl. Phys. **B496**, 103–131 (1997)
22. E.J. Chun, J.E. Kim, H.P. Nilles, A Natural solution of the mu problem with a composite axion in the hidden sector. Nucl. Phys. **B370**, 105–122 (1992)
23. L.E. Ibanez, Hierarchy of quark-lepton masses in orbifold superstring compactification. Phys. Lett. **B181**, 269–272 (1986)
24. H.P. Nilles, M. Olechowski, S. Pokorski, Does a radiative generation of quark masses provide us with the correct mass matrices? Phys. Lett. **B248**, 378–386 (1990)
25. G. Aldazabal, S. Franco, L.E. Ibanez, R. Rabadan, A.M. Uranga, Intersecting brane worlds. J. High Energy Phys. **2**, 047 (2001)
26. G. Aldazabal, S. Franco, L.E. Ibanez, R. Rabadan, A.M. Uranga, $D = 4$ chiral string compactifications from intersecting branes. J. Math. Phys. **42**, 3103–3126 (2001)
27. T. Kobayashi, S. Raby, R.-J. Zhang, Constructing 5-D orbifold grand unified theories from heterotic strings. Phys. Lett. **B593**, 262–270 (2004)
28. S. Forste, H.P. Nilles, P.K.S. Vaudrevange, A. Wingerter, Heterotic brane world. Phys. Rev. **D70**, 106008 (2004)
29. W. Buchmuller, K. Hamaguchi, O. Lebedev, M. Ratz, Dual models of gauge unification in various dimensions. Nucl. Phys. **B712**, 139–156 (2005)
30. P. Hosteins, R. Kappl, M. Ratz, K. Schmidt-Hoberg, Gauge-top unification. J. High Energy Phys. **7**, 029 (2009)
31. M. Blaszczyk, S.G. Nibbelink, M. Ratz, F. Ruehle, M. Trapletti, P.K.S. Vaudrevange, A Z2 × Z2 standard model. Phys. Lett. **B683**, 340–348 (2010)
32. I. Antoniadis, J.R. Ellis, J.S. Hagelin, D.V. Nanopoulos, GUT model building with fermionic four-dimensional strings. Phys. Lett. **B205**, 459–465 (1988)
33. I. Antoniadis, J.R. Ellis, J.S. Hagelin, D.V. Nanopoulos, An improved SU(5) × U(1) model from four-dimensional string. Phys. Lett. **B208**, 209–215 (1988). [Addendum: Phys. Lett. **B213**, 562 (1988)]
34. A.E. Faraggi, D.V. Nanopoulos, K.-J. Yuan, A standard like model in the 4D free fermionic string formulation. Nucl. Phys. **B335**, 347–362 (1990)
35. A.E. Faraggi, A New standard-like model in the four-dimensional free fermionic string formulation. Phys. Lett. **B278**, 131–139 (1992)
36. A.E. Faraggi, Construction of realistic standard-like models in the free fermionic superstring formulation. Nucl. Phys. **B387**, 239–262 (1992)
37. A.E. Faraggi, Z(2) × Z(2) Orbifold compactification as the origin of realistic free fermionic models. Phys. Lett. **B326**, 62–68 (1994)
38. P. Athanasopoulos, A.E. Faraggi, S. Groot Nibbelink, V.M. Mehta, Heterotic free fermionic and symmetric toroidal orbifold models. J. High Energy Phys. **4**, 038 (2016).
39. J.E. Kim, A model of flavor unity. Phys. Rev. Lett. **45**, 1916 (1980)
40. J.E. Kim, Flavor unity in SU(7): low mass magnetic monopole, doubly charged lepton, and $Q = 5/3, -4/3$ quarks. Phys. Rev. **D23**, 2706 (1981)
41. J.E. Kim, Towards unity of families: anti-SU(7) from $Z_{12I}$ orbifold compactification. J. High Energy Phys. **6**, 114 (2015)
42. J.P. Derendinger, J.E. Kim, D.V. Nanopoulos, Anti-SU(5). Phys. Lett. **B139**, 170–176 (1984)
43. S. M. Barr, A new symmetry breaking pattern for SO(10) and proton decay. Phys. Lett. **B112**, 219–222 (1982)
44. I. Antoniadis, J. R. Ellis, J.S. Hagelin, D.V. Nanopoulos, The flipped SU(5) × U(1) string model revamped. Phys. Lett. **B231**, 65–74 (1989)
45. G.L. Kane, M.E. Peskin, A constraint from B decay on models with no T quark. Nucl. Phys. **B195**, 29–38 (1982)

46. S.B. Giddings, A. Strominger, Axion induced topology change in quantum gravity and string theory. Nucl. Phys. **B306**, 890–907 (1988)
47. X.G. Wen, E. Witten, World sheet instantons and the Peccei–Quinn symmetry. Phys. Lett. **B166**, 397–401 (1986)
48. K.-W. Choi, D.B. Kaplan, A.E. Nelson, Is CP a gauge symmetry? Nucl. Phys. **B391**, 515–530 (1993)
49. J.E. Kim, J.-H. Kim, B. Kyae, Superstring standard model from Z(12-I) orbifold compactification with and without exotics, and effective R-parity. J. High Energy Phys. **6**, 034 (2007)
50. S. Weinberg, Supersymmetry at ordinary energies. 1. Masses and conservation laws. Phys. Rev. **D26**, 287 (1982)
51. R. Dermisek, A. Mafi, S. Raby, SUSY GUTs under siege: proton decay. Phys. Rev. **D63**, 035001 (2001)
52. S. Weinberg, Baryon and lepton nonconserving processes. Phys. Rev. Lett. **43**, 1566–1570 (1979)
53. J.E. Kim, H.P. Nilles, The mu problem and the strong CP problem. Phys. Lett. **138B**, 150–154 (1984)
54. G.F. Giudice, A. Masiero, A natural solution to the mu problem in supergravity theories. Phys. Lett. **B206**, 480–484 (1988)
55. T. Kobayashi, O. Lebedev, Heterotic Yukawa couplings and continuous Wilson lines. Phys. Lett. **B566**, 164–170 (2003)
56. C.A. Baker et al., An Improved experimental limit on the electric dipole moment of the neutron. Phys. Rev. Lett. **97**, 131801 (2006)
57. J.E. Kim, G. Carosi, Axions and the strong CP problem. Rev. Mod. Phys. **82**, 557–602 (2010)
58. J.E. Kim, Light pseudoscalars, particle physics and cosmology. Phys. Rept. **150**, 1–177 (1987)
59. J.E. Kim, Weak interaction singlet and strong CP invariance. Phys. Rev. Lett. **43**, 103 (1979)
60. R.D. Peccei, H.R. Quinn, CP conservation in the presence of instantons. Phys. Rev. Lett. **38**, 1440–1443 (1977). [328 (1977)]
61. K. Choi, How precisely can one determine $\mu_u/\mu_d$? Nucl. Phys. **B383**, 58–72 (1992)
62. K. Choi, C.W. Kim, W.K. Sze, Mass renormalization by instantons and the strong CP problem. Phys. Rev. Lett. **61**, 794 (1988)
63. D.B. Kaplan, A.V. Manohar, Current mass ratios of the light quarks. Phys. Rev. Lett. **56**, 2004 (1986)
64. A.E. Nelson, Naturally weak CP violation. Phys. Lett. **B136**, 387–391 (1984)
65. S.M. Barr, Solving the strong CP problem without the Peccei–Quinn symmetry. Phys. Rev. Lett. **53**, 329 (1984)
66. G. Lazarides, C. Panagiotakopoulos, Q. Shafi, Phenomenology and cosmology with superstrings. Phys. Rev. Lett. **56**, 432 (1986)
67. C. Jarlskog, Commutator of the quark mass matrices in the standard electroweak model and a measure of maximal CP violation. Phys. Rev. Lett. **55**, 1039 (1985)
68. C. Jarlskog, A basis independent formulation of the connection between quark mass matrices, CP violation and experiment. Z. Phys. **C29**, 491–497 (1985)
69. J.E. Kim, D.Y. Mo, S. Nam, Final state interaction phases obtained by data from CP asymmetries. J. Korean Phys. Soc. **66**(6), 894–899 (2015)
70. L.M. Krauss, F. Wilczek, Discrete gauge symmetry in continuum theories. Phys. Rev. Lett. **62**, 1221 (1989)
71. G. 't Hooft, Renormalizable Lagrangians for massive Yang–Mills fields. Nucl. Phys. **B35**, 167–188 (1971). [201 (1971)]
72. J.E. Kim, Anomalous gauge-U(1), 't Hooft mechanism, and "invisible" QCD axion from string. PoS **CORFU2017**, 202 (2018)
73. J.E. Kim, B. Kyae, S. Nam, The anomalous $U(1)_{anom}$ symmetry and flavors from an SU(5) × SU(5)$'$ GUT in $\mathbf{Z}_{12-I}$ orbifold compactification. Eur. Phys. J. **C77**(12), 847 (2017).
74. J.E. Kim, S. Nam, Y.K. Semetzidis, Fate of global symmetries in the universe: QCD axion, quintessential axion and trans-Planckian inflaton decay-constant. Int. J. Mod. Phys. **A33**(3), 1830002 (2018)

75. P. Sikivie, Of axions, domain walls and the early universe. Phys. Rev. Lett. **48**, 1156–1159 (1982)
76. G. Lazarides, Q. Shafi, Axion models with no domain wall problem. Phys. Lett. **B115**, 21–25 (1982)
77. K. Choi, J.E. Kim, Domain walls in superstring models. Phys. Rev. Lett. **55**, 2637 (1985)
78. J.E. Kim, Axionic domain wall number related to U(1)$_{anom}$ global symmetry. Phys. Lett. **B759**, 58–63 (2016)
79. E. Witten, Cosmic superstrings. Phys. Lett. **B153**, 243–246 (1985)
80. K. Choi, J.E. Kim, Compactification and axions in E(8) × E(8)-prime superstring models. Phys. Lett. **B165**, 71–75 (1985)
81. K. Choi, Axions and the strong CP problem in M theory. Phys. Rev. **D56**, 6588–6600 (1997)
82. M. Dine, N. Seiberg, X.G. Wen, E. Witten, Nonperturbative effects on the string world sheet. Nucl. Phys. **B278**, 769–789 (1986)
83. M. Berg, J.P. Conlon, F. Day, N. Jennings, S. Krippendorf, A.J. Powell, M. Rummel, Constraints on axion-like particles from X-ray observations of NGC1275. Astrophys. J. **847**(2), 101 (2017)
84. A.G. Dias, A.C.B. Machado, C.C. Nishi, A. Ringwald, P. Vaudrevange, The quest for an intermediate-scale accidental axion and further ALPs. J. High Energy Phys. **6**, 37 (2014)
85. O. Lebedev, The CKM phase in heterotic orbifold models. Phys. Lett. **B521**, 71–78 (2001)
86. S. Khalil, O. Lebedev, S. Morris, CP violation and dilaton stabilization in heterotic string models. Phys. Rev. **D65**, 115014 (2002)
87. M.K. Gaillard, J. Giedt, More modular invariant anomalous U(1) breaking. Nucl. Phys. **B643**, 201–228 (2002)
88. J.E. Kim, H.P. Nilles, Symmetry principles toward solutions of the mu problem. Mod. Phys. Lett. **A9**, 3575–3584 (1994)
89. I. Antoniadis, E. Gava, K.S. Narain, T.R. Taylor, Effective mu term in superstring theory. Nucl. Phys. **B432**, 187–204 (1994)
90. J.A. Casas, C. Munoz, A Natural solution to the mu problem. Phys. Lett. **B306**, 288–294 (1993)

# String Unification

# 14

In this chapter, we discuss grand unified theories (GUTs) from the $E_8 \times E_8'$ heterotic string and present the resultant phenomenological implications. We take concrete example a flipped SU(5) model on $\mathbb{Z}_{12-I}$ orbifold. In Chaps. 7–12, we discussed in length the theoretical framework for orbifolded superstrings.

## 14.1 Requirements for GUTs

In Chap. 13, required viewpoints were presented for the various aspects of standard-like models from orbifold compactification. These are required also in GUTs from string compactification. Here, we list them again, adding more requirements by extending standard-like models to GUTs.

The requirement for three families is the same. GUTs differ from standard-like models basically in two aspects, one is a merit on charge quantization and the other is a shortcoming arising from spontaneous gauge symmetry breaking. In ordinary GUTs, the charge quantization, $Q_{em}(p) = -Q_{em}(e)$, has been considered to be a merit that cannot be false. Before the advent of gauge theories, the hypothesis on conserved vector current (CVC) was considered to be the basis for this charge quantization. Certainly, loop effects do not alter once the charge quantization relation is given. GUTs give this charge quantization relation by not requiring extra charged particles in the first family. But string compactification introduces exotics and the rationale for the charge quantization is not so strong. Nevertheless, there still exists the charge quantization in string compactification even with exotics, in the sense that exotics have definite fractional charges. For spontaneous gauge symmetry breaking, there appears the gauge hierarchy problem in GUTs.

- If a supersymmetric standard model(SSM) is obtained via the interim GUTs, we should pay attention to the requirements added for GUTs, especially on GUTs with $r \geq 5$ and the doublet–triplet splitting problem. If a SSM is obtained via

© Springer Nature Switzerland AG 2020
K.-S. Choi, J. E. Kim, *Quarks and Leptons From Orbifolded Superstring*,
Lecture Notes in Physics 954, https://doi.org/10.1007/978-3-030-54005-0_14

interim GUTs, it goes without saying that the model unifies gravity with all the elementary particle forces.

- When we consider GUTs, the colored scalar partner of Higgs doublets $H_u$ and $H_d$ can trigger proton decay also. Since the R-parities of the members in **5** and $\bar{\mathbf{5}}$ are the same, the dimension-3 superpotential term is managed to be forbidden, but dimension-4 term cannot be forbidden. Therefore, the color-triplet partners of the Higgs doublets must be superheavy such that proton lifetime bound is satisfied. That is the reason we consider the doublet–triplet splitting.

- *Adjoint problem*: The affine Lie algebra or Kac–Moody algebra with level $k = 1$ does not allow an adjoint representation as discussed in Chap. 12. Thus, to break a string derived GUT group from level 1 algebra down to the SM, one needs a GUT group with rank $r \geq 5$ since the VEV of an antisymmetric representation needed to break the GUT to the SM reduces the rank.

- *Proton longevity problem*: In GUT the problem related to rapid nucleon decay appears in a different angle. The gauge bosons of GUT that is lying outside the Standard Model group can mediate dimension via dimension six operator. Also, if we embed Higgs bosons in unified multiplets, colored partners can also give rise to nucleon decays.

- *E-chain of unification*: We restricted our discussion to the $E_8 \times E_8'$ heterotic string, chiefly because breaking the chain of $E_8$ down to the SU(5) or SU(3)$^3$ goes through the intermediate **27** of $E_6$. This is an important observation since the fifteen chiral fields (2.8) of the standard model(SM) are contained in the fundamental representation **27** of $E_6$ or in the spinor representation **16** of SO(10). By restricting ourselves to only this chain, we can automatically achieve the correct charge assignments, coming from **16** of SO(10). On the other hand, if the intermediate step cannot contain **27** of $E_6$, it is most probable that exotically charged particles(exotics) would appear.[1]

  This argument utilizes matter only from the untwisted sector. For SO(32) heterotic string, we may obtain spinorial representation for SO(10) unification group [1, 2]. Recently, SU(9)×SU(5) GUT was obtained from SO(32) heterotic string via $\mathbf{Z}_{12-I}$ orbifold [3].

## 14.2 GUTs from $\mathbb{Z}_{12-I}$ Orbifold

In early 1980s, consideration of unification discussed in Chap. 2 led to the SUSY SU(5) [4,5] which needed an adjoint Higgs representation. From 10D string models, it is difficult to acquire such an adjoint representation. So, to break a GUT to the SM, a SUSY anti-SU(5) is needed with the Higgs spectrum **10** $\oplus$ $\overline{\mathbf{10}}$ supplied. Namely,

---

[1]Exotics can come in two varieties: chiral exotics and vectorlike exotics. The more dangerous chiral exotics has the problem to be observed below the electroweak scale.

in string compactification the gauge coupling unification with SUSY suggested in simple group SU(5) [6] is modified to anti-SU(5) [7].[2]

In this section, we construct a $\mathbb{Z}_{12}$ orbifold model in detail. The chief motivation considering this complicated example is to exercise a model building with many possible ingredients we discussed in the previous chapters.

The point group $\mathbb{Z}_{12-I}$ takes the shift vector of Table 3.4 as

$$\phi = \left( \tfrac{5}{12} \ \tfrac{4}{12} \ \tfrac{1}{12} \right). \tag{14.1}$$

Let us look for anti-SU(5) models on it. First, we introduce a method of using Dynkin diagrams. Next, we discuss an example with a Wilson line not to allow some degeneracy of matter fields.

### 14.2.1 Without a Wilson Line

There are two interesting $E_8$ shift vectors [8] allowing SU(5) groups, which are shown below as Table 14.1. For an anti-SU(5), we need just SU(5) since there appear many U(1)s to produce $U(1)_X$ of anti-SU(5) in models from string compactification.

We illustrate Case (i) of Table 14.1 to familiarize with the Dynkin diagram technique. Let us try to include SU(5). For a possibility of three families, let us also include SU(3).[3] From the Dynkin diagram technique, we find that one possibility is choosing $a_4 + 2a_7 + a_8 = 12$ where $a_i$s are the Coxeter labels of simple roots $\alpha_i$. In Fig. 14.1, we strike out $\alpha_4, \alpha_7,$ and $\alpha_8$, and obtain the remaining SU(5) and SU(3) Dynkin diagrams. Now, let us obtain the shift vector. Choosing the shift vector as $V = \frac{1}{12}(\Lambda_4 + 2\Lambda_7 + \Lambda_8)$ (See Appendix A), we have a $\mathbb{Z}_{12}$ orbifold with

$$v = \tfrac{1}{12} \left( \tfrac{21}{2} \ \tfrac{3}{2} \ \tfrac{3}{2} \ \tfrac{3}{2} \ \tfrac{3}{2} \ \tfrac{1}{2} \ \tfrac{1}{2} \ \tfrac{1}{2} \right) (\cdots).$$

By Weyl transformations, we can transform $v$ to

$$v' = \tfrac{1}{12} \left( \tfrac{21}{2} \ \tfrac{3}{2} \ \tfrac{3}{2} \ \tfrac{3}{2} \ \tfrac{1}{2} \ \tfrac{1}{2} \ \tfrac{1}{2} \ \tfrac{3}{2} \right) (\cdots)$$

$$\equiv \tfrac{1}{12} \left( \tfrac{21}{2} \ \tfrac{3}{2} \ \tfrac{3}{2} \ \tfrac{3}{2} \ \tfrac{1}{2} \ \tfrac{1}{2} \ \tfrac{1}{2} \ \tfrac{51}{2} \right) (\cdots),$$

where a lattice shift is applied in the last line. Now let us Weyl-reflect $v'$ with respect to the plane orthogonal to $\alpha_7$ with $\alpha_7 \cdot v' = \tfrac{15}{12}$,

$$v'' = v' - 2\tfrac{\alpha_7 \cdot v'}{\alpha_7^2}\alpha_7 = v' - \tfrac{15}{12}\alpha_7$$

---

[2]Instead of flipped SU(5), here we use the word "anti-SU(5)" to stress the need for the antisymmetric representation $\mathbf{10} \oplus \overline{\mathbf{10}}$ for the GUT breaking.

[3]The SU(3) degeneracy is not broken at this level. It must be broken spontaneously.

**Table 14.1** Two shift vectors for SU(5)

|      | $E_8$ shift | $E_8'$ shift | 4D gauge group |
|------|-------------|--------------|-----------------|
| (i)  | $(\frac{1}{4}\ \frac{1}{4}\ \frac{1}{4}\ -\frac{1}{4}\ \frac{1}{2}\ \frac{1}{3}\ \frac{1}{3}\ \frac{1}{3})$ | $(\cdots)$ | $\mathrm{SU}(5)\times\mathrm{SU}(3)\times\mathrm{U}(1)^2 \times \cdots$ |
| (ii) | $(\frac{1}{2}\ \frac{5}{12}\ \frac{5}{12}\ \frac{5}{12}\ \frac{1}{6}\ \frac{1}{12}\ 0\ 0)$ | $(\cdots)$ | $\mathrm{SU}(5)\times\mathrm{SU}(3)\times\mathrm{U}(1)^2 \times \cdots$ |

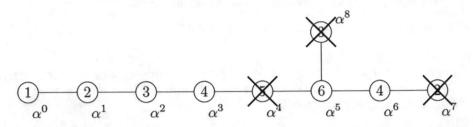

**Fig. 14.1** The SU(5)×SU(3) subgroup of $E_8$. The Coxeter labels are shown inside circles. The Coxeter labels of $\alpha_{4,7,8}$ are 5, 2, and 3, respectively

$$= \tfrac{1}{12}(3\ 9\ 9\ 9\ 8\ 8\ 8\ 18)(\cdots)$$

$$= \tfrac{1}{12}(3\ -3\ -3\ -3\ -4\ -4\ -4\ -6)(\cdots).$$

Weyl reflections of the above shift lead to Case (i) up to extra negative signs. They are again removed by Weyl reflections. Thus, let us consider the following for Case (i) of Table 14.1, with the $E_8'$ shift included,

$$V = \left(\tfrac{1}{4}\ \tfrac{1}{4}\ \tfrac{1}{4}\ \tfrac{-1}{4}\ \tfrac{1}{2}\ \tfrac{1}{3}\ \tfrac{1}{3}\ \tfrac{1}{3}\right)\left(\tfrac{1}{12}\ \tfrac{1}{12}\ \tfrac{1}{3}\ 0\ 0\ 0\ 0\ 0\right), \qquad (14.2)$$

which satisfies the modular invariance condition of Eq. (7.50),

$$12(V^2 - \phi^2) = 8 \equiv 0 \mod 2. \qquad (14.3)$$

### Gauge Group from Untwisted Sector

From the above Dynkin diagram construction of the shift vector, it is obvious that SU(5)×SU(3) results from the first $E_8$. Here, let us observe this explicitly. Among the original $E_8 \times E_8'$ roots $P$ with $P^2 = 2$, we seek the ones satisfying (7.53)

$$P \cdot V = 0 \qquad \mod 1. \qquad (14.4)$$

These form the $E_8$ gauge multiplets shown in Table 14.2.

In Table 14.2, we have the same convention as in the previous chapter. There are 6 momentum-winding states (nonzero roots) in the first row and adding two oscillators we have the adjoint 8 of SU(3). The rest of Table 14.2 has 20 nonzero roots which form the adjoint representation of SU(5) together with four oscillators.

**Table 14.2** Root vectors in untwisted sector satisfying $P \cdot V = 0$

| Vector | Number of states | Representation |
|---|---|---|
| $(0\ 0\ 0\ 0\ 0\ \underline{1\ -1}\ 0)$ | 6 | **8** of SU(3) |
| $(\underline{1\ -1\ 0}\ 0\ 0\ 0\ 0\ 0)$ | 6 | **24** of SU(5) |
| $(\underline{1\ 0\ 0}\ 1\ 0\ 0\ 0\ 0)$ | 3 | |
| $(\underline{-1\ 0\ 0}\ -1\ 0\ 0\ 0\ 0)$ | 3 | |
| $(+\ +\ +\ +\ +\ +\ +\ +)$ | 1 | |
| $(-\ -\ -\ -\ -\ -\ -\ -)$ | 1 | |
| $(\underline{-\ +\ +}\ -\ +\ +\ +\ +)$ | 3 | |
| $(\underline{-\ -\ +}\ +\ -\ -\ -\ -)$ | 3 | |

The underlined entries allow permutations. The $+$ and $-$ in the spinor part denote $\frac{1}{2}$ and $-\frac{1}{2}$, respectively

The remaining 2 oscillators out of 8 generate $U(1)^2$. These, coming from $E_8$, form the adjoint representations of $SU(5) \times SU(3) \times U(1)^2$.

Let us take two simple roots of SU(3) as

$$\alpha_a = (0\ 0\ 0\ 0\ 0\ 1\ -1\ 0\ ), \quad \alpha_b = (0\ 0\ 0\ 0\ 0\ 0\ 1\ -1) \tag{14.5}$$

and four simple roots of SU(5) as

$$\begin{aligned}
\alpha_1 &= (\ 1\ -1\ \ 0\ \ 0\ 0\ 0\ 0\ 0\ ) \\
\alpha_2 &= (\ 0\ \ 1\ \ -1\ 0\ 0\ 0\ 0\ 0\ ) \\
\alpha_3 &= (\ 0\ \ 0\ \ 1\ \ 1\ 0\ 0\ 0\ 0\ ) \\
\alpha_4 &= (-\ -\ \ -\ -\ -\ -\ -\ -\ ).
\end{aligned} \tag{14.6}$$

Then, the highest weights of some representations are

$$\begin{aligned}
\mathbf{3}: &(0\ 0\ \ 0\ \ 0\ 0\ 1\ 0\ \ 0\ ) \\
\mathbf{\bar{3}}: &(0\ 0\ \ 0\ \ 0\ 0\ 0\ 0\ -1) \\
\mathbf{24}: &(+\ -\ \ -\ +\ -\ -\ -\ \ -\ ) \\
\mathbf{5}: &(1\ 0\ \ 0\ \ 1\ 0\ 0\ 0\ \ 0\ ) \\
\mathbf{\bar{5}}: &(-\ -\ \ -\ -\ -\ -\ -\ \ -\ ) \\
\mathbf{10}: &(+\ +\ +\ +\ -\ -\ -\ \ -\ ) \\
\mathbf{\overline{10}}: &(0\ 0\ -1\ 1\ 0\ 0\ 0\ \ 0\ )
\end{aligned} \tag{14.7}$$

In the same way, we obtain the gauge group $SO(10) \times SU(2) \times U(1)^2$ from the other $E_8$ sector.

## Matter from Untwisted Sector

From the untwisted sector, the massless fields satisfy $P^2 = 2$. Those with $P \cdot V \equiv 0$ mod 1 are the gauge multiplets. The massless matter fields are those with

$$P \cdot V \equiv \frac{k}{12} \mod 1, \quad k = 1, 2, \ldots, 11.$$

We will consider only $k = 1, 2, \ldots, 6$, since the $\mathscr{CTP}$ conjugates of $k$th twisted sectors appear in $12 - k$th twisted sectors. We obtain the following spectrum.

| Sector | Highest weight $P$ | Representation |
|---|---|---|
| U | $(0\,0\,0\,0\,0\,0\,0\,0)(1\,0\,0\,1\,0\,0\,0\,0)$ | $(\mathbf{1}, \mathbf{1}; \mathbf{10}, \mathbf{2})$ |
| | $(1\,0\,0\,0\,0\,1\,0\,0)(0\,0\,0\,0\,0\,0\,0\,0)$ | $(\mathbf{10}, \mathbf{3}; \mathbf{1}, \mathbf{1})$ |
| | $(0\,0\,0\,0\,0\,0\,0\,0)(0\,0\,1\,1\,0\,0\,0\,0)$ | $(\mathbf{1}, \mathbf{1}; \mathbf{10}, \mathbf{1})$ |
| | $(0\,0\,0\,1\,0\,1\,0\,0)(0\,0\,0\,0\,0\,0\,0\,0)$ | $(\bar{\mathbf{5}}, \mathbf{3}; \mathbf{1}, \mathbf{1})$ |
| | $(0\,0\,0\,0\,0\,0\,0\,0)(0\,-1\,-1\,0\,0\,0\,0\,0)$ | $(\mathbf{1}, \mathbf{1}; \mathbf{1}, \mathbf{2})$ |
| | $(+\,+\,+\,+\,+\,+\,-\,-)(0\,0\,0\,0\,0\,0\,0\,0)$ | $(\bar{\mathbf{5}}, \mathbf{3}; \mathbf{1}, \mathbf{1})$ |
| | $(0\,0\,0\,0\,0\,0\,0\,0)(-\,-\,+\,+\,+\,+\,+\,+)$ | $(\mathbf{1}, \mathbf{1}; \mathbf{16}, \mathbf{1})$ |
| | $(+\,+\,+\,-\,-\,+\,-\,-)(0\,0\,0\,0\,0\,0\,0\,0)$ | $(\mathbf{1}, \mathbf{3}; \mathbf{1}, \mathbf{1})$ |

The chirality is discussed in Sect. 7.4.1. The massless states in the Neveu–Schwarz sectors are $s = (\pm 1, 0, 0, 0)$ and we may obtain the corresponding Ramond sector field by adding the $r$ vector as $s_R = s_{NS} + r$. Below, we discuss the Ramond states. The left and the right mover states generate phases under the point group transformation.

Let us define $\alpha = e^{2\pi i/12}$. For $k = 1$ or $P \cdot V = \frac{1}{12}$, the left movers obtain a phase $\alpha$. We need an extra phase $\alpha^{-1}$ from the right movers, which is accomplished by $e^{-2\pi i \tilde{s} \cdot \phi}$ where $\phi = \frac{1}{12}(5\,4\,1)$ and $s = (\ominus + - +)$. It is left-handed (See Table 14.5). The four dimensional chirality is read of from the $s_0$ components of right mover, denoted as $(s_0, s_1, s_2, s_3)$. We make it a convention for $s = -\frac{1}{2}$ to left-handed. For $k = 4$, $s = (\ominus + + -)$ provides the needed $\alpha^{-4}$; thus it is left-handed. For $k = 5$, $s = (\oplus + + +)$ provides $\alpha^{-5}$; thus it is right-handed.

$\alpha$ from the right movers is provided by $s = (\oplus - + -)$ which thus will couple to $k = 11$. It is right-handed. $\alpha^4$ from the right movers is provided by $s = (\oplus - - +)$ which will couple to $k = 8$. It is right-handed. $\alpha^5$ from the right movers is provided by $s = (\oplus - - -)$ which will couple to $k = 7$. It is left-handed. These give the antiparticle spectra of those obtained in the previous paragraph.

So far we considered six chirality operators from $\mathbf{8}$ of the Ramond sector. There are two more chirality operators: $s = (\oplus + + +)$ and $s = (\ominus - - -)$. These from the right movers do not provide any phase; so they must couple to the $k = 0$ (or $k = 12$) sector. In fact we considered them already in obtaining the gauge sector which is chosen as left-handed through $s = (\ominus - - -)$.

**Table 14.3** Summary of massless matter from the untwisted sector

| $P \cdot V$ | $[SU(5) \times SU(3)]_{\text{chirality}}$ |
|---|---|
| $\frac{5}{12}$ | $(\mathbf{10}, \mathbf{3})_L$ |
| $\frac{4}{12}$ | $(\bar{\mathbf{5}}, \mathbf{3})_L$ |
| $\frac{1}{12}$ | $(\bar{\mathbf{5}}, \mathbf{3})_L \oplus (\mathbf{1}, \mathbf{3})_L$ |

Table 14.3 summarizes matter from the untwisted sector. Note that for $P \cdot V = \frac{5}{12}$ the original R-handed field is changed to the charge conjugated L-handed field in Eq. (14.3).

**Twisted Sector**

The $\mathbb{Z}_{12-\mathrm{I}}$ with the twist vector $\phi = \frac{1}{12}(5\ 4\ 1)$ has three fixed points in the prime, i.e., $\theta^1$-, $\theta^2$- and $\theta^5$- twisted sectors. This is because it has three fixed points in the second torus since it is the same as $\mathbb{Z}_3$, and for the first and the third torus the origin is the only fixed point. For the other non-prime-order twists such as $k = 4$ and 6, counting the number of massless states involves a more complicated nonvanishing projector $P_k$, Eq. (8.59).

The masslessness condition for left movers is

$$\frac{(P + kV)^2}{2} + \tilde{N} + \tilde{c} = 0 \tag{14.8}$$

with the zero point energy $\tilde{c}$ given in (7.15). In each twisted sector, $\tilde{c}$ has the value

$$2\tilde{c} = \begin{cases} \frac{210}{144}, \ k = 1; & \frac{192}{144}, \ k = 4 \\[2mm] \frac{216}{144}, \ k = 2; & \frac{210}{144}, \ k = 5 \\[2mm] \frac{234}{144}, \ k = 3; & \frac{216}{144}, \ k = 6. \end{cases} \tag{14.9}$$

On the other hand, the right mover masslessness condition leads to

$$2N + (s + k\phi_s)^2 = 2c = 2\tilde{c} - 1 \tag{14.10}$$

which are

$$2c = \tfrac{11}{24}, \ \tfrac{1}{2}, \ \tfrac{5}{8}, \ \tfrac{1}{3}, \ \tfrac{11}{24}, \ \tfrac{1}{2}, \tag{14.11}$$

for $k = 1, 2, \ldots, 6$, respectively.

The number of zero modes in the twisted sector are given by projector

$$P_m = \frac{1}{N} \sum \tilde{\chi}(\theta^m, \theta^n) \Delta_{(\theta^m)}{}^n, \tag{14.12}$$

where $\Delta$ is the phase of the specified twisted sector and $\tilde{\chi}(\theta^i, \theta^j)$ is the number of simultaneous fixed points on nontrivial tori (8.62) of the internal space both given in Chap. 8. Table 14.4 summarizes matter from the twisted sector. In the $\theta^6$ sector, we omitted, except in the first row, hidden sector representations $(-1^2 - 2\,0^5)$ and $(0^2 - 2\,0^5)$ which provide $\mathscr{CPT}$ conjugate to each other.

The generalized GSO projector (14.12) gives the multiplicity. For the prime orbifold $\mathbb{Z}_3$, the multiplicity is just $\frac{1}{3}(1 + \Delta + \Delta^2)$ which can be either 1 (for $\Delta = 1$) or 0 (for $\Delta = e^{\pm 2\pi i/3}$). So in Sect. 7.4.2 it was sufficient to count those with the vanishing phase. But for non-prime orbifolds such as $\mathbb{Z}_{12}$, the multiplicity (14.12) is nonvanishing even if $\Delta$ were not 1. Only for $\theta^1$, $\theta^2$, and $\theta^5$ twists, the multiplicity is given by the vanishing $\Delta$ phase.

Since the calculation is straightforward even though tedious, here we show only the $\theta^4$ twist in detail. The case of $\theta^4$ includes all possible complications one can anticipate. The $\theta^4$ twist vectors are

$$4\phi = \left(\tfrac{2}{3}, \tfrac{1}{3}, \tfrac{1}{3}\right)$$

$$4V \equiv \left(1\,1\,1\,-1\,2\,\tfrac{4}{3}\,\tfrac{4}{3}\,\tfrac{4}{3}\right)\left(\tfrac{1}{3}\,\tfrac{1}{3}\,\tfrac{4}{3}\,0\,0\,0\,0\,0\right). \tag{14.13}$$

In the $\theta^4$-twisted sector, the mass shell condition (14.8) becomes $(P + 4V)^2 + 2\tilde{N} = 2\tilde{c} = \tfrac{4}{3}$. We have solutions constituting the representation $(\mathbf{1}, \bar{\mathbf{3}})_L$ in Table 14.5. One of the components is $\tilde{N}_L = 0$, $P = (0^5 \sim \underline{-2} \sim -1 \sim -1)(0 \sim 0 \sim -2 \sim 0^5)$ is massless.

Now consider the masslessness condition for the right movers. From (14.11), we need $(s + 4\phi)^2 + 2N = 1/3$. We obtain $\mathscr{N} = 0$, $s = (\ominus\,\tfrac{-3}{2}\,\tfrac{-3}{2}\,-)$. It shows that the chirality is left-handed.

The GSO projection is given when combined with the right movers. We have for the above $P$ and $s$,

$$(P + 4V) \cdot V - (s + 4\phi) \cdot \phi - \frac{4}{2}\left(V^2 - \phi^2\right) = \frac{3}{4}. \tag{14.14}$$

We use the number of simultaneous fixed points (8.62),

$$\tilde{\chi}(\theta^4, 1) = 27, \quad \tilde{\chi}(\theta^4, \theta^1) = 3, \quad \tilde{\chi}(\theta^4, \theta^2) = 3, \quad \tilde{\chi}(\theta^4, \theta^1) = 3,$$

$$\tilde{\chi}(\theta^4, \theta^4) = 27, \quad \tilde{\chi}(\theta^4, \theta^5) = 3, \quad \tilde{\chi}(\theta^4, \theta^6) = 3, \quad \tilde{\chi}(\theta^4, \theta^7) = 3, \quad (14.15)$$

$$\tilde{\chi}(\theta^4, \theta^8) = 27, \quad \tilde{\chi}(\theta^4, \theta^9) = 3, \quad \tilde{\chi}(\theta^4, \theta^{10}) = 3, \quad \tilde{\chi}(\theta^4, \theta^{11}) = 3.$$

Therefore, we obtain

$$P_{\theta^4} = \tfrac{1}{12}\left\{3(1 + \Delta^4 + \Delta^8) \cdot \left[8 + 1 + \Delta(1 + \Delta + \Delta^2)\right]\right\} \tag{14.16}$$

which becomes 9, 6, 6 for $\Delta = 1, -1, \pm i$, respectively, and 0 for the other cases.

We can calculate the spectrum in the same way. The result is shown in Table 14.4.

**Table 14.4** Matter fields from twisted sectors. All the states are L-handed in four dimensions

| Sector | Highest weight $(P + kV)$ | Representation |
|---|---|---|
| U | $(0\,0\,0\,0\,0\,0\,0\,0)(1\,0\,0\,1\,0\,0\,0\,0)$ | $(\mathbf{1}, \mathbf{1}; \mathbf{10}, \mathbf{2})$ |
| | $(1\,0\,0\,0\,0\,1\,0\,0)(0\,0\,0\,0\,0\,0\,0\,0)$ | $(\mathbf{10}, \mathbf{3}; \mathbf{1}, \mathbf{1})$ |
| | $(0\,0\,0\,0\,0\,0\,0\,0)(0\,0\,1\,1\,0\,0\,0\,0)$ | $(\mathbf{1}, \mathbf{1}; \mathbf{10}, \mathbf{1})$ |
| | $(0\,0\,0\,1\,0\,1\,0\,0)(0\,0\,0\,0\,0\,0\,0\,0)$ | $(\mathbf{\bar{5}}, \mathbf{3}; \mathbf{1}, \mathbf{1})$ |
| | $(0\,0\,0\,0\,0\,0\,0\,0)(0\,-1\,-1\,0\,0\,0\,0\,0)$ | $(\mathbf{1}, \mathbf{1}; \mathbf{1}, \mathbf{2})$ |
| | $(+\,+\,+\,+\,+\,+\,-\,-)(0\,0\,0\,0\,0\,0\,0\,0)$ | $(\mathbf{\bar{5}}, \mathbf{3}; \mathbf{1}, \mathbf{1})$ |
| | $(0\,0\,0\,0\,0\,0\,0\,0)(-\,-\,+\,+\,+\,+\,+\,+)$ | $(\mathbf{1}, \mathbf{1}; \mathbf{16}, \mathbf{1})$ |
| | $(+\,+\,+\,-\,-\,+\,-\,-)(0\,0\,0\,0\,0\,0\,0\,0)$ | $(\mathbf{1}, \mathbf{3}; \mathbf{1}, \mathbf{1})$ |
| T1 | $\frac{1}{12}(3\,3\,3\,9\,-6\,4\,4\,4)(1\,1\,4\,0\,0\,0\,0\,0)$ | $3(\mathbf{\bar{5}}, \mathbf{1}; \mathbf{1}, \mathbf{1})$ |
| | $\frac{1}{12}(-3\,-3\,-3\,3\,12\,-2\,-2\,-2)(1\,1\,4\,0\,0\,0\,0\,0)$ | $3(\mathbf{1}, \mathbf{1}; \mathbf{1}, \mathbf{1})$ |
| | $\frac{1}{12}(3\,3\,3\,-3\,-6\,4\,4\,-8)(1\,1\,4\,0\,0\,0\,0\,0)$ | $3(\mathbf{1}, \mathbf{\bar{3}}; \mathbf{1}, \mathbf{1})$ |
| | $\frac{1}{12}(-3\,-3\,-3\,3\,0\,10\,-2\,-2)(1\,1\,4\,0\,0\,0\,0\,0)$ | $3(\mathbf{1}, \mathbf{3}; \mathbf{1}, \mathbf{1})$ |
| | $\frac{1}{12}(9\,-3\,-3\,3\,0\,-2\,-2\,-2)(1\,1\,4\,0\,0\,0\,0\,0)$ | $3(\mathbf{5}, \mathbf{1}; \mathbf{1}, \mathbf{1})$ |
| T2 | $\frac{1}{12}(0\,0\,0\,0\,-6\,2\,2\,2)(2\,2\,-4\,1\,2\,0\,0\,0\,0)$ | $3(\mathbf{1}, \mathbf{1}; \mathbf{10}, \mathbf{1})$ |
| | $\frac{1}{12}(0\,0\,0\,0\,-6\,2\,2\,2)(2\,-10\,-4\,0\,0\,0\,0)$ | $3(\mathbf{1}, \mathbf{1}; \mathbf{1}, \mathbf{2})$ |
| | $\frac{1}{12}(0\,0\,0\,0\,-6\,2\,2\,2)(2\,2\,8\,0\,0\,0\,0\,0)$ | $6(\mathbf{1}, \mathbf{1}; \mathbf{1}, \mathbf{1})$ |
| T3 | $\frac{1}{12}(9\,-3\,-3\,3\,-6\,0\,0\,0)(3\,-9\,0\,0\,0\,0\,0\,0)$ | $2(\mathbf{5}, \mathbf{1}; \mathbf{1}, \mathbf{2})$ |
| | $\frac{1}{12}(3\,3\,3\,-3\,0\,6\,-6\,-6)(3\,-9\,0\,0\,0\,0\,0\,0)$ | $(\mathbf{1}, \mathbf{3}; \mathbf{1}, \mathbf{2})$ |
| | $\frac{1}{12}(-3\,-3\,-3\,3\,6\,0\,0\,0)(3\,3\,12\,0\,0\,0\,0\,0)$ | $(\mathbf{1}, \mathbf{1}; \mathbf{1}, \mathbf{1})$ |
| | $\frac{1}{12}(-3\,-3\,-3\,3\,6\,0\,0\,0)(3\,3\,0\,12\,0\,0\,0\,0)$ | $(\mathbf{1}, \mathbf{1}; \mathbf{10}, \mathbf{1})$ |
| | $\frac{1}{12}(-3\,-3\,-3\,3\,6\,0\,0\,0)(0\,3\,3\,-12\,0\,0\,0\,0)$ | $2(\mathbf{1}, \mathbf{1}; \mathbf{1}, \mathbf{1})$ |
| | $\frac{1}{12}(-3\,-3\,-3\,3\,6\,0\,0\,0)(3\,-9\,0\,0\,0\,0\,0\,0)$ | $2(\mathbf{1}, \mathbf{1}; \mathbf{1}, \mathbf{2})$ |
| T4 | $\frac{1}{12}(0\,0\,0\,0\,0\,4\,4\,-8)(4\,-8\,4\,0\,0\,0\,0\,0)$ | $6(\mathbf{1}, \mathbf{\bar{3}}; \mathbf{1}, \mathbf{2})$ |
| | $\frac{1}{12}(0\,0\,0\,0\,0\,4\,4\,-8)(4\,4\,-8\,0\,0\,0\,0\,0)$ | $6(\mathbf{1}, \mathbf{\bar{3}}; \mathbf{1}, \mathbf{1})$ |
| T6 | $\frac{1}{12}(6\,-6\,-6\,6\,0\,0\,0\,0)(-6\,-6\,0\,0\,0\,0\,0\,0)$ | $2(\mathbf{5}, \mathbf{1}; \mathbf{1}, \mathbf{1})$ |
| | $\frac{1}{12}(6\,-6\,-6\,6\,0\,0\,0\,0)(6\,6\,0\,0\,0\,0\,0\,0)$ | $4(\mathbf{5}, \mathbf{1}; \mathbf{1}, \mathbf{1})$ |
| | $\frac{1}{12}(6\,6\,6\,6\,0\,0\,0\,0)(-6\,-6\,0\,0\,0\,0\,0\,0)$ | $3(\mathbf{\bar{5}}, \mathbf{1}; \mathbf{1}, \mathbf{1})$ |
| | $\frac{1}{12}(6\,6\,6\,6\,0\,0\,0\,0)(6\,6\,0\,0\,0\,0\,0\,0)$ | $2(\mathbf{\bar{5}}, \mathbf{1}; \mathbf{1}, \mathbf{1})$ |
| | $\frac{1}{12}(0\,0\,0\,0\,-6\,6\,6\,-6)(-6\,-6\,0\,0\,0\,0\,0\,0)$ | $4(\mathbf{1}, \mathbf{\bar{3}}; \mathbf{1}, \mathbf{1})$ |
| | $\frac{1}{12}(0\,0\,0\,0\,-6\,6\,6\,-6)(6\,6\,0\,0\,0\,0\,0\,0)$ | $2(\mathbf{1}, \mathbf{\bar{3}}; \mathbf{1}, \mathbf{1})$ |
| | $\frac{1}{12}(0\,0\,0\,0\,6\,6\,-6\,-6)(-6\,-6\,0\,0\,0\,0\,0\,0)$ | $2(\mathbf{1}, \mathbf{3}; \mathbf{1}, \mathbf{1})$ |
| | $\frac{1}{12}(0\,0\,0\,0\,6\,6\,-6\,-6)(6\,6\,0\,0\,0\,0\,0\,0)$ | $3(\mathbf{1}, \mathbf{3}; \mathbf{1}, \mathbf{1})$ |
| T7 | $\frac{1}{12}(3\,3\,3\,-3\,0\,-2\,-2\,-2)(7\,-5\,-8\,0\,0\,0\,0\,0)$ | $3(\mathbf{1}, \mathbf{1}; \mathbf{1}, \mathbf{2})$ |
| | $\frac{1}{12}(-3\,-3\,-3\,3\,-6\,4\,4\,4)(-5\,-5\,4\,0\,0\,0\,0\,0)$ | $3(\mathbf{1}, \mathbf{1}; \mathbf{1}, \mathbf{1})$ |
| | $\frac{1}{12}(3\,3\,3\,-3\,0\,-2\,-2\,-2)(7\,7\,4\,0\,0\,0\,0\,0)$ | $3(\mathbf{1}, \mathbf{1}; \mathbf{1}, \mathbf{1})$ |
| | $\frac{1}{12}(3\,3\,3\,-3\,0\,-2\,-2\,-2)(-5\,-5\,4\,0\,0\,0\,0\,0)$ | $6(\mathbf{1}, \mathbf{1}; \mathbf{1}, \mathbf{1})$ |
| T9 | $\frac{1}{12}(3\,3\,3\,9\,6\,0\,0\,0)(9\,-3\,0\,0\,0\,0\,0\,0)$ | $(\mathbf{\bar{5}}, \mathbf{1}; \mathbf{1}, \mathbf{2})$ |
| | $\frac{1}{12}(-3\,-3\,-3\,3\,0\,6\,6\,-6)(9\,-3\,0\,0\,0\,0\,0\,0)$ | $2(\mathbf{1}, \mathbf{\bar{3}}; \mathbf{1}, \mathbf{2})$ |
| | $\frac{1}{12}(3\,3\,3\,-3\,-6\,0\,0\,0)(-3\,-3\,12\,0\,0\,0\,0\,0)$ | $(\mathbf{1}, \mathbf{1}; \mathbf{1}, \mathbf{1})$ |
| | $\frac{1}{12}(3\,3\,3\,-3\,-6\,0\,0\,0)(-3\,-3\,0\,12\,0\,0\,0\,0)$ | $2(\mathbf{1}, \mathbf{1}; \mathbf{10}, \mathbf{1})$ |
| | $\frac{1}{12}(3\,3\,3\,-3\,-6\,0\,0\,0)(-3\,-3\,-12\,0\,0\,0\,0\,0)$ | $(\mathbf{1}, \mathbf{1}; \mathbf{1}, \mathbf{1})$ |
| | $\frac{1}{12}(3\,3\,3\,-3\,-6\,0\,0\,0)(9\,-3\,0\,0\,0\,0\,0\,0)$ | $3(\mathbf{1}, \mathbf{1}; \mathbf{1}, \mathbf{2})$ |

**Table 14.5** $+$ and $-$ denote $+\frac{1}{2}$ and $-\frac{1}{2}$, respectively, and $\oplus(\ominus)$ is R(L)-handed

| $\frac{1}{2}\chi$ | $\tilde{s} = (r + \tilde{\omega})$ | $\tilde{s}\cdot\phi_s$ | $(\tilde{s}+\tilde{\phi})\cdot\phi_s$ | $\Delta$ phase | Multiplicity |
|---|---|---|---|---|---|
| $\oplus$ | $+\ +\ +$ | $\frac{5}{12}$ | $\frac{19}{12}$ | $-\frac{1}{12}\cdot 2\pi$ | 0 |
| $\oplus$ | $+\ -\ -$ | $0$ | $\frac{14}{12}$ | $\frac{1}{3}\cdot 2\pi$ | 0 |
| $\oplus$ | $-\ +\ -$ | $-\frac{1}{12}$ | $\frac{13}{12}$ | $\frac{5}{12}\cdot 2\pi$ | 0 |
| $\oplus$ | $-\ -\ +$ | $-\frac{4}{12}$ | $\frac{10}{12}$ | $\frac{2}{3}\cdot 2\pi$ | 0 |
| $\ominus$ | $-\ -\ -$ | $-\frac{5}{12}$ | $\frac{9}{12}$ | $\frac{3}{4}\cdot 2\pi$ | 6 |
| $\ominus$ | $-\ +\ +$ | $0$ | $\frac{14}{12}$ | $\frac{1}{3}\cdot 2\pi$ | 0 |
| $\ominus$ | $+\ -\ +$ | $\frac{1}{12}$ | $\frac{15}{12}$ | $\frac{1}{4}\cdot 2\pi$ | 6 |
| $\ominus$ | $+\ +\ -$ | $\frac{4}{12}$ | $\frac{18}{12}$ | $0\cdot 2\pi$ | 9 |

Note that $(P + \tilde{V})\cdot V - \frac{1}{2}m(V^2 - \phi_s^2) = -\frac{1}{2} \to \frac{1}{2}$

Note that in the $\theta_{(6)}$ twisted sector, states are self-conjugate under the $\mathscr{CTP}$, which is always the case in the $\theta_{(N)}$ twisted sector in a $\mathbb{Z}_{2N}$ orbifold model. In calculating the multiplicity, we have to consider the $\mathbb{Z}_2$ symmetry in the $\theta_{(N)}$ twisted sector, i.e. dividing by 2. In our case of $\theta_{(6)}$ twisted sector, there appears another factor 2 coming from the hidden sector multiplicity 2. It is shown only in the first line of the $\theta_{(6)}$ twisted sector of the twisted sector table. In the $\theta^6$ twisted sector, the $\mathscr{CTP}$ conjugate of an L mover should be an R mover. Still the spectrum can be chiral because the projection may select one chirality.

In Table 14.6 we summarize the observable sector fields except singlets. Adding the representations from the untwisted sector, Table 14.3, we can easily check that there do not exist any non-Abelian anomalies of SU(5) and SU(3). Even though we have not shown explicitly, there does not exist any gauge anomalies except for one U(1) anomaly which is cancelled by the Green–Schwarz mechanism.

## 14.2.2 A Model with a Wilson Line

Not to introduce any degeneracy, we must employ the full information on the Wilson lines. So, instead of introducing SU(5) directly, we first introduce SO(10) from the shift vector.

**Table 14.6** Summary of massless observable matter from the twisted-sectors, all in the L-handed. The multiplicity may come from either the number of fixed points or non-Abelian charge.

| Sector | Representation |
|---|---|
| T1 | $3(\bar{\mathbf{5}}, \mathbf{1}) + 3(\mathbf{5}, \mathbf{1}) + 3(\mathbf{1}, \mathbf{3}) + 3(\mathbf{1}, \bar{\mathbf{3}})$ |
| T2 | no observable matter |
| T3 | $4(\mathbf{5}, \mathbf{1}) + 2(\mathbf{1}, \mathbf{3})$ |
| T4 | $18(\mathbf{1}, \bar{\mathbf{3}})$ |
| T6 | $6(\mathbf{5}, \mathbf{1}) + 5(\bar{\mathbf{5}}, \mathbf{1}) + 6(\mathbf{1}, \bar{\mathbf{3}}) + 5(\mathbf{1}, \mathbf{3})$ |
| T7 | no observable matter |
| T9 | $2(\bar{\mathbf{5}}, \mathbf{1}) + 4(\mathbf{1}, \bar{\mathbf{3}})$ |

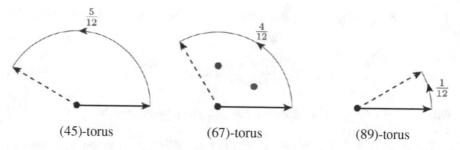

(45)-torus                    (67)-torus                    (89)-torus

**Fig. 14.2** Fixed points in $\mathbb{Z}_{12-\mathrm{I}}$ orbifold. The three fixed points are distinguished in the (67)-torus by the Wilson lines

$$V = \left(0, 0, 0, 0, 0; \tfrac{-1}{6}, \tfrac{-1}{6}, \tfrac{-1}{6}\right) \left(0, 0, 0, 0, 0; \tfrac{1}{4}, \tfrac{1}{4}, \tfrac{-2}{4}\right)'. \qquad (14.17)$$

Then we turn on Wilson line to break this group down to SU(5). Among the lattices of the $\mathbb{Z}_{12-\mathrm{I}}$ orbifold, shown in Fig. 14.2, we have SU(3) lattice of order three orbifold along the (67) direction. We turn on a Wilson line along this direction

$$a = \left(\tfrac{2}{3}, \tfrac{2}{3}, \tfrac{2}{3}, \tfrac{2}{3}, \tfrac{2}{3}; 0, \tfrac{-2}{3}, \tfrac{2}{3}\right) \left(\tfrac{2}{3}, \tfrac{2}{3}, \tfrac{2}{3}, \tfrac{2}{3}, 0; \tfrac{-2}{3}, 0, 0\right)'. \qquad (14.18)$$

The resulting gauge group is SU(5)$\times$SU(5)$'\times$SU(2)$'\times$U(1)$^7$ [9].

To remove degeneracies completely toward a successful flavor phenomenology, all fixed points should be distinguished. Because the number of fixed points is minimum in $\mathbb{Z}_{12-\mathrm{I}}$ orbifold as shown in the first table in Appendix A, it can be considered as the simplest example in phenomenological application. Indeed, Model (14.18) contains possible ingredients to answer the questions listed in the beginning of this chapter.

Since the method to obtain massless spectra was considered in detail in Sect. 14.2.1, in this subsection we restrict just to three twisted sectors $T_4^0$ (for matter fields),[4] $T_3$ (for $\mathbf{10}_{-1} \oplus \overline{\mathbf{10}}_{+1}$), and $T_1^{0,+,-}$ (for the hidden SU(5)$'$ fields) to acquaint with the case of the Wilson line addition.

Let us recapitulate the generalized GSO projection presented in Sect. 9.4.3. The phase associated with the orbifold is

$$\Theta_k = \sum_i (N_i^L - N_i^R)\hat{\phi}_i + (P + kV_f)\cdot V_f - (s + k\phi)\cdot\phi - \frac{k}{2}(V^2 - \phi^2), \qquad (14.19)$$

---

[4]Sectors in $\theta_{(4)}$ are distinguished by Wilson lines: $T_4^0 = V$, $T_4^+ = V + a$, $T_4^- = V - a$. The visible sector families arise in $T_4^0$.

where $\hat{\phi}_j = \phi_j$ and $\hat{\phi}_{\bar{j}} = -\phi_j$. Concretely, the multiplicity in the twisted sector $T_k^f$ is[5]

$$\mathscr{P}_k(f) = \frac{1}{12 \cdot 3} \sum_{l=0}^{11} \tilde{\chi}(\theta^k, \theta^l) e^{i\, 2\pi l\, \Theta_k}, \tag{14.20}$$

where $f(= \{f_0, f_+, f_-\})$ denote twisted sectors associated with $kV_f = kV$, $k(V + a), k(V - a)$.

For $k = 0, 3, 6, 9$, $\mathscr{P}_k(f_0) = \mathscr{P}_k(f_+) = \mathscr{P}_k(f_-)$. Thus we only need to calculate one of them, say $\mathscr{P}_k(f_0)$ and multiply three. We require in addition,

$$P \cdot a = 0 \text{ mod integer in the sector,}$$

$$12(P + V_k^{0,+,-}) \cdot a = 0 \text{ mod integer in } k = 3, 6, 9 \text{ sectors.} \tag{14.21}$$

Note that the four entry $s$ and the three entry $\tilde{s}$ with the relation $s = (\oplus \text{or} \ominus; \tilde{s})$ such that $\oplus$ or $\ominus$ is chosen to make the total number of minus signs even. For the subsector $f = 0$, i.e. for $T_0^k$ (useful for $T_4^0$ and $T_1^0$), from the masslessness condition, $2\tilde{c}_k = \sum_i (N_i^L)\hat{\phi}_i + P \cdot V + \frac{k}{2}V^2$, we have

$$\Theta_k^0 = \sum_i (N_i^L - N_i^R)\hat{\phi}_i + P \cdot V - s \cdot \phi + \frac{k}{2}(V^2 - \phi^2) \tag{14.22}$$

$$= \sum_i (N_i^L - N_i^R)\hat{\phi}_i + P \cdot V - s \cdot \phi + \frac{k}{12}. \tag{14.23}$$

**Twisted Sector $T_4^0$**  In the sector $\theta_{(4)}$, the degeneracy factors of Appendix A are

$$\tilde{\chi}(\theta^4, \theta^j) = \begin{cases} j = & 0, 1, 2, 3, \quad 4, 5, 6, 7, \quad 8, 9, 10, 11 \\ & 27, 3, 3, 3, \; 27, 3, 3, 3, \; 27, 3, \;\; 3, 3. \end{cases} \tag{14.24}$$

With $\phi$ of Eq. (14.1) for right movers, note that $4\phi$ is $\left(\frac{20}{12}, \frac{16}{12}, \frac{4}{12}\right)$ which is equivalent to $\left(\frac{2}{3}, \frac{1}{3}, \frac{1}{3}\right)$. Locally its $\mathbb{Z}_3$ twist. The mass shell condition is $s_0^2 + (\tilde{s} + 4\phi)^2 = -2c = \frac{1}{3}$, i.e. $(\tilde{s} + 4\phi)^2 = \frac{1}{12}$ with $(s_0)^2 = \frac{1}{4}$, which is satisfied by $\tilde{s} = (- - -)$ and $s_0 = -$.

For matter fields in $T_4^0$ without oscillators, we insert $P \cdot V = -\frac{1}{4}$ for $k = 4$ in Eq. (14.23),

$$\Theta_4^0(\text{matter}) = -\tilde{s} \cdot \phi + \frac{1}{12}. \tag{14.25}$$

---

[5]In Eq. (14.20), if we use the denominator 12 instead of 36, then we must use 9 1 1 1 9 1 1 1 9 1 1 1 instead of 27 3 3 3 27 3 3 3 27 3 3 3 in the $k = 4$ row of $Z_{12-I}$ for the table $\tilde{\chi}(\theta^k, \theta^l)$ in Appendix A.

So, we choose $s = (\ominus; \tilde{s})$, i.e. it is L-handed, and obtain $\Theta_4^0(\text{matter}) = \frac{1}{2}$. Now, the multiplicity given in Eq. (14.20) is

$$
\mathcal{P}_4^0 = \frac{1}{36}\Big(27 \cdot e^{i\pi \cdot 0} + 3 \cdot e^{i\pi \cdot 1} + 3 \cdot e^{i\pi \cdot 2} + 3 \cdot e^{i\pi \cdot 3} + 27 \cdot e^{i\pi \cdot 4}
$$
$$
+ 3 \cdot e^{i\pi \cdot 5} + 3 \cdot e^{i\pi \cdot 6}
$$
$$
+ 3 \cdot e^{i\pi \cdot 7} + 27 \cdot e^{i\pi \cdot 8} + 3 \cdot e^{i\pi \cdot 9} + 3 \cdot e^{i\pi \cdot 10} + 3 \cdot e^{i\pi \cdot 11}\Big)
$$
$$
= 2. \tag{14.26}
$$

Thus, two families appear from $T_4^0$. GUTs are simple in counting families.

In $T_4^0$, consider $P = (\underline{1\,0\,0\,0}; 1\,1\,1)(0^8)'$, satisfying $(P+4V)^2 + 2\tilde{c} = 0$, which gives $P \cdot V = -\frac{1}{2}$. It has the quantum number of $H_u$. We insert $P \cdot V = -\frac{1}{2}$ for $k = 4$ in Eq. (14.23) to obtain

$$
\Theta_4^0(\text{Higgs}) = -\tilde{s} \cdot \phi + \frac{-2}{12}. \tag{14.27}
$$

Since $4\phi$ is $(\frac{2}{3}, \frac{1}{3}, \frac{1}{3})$, the masslessness condition is the same as above, $s = (\ominus; -\,-\,-)$, i.e. it is left-handed (L-handed), and we obtain $\Theta_4^0(\text{Higgs}) = \frac{1}{4}$. Thus, the multiplicity of $H_u$ is given by Eq. (14.20)

$$
\mathcal{P}_4^0 = \frac{1}{36}\Big(27 \cdot e^{i\pi \cdot 0} + 3 \cdot e^{i\pi \cdot \frac{1}{2}} + 3 \cdot e^{i\pi \cdot \frac{2}{2}} + 3 \cdot e^{i\pi \cdot \frac{3}{2}} + 27 \cdot e^{i\pi \cdot \frac{4}{2}} + 3 \cdot e^{i\pi \cdot \frac{5}{2}}
$$
$$
+ 3 \cdot e^{i\pi \cdot \frac{6}{2}} + 3 \cdot e^{i\pi \cdot \frac{7}{2}} + 27 \cdot e^{i\pi \cdot \frac{8}{2}} + 3 \cdot e^{i\pi \cdot \frac{9}{2}} + 3 \cdot e^{i\pi \cdot \frac{10}{2}} + 3 \cdot e^{i\pi \cdot \frac{11}{2}}\Big)
$$
$$
= 2. \tag{14.28}
$$

Again, we obtain $\mathcal{P} = 2$. These were used for the Higgs fields in [10], but Higgs from $T_6$ was employed in [11].

**Twisted Sector $T_3$: $10b_{-1} + 10_{+1}$**  The degeneracy factors in $T_3$ are

$$
\tilde{\chi}(\theta^3, \theta^j) = \begin{cases} j = 0,\ 1,\ 2,\ 3,\ 4,\ 5,\ 6,\ 7,\ 8,\ 9,\ 10,\ 11 \\ \quad 4,\ 1,\ 1,\ 4,\ 1,\ 1,\ 4,\ 1,\ 1,\ 4,\ \ 1,\ \ 1 \end{cases}. \tag{14.29}
$$

From Eq. (14.19), the generalized GSO phase is

$$
\Theta_3 = \sum_i \left(N_i^L - N_i^R\right) \hat{\phi}_i - \tilde{s} \cdot \phi + P \cdot V + \frac{1}{4}. \tag{14.30}
$$

We have the weight vectors

$$\overline{\mathbf{10}}_{-1} : P + 3V = (+++--; 0^3)(0^5; \tfrac{-1}{4}, \tfrac{-1}{4}, \tfrac{+1}{2})', \quad P \cdot V = \tfrac{1}{4} \qquad (14.31)$$

$$\mathbf{10}_1 : P + 3V = (++---; -1-1-1)(0^5; \tfrac{7}{4}, \tfrac{7}{4}, -\tfrac{14}{4})', \quad P \cdot V = -\tfrac{1}{4}. \qquad (14.32)$$

For these, $(P + 3V) \cdot a = 0$ mod integer. We can also use Eq. (14.30) to obtain multiplicities for massless $\overline{\mathbf{10}}_{-1}$ and $\mathbf{10}_1$,

| $s$ | $(N_i^L - N_i^R)\hat{\phi}_i$, | $\tilde{s} \cdot \phi$, | $P \cdot V$, | $\Theta_3$, | Multiplicity |
|---|---|---|---|---|---|
| $(\oplus\, \vert + --)$ : | $0,$ | $\tfrac{0}{12},$ | $\tfrac{+1}{4}(\overline{\mathbf{10}}_{-1})$ | $\tfrac{+6}{12},$ | $0$ |
| $(\oplus\, \vert - +-)$ : | $0,$ | $\tfrac{-1}{12},$ | $\tfrac{+1}{4}(\overline{\mathbf{10}}_{-1})$ | $\tfrac{+7}{12},$ | $0$ |
| $(\oplus\, \vert - -+)$ : | $0,$ | $\tfrac{-4}{12},$ | $\tfrac{+1}{4}(\overline{\mathbf{10}}_{-1})$ | $\tfrac{-2}{12},$ | $0$ |
| $(\ominus\, \vert - ++)$ : | $0,$ | $\tfrac{0}{12},$ | $\tfrac{+1}{4}(\overline{\mathbf{10}}_{-1})$ | $\tfrac{+6}{12},$ | $0$ |
| $(\ominus\, \vert + -+)$ : | $0,$ | $\tfrac{+1}{12},$ | $\tfrac{+1}{4}(\overline{\mathbf{10}}_{-1})$ | $\tfrac{+5}{12},$ | $0$ |
| $(\ominus\, \vert + +-)$ : | $0,$ | $\tfrac{+4}{12},$ | $\tfrac{+1}{4}(\overline{\mathbf{10}}_{-1})$ | $\tfrac{+2}{12},$ | $0$ |

(14.33)

| $s$ | $(N_i^L - N_i^R)\hat{\phi}_i$, | $\tilde{s} \cdot \phi$, | $P \cdot V$, | $\Theta_3$, | Multiplicity |
|---|---|---|---|---|---|
| $(\oplus\, \vert + --)$ : | $0,$ | $\tfrac{0}{12},$ | $\tfrac{-1}{4}(\mathbf{10}_1)$ | $\tfrac{0}{12},$ | $2$ |
| $(\oplus\, \vert - +-)$ : | $0,$ | $\tfrac{-1}{12},$ | $\tfrac{-1}{4}(\mathbf{10}_1)$ | $\tfrac{+1}{12},$ | $0$ |
| $(\oplus\, \vert - -+)$ : | $0,$ | $\tfrac{-4}{12},$ | $\tfrac{-1}{4}(\mathbf{10}_1)$ | $\tfrac{+4}{12},$ | $1$ |
| $(\ominus\, \vert - ++)$ : | $0,$ | $\tfrac{0}{12},$ | $\tfrac{-1}{4}(\mathbf{10}_1)$ | $\tfrac{0}{12},$ | $2$ |
| $(\ominus\, \vert + -+)$ : | $0,$ | $\tfrac{+1}{12},$ | $\tfrac{-1}{4}(\mathbf{10}_1)$ | $\tfrac{-1}{12},$ | $0$ |
| $(\ominus\, \vert + +-)$ : | $0,$ | $\tfrac{+4}{12},$ | $\tfrac{-1}{4}(\mathbf{10}_1)$ | $\tfrac{-4}{12},$ | $1$ |

(14.34)

$P \cdot V = +\tfrac{1}{4}$ does not lead to a massless field but $P \cdot V = -\tfrac{1}{4}$ leads to massless pair, L-handed $\mathbf{10}_1$ and R-handed $\mathbf{10}_1$. Because the quantum numbers of $\mathbf{10}_1$ are the opposite of those of $\overline{\mathbf{10}}_{-1}$, the R-handed first and third rows in (14.34) on $\mathbf{10}_1$ can be called L-handed $\overline{\mathbf{10}}_{-1}$. Thus, we obtain three L-handed pairs $3(\overline{\mathbf{10}}_{-1} \oplus \mathbf{10}_1)$ from $T_3$.

**Twisted Sector $T_1^0$: Chiral Matter $\overline{\mathbf{10}}'$ of Hidden SU(5)$'$**  In $T_1$, we have

$$\tilde{\chi}(\theta^1, \theta^j) = \begin{cases} j = 0,\ 1,\ 2,\ 3,\ 4,\ 5,\ 6,\ 7,\ 8,\ 9,\ 10,\ 11 \\ \ \ \ \ 3,\ 3,\ 3,\ 3,\ 3,\ 3,\ 3,\ 3,\ 3,\ 3,\ \ 3,\ \ 3 \end{cases}. \qquad (14.35)$$

In $T^1$, we have from (14.23),

$$\Theta_1 = \sum_i (N_i^L - N_i^R)\hat{\phi}_i - \tilde{s} \cdot \phi + P \cdot V + \frac{1}{12}. \tag{14.36}$$

**State 10$b'$** Consider

$$P = (0^5; \left(\frac{-1}{6}\right)^3)\left(\underline{-1\,0^3}\,0; \frac{+1}{4}\,\frac{-1}{4}\,\frac{+1}{2}\right)' \oplus \left(0^5; \left(\frac{-1}{6}\right)^3\right)$$

$$\times \left(\frac{1}{2}\,\frac{1}{2}\,\frac{-1}{2}\,\frac{-1}{2}\,0; \frac{+1}{4}\,\frac{-1}{4}\,\frac{1}{2}\right)', \tag{14.37}$$

for which $P \cdot V$ is $\frac{-2}{12}$. Without oscillators, the masslessness condition is satisfied for

$$
\begin{array}{cccccc}
s & (N_i^L - N_i^R)\hat{\phi}_i, & \tilde{s} \cdot \phi, & P \cdot V, & \Theta_1, & \text{Multiplicity} \\
(\oplus|-+-): & 0, & \frac{-1}{12}, & \frac{-2}{12}, & 0, & 1
\end{array} \tag{14.38}
$$

It is a R-handed field with U(1)$_{\text{KK}}$ charges $Q_{18,20,22} = (-1, +1, -1)$. It is listed in the first row in Table 14.7.

**State $(5, 2)'$** Considering

$$P = (0^5; \frac{-1}{6}\,\frac{-1}{6}\,\frac{-1}{6})(\underline{+1\,0^3}\,0; \frac{+1}{4}\,\frac{-1}{4}\,\frac{+1}{2})'$$

$$(0^5; \frac{-1}{6}\,\frac{-1}{6}\,\frac{-1}{6})(0\,0\,0\,0-1; \frac{+1}{4}\,\frac{-1}{4}\,\frac{+1}{2})', \tag{14.39}$$

we obtain an SU(5)$'$ quintet **5**$'$. Another quintet is

$$\left(0^5; \frac{-1}{6}\,\frac{-1}{6}\,\frac{-1}{6}\right)\left(\underline{\frac{+1}{2}\,\frac{-1}{2}\,\frac{-1}{2}\,\frac{-1}{2}}\,0; \frac{+1}{4}\,\frac{-1}{4}\,\frac{+1}{2}\right)'$$

$$\left(0^5; \frac{-1}{6}\,\frac{-1}{6}\,\frac{-1}{6}\right)\left(\underline{\frac{+1}{2}\,\frac{+1}{2}\,\frac{+1}{2}\,\frac{+1}{2}}\,0; \frac{+1}{4}\,\frac{-1}{4}\,\frac{+1}{2}\right)'. \tag{14.40}$$

For these, $P \cdot V$ is $\frac{-2}{12}$. These are listed as $(\bar{\mathbf{5}}', \mathbf{2}')$ in Table 14.7.

**Table 14.7** U(1) charges of SU(5)' fields

| $R_X(T_1^{0,+})_{\oplus,\ominus}$ | State$(P+kV^{0,\pm})$ | $Q_R$ | $Q_{18}$ | $Q_{20}$ | $Q_{22}$ |
|---|---|---|---|---|---|
| $10b'_0(T_1^0)_R$ | $(0^5; \frac{-1}{6}\frac{-1}{6}\frac{-1}{6})(-1\,0^3\,0; \frac{+1}{4}\frac{-1}{4}\frac{+1}{2})'$ | $+1$ | $-1$ | $+1$ | $-1$ |
| | $(0^5; \frac{-1}{6}\frac{-1}{6}\frac{-1}{6})(\frac{+1}{2}\frac{+1}{2}\frac{-1}{2}\frac{-1}{2}0; \frac{+1}{4}\frac{-1}{4}\frac{+1}{2})'$ | | | | |
| $(5', 2')_0(T_1^0)_R$ | $(0^5; \frac{-1}{6}\frac{-1}{6}\frac{-1}{6})(+1\,0^3\,0; \frac{+1}{4}\frac{-1}{4}\frac{+1}{2})'_\uparrow$ | $+1$ | $-1$ | $+1$ | $-1$ |
| | $(0^5; \frac{-1}{6}\frac{-1}{6}\frac{-1}{6})(0\,0\,0\,0-1; \frac{+1}{4}\frac{+1}{4}\frac{-1}{2})''_\downarrow$ | | | | |
| | $(0^5; \frac{-1}{6}\frac{-1}{6}\frac{-1}{6})(\frac{+1}{2}\frac{-1}{2}\frac{-1}{2}\frac{-1}{2}\frac{-1}{2}; \frac{-1}{4}\frac{-3}{4}0)''_\uparrow$ | | | | |
| | $(0^5; \frac{-1}{6}\frac{-1}{6}\frac{-1}{6})(\frac{+1}{2}\frac{+1}{2}\frac{+1}{2}\frac{+1}{2}\frac{-1}{2}; \frac{+3}{4}\frac{+1}{4}0)''_\downarrow$ | | | | |
| $5b'_0(T_1^+)_R$ | $(0^5; \frac{-1}{6}\frac{-1}{6}\frac{-1}{6})(\frac{-1}{2}\frac{+1}{2}\frac{+1}{2}\frac{+1}{2}0; \frac{+1}{4}\frac{-1}{4}\frac{-1}{2})'$ | $0$ | $+1$ | $+1$ | $+1$ |
| | $(0^5; \frac{-1}{6}\frac{-1}{6}\frac{-1}{6})(0\,0\,0\,0\frac{+1}{2}; \frac{-1}{4}\frac{+1}{4}0)'$ | | | | |
| $5b'_{-5/3}(T_1^+)_R$ | $((\frac{+1}{6})^5; \frac{-1}{3}\frac{+1}{3}0)(\frac{-5}{6}\frac{+1}{6}\frac{+1}{6}\frac{+1}{6}\frac{+1}{6}; \frac{-1}{2}\frac{-1}{4}\frac{-1}{4}0)'$ | $-4$ | $-1$ | $-1$ | $+1$ |
| | $((\frac{+1}{6})^5; \frac{-1}{3}\frac{-1}{3}0)(\frac{+1}{3}\frac{+1}{3}\frac{+1}{3}\frac{+1}{3}0; \frac{-1}{12}\frac{+1}{4}\frac{-1}{2})'$ | | | | |

SU(2) doublets are denoted as up and down arrows

## Twisted Sector $T_1^+$

**State $5b'_0$** In the sector $T_1^+$, we consider

$$V + a = \left(\frac{2}{3}, \frac{2}{3}, \frac{2}{3}, \frac{2}{3}, \frac{2}{3}; \frac{-1}{6}, \frac{-5}{6}, \frac{+1}{2}\right)\left(\frac{2}{3}, \frac{2}{3}, \frac{2}{3}, \frac{2}{3}, 0; \frac{-5}{12}, \frac{+1}{4}, \frac{-1}{2}\right)' \tag{14.41}$$

for which we consider the phase (14.19),

$$\Theta_1^+ = \sum_i (N_i^L - N_i^R)\hat{\phi}_i - \tilde{s}\cdot\phi + P\cdot V^+ + \frac{1}{2}\cdot\frac{62}{12}. \tag{14.42}$$

With $P = \left(0^5; \frac{-1}{6}\frac{-1}{6}\frac{-1}{6}\right)\left(\frac{-1}{2}\frac{+1}{2}\frac{+1}{2}\frac{+1}{2}0; \frac{+1}{4}\frac{-1}{4}\frac{-1}{2}\right)'$, we obtain $P\cdot V^+ = \frac{+10}{12}$. An ingenious lattice shift of the vector $P + V_1^+ = \left(\frac{2}{3}, \frac{2}{3}, \frac{2}{3}, \frac{2}{3}, \frac{2}{3}; \frac{-1}{3}, -1, \frac{1}{3}\right)$ $\left(\frac{1}{6}, \frac{7}{6}, \frac{7}{6}, \frac{7}{6}, \frac{-1}{6}, 0, -1\right)'$ gives $(P + V_1^+)^2 = \frac{210}{144}$, and the magnitude of the masslessness condition is satisfied. Then, we obtain the following multiplicity:

$$\begin{array}{ccccc}
s & (N_i^L - N_i^R)\hat{\phi}_i, & \tilde{s}\cdot\phi, & P\cdot V, & \Theta_1, \text{ Multiplicity} \\
(\oplus| + ++): & 0, & \frac{+5}{12}, & \frac{+10}{12}, & 0, \quad 1
\end{array} \tag{14.43}$$

It is a R-handed field with $U(1)_{KK}$ charges $Q_{18,20,22} = (+1, +1, +1)$. It is listed in Table 14.7.

**State $\overline{5}'_{-5/3}$** Similarly, we obtain a charged quintet $\overline{5}'_{-5/3}(T_1^+)_R$ from the sector $T_1^+$.

### 14.2.3 Doublet–Triplet Splitting

With the simplest and most widely used level $k = 1$, gauge couplings are unified. If one goes beyond $k = 1$, then the resulting spectrum must be much more abundant and would not be a minimal model. Models with differing $k_i$ do not unify coupling constants. Therefore, it has been hoped that a reasonable SSM can be obtained at the level $k = 1$. In this spirit, the initial motivation was to obtain the factor group SU(3)×SU(2) directly by compactification.

But GUTs are attractive in many ways. The Higgs doublets $H_u$ and $H_d$ contained in some representations of a GUT model must accompany their color partners. As commented earlier, low scale colored scalars are dangerous since they can easily make proton decay very fast. So the color-triplet partners of $H_u$ and $H_d$ must be made superheavy. This doublet–triplet splitting problem in SUSY SU(5) was addressed in field theoretic orbifold in Chap. 4, with the discrete group $\mathbb{Z}_2 \times \mathbb{Z}_2'$ in 5D space. In 5D, the internal space $y$ is split into two branes (at $y = 0$ and $\pi R$) and the bulk in between. At the $y = 0$ brane, the full SU(5) is respected. At the $y = \pi R$ brane, the discrete projectors $P$ associated with $\mathbb{Z}_2$ and $P'$ associated with $\mathbb{Z}_2'$ are imposed

$$P = \text{diag}(1\,1\,1\,1\,1) \tag{14.44}$$

$$P' = \text{diag}(1\,1\,1\,{-1}\,{-1}). \tag{14.45}$$

The $S_1/\mathbb{Z}_2 \times \mathbb{Z}_2'$ orbifold breaks SU(5) to the SM and Higgs $\mathbf{5}$ and $\bar{\mathbf{5}}$ are split into massless $H_u$ and $H_d$ plus their superheavy color partners [12]. This is the realization of a doublet–triplet splitting in a 5D field theory model.

In string orbifold, the standard-like models have a potential to have it achieved already if there is no colored triplets beyond the SM spectrum. Indeed, Example 3 of Ref. [13] achives the doublet–triplet splitting. But in GUTs from string orbifold, the problem needs to be resolved. We find, however, that in anti-SU(5) models there is a base to address this problem. This is because the symmetry breaking fields $\mathbf{10}_{+1}$ and $\overline{\mathbf{10}}_{-1}$ have nonvanishing VEVs in the (45) direction, $\langle \Sigma_2^{45} \rangle = \langle \Sigma_{1,45}^* \rangle \neq 0$. Consider first the Higgs quintet $\mathbf{5}$. The $\mathbf{10} \cdot \mathbf{10} \cdot \mathbf{5}$ coupling related to $H_d$ is $\mathbf{10}_1^{12}\mathbf{10}_1^{34}H^5$ is vanishing and $H_d$ is massless. On the other hand $\mathbf{10}_1^{12}\mathbf{10}_1^{45}H^3$ leads to a colored Higgsino mass term for $\tilde{\Sigma}_2^{12} \cdot \tilde{H}^3$ to be of order $\langle \mathbf{10}_1^{45} \rangle$. Similarly, for the Higgs quintet $\bar{\mathbf{5}}$, we can make the colored Higgsino mass term for $\tilde{\Sigma}_{1,12}^* \cdot \tilde{\bar{H}}_3$ to be of order $\langle \Sigma_{1,45}^* \rangle$. But, this solution works only if there is no contribution to the Higgsino mass term, $\tilde{H}_u \tilde{H}_d$. If there is a mass term in the superpotential of the form $\mathbf{5} \cdot \bar{\mathbf{5}} = H\bar{H}$, this doublet–triplet solution does not work. This is the issue discussed in Sect. 13.8, and realistic doublet–triplet splitting from GUTs can be easily realized in anti-SU(5) models.

## 14.2.4  U(1) Charges

To distinguish the SM chiral fields among the plethora of fields, one has to identify the weak hypercharge $U(1)_Y$. If the first five entries of $P$ are anti-SU(5) indices, with the first three for color SU(3) and the next two for the weak SU(2), the weak hypercharge is proportional to

$$q_Y \propto (2\,2\,2\,-3\,-3\ n_6\ n_7\ n_8)(n_9, \cdots, n_{16})'. \tag{14.46}$$

If all of $n_6, \ldots, n_{16}$ are zero, then we obtain $\sin^2\theta_W^0 = \frac{3}{8}$. If any of $n_6, \ldots, n_{16}$ is nonzero, then $\sin^2\theta_W^0 < \frac{3}{8}$.

Model (14.18) leads to the rank 9 non-Abelian gauge group SU(5)×SU(5)'× SU(2)'. There are 7 U(1)s. Besides of the above hypercharge, we have six other generators

$$
\begin{aligned}
q_1 &= (0^5;\, 12, 0, 0)\,(0^8)', \\
q_2 &= (0^5;\, 0, 12, 0)\,(0^8)', \\
q_3 &= (0^5;\, 0, 0, 12)\,(0^8)', \\
q_4 &= (0^8)\,(0^4, 0;\, 12, -12, 0)', \\
q_5 &= (0^8)\,(0^4, 0;\, -6, -6, 12)', \\
q_6 &= (0^8)\,(-6, -6, -6, -6, 18;\, 0, 0, 6)'.
\end{aligned}
\tag{14.47}
$$

We can always make only one of them anomalous $U(1)_X$ by change of basis [11],

## 14.2.5  U(1)$_R$ Identification

As discussed in Sect. 2.3, supergravity in 4D has three functions, superpotential $W(z)$, gauge kinetic function $f(z)$, and Kähler potential $K(z, z^*)$. In the global SUSY case, the Lagrangian contains $\int d^2\vartheta\,(W + f(z)(\lambda^a)^2)$ plus $\frac{1}{2}(D^a)^2$ where $\lambda^a$ are chiral gaugino superfields and $(D^a)^2$ is the D-term. Since $(D^a)^2$ is real, we consider just $\int d^2\vartheta\,(W + f(z)(\lambda^a)^2)$ to look for the complex functions. Since the integrand is complex, this can be matched by the complex $d^2\vartheta$, i.e. let there be a symmetry

$$\text{R-symmetry}: \ \vartheta \to e^{i\,\alpha_R}\vartheta. \tag{14.48}$$

For (14.48) to be a symmetry, $W$ and gauginos transform as

$$W \to e^{i\,2\alpha_R}\,W, \tag{14.49}$$

$$\lambda^a \to e^{i\,\alpha_R}\lambda^a. \tag{14.50}$$

This phase transformation is called R-symmetry, U(1)$_R$. The R-charge of the $\vartheta$'s 0$^{\mathrm{th}}$ power component of a superfield is usually used as the R-charge of the superfield. For example, gauginos carry 1 unit of R-charge and gauge fields carry 0 unit of R-charge. If a chiral field $z$ carries R-charge $R_z$, its fermionic partner carries R-charge $R_z - 1$.

For U(1)$_R$ in supergravity, we consider gravitino also. The Kähler potential $K(z, z^*)$ contains a physical parameter describing gravitino mass. So, in supergravity SUSY is described in the broken phase and U(1)$_R$ is effectively broken. The degree of breaking is from the knowledge of SUSY breaking scale.

Since matter fields $z$ and $\lambda^a$ carry U(1)$_R$ charge, its origin can be from U(1)$_{EE}$ and U(1)$_{KK}$, and in general it can be a linear combination of them,

$$Q_R = \sum_i l_i Q_{EE}^i + l_{18} Q_{KK}^{18} + l_{20} Q_{KK}^{20} + l_{18} Q_{KK}^{22}, \tag{14.51}$$

where $U_{KK}^{18} - U_{KK}^{20}$ are U(1)'s from U(1)$_{KK}$.[6]

## 14.2.6  Discrete Symmetry $\mathbb{Z}_{4R}$

In Sect. 13.5.2, we discussed a need for introducing a kind of R-parity toward proton longevity, forbidding the dimension-5 operator Eq. (13.37). This has to be considered together with the neutrino mass generation since the leading neutrino mass operator appears at the same dimension-5 level [14]. At the dimension-5 level, the $\mu$ term also appears [15]. So, these two operators must be considered together with the $\mu$ problem.

In string compactification, U(1)$_{\mathrm{anom}}$ appears frequently. In this regard, Lee et al. considered the anomaly coefficients in SUSY field theory and found that U(1)$_{EE}$ alone cannot be used to suppress the $\mu$ term [16]. We adopt their conclusion on $\mathbb{Z}_{4R}$ that the needed R-parity is a subgroup of a U(1)$_R$ symmetry. So, let our U(1)$_R$ be a linear combination of U(1)$_{EE}$ and U(1)$_{KK}$.

In supergravity, an intermediate scale $M_I$ generates the electroweak scale $v_{ew}$ as $\sim M_I^2/M_P$. Then, the $\mu$ term and $v_{ew}$ can be economically generated at the same scale [17]. Also, for a multiple appearance of Higgs pairs, the democratic mass matrix, by some kind of fine tuning, always guarantees at least one massless pair of Higgs doublets [18]. So, we may consider the cases of discrete groups $\mathbb{Z}_4, \mathbb{Z}_6, \mathbb{Z}_8$, and $\mathbb{Z}_{12}$ of Ref. [19].

Let us consider the following operators, relevant for the dimension-5 proton decay and neutrino mass operators in an anti-SU(5) model,

$$W^{\Delta B} \equiv \overline{10}\,\overline{10}\,\overline{10}\,5 \tag{14.52}$$

---

[6]Even if the KK U(1)'s are projected out, the U(1)'s from U(1)$_{EE}$ are affected by these. In a sense, the other combination is considered to be projected out and Eq. (14.51) survives.

$$W^{\nu \, \text{mass}} \equiv \mathbf{5} \, \mathbf{5} \, \overline{\mathbf{5}}_{H_u} \, \overline{\mathbf{5}}_{H_u},$$                      (14.53)

where the subscripts $m$ and $H$ denote matter fields and Higgs fields, respectively. If an operator is present in the superpotential, $U(1)_{KK}$ transformations of the fields of an operator is cancelled by the transformation of the anti-commuting variable $\vartheta$. Under certain normalization, the superpotential is required to have $+2$ units of the $U(1)_{KK}$ charge.

Consider the bullet of Fig. 14.3 whose charge is 0. Since the rotation angle of variable $\vartheta$ can be taken as the negative of the previous transformation, $-2$ units of the $U(1)_{KK}$ charge must be allowed also as illustrated in Fig. 14.3. Let both charge $\pm 2$ scalars develop VEVs, but the charge 0 scalars do not. In this case, we have a $\mathbb{Z}_4$ symmetry $-2 \equiv +2$, i.e. minimally we require $\mathbb{Z}_{4R}$ symmetry when we consider the global transformation of $\vartheta$. The $\mathbb{Z}_{4R}$ quantum numbers can be labelled as those in green color, and the black number assignment is identical to those of green colors.

Under $\mathbb{Z}_{4R}$, the superpotential $W$ leading to proton decay operator and the $\mu$ term are required to carry $+4 \equiv 0$ units which are then forbidden by $U(1)_R$, and the superpotential for neutrino mass operator carries $+2$ units which is allowed by $U(1)_R$. These can be satisfied with the matter charges $+1$ and the Higgs charges 0, for example. In string compactification, the realization may be more complex because one must take into account the sectors where these fields appear.

The R-parity $\mathbb{Z}_{4R}$ is a discrete subgroup of $U(1)_R$,

$$U(1)_R \subset U(1)_{EE} \otimes U(1)_{KK}$$                      (14.54)

which is obtained by VEVs of $U(1)_R$ charge $4n \cdot$ (integer) fields, including $n = 0$ and $\pm 1$. If any $n = 0$ fields do not develop a VEV, then $(4n + 2) \cdot$ (integer) fields, including $n = 0$ and $\pm 1$, will do the job. In any case, superpotential carries $+2$ (modulo 4) units of $U(1)_R$ charge. On the other hand, the integrand under $d^2\vartheta \, d^2\bar{\vartheta}$ carries $+4$ (modulo 4) units of $U(1)_R$ charge.

The scale of $\mu$ can be considered in the model, forbidding the operator of Eq. (14.52) but allowing the operator of Eq. (14.53). These are checked for Model (14.18) in Ref. [11].

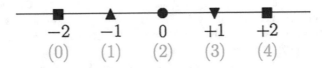

**Fig. 14.3** $\mathbb{Z}_{4R}$ quantum numbers in the region $[-2, +2]$. Numbers in the region $[0, 4]$ are shown in the brackets

### 14.2.7 A $\mathbb{Z}_6$ Orbifold Model

We present an example model of Ref. [20]. It is constructed on a $T^6/\mathbb{Z}_{6-\mathrm{II}}$ orbifold with shift vectors

$$V = \frac{1}{6}(2 \;-3 \;-3\,0\,0\,0\,0\,0)(3 \;-1 \;-3 \;-3 \;-3 \;-3 \;-3\,3)$$

$$a_1 = \frac{1}{6}(\tfrac{3}{2} \;-\tfrac{3}{2} \;-\tfrac{3}{2} \;-\tfrac{3}{2} \;-\tfrac{3}{2}\,\tfrac{3}{2}\,\tfrac{3}{2}\,\tfrac{3}{2})(6 \;-6 \;-15 \;-9 \;-3 \;-15 \;-9\,9)$$

$$a_3 = \frac{1}{6}(-3 \;-3\,1\,1\,1\,1\,1\,1)(0\,0\,0\,0\,0\,0\,0). \tag{14.55}$$

Here, $a_1$ is order two and $a_3$ is order six. The resulting gauge group is $SU(3) \times SU(2) \times U(1) \times SO(8) \times SU(2) \times U(1)^7$ where the first three corresponds to the SM groups.

### 14.2.8 Other Unified Models

There is a unified model based on the Pati–Salam group [21]. It takes the $\mathbb{Z}_6$ orbifold, to insert a $\mathbb{Z}_2$ symmetry so that the recent 5D field theoretic orbifolds using $\mathbb{Z}_2$ discussed in Chap. 5 are derivable from string construction. Here, different scales of tori are necessarily introduced since one is trying to achieve intermediate 5D physics.

### Exercises

▶ **Exercise 14.1** Show that the states with nonvanishing multiplicities in Table 14.5 do not satisfy the mass shell condition $M_R^2 = 0$ but $s = \left(\ominus, \frac{-3}{2}, \frac{-3}{2}, -\right)$ satisfies the mass shell condition.

▶ **Exercise 14.2** Prove that the chirality of the $\theta^1$-twisted sector in Sect. 14.2.1 is $-1$, i.e. left-handed.

### References

1. H. Abe, T. Kobayashi, H. Otsuka, Y. Takano, Realistic three-generation models from SO(32) heterotic string theory. J. High Energy Phys. **9**, 056 (2015)
2. K.-S. Choi, S.G. Nibbelink, M. Trapletti, Heterotic SO(32) model building in four dimensions. J. High Energy Phys. **12**, 063 (2004)
3. J. E. Kim, Grand unfication models from SO(32) heterotic string. e-print: 2008.00367 [hep-th]
4. S. Dimopoulos, H. Georgi, Softly Broken Supersymmetry and SU(5). Nucl. Phys. **B193**, 150–162 (1981)
5. N. Sakai, Naturalness in supersymmetric guts. Z. Phys. **C11**, 153 (1981)

6. S. Dimopoulos, S. Raby, F. Wilczek, Supersymmetry and the scale of unification. Phys. Rev. **D24**, 1681–1683 (1981)
7. J.P. Derendinger, J.E. Kim, D.V. Nanopoulos, Anti-SU(5). Phys. Lett. **B139**, 170–176 (1984)
8. Y. Katsuki, Y. Kawamura, T. Kobayashi, N. Otsubo, Y. Ono, K. Tanioka, *Tables in* $\mathbf{Z}_N$ *orbifold models*. Kanazawa University preprint DPKU-8904 (1989)
9. J.-H. Huh, J.E. Kim, B. Kyae, SU(5)(flip) $\times$ SU(5)-prime from Z(12-I). Phys. Rev. **D80**, 115012 (2009)
10. J.E. Kim, Theory of flavors: string compactification. Phys. Rev. **D98**(5), 055005 (2018)
11. J.E. Kim, R-parity from string compactification. Phys. Rev. **D99**(9), 093004 (2019)
12. Y. Kawamura, Triplet doublet splitting, proton stability and extra dimension. Prog. Theor. Phys. **105**, 999–1006 (2001)
13. L.E. Ibanez, J.E. Kim, H.P. Nilles, F. Quevedo, Orbifold compactifications with three families of SU(3) $\times$ SU(2) $\times$ U(1)**n. Phys. Lett. **B191**, 282–286 (1987)
14. S. Weinberg, Baryon and lepton nonconserving processes. Phys. Rev. Lett. **43**, 1566–1570 (1979)
15. J.E. Kim, H.P. Nilles, The mu problem and the strong CP problem. Phys. Lett. **B138**, 150–154 (1984)
16. H.M. Lee, S. Raby, M. Ratz, G.G. Ross, R. Schieren, K. Schmidt-Hoberg, P.K.S. Vaudrevange, A unique $\mathbb{Z}_4^R$ symmetry for the MSSM. Phys. Lett. **B694**, 491–495 (2011)
17. J.E. Kim, A common scale for the invisible axion, local SUSY GUTs and Saxino decay. Phys. Lett. **136B**, 378–382 (1984)
18. J.E. Kim, B. Kyae, S. Nam, The anomalous $U(1)_{\text{anom}}$ symmetry and flavors from an SU(5) $\times$ SU(5) $'$ GUT in $\mathbf{Z}_{12-I}$ orbifold compactification. Eur. Phys. J. **C77**(12), 847 (2017)
19. H.M. Lee, S. Raby, M. Ratz, G.G. Ross, R.S.K. Schmidt-Hoberg, P.K.S. Vaudrevange, Discrete R symmetries for the MSSM and its singlet extensions. Nucl. Phys. **B850**, 1–30 (2011)
20. O. Lebedev, H.P. Nilles, S. Raby, S. Ramos-Sanchez, M. Ratz, P.K.S. Vaudrevange, A. Wingerter, A mini-landscape of exact MSSM spectra in heterotic orbifolds. Phys. Lett. B **645**(1), 88–94 (2007)
21. T. Kobayashi, S. Raby, R.-J. Zhang, Constructing 5-D orbifold grand unified theories from heterotic strings. Phys. Lett. **B593**, 262–270 (2004)

# Smooth Compactification

# 15

An important class of manifolds permitting $\mathcal{N} = 1$ supersymmetry in four dimension are the Calabi–Yau manifolds [1]. For a gauge hierarchy solution supersymmetry is desirable. For the chiral nature of the SM fermions it is restricted to minimal supersymmetry.

Orbifold may be regarded as the singular limit of Calabi–Yau manifold. We have a standard mathematical tool for resolving singularity to recover a smooth manifold geometry. In defining the topology, the resolution is necessary. For instance, the Euler number of an orbifold is defined by that of the resolved orbifold. We may understand this procedure in many ways. We construct explicit resolution in the non-compact geometry. Also, we can more easily deal with the geometry algebraically. The latter approach is powerful enough since most quantities are topological.

We take inductive approach: after dealing with concrete examples, we generalize the results in more abstract mathematical objects without rigorous proofs. Many features of Calabi–Yau manifold are shared by orbifold. A detailed treatment is given in Refs. [1–3], especially on K3 manifold in Ref. [4], differential geometry and index theorem in Ref. [5], and on special holonomy in Refs. [6, 7]. For differential geometry and algebric geometry, we refer to [3, 8]. For toric geometry, see [9–11].

## 15.1  Calabi–Yau Manifold

We study dimensional reduction further by considering again the ten dimensional effective action (11.1). To examine the condition of supersymmetry breaking, we need to look at the transformation properties of fermions, the gravitino $\psi_M$, gaugino $\chi^a$, and dilatino $\lambda$. It is because their superpartners, the scalars, can develop VEV to break supersymmetry. We present them schematically

$$\delta\psi_M = \left(\partial_M + \frac{1}{4}(\omega_{MNP} - \frac{1}{2}H_{MNP})\Gamma^{NP}\right)\epsilon^{(10)}\left(x^M\right) \tag{15.1}$$

© Springer Nature Switzerland AG 2020
K.-S. Choi, J. E. Kim, *Quarks and Leptons From Orbifolded Superstring*,
Lecture Notes in Physics 954, https://doi.org/10.1007/978-3-030-54005-0_15

$$\delta \lambda^A = \Gamma^{MN} F^A_{MN} \epsilon^{(10)} \left( x^M \right) \tag{15.2}$$

$$\delta \chi = \left( \Gamma^M \partial_M \Phi - \frac{1}{12} \Gamma^{MNP} H_{MNP} \right) \epsilon^{(10)} \left( x^M \right), \tag{15.3}$$

up to four fermion terms. Here, the gamma matrices are antisymmetrized, e.g. $\Gamma^{PQ} = \Gamma^{[P} \Gamma^{Q]}$. Also, $\epsilon^{(10)}$ is the SUSY parameter which is Majorana–Weyl fermion in ten dimensions. In this book, we set $H_{MNP} = 0$, $\Phi = $ constant and $G_{\mu\nu} = \eta_{\mu\nu}$.

Throughout this book, we have used implicitly the notion of complex manifolds. It is more than a convenience, since the holonomy group $SU(n)$ and its subgroup, the point group of the orbifold, act on complex representations.

### 15.1.1 Geometry Breaks Supersymmetry

In the setup $H_{MNP} = 0$, the conditions for the gravitino reduce to

$$\nabla_M \epsilon^{(10)} \left( x^M \right) = 0. \tag{15.4}$$

That is, the covariantly constant spinor equation with respect to the metric connection. It is purely a *geometric* condition.

Next, let us compactify six dimensions. We can decompose the parameter $\epsilon^{(10)}$ under the spacetime group $SO(6) \times SO(1, 3) \subset SO(1, 9)$

$$\epsilon^{(10)} \left( x^M \right) = \eta \left( y^m \right) \otimes \epsilon \left( x^\mu \right) \oplus \eta^* \left( y^m \right) \otimes \epsilon^* \left( x^\mu \right), \tag{15.5}$$

where $\eta(y^m), \epsilon(x^\mu)$ are, respectively, 6D and 4D SUSY parameters. The ten dimensional chirality operator is decomposed into four and six dimensional ones

$$\Gamma^{(10)} = \Gamma^{(4)} \Gamma^{(6)}, \quad \Gamma^{(6)} = \Gamma^4 \Gamma^5 \cdots \Gamma^9. \tag{15.6}$$

Once we fix the ten dimensional one, the chiralities of six dimensions and four dimensions are correlated

$$\Gamma^{(4)} \epsilon^{(10)} = \Gamma^{(6)} \epsilon^{(10)} = \pm \epsilon^{(10)}. \tag{15.7}$$

This amounts to the branching

$$\mathbf{16} \rightarrow (\mathbf{4}, \mathbf{2}) + (\bar{\mathbf{4}}, \bar{\mathbf{2}}),$$

and the chiralities are correlated. The number of invariant components of $\mathbf{4}$ leaves the same number of 4D supersymmetries.

We define *holonomy* of a manifold $M$ as the rotation group of the spinor $\eta(y^m)$ under arbitrary transportaion with the metric connection, around a closed loop. A general holonomy of $d$ dimensional real manifold is $SO(d)$. We consider *special* holonomy as a proper subgroup of $SO(d)$. For instance, if the holonomy group is $SU(3) \subset SO(6) \simeq SU(4)$, a general transportation rotates only three components of $\eta(y^m)$ in a suitable basis. Then, we always have one invariant component

$$4 \rightarrow 3 + 1, \tag{15.8}$$

and one-fourth of the SUSY is preserved. Starting from heterotic string, we have $\mathcal{N} = 4/4 = 1$ SUSY in 4D and this is what we require for the chirality. A manifold admitting an $SU(n)$ holonomy is called Calabi–Yau manifold.

Also note that this unbroken supersymmetry condition, or the special holonomy condition, is just continuous version of Eq. (3.73). To study holonomy, we review properties of complex manifolds.

### 15.1.2 Complex Manifold

We first complexify the spacetime. In flat space, complex coordinates are naturally introduced as

$$z^a \equiv \frac{1}{\sqrt{2}} \left( y^{2a-1} + iy^{2a} \right), \quad z^{\bar a} \equiv \overline{z^a} \equiv \frac{1}{\sqrt{2}} \left( y^{2a-1} - iy^{2a} \right). \tag{15.9}$$

Without introducing imaginary number, we can also complexify real dimensions. The block-diagonal matrix defined on a point

$$J = \begin{pmatrix} 0 & -1 & & & \\ 1 & 0 & & & \\ & & 0 & -1 & \\ & & 1 & 0 & \\ & & & & \ddots \end{pmatrix} \tag{15.10}$$

satisfies $J^2 = -\mathbf{1}_n$. We can verify that it is diagonalizable to $J^a{}_b = i\delta^a{}_b$, $J^{\bar a}{}_{\bar b} = -i\delta^a{}_b$ in the complex basis $\{z^a, z^{\bar a}\}$. This complexification is nothing but decomposition of an $SO(2n)$ vector representation into an $SU(n)$ fundamental representation. Every vector, spinorial, and antisymmetric representation of $SO(2n)$ can be decomposed under suitable tensor products of fundamental $\mathbf{n}$ and anti-fundamental $\bar{\mathbf{n}}$ of $SU(n)$. The example of $SO(10) \rightarrow SU(5)$ is presented in Eq. (5.86) in Chap. 5 and below.

We may think of a tensor $J(x^m)$, called *almost complex structure*, that has the same components $J(x^m) = J$ at a point $x^m$. In terms of holonomy singlet spinor

$\eta(y)$ in (15.4),

$$J^m{}_n(y) = G^{mp}\eta^\dagger(y)\Gamma_{pn}\eta(y). \tag{15.11}$$

We can check that this is the desired complex structure from the singlet property of $\eta$. At *one* point, every real, traceless, and SU(3) invariant tensor can be always adjusted to have the form (15.10). The problem is whether we can do this at *all* the other points in a local neighborhood. It would be nice to think about a similar consideration about flatness. In Riemannian geometry, we have a criterion, where a manifold is locally flat: it is so if its Riemann tensor vanishes. For the complex structure, we have a similar counterpart, called *Nijenhuis* tensor

$$N^p{}_{mn} \equiv J^q{}_m \left(\partial_q J^p{}_n - \partial_n J^p{}_q\right) - J^p{}_n \left(\partial_q J^p{}_m - \partial_m J^p{}_q\right).$$

If this vanishes, then the almost complex structure becomes (15.10) by the Newlander–Nirenberg theorem [8]. Not every almost complex manifold admits a complex structure, e.g. $S^4$ does not admit a complex structure [6].

**Kähler Manifold**

The metric of a form $G_{ab} = G_{\bar{a}\bar{b}} = 0$ always gives a positive definite norm. Then, the manifold is called *Hermitian*.

We can construct two forms out of metric by contracting the complex structure $k_{ij} = G_{ik}J^k{}_j$. This makes a metric tensor into a differential-(1, 1)-form called the Kähler form. In the complex basis, it is

$$k_{a\bar{b}} = -iG_{a\bar{b}} = -k_{\bar{b}a}, \quad k_{ab} = k_{\bar{a}\bar{b}} = 0. \tag{15.12}$$

When the metric is Hermitian and the Kähler form is closed,

$$dk = 0, \tag{15.13}$$

where d is the exterior derivative, and the manifold is called *Kähler manifold*. Since $d = \partial + \bar{\partial}$, it implies

$$\partial k = \bar{\partial} k = 0. \tag{15.14}$$

Since $k$ is a total derivative in the holomorphic and antiholomorphic derivative, $G$ of (15.12) is obtained from a single function $\mathcal{K}$ called *Kähler potential*

$$G_{a\bar{b}} = -\partial_a \partial_{\bar{b}} \mathcal{K}. \tag{15.15}$$

This is an important feature. As a consequence, the nonvanishing connection is either purely holomorphic or antiholomorphic

$$\Gamma^a_{bc} = G^{a\bar{d}}\partial_b G_{c\bar{d}}, \quad \Gamma^{\bar{a}}_{\bar{b}\bar{c}} = G^{\bar{a}d}\partial_{\bar{b}} G_{\bar{c}d}. \tag{15.16}$$

This means that a holomorphic (or an antiholomorphic) index transforms again into a holomorphic (or an antiholomorphic) one. Thus, an $n$-(complex) dimensional Kähler manifold admits a U($n$) holonomy.

## Calabi–Yau Manifold

We want SU($n$) holonomy from the discussion in Sect. 15.1.1. Since we have a U($n$) holonomy in the Kähler manifold, we may seek a further condition making the U(1) part trivial.

Note that the U(1) generator of SO($2n$) is nothing but the almost complex structure $J$ in (15.11). Therefore, given an SO($2n$) generator $M$, we can extract the U(1) part as Tr $JM$. Viewing the Riemann tensor as a matrix $R_{nq}$ with the local SO($2n$) rotational indices

$$R^m{}_{npq} = (R_{nq})^m{}_p, \tag{15.17}$$

we see that the Ricci tensor is the U(1) part. So, the Ricci-flat condition

$$R_{mn} = 0 \tag{15.18}$$

makes the U(1) part trivial. The first Chern class is defined as the Ricci tensor with a suitable normalization. Thus, the Ricci flatness means vanishing first Chern class. A Kähler manifold with vanishing first Chern class is called *Calabi–Yau manifold*. In other words, the Calabi–Yau (CY) manifold admits an SU($n$) holonomy.

In the Calabi–Yau manifold, we have invariant tensor $\eta(y^m)$ in (15.5). We can show an $(n, 0)$ form

$$\Omega_{m_1 \ldots m_n} = \eta^\top(y) \Gamma_{m_1 \ldots m_n} \eta(y). \tag{15.19}$$

We can show that this form is covariantly constant

$$\bar{\partial} \Omega = 0. \tag{15.20}$$

In this standard, the torus $T^n$ satisfies the above because it trivially satisfies $R_{mnpq} = 0$ everywhere. So, most of the time we demand a stronger definition that the holonomy should not bigger than SU($n$). This means that the manifold should not be reduced to the product involving a sub-torus.

## 15.1.3 Mode Expansion

We need harmonic forms, zero-mode solutions to harmonic equation. We briefly summarize the Hodge theory on classifying harmonic forms and its geometric dual.

## Homology and Cohomology

We are mainly interested in the mode expansion and we need harmonic forms with zero eigenvalues. They have redundancy under gauge transform. Cohomology provides a good way of counting them.

We first consider a submanifold and introduce a boundary operator $\delta$. We say a submanifold $C$ is closed if it has no boundary, or annihilated by the operator $\delta$, $\delta C = 0$. The boundary $\delta D$ of a submanifold $D$ is called exact. It has no boundary $\delta \delta D = 0$, and therefore $\delta^2 = 0$ and the boundary operator is said nilpotent. A *homology* class of $p$-cycles $H_p$ is defined as

$$H_p = \frac{\text{closed } p\text{-chain}}{\text{exact } p\text{-chain}}. \tag{15.21}$$

We may deform a cycle to another if they form the boundary of a higher dimensional chain.

There is a similar notion in differential forms. For the real differential operator d, $F$ is closed form if d $F = 0$ and is exact if it is expressed as $F = $ d $A$. Every exact form is closed, and d is nilpotent d$^2 = 0$. A closed form is locally exact by Poincaré lemma [8], but not globally. We can define the cohomology class of $p$-form $H^p$ as

$$H^p = \frac{\text{closed } p\text{-form}}{\text{exact } p\text{-form}}. \tag{15.22}$$

Namely, it is the closed class of differential $p$-forms, not overcounting those connected by gauge transformation, $F_{[p]} \sim F_{[p]} + $ d $\Lambda_{[p-1]}$. We call its dimension Betti number $b^p$. Since all the points are homologous and there is only one volume form, we have $b_0 = b_m = 1$.

It is de Rham's theorem [9] that establishes the one-to-one correspondence $H^p(M) \simeq H_p(M)$ between a $p$-cycle $c_{[p]}$ and a $p$-form $\omega_{[p]}$ because there is a unique real number for the integral

$$\int_{C_p} \omega_{[p]}. \tag{15.23}$$

Their dimensions are the same as well $b^p = b_p$. Finally, in $m$ dimensions, we have Poincaré duality $H_p(M) = H_{m-p}(M)$ between $p$-form $\omega_{[p]}$ and $(m - p)$-cycles $C_{[m-p]}$

$$\int_{C_{[p]}} \omega_{[p]} = \int_M \omega_{[m-p]} \wedge \omega_{[p]}. \tag{15.24}$$

Thus we may define Poincaré dual form $\omega_{[m-p]}$ of $\omega_{[p]}$. Thus we have $b^p = b^{m-p}$.

This also naturally defines the intersection number between homology cycles

$$C_{[p]} \cdot C_{[m-p]} = \int_M \omega_{[m-p]} \wedge \omega_{[p]}. \tag{15.25}$$

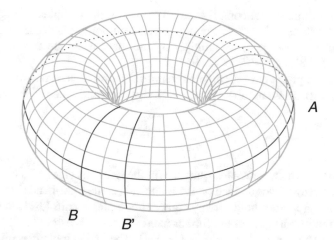

**Fig. 15.1** Homology one-cycles on a two-torus. The cycles $A$ and $B$ are different, but $B$ and $B'$ are homologous

We should assume that the two cycles should intersect transversally: the intersection should have lower dimension than that of each. Consider a two-torus and two different 1-cycles $A$, $B$ on it. Their transversal intersections are points. We deform one of the cycles and make them intersect more times. However, this makes the intersection number invariant if we add $(+1)$ if they meet in the right-hand rule and $(-1)$ in the left. We may show that this is invariant under the deformation [13] (Fig. 15.1).

We have complex generalization of the cohomology, called Dolbeault cohomology. We have complex differential operators $\partial$ and $\bar{\partial}$ generate holomorphic and antiholomorphic indices. They are nilpotent $\partial^2 = \bar{\partial}^2 = 0$. Naturally, we can think of the $(p, q)$ form having $p$ and $q$ holomorphic and antiholomorphic indices, respectively, and similarly the closed and exact forms. We define $p, q$ cohomology class as

$$H_{\bar{\partial}}^{p,q} = \frac{\partial\text{-closed } (p, q)\text{-form}}{\partial\text{-exact } (p, q)\text{-form}}. \tag{15.26}$$

In this case also, we have the de Rham counterpart homology $H_{p,q}$. Therefore, $h^{p,q} \equiv \dim H^{p,q}$ is the same as $h_{p,q} \equiv \dim H_{p,q}$. As in the de Rham cohomology, we have $h^{0,0} = h^{n,n} = 1$.

The complex conjugation takes

$$H^{p,q} \simeq H^{q,p}. \tag{15.27}$$

This is isomorphism and so the resulting dimensions are the same $h^{p,q} = h^{q,p}$. We have Serre duality, a generalized Hodge star in Dolbeault cohomology [14, 15]

$$H^{p,q} \simeq H^{n-q,n-p}, \tag{15.28}$$

which is in $n$ dimensional complex manifold. Using the complex conjugation, we also have $H^{p,q} \simeq H^{n-p,n-q}$. Finally, we have holomorphic duality converting $(p,0)$-form into $(n-p,0)$ by multiplying the three form and take the Poincare dual. Thus, $H^{p,0} \simeq H^{n-p,0}$.

An invariant property of a complex manifold is the dimensions $h^{p,q}$ (which is not topological). These numbers form the Hodge diamond. The Euler number is

$$\chi = \sum_{p,q} (-1)^{p+q} h^{p,q}. \tag{15.29}$$

We have connected the geometric quantity (in the left-hand side) with characteristic quantities of gauge fields defined on the manifold (in the right-hand side).

The above properties hold for any compact complex manifolds. If the manifold is Calabi–Yau, some numbers are further fixed. Since a CY $n$-fold has the nowhere-vanishing $(n,0)$ form, we have $h^{n,0} = h^{0,n} = 1$. The first homology is related to the fundamental class of the topology. It means that the number of 1-cycle is the same as the number of non-contractable cycle. We strictly require the holonomy of the Calabi–Yau manifold $SU(n)$, not a subgroup, and we have no torus factor $b_1 = 0$. Thus, $h^{1,0} = h^{0,1} = 0$. Using the holomorphic duality, we have $h^{n-1,0} = h^{1,0} = h^{0,n-1} = 0$.

### K3 Manifold
The simplest nontrivial Calabi–Yau manifold has two complex dimensions. The only nontrivial Hodge number is $h^{1,1} = 20$. There are many way to obtain this. For instance, we may write the algebraic equation and impose the vanishing first Chern class condition. We may do this for just one case, thanks to the theorem that there is a unique manifold called K3, in the sense that all the K3 manifolds are diffeomorphic to each other, that is, they are connected by maps preserving differentiation structure [4].

$$
\begin{matrix}
 & & h^{0,0} & & & & & & 1 & & \\
 & h^{1,0} & & h^{0,1} & & & & 0 & & 0 & \\
h^{2,0} & & h^{1,1} & & h^{0,2} & = & 1 & & 20 & & 1. \\
 & h^{2,1} & & h^{1,2} & & & & 0 & & 0 & \\
 & & h^{2,2} & & & & & & 1 & &
\end{matrix}
\tag{15.30}
$$

Therefore, we have $b^2 = 22$ and the Euler number is 24. The harmonic forms $\omega^i$ in $H^{1,1}(K3)$ have intersection number [4]

$$\int \omega^i \wedge \omega^j = A^{ij}, \tag{15.31}$$

where $A^{ij}$ is the same as the inverse metric of the lattice $\Gamma_8 \times \Gamma_8 \times \Gamma_{1,1} \times \Gamma_{1,1} \times \Gamma_{1,1}$.

**Calabi–Yau Threefold**

Now, we turn to threefold. Nontrivial numbers are $h^{2,1} = h^{1,2}, h^{1,1} = h^{2,2}$. They are summarized in the Hodge diamond

$$
\begin{array}{ccccccc}
 & & & h^{0,0} & & & \\
 & & h^{1,0} & & h^{0,1} & & \\
 & h^{2,0} & & h^{1,1} & & h^{0,2} & \\
h^{3,0} & & h^{2,1} & & h^{1,2} & & h^{0,3} \\
 & h^{3,1} & & h^{2,2} & & h^{1,3} & \\
 & & h^{3,2} & & h^{2,3} & & \\
 & & & h^{3,3} & & &
\end{array}
=
\begin{array}{ccccccc}
 & & & 1 & & & \\
 & & 0 & & 0 & & \\
 & 0 & & h^{1,1} & & 0 & \\
1 & & h^{2,1} & & h^{2,1} & & 1 \\
 & 0 & & h^{1,1} & & 0 & \\
 & & 0 & & 0 & & \\
 & & & 1 & & &
\end{array}
. \tag{15.32}
$$

Since Kähler form $k_{a\bar{b}}$ is (1, 1) form, we have at least $h^{1,1} \geq 1$. In many cases, we have $h^{1,1} = 1$.

The elements in $H^{2,1}$ correspond to the number of parameters describing deformations of complex structure $G_{ij}\delta J^j{}_k$, as in (15.11). The $H^{1,1}$ elements describe the deformations of the Kähler structure $\delta J_{a\bar{b}}$.

Finally, there is *mirror symmetry* relating a string theory on Calabi–Yau manifold $X$ and another one on mirror manifold $\check{X}$, whose Hodge numbers $h^{1,2}$ and $h^{1,1}$ are exchanged .

## 15.2 Standard Embedding

As in the orbifold case, we break the group by relating a holonomy action in the compact space (a generalized point group action of the orbifold case) with a background gauge group. The space action of Calabi–Yau threefold belongs to the SU(3) holonomy.

Let us choose the *standard embedding*, that is, to identify the spatial rotation with that in the group space. Understanding the Riemann tensor as the field strength for the SO($n$) transformation as in (15.17), we take

$$
\overline{R}_{mn} = \overline{F}_{mn}. \tag{15.33}
$$

This naturally breaks $E_8 \times E_8$ into the commutant to SU(3)

$$
E_6 \times E_8.
$$

### 15.2.1 Mode Expansion

The SU(3) part does not survive. Nevertheless, it is useful to consider taking into account the SU(3) part via branching. The $(\mathbf{248}, \mathbf{1}) + (\mathbf{1}, \mathbf{248})$ branches into

$$(Q_a)_b^x : (\mathbf{3}, \mathbf{27}, \mathbf{1}), \quad (Q_{\bar{a}})_{\bar{b}}^{\bar{x}} : (\bar{\mathbf{3}}, \overline{\mathbf{27}}, \mathbf{1}), \quad (Q_a)_{\bar{b}} : (\mathbf{8}, \mathbf{1}, \mathbf{1}),$$
$$A_\mu^A : (\mathbf{1}, \mathbf{78}, \mathbf{1}) + (\mathbf{1}, \mathbf{1}, \mathbf{248}). \tag{15.34}$$

Again, the first entries pretend as representations of the gauge group but are of the SU(3) holonomy group. Their indices are attached, in addition to the Lorentz index $M$. Since $\mathbf{3}$ and $\mathbf{27}$ are correlated, we suppressed indices of the latter.

Now, we make the Klein reduction of fields on the Calabi–Yau manifold. The basic strategy is the same as in the toroidal and orbifold case discussed in Sect. 5.1. We can decompose the Laplacian and the Dirac operator into our 4D spacetime part $(x^\mu)$ and the 6D internal space part $(y^m)$

$$\nabla_M \nabla^M = \partial_\mu \partial^\mu + \nabla_m \nabla^m, \tag{15.35}$$

$$\Gamma_M \nabla^M = \Gamma_\mu \partial^\mu + \Gamma_m \nabla^m, \tag{15.36}$$

where the covariant derivative $\nabla_m$ contains the gauge and metric connections. Then, we mode expand fields $\phi$ in terms of the complete eigenstates of the harmonic operators in the internal space[1]

$$\phi = \sum \phi_4(x^\mu) \omega(y^i). \tag{15.37}$$

Note however that unlike in Sect. 5.1, we expand here only zero eigenstates. Here, the harmonic forms $\omega$ are the solution of the Laplace equation, $(\partial + \bar{\partial})^2 \omega = 0$, or equivalently

$$\partial \omega = \bar{\partial} \omega = 0. \tag{15.38}$$

The harmonic forms are the bases of homology cycle $H^{p,q}$. All of them have the zero eigenvalues, and we may have multiple number of zero modes.

Now, let us consider the standard embedding. Consider first the field transforming as $(\bar{\mathbf{3}}, \overline{\mathbf{27}})$. As in (15.12), the one represented as $A_{i,\bar{b}}$ is a $(1,1)$ form; therefore, we have a similar mode expansion

$$Q_{a,\bar{b}} = \sum_{p=1}^{h^{1,1}} Q^{[p]}(x^\mu) \omega_{a\bar{b}}^{[p]}(y^i). \tag{15.39}$$

---

[1]In general, $\Gamma_m \partial^m$ and $\Gamma_\mu \partial^\mu$ do not commute. However, we can multiply a suitable gamma matrix to make them commute. Even, for the consideration of the zero modes of $\Gamma_m \partial^m$, the eigenvalue vanishes and so there is no such ambiguity.

Here, $Q^{[p]}$ has no spatial index. In four dimensions, we have $h^{1,1}$ hypermultiplets
**27**. The "Kähler" moduli $T_{a\bar{b}}$ of (3.77) are also of (1,1) form; therefore, it has the
same mode expansion and the same number of zero modes.

Next, consider (**3**, **27**), denoted by $(Q_a)_{bx}$. This is not a differential form because
the indices $a$ and $b$ are symmetric. We can make it a differential form using the
covariantly constant (3, 0) tensor

$$Q_{a,\bar{d}\bar{e}} = Q_{a,b} G^{b\bar{c}} \Omega_{\bar{c}\bar{d}\bar{e}}. \tag{15.40}$$

Then, we have the following mode expansion:

$$Q_{a,\bar{d}\bar{e}} = \sum_{p=1}^{h^{1,2}} Q^{[p]}(x^{\mu}) \omega^{[p]}_{a\bar{d}\bar{e}}(y^i). \tag{15.41}$$

Therefore, the number of hypermultiplets (**3**, **27**)s is $h^{1,2}$. The same argument
applies to the complex structure moduli $U_{ab}$, yielding the same number of zero
modes.

Since we have $h^{1,0} = 0$, there are no zero modes for $Q_{a,x}$.

### 15.2.2 Number of Generations

We know that **27** of $E_6$ houses one complete family, including a pair of Higgs
doublets, as discussed in 2.24. Because of the Survival Hypothesis, every pair of
**27** and $\overline{\mathbf{27}}$ can form a mass term around the unification scale; thus, they decouple at
low energy if there is no symmetry forbidding it to happen. Thus, the net number of
families is the difference $|h^{2,1} - h^{1,1}|$. It is remarkable to see, by the definition of
Euler number $\chi$ in Eq. (15.29),

$$\text{Number of families}: \quad |h^{2,1} - h^{1,1}| = \frac{|\chi|}{2}. \tag{15.42}$$

The number of families reduces to the *topological property*. This relies on the
assumptions that (1) the Survival Hypothesis (SH) holds,[2] (2) one representation,
here **27**, houses a complete family, and (3) as seen in (15.34), the field transforming
as **27** carries the internal space index $a$ of the special holonomy SU(3). In particular,
the last condition is violated in the presence of Wilson lines. A typical construction
of Calabi–Yau manifold gives a very large Euler number of order 100, so that we
need a further sophistication. The one constructed by Tian and Yau gives $|\chi| = 6$ to
allow three families [16].

---

[2]One may envision that this SH is evaded by a stringy calculation, for example, in cases where
a mass term cannot be formed. But, massless $\overline{\mathbf{27}}$ must be made heavy around or above the
electroweak scale; so, the number of chiral families is still $\frac{|\chi|}{2}$.

### 15.2.3 Wilson Lines

With the embedding (15.33), we do not have a realistic model, for the same reason
as in the orbifold case. We need a further symmetry breaking. Still there is another
object, Wilson line, related with non-contractible cycles dealt in Sect. 7.5.

Unfortunately, Calabi–Yau manifold has no isometry. If we have a Killing vector
$\nabla_{(m}\xi_{n)} = 0$, then the Eq. (15.73) tells us that this vector should be harmonic form.
This contradict that there is no harmonic 1-form $h^{1,0} = h^{0,1} = 0$. However, it is
possible to have a discrete symmetry with which we may mod out the manifold.
This enables us to turn on the Wilson lines and at the same time reduce the Euler
number that is related to the number of generations.

We generalize the Wilson line $U$ as the object associating one-cycle (instead
of translational symmetry that the torus possesses) with the action in the group
space. As in the orbifold case, the Wilson line should satisfy the consistency
condition (15.53). This condition is exhausted by the identification of the SU(3)
in the group space and the holonomy group, leaving the field strength $F$ of the other
group $E_6$ being zero. This is exactly the situation, where the Wilson line emerges.

To be compatible with the standard embedding, we do not touch on the SU(3)
subgroup associated with the holonomy group, but consider the unbroken $E_6$. As an
example, let us work in the $SU(3)_C \times SU(3)_L \times SU(3)_R$ basis which is one of the
maximal subgroups of $E_6$. Consider a matrix acting on the generator of this group.
Specifically, we choose $U$ of (7.83) as

$$U = \mathbf{1}_3 \otimes \begin{pmatrix} \beta & & \\ & \beta & \\ & & \beta^{-2} \end{pmatrix} \otimes \begin{pmatrix} \gamma & & \\ & \delta & \\ & & (\gamma\delta)^{-1} \end{pmatrix}, \tag{15.43}$$

where $\alpha, \beta, \ldots$ are the unit roots of 1. Since only the commuting generators
of (15.34) survive, this breaks $E_6$ down to

$$SU(3) \times SU(2) \times U(1)^3.$$

It contains the desired standard model group, with extra two U(1) groups. The order
of Wilson line should not exceed that of the orbifolding $V$.

We may go on to turn on more Wilson lines and choose another CY manifold to
find a realistic model. There have been endeavors along this line [17].

## 15.3    General Embedding

We want to turn on more general background gauge field which may lead to
realistic gauge group like GUT or the SM groups. If the background gauge field
belongs to larger gauge group, we can obtain smaller unbroken group. We study the
requirement for this gauge bundle and construct more realistic models.

## 15.3.1 Background Gauge Field

First, we study the condition for the background gauge field.

### Hermitian Yang–Mills Equations

So far, we have studied the conditions imposed on the geometry of compact manifold $\mathcal{M}$. Taking into account gauge fields on $\mathcal{M}$, we need more supersymmetric conditions on these. The condition for gaugino (15.2) reads

$$\Gamma^{mn} F_{mn}\epsilon = \left(\Gamma^{ab} F_{ab} + \Gamma^{\bar{a}\bar{b}} F_{\bar{a}\bar{b}} + 2\Gamma^{a\bar{b}} F_{a\bar{b}}\right)\epsilon = 0. \tag{15.44}$$

Due to Hodge, we decomposed the two-form $F_{mn}$ into, respectively, $(2, 0)$, $(0, 2)$ and $(1, 1)$ parts. Since the Kähler manifold does not mix holomorphic and antiholomorhic indices, each term should separately vanish

$$F_{ab} = F_{\bar{a}\bar{b}} = 0, \tag{15.45}$$

$$g^{a\bar{b}} F_{a\bar{b}} = 0. \tag{15.46}$$

The first equation in (15.45) states that locally the field strength is a pure gauge

$$A_{\bar{a}} = -iU^{-1}\partial_{\bar{a}}U, \tag{15.47}$$

with a gauge transformation matrix $U(z^a, \bar{z}^{\bar{a}})$. Globally, the transition function is holomorphic

$$A_{\bar{a}} \to U A_{\bar{a}} U^{-1} + iU^{-1}\partial_{\bar{a}}U \tag{15.48}$$

among different patches, for which we have the stanadrd patchability in the manifold. A similar antiholomorphicity holds for $A_a$ hence $F_{\bar{a}\bar{b}}$. In the standard embedding of Sect. 15.1.3, this is automatically satisfied since we have identified the gauge connection with the spin connection which is holomorphic.

The second Eq. (15.46) is called Hermitian Yang–Mills equation. This generalizes the self-dual equation for instantons in four dimensions. So, we call the group that the background gauge field belongs to structure group. To be specific, consider the case where $M$ is a Kähler manifold of complex dimension $n$, with the Kähler form $J$. We can show that, for a U(1) part, it can be converted into

$$\text{tr } F \wedge J \wedge J = 0. \tag{15.49}$$

Locally, we can always make this vanish with a suitable choice of $g_{a\bar{b}}$, globally not. Noting that $c_1(V) \equiv i \text{ tr } F/(2\pi)$, the global condition is that the integral

$$\int_M c_1(V) \wedge J \wedge J \equiv \deg(V) \tag{15.50}$$

over the compact manifold is a topological invariant, and easily one might see that $F$ is a topological invariant, the first Chern class. We call it degree because, in the case of $\dim_{\mathbb{C}} M = 2$, the algebraic degree of the defining equation is dual to that of the line bundle.

We also define slope [18, 19]

$$\mu(V) \equiv \frac{\deg(E)}{\text{rank}(E)}. \tag{15.51}$$

A vector bundle $E$ is stable,[3] if and only if for any subsheaf $S$ of $E$ satisfy the inequality

$$\mu(S) \leq \mu(V). \tag{15.52}$$

Sometimes, the stable condition means the one without equality, while this relaxed condition is called semistable. By the theorem of Donaldson and that of Uhlenbeck and Yau [6, 22, 23], any stable bundle of satisfying (15.45) also satisfy (15.46) and vice versa. This is analogous condition for D-flatness. Roughly speaking, the stability condition means the vector bundle is not reducible [19].

The consistency condition comes from the Bianchi identity for the antisymmetric tensor field $H$ (11.133)

$$\text{Tr}_{\text{v}}\, \overline{R}^2 - \text{Tr}_{\text{v}}\, \overline{F}^2 = 0. \tag{15.53}$$

Here, $\text{Tr}_{\text{v}}$ is the trace over the vector representation, and as before we mean the product of differential forms as the wedge product. If we have semisimple groups $E_8 \times E_8$, we mean $\text{Tr}_{\text{v}}\, \overline{F}^2 = \text{Tr}_{\text{v}}\, \overline{F}_1^2 + \text{Tr}_{\text{v}}\, \overline{F}_2^2$, with the background field strengths of the two $E_8$s. We may convert this trace with that for the adjoint using $\text{Tr}_{\text{v}} = \frac{1}{g^\vee}\text{Tr}_{\text{adj}}$, where $g^\vee$ is the dual Coxeter number of the group $g$, displayed in Table 12.2.

The above conditions (15.33) naturally remind us of the relation between the space twist and the shift vector (7.50). It also follows that (15.53) corresponds to the modular invariance condition (7.50). Certainly, this Calabi–Yau case is the more general and continuous version. The correspondence is studied in Refs. [24, 25].

### 15.3.2 Spectrum

We can rephrase the above discussion in terms of characteristic classes, following Ref. [26]. For a choice of the structure group $G$, the unbroken group is its commutant $H$ in $E_8 \times E_8$. All the generators in $H$ commute with all the generators in $G$.

---

[3]This is called $\mu$-stability in the literature, because the ratio is called $\mu$-slope . There is another kind of stabilities, which are mostly equivalent [20, 21].

Considering the breaking $E_8 \to G \times H$, the adjoint branches as

$$\mathbf{248} \to (\mathbf{adj}_H, \mathbf{1}) + (\mathbf{1}, \mathbf{adj}_G) + \sum_{\mathbf{R}} (\mathbf{R}, \mathbf{R}'). \tag{15.54}$$

Although the representations of $G$ do not survive, they serve tracking tools for the number of zero modes of $H$.

The Bianchi identity (15.53) is expressed as

$$\mathrm{ch}_2(\overline{F}) - \mathrm{ch}_2(\overline{R}) = 0. \tag{15.55}$$

For the Calabi–Yau, we have $c_1(\overline{F}) = 0$, which we assume in the sequel, thus $\mathrm{ch}_2(F) = -c_2(F)$.

The ten dimensional gaugino satisfies Dirac equation and dimensionally reduced to four dimensional chiral fermions.

The zero mode is by definition the eigenstates of the Dirac operator $\slashed{D}$ in the gauge background of $G$. The solution of the Dirac operator $\slashed{D}$ is a charged wavefunction transforming as vector bundle and at the same spacetime tensors with Lorentz indices.

The cohomology $H^r(X, \mathcal{V})$ contains harmonic $r$-forms having the value in $\mathcal{V}$. If $\mathbf{R}'$ is a rank $r$ totally antisymmetric tensor product $\wedge^r \mathcal{V}$, then the zero mode belongs to the cohomology $H^r(X, \wedge \mathcal{V})$. We expand the spinors in the harmonic basis of $H^k(X, \mathcal{V})$. We have seen that the $\mathbf{3}$ part of the field $(\mathbf{3}, \mathbf{27})$ transforms as the vector bundle $\mathcal{V}$ under the structure group SU(3). The four dimensional zero modes become

$$H^0(X, \mathcal{O}_X) \otimes \mathbf{78}, H^1(X, \mathrm{ad}\,\mathcal{V}) \otimes \mathbf{1},$$
$$H^1(X, \mathcal{V}) \otimes \mathbf{27}, \quad H^1(X, \mathcal{V}^*) \otimes \overline{\mathbf{27}}.$$

We learned that in the standard embedding we identified the background gauge field with background connection. Thus, $\mathcal{V}$ adds the indices of $(1, 1)$-form as in (15.40), and $\mathcal{V}^*$ is adds one antiholomorphic index as in (15.39)

$$H^1(X, \mathcal{V}) \simeq H^{2,1}(X), \quad H^1(X, \mathcal{V}^*) \simeq H^{1,1}(X). \tag{15.56}$$

Thus, the number of generations is given by the difference of their dimensions

$$-n_{27} + n_{\overline{27}} = -h^{2,1} + h^{1,1}, \tag{15.57}$$

which is obtained in (15.42).

We also consider a nonstandard embedding. Taking the structure group $SU(5)_\perp \times E_8$, the unbroken group is the commutant SU(5). We have branching, under SU(5) $\times$ $SU(5)_\perp$,

$$\mathbf{248} \to (\mathbf{24}, \mathbf{1}) + (\mathbf{1}, \mathbf{24}) + (\mathbf{10}, \mathbf{5}) + (\overline{\mathbf{5}}, \mathbf{10}) + (\overline{\mathbf{10}}, \mathbf{5}) + (\mathbf{5}, \overline{\mathbf{10}}). \tag{15.58}$$

Now, focus on the structure group $SU(5)_\perp$. The vector bundle $\mathscr{V}$ transforms as **5**, and the tensor product **10** transforms as $\mathscr{V} \wedge \mathscr{V}$, which we also denote as $\wedge^2 \mathscr{V}$. The spectrum and the corresponding number of zero modes are

$$
\begin{aligned}
H^0(X, \mathscr{O}_X) \otimes \mathbf{24}, && n_{24} &= 1, \\
H^1(X, \mathrm{ad}\,\mathscr{V}) \otimes \mathbf{1}, && n_1 &= h^1(\mathrm{ad}\,\mathscr{V}), \\
H^1(X, \wedge^2\mathscr{V}) \otimes \mathbf{5}, && n_5 &= h^1(\wedge^2\mathscr{V}), \\
H^1(X, \wedge^2\mathscr{V}^*) \otimes \mathbf{\bar{5}}, && n_{\bar{5}} &= h^1(\wedge^2\mathscr{V}^*), \\
H^1(X, \mathscr{V}) \otimes \mathbf{10}, && n_{10} &= h^1(\mathscr{V}), \\
H^1(X, \mathscr{V}^*) \otimes \mathbf{\overline{10}}, && n_{\overline{10}} &= h^1(\mathscr{V}^*).
\end{aligned}
$$

In this nonstandard embedding, we cannot relate the vector bundle-valued cohomology with that of ordinary cohomology. So, we have to have the number of generations as

$$
\begin{aligned}
-n_{10} + n_{\overline{10}} - n_{\bar{5}} + n_5 &= -h^1(\mathscr{V}) + h^1(\mathscr{V}^*) - h^1(\wedge^2\mathscr{V}^*) + h^1(\wedge^2\mathscr{V}) \\
&= 2h^0(\mathscr{V}) - 2h^1(\mathscr{V}) + 2h^2(\mathscr{V}) - 2h^3(\mathscr{V}),
\end{aligned}
\tag{15.59}
$$

where we used $h^k(\mathscr{V}^*) = -h^k(\mathscr{V})$ and inserted $h^0(\mathscr{V}) = h^3(\mathscr{V}) = 1$.

### 15.3.3  Index Theorem

We have seen that the number of generation is the half of the Euler number. Euler number is a *topological invariant* in the sense that it is invariant under continuous change of geometry. De Rham's theorem relates this to cohomology: the zero eigenstates of an operator, in this case Dirac. Can we have also topological invariant from the operator side? The affirmative answer comes from the index theorem by Atiyah and Singer.

We consider Dirac operator in the internal space. We saw in (15.7) that the chirality of six and four dimensional spinors is correlated. We know that the Klein–Gordon operator $H = \nabla_m \nabla^m$ in (15.35) is the square of the Dirac operator $i\slashed{D} \equiv \Gamma_m \nabla^m$ in (15.36). We check

$$
[\Gamma^{(6)}, H] = 0, \quad \{\Gamma^{(6)}, \slashed{D}\} = 0.
$$

This means that the operator $i\slashed{D}$ flips the chirality of the eigenstate $\phi$ and the resulting $i\slashed{D}\phi$ is also the eigenstates of $H$. Note that this argument only applies to nonzero eigenstate of $\slashed{D}$ since the zero mode is annihilated. However, the following quantity, called *index*, as difference of the numbers of zero eigenstates of left and right movers,

$$\text{index}_{\mathbf{R}'} \, i \slashed{D} = n_{\mathbf{R}'} - n_{\overline{\mathbf{R}}'} \tag{15.60}$$

is invariant. Here, the subscript means that the covariant derivative is with respect to the representation $\mathbf{R}'$ of the structure group. In the previous example, it is the SU(3) structure group. We also observe that in four dimensions, charge conjugation changes $\mathbf{R}'$ to its conjugate $\overline{\mathbf{R}}'$ and exchange chirality as well, therefore $n_+ \leftrightarrow n_-$ and $\text{index}_{\overline{\mathbf{R}}'} \slashed{D} = -\text{index}_{\mathbf{R}'} \slashed{D}$. It follows that the real representation is not counted, but only complex representation **3**, and thus **27** (correlated!) is counted.

Using *Atiyah–Singer index theorem*, we can calculate the index in terms of a characteristic classes [27, 28]. For an "elliptic" operator $D$ on $m$ dimensional complex manifold with covariant derivative taking values in $\mathscr{V}$, we have [5, 8]

$$\text{ind}(\mathscr{V}, D) = (-1)^{m(m+1)/2} \int_M \text{ch}\left(\oplus_r (-1)^r E_r\right) \frac{\text{Td}(TM^{\mathbb{C}})}{e(TM)}\bigg|_{\text{vol}}. \tag{15.61}$$

The quantities in the integrand are to be defined soon. Only the $2n$-forms will be extracted for the manifold $M$ of the $n$ complex dimensions.

Let us apply this to Dirac operator. From the commutation relation of gamma matrices (4.16), we may associate the Gamma matrices with creation and annihilation operators [3]

$$d = \Gamma^{\tilde{a}+} D_a, \quad d^* = \Gamma^{\tilde{a}-} D_a, \tag{15.62}$$

where $\Gamma^{\tilde{a}\pm}$ are ladder operators for the spinorial states introduced in (4.16). From this, we can verify that the Dirac operator $i\slashed{D} = i\Gamma^m \partial_m$ is nilpotent. The zero modes belong to the cohomology $H^r(X, \mathscr{V})$. That is, it contains the complete zero eigenstates under the Dirac operators. Like the Euler characteristic (15.29), we may also define an alternating sum of the dimensions of the homologies

$$\chi(\mathscr{V}) \equiv \sum_{r=0}^{6} (-1)^r b^r(\mathscr{V}) = \sum_{p,q} (-1)^{p+q} h^{p,q}(\mathscr{V}). \tag{15.63}$$

For example, considering the SU($n$) vector bundle on Calabi–Yau threefold, we have $c_1(\mathscr{V}) = c_1(M) = 0$. So,

$$\int_M \text{ch}(\mathscr{V}) \, \text{Td}(M) = \int_M \left(n - c_2(\mathscr{V}) + \frac{1}{2} c_3(\mathscr{V})\right) \left(1 + \frac{1}{12} c_2(M)\right)$$
$$= \frac{1}{2} \int_M c_3(\mathscr{V}) = \frac{1}{2} \chi(\mathscr{V}), \tag{15.64}$$

where in the last line we evaluate six-forms.

In the standard embedding, we identified the vector bundle with the tangent bundle of the manifold, and we have $c_r(\mathscr{V}) = c_r(M)$ and $\chi(\mathscr{V}) = \chi$, as is well-known. As expected, this result is exactly the same as in the orbifold case we discussed in Sect. 7.4. For the standard embedding case of Sect. 7.4, being Calabi–Yau manifold, tracking the Hodge numbers gives $h^{1,1} = 72, h^{2,1} = 0$. Therefore, we have the correct number of families, $|\chi|/2 = 36$.

Also, it makes general enough statements that the origin of chirality in four dimensions is geometry and/or background gauge fields.

## Characteristic Classes

We have seen that many quantities like the number of generalization are topological and obtained by generalization of Euler theorem. At the heart lies index theorem, which states that the number of zero modes of a given operator is given by topological number, which is given by *characteristic class*. An integer quantized number of zero modes minus anti-zero modes is given by integral of a polynomial made of gauge-invariant field strengths.

The building block is the Chern class

$$c(V) = \det\left(1 + \frac{iF}{2\pi}\right),\tag{15.65}$$

which is the generating function for the $r$th Chern classes

$$c(V) \equiv \sum c_r(V),\tag{15.66}$$

where

$$c_0(V) = 1$$

$$c_1(V) = \frac{iF}{2\pi}$$

$$c_2(V) = \frac{1}{2}\frac{1}{(2\pi)^2}\left(\operatorname{tr} iF \wedge iF - \operatorname{tr}(iF) \wedge \operatorname{tr}(iF)\right),$$

$$\vdots$$

$$c_n(V) = \frac{1}{(2\pi)^n}\det(iF).$$

Note that it is generalization of monopole and instanton numbers.

We may compute anomaly polynomial using index theorem [27, 29, 30].

- Gauge anomaly polynomial counts the number of zero modes of the Dirac operator

$$\operatorname{ind} i\,\slashed{D}_{m+2} = \int \operatorname{ch}_{l+1}(F),\tag{15.67}$$

where ch$_i$ is called Chern character

$$\mathrm{ch}(F) \equiv \mathrm{tr} \exp \frac{iF}{2\pi} = \sum \frac{1}{j!} \mathrm{tr} \left( \frac{iF}{2\pi} \right)^j . \tag{15.68}$$

We can further define $n$th Chern characters as

$$\mathrm{ch}(F) \equiv \sum \mathrm{ch}_j(F),$$

where

$$\mathrm{ch}_0 = \mathrm{rank}\, g$$

$$\mathrm{ch}_1 = c_1$$

$$\mathrm{ch}_2 = \frac{1}{2} \left( c_1^2 - 2c_2 \right)$$

$$\mathrm{ch}_3 = \frac{1}{6} \left( c_1^3 - 3c_1 c_2 + 3c_3 \right).$$

- Gravitational anomaly for Weyl fermions is generated by Dirac genus [27, 29, 30]

$$\hat{I}_{1/2}(R) = \hat{A}(R) = \prod_{i=1}^{2k+1} \frac{\frac{1}{2}x_i}{\sinh \frac{1}{2}x_i} . \tag{15.69}$$

$$\hat{A}(\mathcal{V}) = 1 - \frac{1}{24} p_1(\mathcal{V}) + \frac{1}{5760} \left( 7p_1^2 - 4p_2 \right)(V) + \ldots,$$

where

$$p_j(\mathcal{V}) = (-1)^j c_{2j}.$$

- A gravitino has spin 3/2, which is obtained by tensor product between spin 1/2 and vector. However, we also need to extract spin 1/2 component. The resulting gravitational anomaly is generated by

$$\hat{I}_{3/2}(R) = \hat{A}(R) \left( 2 \sum_{j=1}^{2k+1} \cosh(x_j) - 1 \right) \tag{15.70}$$

- The rank-two antisymmetric tensor anomaly is generated by Hirzebruch $L$-polynomial

$$L(x) = \prod_{j=1}^{k} \frac{x_j}{\tanh x_j} . \tag{15.71}$$

- The zero modes of Dirac operator are counted by Todd class

$$\mathrm{Td}(E)=\prod_{1}^{n}\frac{x_i}{1-e^{-x_i}}=1+\frac{1}{2}c_1(E)+\frac{1}{12}(c_1(E)^2+c_2(E))+\frac{1}{24}c_1(E)c_2(E)+\cdots .$$

(15.72)

## 15.4   Relation to Orbifold

We review the relation between orbifold and smooth Calabi–Yau manifold. They are related to blow-ups.

The four dimensional manifolds $T^4/\mathbb{Z}_N$ with $N = 2, 3, 4, 6$ are related to the K3 manifold by "blowing-up" the singularities. This should be due to the uniqueness of K3. Naively, this correspondence is understood by excising around the apex of the cone and replacing it with a "well-patched" smooth manifold, that we will see shortly (Fig. 15.2).

**Blowing Up**

In this four dimensional space, we can solve the Ricci-flat constraint. From (15.4), the integrability condition becomes

$$[\nabla_m, \nabla_n]\eta = \frac{1}{4}R_{mn}{}^{pq}\Gamma_{pq}\epsilon = 0.$$

(15.73)

It is certain that if not Ricci–Flat there is an additional piece proportional to $\mathrm{Tr}\, J R$ in (15.18). Since a chiral spinor satisfies $\Gamma\epsilon = \epsilon$, the constraint becomes the self-dual equation

$$R_{mnpq} = \tfrac{1}{2}\epsilon^{abcd} R_{mn}{}^{pq}.$$

(15.74)

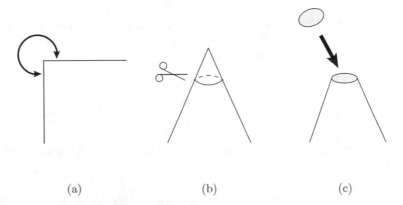

(a)                              (b)                              (c)

**Fig. 15.2**  Blow-up of orbifold singularity (**a**). This is performed (**b**) by cutting around the singular fixed points and (**c**) replacing them with a smooth manifold, which is also modded out by the same $\mathbb{Z}_2$

We may check that the Eguchi–Hanson space with the following metric is a solution [31, 32],

$$ds^2 = \left(1 - \left(\frac{a}{r}\right)^4\right)^{-1} dr^2 + r^2 \left(1 - \left(\frac{a}{r}\right)^4\right)\sigma_3^2 + r^2 \left(\sigma_1^2 + \sigma_2^2\right), \qquad (15.75)$$

where we parameterized the line element by radial variable $r > 0$ and these Euler angles $(\theta, \phi, \psi)$, with $0 \leq \theta < \pi, 0 \leq \phi < 2\pi, 0 \leq \psi < 4\pi$,

$$\sigma_1 = -\sin\psi d\theta + \cos\psi \sin\theta d\phi$$
$$\sigma_2 = \cos\psi d\theta + \sin\psi \sin\theta d\phi \qquad (15.76)$$
$$\sigma_3 = d\psi + \cos\theta d\phi.$$

This looks like a strange convention, but we can easily check that this is a differential form version of the Pauli matrix satisfying Maurer–Cartan equation $\sigma_a \wedge \sigma_b = 2\epsilon_{abc} d\sigma_c$. We verify that $\sigma_1^2 + \sigma_2^2$ is the metric of two-sphere $d\Omega_2^2 = d\theta^2 + \sin^2\theta d\phi^2$ and $\sigma_1^2 + \sigma_2^2 + \sigma_3^2$ is that of three-sphere $S^3$.

The metric (15.75) has one parameter $a$ corresponding to the radius of two-cycle. Let us look at the local property near $r = a$. Parameterizing $r \equiv a + \epsilon^2/a$ we have

$$ds^2 = d\epsilon^2 + \epsilon^2(d\psi + \cos\theta d\phi)^2 + a^2 d\Omega_2^2 \qquad (15.77)$$

in the limit $\epsilon \to 0$. Locally (neglecting $d\phi$) we see that the space looks like $R^2(\epsilon, \psi) \times S^2(\theta, \phi)$ whose Euler number is that of the two-sphere. However, the angular variable $\psi$ has wrong periodicity $4\pi$ instead of $2\pi$. This becomes smooth at $r = a$, if we mod out by a $\mathbb{Z}_2$ action $\psi \to 4\pi - \psi$. In the original coordinates, this action identifies the antipodal points to make $S^3/\mathbb{Z}_2$. This is the original $\mathbb{Z}_2$ action on $\mathbb{C}^2/\mathbb{Z}_2$ orbifold. Also, the curvature drops as $1/r$ as $r \to \infty$, and the resulting geometry is asymptotically locally Euclidian (ALE). Therefore, this space is well-glued that we can cut out the small cone around the fixed point of $\mathbb{C}^2/\mathbb{Z}_2$ and replace it with this space. A more generalized series of solutions are found, providing blow-up $\mathbb{C}2/\Gamma$ singularities, where $\Gamma$ is the discrete subgroup of SU(2).

**Euler Number**

Here, we focus on the case of $T^4/\mathbb{Z}_2$ orbifold. The easiest criterion for the equivalence comes from topology. The Euler number becomes [1]

$$\chi = \frac{\chi(M) - \chi(F)}{N} + \chi(N), \qquad (15.78)$$

with the notation to be understood as the following. The original two-torus is everywhere flat and thus has the Euler number $\chi(M) = 0$. We cut out discs, which has Euler number $\chi(D^2) = 1$, at sixteen fixed points. Because of modding with $\mathbb{Z}_N$, the region for integration to calculate the topological number on the orbifold is reduced to $1/N$ of the torus integration region, and thus it is divided by $N$. $N = 2$ is the order of orbifold action. Shortly, we will see that the patched region corresponds to a smooth manifold of sphere with Euler number 2; so $\chi(N) = 2$. We therefore have

$$\chi(T^4/\mathbb{Z}_2) = 16\chi(\mathbb{C}^2/\mathbb{Z}_2) = 16\left(-\tfrac{1}{2} + 2\right) = 24,$$

in accord with (15.30). For non-prime $N$, we need more than one resolution. For instance, consider $T^4/\mathbb{Z}_4$ that we have seen in Sect. 9.1. It has four order 4 fixed points. Not overcounting these, among twelve order two fixed points, we have six invariant combinations. So effectively, we have six $\mathbb{Z}_2$ fixed points. The total Euler number is

$$\chi\left(T^4/\mathbb{Z}_4\right) = 6\left(-\tfrac{1}{2} + 2\right) + 4\left(-\tfrac{1}{4} + 4\right) = 24,$$

In general, for $\mathbb{Z}_\ell$ fixed point, we may generalize the above to define

$$\chi\left(\mathbb{C}^2/\mathbb{Z}_\ell\right) = -\frac{1}{\ell} + \ell. \tag{15.79}$$

Putting appropriate number $d_j$ of the $\mathbb{Z}_{N/j}$ fixed points, we may verify that

$$24\chi\left(T^4/\mathbb{Z}_N\right) = \sum_{j \in N/\mathbb{Z}} d_j \chi\left(\mathbb{C}^2/\mathbb{Z}_{N/j}\right) = 24. \tag{15.80}$$

We may understand the Euler number as follows. The trivial zero cycle (point) and four cycle (volume) determines, respectively, $h^{0,0} = h^{2,2} = 1$. We have $4 \cdot 3/2 = 6$ independent two-cycles, dual to $d\,y^m \wedge d\,y^n$. Also, each blow-up contributes as $H^{1,1}$ cycles, so there are 16 more elements in $H^{1,1}$. Therefore, there are 24 relevant parameters in total, which coincides with the Euler number we obtained above.

## 15.5   Algebraic Description

We describe the above orbifolds and their resolutions using algebraic geometry. In addition to topological numbers, we may also calculate the size of the Yukawa coupling from intersection theory.

Usually, the Calabi–Yau manifold is described by *algebraic variety*, that is, a (set of) polynomials in the projective spaces and, with them, the shape deformations are readily countable [3]. They are all Kähler and the metric is well-known [14].

Thus, if we have vanishing first Chern class, we have Calabi–Yau manifold. It is known that if the order of the defining equation, which should be homogeneous, is $n + 2$, we have $c_1 = 0$ [15]. Thus, we may describe Calabi–Yau $n$-fold using degree $n + 2$ homogeneous polynomial in $\mathbb{CP}^{n+1}$.

### 15.5.1 $A, D, E$ Singularity

The local geometry of a fixed point of the $T^4/\mathbb{Z}_N$ orbifold is a non-compact Calabi–Yau twofold $\mathbb{C}^2/\mathbb{Z}_N$. Taking the local coordinate $(a, b)$ where the singularity lies at $(0, 0)$, we have the point group action

$$\begin{pmatrix} z_1 \\ z_2 \end{pmatrix} \to \begin{pmatrix} \alpha & 0 \\ 0 & \alpha^{-1} \end{pmatrix} \begin{pmatrix} z_1 \\ z_2 \end{pmatrix}, \quad \alpha = e^{2\pi i/N}. \tag{15.81}$$

Faithful coordinates that are invariant under the transformation are

$$x = z_1^N, \quad y = z_2^N, \quad z = z_1 z_2,$$

satisfying the equation

$$P(x, y, z) \equiv xy - z^N = 0. \tag{15.82}$$

This is regarded as defining equation for the in $\mathbb{C}^2/\mathbb{Z}_N$ in a $\mathbb{C}^3$ with the coordinates. We know this geometry is singular because the gradient is zero at the singular point, so that the slope at this point is not well-defined

$$P(0, 0, 0) = 0, \quad (\partial_x P, \partial_y P, \partial_z P)|_{(0,0,0)} = (0, 0, 0), \quad N \geq 2.$$

### 15.5.2 Resolution

We resolve $C^2/\mathbb{Z}_2$ orbifold singularity following Refs. [33, 34].

We deal with described except $(z_1, z_2) = 0$ by the twist $(\frac{1}{2}, \frac{1}{2})$. We resolve this singularity by introducing a new coordinate $x$ and impose scaling

$$\mathbb{C}^* : (z_1, z_2, x) \sim \left( \lambda^1 z_1, \lambda^1 z_2, \lambda^{-2} x \right). \tag{15.83}$$

Under the monodromy $x \to e^{-2\pi i} x$, at the patch $\lambda^{-2} x = 1$, the coordinate acquires the phase as in the $T^4/\mathbb{Z}_2$ orbifold.

Formally, we defined the resolution

$$\text{Res}\left( \mathbb{C}^2/\mathbb{Z}_2 \right) = \left( \mathbb{C}^2 - \{(0, 0)\} \right)/\mathbb{C}^*. \tag{15.84}$$

We remove the singular point, introduce extra coordinates, and mod out appropriately. Then, the non-singular points in the original space remain the same. The previously singular point is accessed by $x = 0$, then we have $\mathbb{P}^1$ described by

$$(z_1, z_2) \sim (\lambda z_1, \lambda z_2). \tag{15.85}$$

This is what we meant by replacing a singular geometry by $\mathbb{P}^1$. We name this an *exceptional* divisor

$$E \equiv \{x = 0\} \tag{15.86}$$

because it newly appeared after blow-up. This is to be contrasted with ordinary divisors

$$D_i \equiv \{z_i = 0\}, \quad i = 1, 2, \tag{15.87}$$

which are present before blowing up. The blown-up geometry has the same dimension. The location in the exceptional divisor depends on the approaching angle in the bulk space.

The defining functions, $z_1$ for $D_1$ and $z_2$ for $D_2$, are related by rational function: indeed $z_1/z_2$ is the rational function on the $\mathbb{C}^2$. This is tracked by the scaling in (15.85). We also have scaling $1 : -2$ between $D_1, E$ or $D_2, E$. Thus, we have linear equivalence relations $\sim$

$$D_1 \sim D_2, \quad 2D_i + E \sim 0. \tag{15.88}$$

After blowing-up we remove the origin. Thus, the original divisor $\{z = 0\}$ does not contain the origin. We do proper transform, meaning that we fill up the origin to make $D_1$ again closed. At this point, each $D_i, i = 1, 2$ intersects $E$ at one point

$$D_1 \cdot E = D_2 \cdot E = 1. \tag{15.89}$$

Using the linear equivalence relations (15.88), we obtain the self-intersection numbers

$$D_1^2 = D_2^2 = -\frac{1}{2}, \quad E^2 = -2. \tag{15.90}$$

In what follows, we may denote divisors, line bundles, and the first Chern class using the same notation. It is known that for every divisor there is a dual line bundle. So, we denote the line bundle and the field strength with the same symbol. For example, have the vanishing first Chern class

$$c_1(T) = E + D_1 + D_2 \sim 0, \tag{15.91}$$

where we used the relation (15.88). We also find the Euler number is

$$c_2(T) = D_1 \cdot D_2 + E \cdot D_1 + E \cdot D_2 = -\frac{1}{2} + 1 + 1 = \frac{3}{2}, \tag{15.92}$$

which agrees with (15.78)

By associating the spin connection with the vector bundle, we take

$$F = D_1 V^I H_I \tag{15.93}$$

That is, the shift vector is associated with the line bundle dual to $D_1$. The Bianchi identity becomes

$$\frac{1}{2} V^2 = \int \text{tr} F_V^2 = \int \text{tr} R^2 = \frac{1}{2} 6.$$

Recall that the components of $V$ are half-integers. This $V$ is a local shift vector at the fixed point of the $\mathbb{C}^2/\mathbb{Z}_2$ orbifold. We may construct the full orbifold $T^4/\mathbb{Z}_2$ by patching sixteen local orbifolds, as done in Sect. 15.4.

## Dimensional Reduction

In low dimensional theory, the condition

$$dH = 0$$

provides the global consistency condition guaranteeing anomaly cancellation. Under background gauge field $F$ and geometry $R$, we may consider fluctuations obtain four dimensional anomaly polynomial in lower dimensions. We may show that if one satisfies the Bianchi identity (11.133), we may have anomaly-free spectrum. We take an example of SO(32) heterotic string on $\mathbb{C}^3/\mathbb{Z}^3$ with the shift vector [33] Eq. (11.133) that looks similar to modular invariance condition. We will see the connection in Chap. 15.

$$V = \frac{1}{3} \left( 0^{n_0} \, 1^{n_1}, 2^{n_2} \right).$$

We have the modular invariance condition

$$V^2 = 6. \tag{15.94}$$

The anomaly polynomial is expanded

$$\hat{I} = (i\mathscr{F})^3 \left( -\frac{1}{36} \text{tr}[H_V^3 (iF)^3] + \frac{1}{9 \cdot 32} \text{tr}[H_V^3 iF] \text{tr} R^2 \right)$$
$$+ 2i\mathscr{F} \, \text{tr} \mathscr{R}^2 \left( \frac{1}{9 \cdot 32} \text{tr}[H_V (iF)^3] - \frac{1}{9 \cdot 256} \text{tr}[H_V iF] \text{tr} R^2 \right). \tag{15.95}$$

We have

$$\hat{I}^6 = \frac{1}{6} \text{tr} \left( \frac{1}{2} N_V (iF)^3 \right) - \frac{1}{48} \text{tr} \left( \frac{1}{2} N_V iF \right) \text{tr} R^2, \tag{15.96}$$

where we have the number operator

$$N_V = \frac{1}{6}\left(-\frac{1}{3}H_V^2 + 1\right)H_V.$$

### 15.5.3 Classification

The geometry (15.82) we have just analyzed is called the $A_1$ singularity. We will see soon that the resolved geometry has the same connectedness structure as the Dynkin diagram, so that we name the singularity with the same name of the $A_{n-1}$ algebra. It is generalized to $\mathbb{C}^2/\Gamma$, where $\Gamma$ is a discrete subgroup of the holonomy group SU(2).

We need the resolution that preserves the vanishing the first Chern class. This non-discrepancy is called crepant. The crepancy is converted [34] to the condition that the adjacent blown-up spheres $E_i$, $E_{i+1}$ should intersect at one point and each sphere should have the self-intersection

$$E_i \cdot E_{i+1} = 1, \quad E_i^2 = -2. \tag{15.97}$$

The self-intersection number is defined as in (15.90). If we blow up $A, D, E$ singularity, we may show that the self-intersection of the $\mathbb{P}^1$ divisor is $-2$. The two-spheres from the blow-up has the same intersection structure to a Lie algebra. This is known as the McKay correspondence. Thus, the resolution should be simply laced. that is, $A, D, E$ type. The intersection matrix is the same as the Cartan matrix of the Lie algebra. The resolved equation is displayed in Table 15.1. There we used the fact that the same geometry is described by the changed coordinates as $xy \to x^2 + y^2$.

**Table 15.1** $A, D, E$ singularities

| Name | Equation | Generator |
|------|----------|-----------|
| $A_n$ | $y^2 = x^2 + z^{n+1}$ | $\begin{pmatrix} \alpha & 0 \\ 0 & \alpha^{-1} \end{pmatrix}$ |
| $D_n$ | $y^2 = x^2 + z^{n-1}$ | $\begin{pmatrix} \alpha^2 & 0 \\ 0 & \alpha^{-2} \end{pmatrix}, \begin{pmatrix} 0 & 1 \\ -1 & 0 \end{pmatrix}$ |
| $E_6$ | $y^2 = x^3 + z^4$ | $D_4, \frac{1}{\sqrt{2}}\begin{pmatrix} \varepsilon^7 & \varepsilon^7 \\ \varepsilon^5 & \varepsilon^7 \end{pmatrix}, \varepsilon = e^{2\pi i/8}$ |
| $E_7$ | $y^2 = x^3 + xz^3$ | $E_6, \begin{pmatrix} \varepsilon & 0 \\ 0 & \varepsilon^7 \end{pmatrix}$ |
| $E_8$ | $y^2 = x^3 + z^5$ | $E_7, -\begin{pmatrix} \eta^3 & 0 \\ 0 & \eta^2 \end{pmatrix}, \frac{1}{\eta^2-\eta^3}\begin{pmatrix} \eta+\eta^4 & 1 \\ 1 & -\eta-\eta^4 \end{pmatrix}, \eta = e^{2\pi i/5}$ |

We may regard each as an algebraic surface in $\mathbb{C}^3$ described by the corresponding equation. They are regarded as the orbifold $\mathbb{C}^2/\Gamma$, where $\Gamma$ is a discrete subgroup of SU(2) generated by the corresponding generators

Two divisors are linearly equivalent if the defining equations for the two divisors are the same up to multiplication of a rational function Shafarevich. The same holds for the dual objects of holomorphic line bundles that are the same if we have holomorphic transition function that is rational function in the common intersection Griffiths and Harris. Linear equivalence relation is similar to homologous relation in most of cases. Well-known difference is that on a sphere $\mathbb{P}^1$ all the points are linearly equivalent, but on a torus $T$ they are not, although they are homologous.

We may also *deform* the singularity to make it smooth

$$xy = (z - c)(z - d).$$

A similar analysis as above shows that it is not singular at any point. While the resolution breaks $A$, $D$, $E$ singularity like the adjoint Higgs mechanism. However, the deformation changes the structure. In terms of the analogy, similar to Higgs mechanism, it breaks the SU(2) algebraic structure to S[U(1) × U(1)], where the unimodular condition is imposed to a diagonal matrix formed by two U(1) generators.

Sometimes, the singularity is not fully resolved by one blow-up. We may further blow up until the geometry becomes completely smooth.

### 15.5.4 Toric Geometry

In the above example, the most important data in the resolution are scaling relation of the coordinates, as in (15.83). It determines the intersection numbers of the cycles in the end. Also, such scalings can be redefined so they are the objects in linear algebra. Thus it is useful to draw these data in space and we may study the resulting geometry. This leads us to toric geometry, using homogeneous coordinates. It also on the graph now only shows the scaling but also the requirement of resolution. We follow the discussion in Ref. [35].

#### $\mathbb{C}^2/\mathbb{Z}_2$ Orbifold

First, we convert the above data of the $\mathbb{C}^2/\mathbb{Z}_2$ resolution, discussed in Sect. 15.5.2, into that of toric geometry. The twist vector $\phi$ gives the condition for the basis vectors $\{v_a = (v_a^i)\}$ defining the toric diagram

$$\phi_a v_a^i = 0 \qquad \text{mod 1.} \tag{15.98}$$

For the $\mathbb{C}_2/\mathbb{Z}_2$, we choose $\phi = (\frac{1}{2}\ \frac{1}{2}\ 0)$, so

$$\frac{1}{2}v_1^i + \frac{1}{2}v_2^i \equiv 0 \qquad \text{mod 1.} \tag{15.99}$$

We find a particular solution

$$v_1 = (2, 0), \quad v_2 = (0, 2). \tag{15.100}$$

**Fig. 15.3**  Toric diagrams for the $\mathbb{C}^2/\mathbb{Z}_2$ orbifold (**a**) and its resolution (**b**). The original geometry with the divisors $D_1$ and $D_2$ is singular so we resolved the geometry by introducing $E$. Here, Figure (**b**) contains two cones with bases $D_1 E$ and $D_2 E$. The collection of cones forms a fan

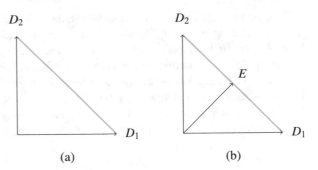

These vectors are going to be related to the ordinary divisors $D_1$, $D_2$ in (15.87), respectively. The resulting diagram, shown in Fig. 15.3. The dimension of toric diagram at this stage should be the same as the complex dimension $d/2$ of the original manifold. A ray made by each vector is semi-infinite line and make an edge. Then, a linear combination with non-negative coefficients makes a face. The collection of faces make the entire diagram, fan.

Blowing up by the exceptional divisor introduced a new coordinate $x$ with the scaling (15.83). Let us make a column matrix containing these powers as

$$Q = \begin{pmatrix} 1 \\ 1 \\ -2 \end{pmatrix}. \tag{15.101}$$

This introduces a new vertex $w$ satisfying

$$v_1 + v_2 - 2w = 0. \tag{15.102}$$

The actual coordinate is

$$\theta : w = \frac{1}{2}v_1 + \frac{1}{2}v_2 = (1, 1). \tag{15.103}$$

We also note that the coefficients of the RHS are the twist vector. It is known that the resulting geometry is smooth if *all the cones* separately span the $\mathbb{Z}^{d/2}$ lattice [9, 10]. We see that the original orbifold $\{v_1, v_2\}$ cannot span the $\mathbb{Z}^2$ lattice because the odd coordinates are not spanned. With the vertex $w$, each cone $\{v_1, w\}$ and $\{v_2, w\}$ spans the entire $\mathbb{Z}^2$ lattice. We saw that the resulting geometry (15.84) is smooth.

We may introduce a matrix summarizing this. First, we make a matrix $P$ by stacking all the vertex vectors (15.100) and (15.103) vertically. We may denote the name of each row by the corresponding divisor. Then, we also make a matrix $Q$ as (15.102) by stacking the coefficient vectors horizontally (in our case, we have only one vector, but in general we have more if we have more scaling relations). Since the number of rows of $P$ and $Q$ matrices are the same, the number of current divisors at hand, then we write a combined matrix

$$(P|Q) = \begin{pmatrix} D_1 & 2 & 0 & | & 1 \\ D_2 & 0 & 2 & | & 1 \\ E & 1 & 1 & | & -2 \end{pmatrix}. \tag{15.104}$$

In particular, the $Q$ submatrix gives us the linear equivalence relation (15.88). We have such relations whenever a linear combination of rows of $Q$ subvectors make another row of $Q$. Most importantly, we can read off the intersection number. In Fig. 15.3b, each cone gives intersection number 1. Thus, we reproduce the intersection number (15.89). Using the matrix (15.104), we may also obtain the relation $D_1 \cdot E = D_2 \cdot E$.

## $\mathbb{C}^3/\mathbb{Z}_{6-\mathrm{I}}$ Orbifold

We apply the construction to a more nontrivial case of $\mathbb{C}^3/\mathbb{Z}_{6-\mathrm{I}}$, which provides the local fixed points of $T^6/\mathbb{Z}_{6-\mathrm{I}}$ orbifold. The orbifold has the scaling

$$\left(z^1, z^2, z^3\right) \rightarrow \left(e^{2\pi i \cdot 4/6} z^1, e^{2\pi i/6} z^2, e^{2\pi i/6} z^3\right).$$

We take the vertices of the toric diagram by solving the equation

$$\frac{4}{6} v_1^i + \frac{1}{6} v_2^i + \frac{1}{6} v_3^i = 0 \mod 1. \tag{15.105}$$

Each vector has three components labelled by $i = 1, 2, 3$. We find a particular solution

$$v_1 = (0, 1, 1), \quad v_2 = (-1, -2, 1), \quad v_3 = (1, -2, 1).$$

They form vertices of the toric diagram and the dimension is three. Instead of drawing the whole cones, we have shown the base plane in Fig. 15.4a. The resulting fan has only one cone.

**Fig. 15.4** Toric diagram for the $\mathbb{C}^3/\mathbb{Z}_{6-\mathrm{I}}$ orbifold (**a**) and its resolution (**b**). Since the space is three dimensional, we need three dimensional diagram; however, all the vertices lie on the $z = 1$ plane

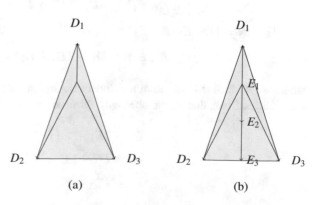

(a)     (b)

We see that the geometry is singular, because not every cone $\{v_1, v_2\}$, $\{v_1, v_3\}$, $\{v_2, v_3\}$ spans $\mathbb{Z}^3$ lattice. In fact, none. We fill in vertices $E_1$, $E_2$, $E_3$ corresponding to the vertices

$$\theta : w_1 = \frac{4}{6}v_1 + \frac{1}{6}v_2 + \frac{1}{6}v_3 = (0, 0, 1), \tag{15.106}$$

$$\theta^2 : w_2 = \frac{2}{6}v_1 + \frac{2}{6}v_2 + \frac{2}{6}v_3 = (0, -1, 1), \tag{15.107}$$

$$\theta^3 : w_3 = \frac{3}{6}v_2 + \frac{3}{6}v_3 = (0, -2, 1). \tag{15.108}$$

Again, we note that these correspond to $\mathbb{Z}_6$, $\mathbb{Z}_3$, $\mathbb{Z}_2$ singularities, respectively. The resulting toric diagram is drawn in Fig. 15.4b.

From this, we may find linear equivalence relations. There are many linear relations that are redundant so we want to find "minimal" relations. We may find them using the method [35, 36]. First, we consider intersections between two cones and find non-common vertices. The fan has six cones. Denoting each cone without mentioning the apex at the origin, we have

$$S_1 = \langle D_3, E_2, E_3 \rangle, \quad S_2 = \langle D_3, E_2, E_1 \rangle, \quad S_3 = \langle D_1, E_1, D_3 \rangle,$$
$$S_4 = \langle D_2, E_2, E_3 \rangle, \quad S_5 = \langle D_2, E_2, E_1 \rangle, \quad S_6 = \langle D_2, E_1, D_1 \rangle. \tag{15.109}$$

Considering $S_3 \cup S_6$, we have non-common divisors $D_2$ and $D_3$. We can find a linear relation from the above definition $4D_1 + D_2 + D_3 - 6E_1$, which always has for the non-common divisors coefficient 1. In this way, we find

$$\begin{aligned}
S_1 \cup S_2 &= \{D_3, \underline{E_1}, E_2, \underline{E_3}\}, \quad & E_1 - 2E_2 + E_3 = 0, \\
S_1 \cup S_4 &= \{\underline{D_2}, \underline{D_3}, E_2, E_3\}, \quad & D_2 + D_3 - 2E_3 = 0, \\
S_2 \cup S_3 &= \{\underline{D_1}, D_3, E_1, \underline{E_2}\}, \quad & D_1 - 2E_1 + E_2 = 0, \\
S_2 \cup S_6 &= \{\underline{D_2}, \underline{D_3}, E_1, E_2\}, \quad & D_2 + D_3 + 2E_1 - 4E_2 = 0, \\
S_3 \cup S_6 &= \{D_1, \underline{D_2}, \underline{D_3}, E_1\}, \quad & 4D_1 + D_2 + D_3 - 6E_1 = 0, \\
S_4 \cup S_5 &= \{D_2, \underline{E_1}, E_2, \underline{E_3}\}, \quad & E_1 - 2E_2 + E_3 = 0, \\
S_5 \cup S_6 &= \{\underline{D_1}, D_2, E_1, \underline{E_2}\}, \quad & D_1 - 2E_1 + E_2 = 0,
\end{aligned} \tag{15.110}$$

where we underlined non-common divisors. Among these relations, we extract the simplest relations, that is, the ones with minimal coefficients. They are

$$D_1 - 2E_1 + E_2 = 0,$$
$$D_2 + D_3 - 2E_3 = 0,$$
$$E_1 - 2E_2 + E_3 = 0,$$

called the Mori generators. We take the coefficients as column vectors

$$Q_1 = (0, 1, 1, 0, 0, -2)^\top,$$

$$Q_2 = (1, 0, 0, -2, 1, 0)^\top,$$

$$Q_3 = (0, 0, 0, 1, -2, 1)^\top,$$

and form the $Q$ matrix by stacking them horizontally. The Mori generators span Mori cones $Q_i$ in the original space, which is the space of the effective curves. By effective we mean that a curve $C$ has nonnegative intersections with all the divisors $C \cdot D \geq 0$ for all the divisors, including exceptional.

Using the above information, we can write down the matrix $(P|Q)$ as above

$$(P|Q) = \begin{pmatrix} D_1 & 0 & 1 & 1 & | & 0 & 1 & 0 \\ D_2 & -1 & -2 & 1 & | & 1 & 0 & 0 \\ D_3 & 1 & -2 & 1 & | & 1 & 0 & 0 \\ E_1 & 0 & 0 & 1 & | & 0 & -2 & 1 \\ E_2 & 0 & -1 & 1 & | & 0 & 1 & -2 \\ E_3 & 0 & -2 & 1 & | & -2 & 0 & 1 \end{pmatrix} \tag{15.111}$$

What we have done is to find the scaling relations between all the ordinary and exceptional coordinates

$$(\mathbb{C}^*)^3 : \left( z^1, z^2, z^3, x^1, x^2, x^3 \right)$$

$$\rightarrow \left( \lambda_2 z^1, \lambda_1 z^2, \lambda_1 z^3, \lambda_2^{-2}, \lambda_3 x^1, \lambda_2 \lambda_3^{-2} x^2, \lambda_1^{-2} \lambda_3 x^3 \right). \tag{15.112}$$

We define the resolution as

$$\text{Res}\left( \mathbb{C}^3 / \mathbb{Z}_{6-\mathrm{I}} \right) = (\mathbb{C}^3 - F)/(\mathbb{C}^*)^3. \tag{15.113}$$

Here, the excluded points form the set

$$F = \{(z_3, x_2) = 0, (z_3, x_1) = 0, (x_1, x_3) = 0, (z_1, z_2) = 0\}.$$

From the $Q$ submatrix, we read off the intersection numbers. Each element gives the intersection between $D_i$ (or $E_i$) and $C_i$. For instance, from $Q_1$ we find $C_1 \cdot D_2 = C_1 \cdot D_3 = 1$, $C_1 \cdot E_3$ and the others vanish. Since $E_1$ and $E_3$ are not adjacent vertices, we know $E_1 \cdot E_3 = 0$. Using linear equivalence relations, the following identification gives the correct intersection numbers:

$$C_1 = E_2 \cdot E_3,$$

$$C_2 = D_3 \cdot E_1 = D_2 \cdot E_1,$$

$$C_3 = D_3 \cdot E_2 = D_2 \cdot E_2,$$

$$E_1 \cdot E_2 = C_1 + 2C_3,$$

$$D_3 \cdot E_1 = C_1 + 4C_2 + 2C_3.$$

We obtain triple intersection numbers as shown in Table 15.2. For instance, $D_3 \cdot E_1 \cdot E_1 = C_1 \cdot E_1 + 4C_2 \cdot E_1 + 2C_3 \cdot E_1 = -6$.

Finally, we comment on triangulation . In this simple example, the six cones are uniquely defined as (15.109), as seen in the diagram, Fig. 15.4b. However, other orbifolds such as $\mathbb{Z}_{6-\mathrm{II}}$ has more than one way to define cones. Its toric diagram is spanned by the vectors

$$v_1 = (0, 1, 1), \quad v_2 = (1, -1, 1), \quad v_3 = (-2, -1, 1),$$

$$w_1 = (0, 0, 1), \quad w_2 = (0, -1, 1), \quad w_3 = (-1, 0, 1), \quad w_4 = (-1, -1, 1),$$
$$\tag{15.114}$$

as drawn in Fig. 15.5. There are five different choices of set of cones. This choice is called *triangulation*, as the diagram clearly shows. Each choice is equally good, giving smooth manifold. However, different triangulation leads different set of intersection numbers, affecting quantitative difference in the resulting four dimensional models (Fig. 15.5).

**Table 15.2** Triple intersection numbers of the resolved $\mathbb{C}^3/\mathbb{Z}_{6-\mathrm{I}}$ orbifold

|               | $D_1$ | $D_2$ | $D_3$ | $E_1$ | $E_2$ | $E_3$ |
|---------------|-------|-------|-------|-------|-------|-------|
| $D_1 \cdot E_1$ | 4     | 1     | 1     | $-6$  | 0     | 0     |
| $D_2 \cdot E_1$ | 1     | 0     | 0     | $-2$  | 1     | 0     |
| $D_2 \cdot E_2$ | 0     | 0     | 0     | 1     | $-2$  | 1     |
| $D_3 \cdot E_1$ | 1     | 0     | 0     | $-2$  | 1     | 0     |
| $D_3 \cdot E_2$ | 0     | 0     | 0     | 1     | $-2$  | 1     |
| $E_1 \cdot E_2$ | 0     | 1     | 1     | 2     | $-4$  | 0     |
| $E_2 \cdot E_3$ | 0     | 1     | 1     | 0     | 0     | $-2$  |

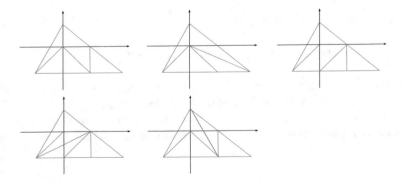

**Fig. 15.5** There are five different triangulations in the toric diagram of the resolved $\mathbb{Z}_{6-\mathrm{II}}$

So far, we have done toric resolution for the non-compact orbifold $\mathbb{C}^{d/2}/\mathbb{Z}_N$. We may study toric geometry on the global orbifold $T^d/\mathbb{Z}_N$ as discussed in Ref. [35].

## 15.6 Dynamics of the Geometry

Recall that geometrical parameters become dynamical fields, or moduli fields. The role of the fields from graviton $G_{mn}$ is evident, but it was not clear what determines the geometry of the fixed points. It turns out that the localized matter field also plays the role of the moduli field, controlling the blown-up geometry. Remarkably, the geometry is caught by gauged linear sigma model (GLSM) [37]. We follow Refs. [38, 39] to discuss the dynamics.

In most of the string compactification, we have (2,0) worldsheet supersymemtry. Let $\theta^+$ be the corresponding Grassmannian variable. Here, we describe the current algebra using 16 complexified fermions. We write the target space variable $z^a$, $a = 1, 2, 3$ and the current algebra $\lambda^A$. The former belongs to a chiral fermion. The current algebra form a Fermi multiplet.

We will also gauge the The $H$-momenta of $s^a$ that are promoted to the $U(1)$ charges for the fields $z^a$.

The two dimensional action

$$S = \int d^2\sigma d^2\theta^+ \left( \frac{i}{4}\overline{\Psi}_a \overline{\mathscr{D}}\Psi^a - \frac{1}{4}\overline{\Lambda_A}\Lambda^A \right.$$
$$\left. + \frac{1}{2e^2}\overline{F}_A F_A \right) + \int d^2\sigma d^2\theta^+ \rho(\Psi)F_A + \text{h.c..} \tag{15.115}$$

The most important scalar potential comes from D-term

$$V = \sum_A \frac{e_A^2}{2} \left( \sum_i q_i|z_i|^2 - b_A \right). \tag{15.116}$$

For each $U(1)_A$, the Fayet–Illiopoulos parameter $b_A$ is given by the complexified axio-dilaton $\rho_A = b_A + i\beta_A$.

We have two global consistency conditions

$$\sum_A Q_A^\alpha Q_B^\alpha = \sum_a q_A^a q_B^a, \quad \text{for all } A, B, \tag{15.117}$$

$$\sum_a q_A^a = 0 \quad \text{for all } A. \tag{15.118}$$

So, not every field can take part in the sigma model.

We consider SO(32) heterotic string compactified on $T^6/\mathbb{Z}_3$ orbifold with a shift vector

$$V = \frac{1}{3}\left(2\,1\,1\,0^{13}\right).$$

Without Wilson lines, we have equally distributed spectrum on 27 fixed points. This means that compactification on $\mathbb{C}^3/\mathbb{Z}_3$ yields the spectrum

$$\frac{1}{27}[3(\mathbf{3}, \mathbf{26})_1 + 3(\mathbf{3}, \mathbf{1})_{-2}] + 3(\overline{\mathbf{3}}, \mathbf{1})_0 + (\mathbf{1}, \mathbf{26})_1 + (\mathbf{1}, \mathbf{1})_{-2}. \qquad (15.119)$$

The fractional multiplicity reflects that the corresponding states come from the bulk. It is shown that we cannot make use of $(\mathbf{1}, \mathbf{1})_{-2}$ or $3(\mathbf{3}, \mathbf{1})_{-2}$ for blowing up, because it cannot satisfy anomaly cancellation condition (15.117) [38]. We may include the fields $(\mathbf{3}, \mathbf{26})_1, (\overline{\mathbf{3}}, \mathbf{1}), (\mathbf{1}, \mathbf{26})_1$.

Finally, we investigate a $\mathbb{C}^3/(\mathbb{Z}_2 \times \mathbb{Z}_2)$ model to see various different partial resolutions [38]. We have seen that some orbifold have different selection of triangulation, giving different intersection numbers. We see that they are all controlled by VEV of fields.

With the shift vectors, $\phi_1 = (0\,\frac{1}{2}\,\frac{1}{2})$, $\phi_2 = (\frac{1}{2}\,0\,\frac{1}{2})$, $\phi_3 = (\frac{1}{2}\,\frac{1}{2}\,0) \equiv \phi_1 + \phi_2$ we consider the standard embedding. We have three twisted sectors and hence there can be up to three resolutions. We have coordinate superfields $\Psi$s. We should introduce three U(1) gauge group. The scalar potential is

$$V = \frac{e_1^2}{2}\left(\frac{|z^2|^2 + |z^3|^2}{2} - |x^1| - b_1\right)^2 + \frac{e_2^2}{2}\left(\frac{|z^1|^2 + |z^3|^2}{2} - |x^2| - b_2\right)^2$$

$$+ \frac{e_3^2}{2}\left(\frac{|z^1|^2 + |z^2|^2}{2} - |x^3| - b_3\right)^2,$$

$$(15.120)$$

where as before $z^a$, $a = 1, 2, 3$ are the coordinates of $\mathbb{C}^3$ associated with the divisors $D_a = \{z^a = 0\}$ and $x^r$, $r = 1, 2, 3$ are those of the exceptional divisors $E_r = \{x^r = 0\}$. The exceptional divisor $E_r$ can exist only when $b_r \geq 0$. It is because otherwise we have $b_r < 0$ and $x_a = 0$ that cannot make the potential $V$ in (15.120) vanish. We may consider a curve, for example, the intersection between $D_1$ and $E_1$. Setting $z^1 = x^1 = 0$, we obtain

$$V = \frac{e_1^2}{2}\left(\frac{|z^2|^2 + |z^3|^2}{2} - b_1\right)^2 + \frac{e_2^2}{2}\left(\frac{|z^3|^2}{2} - |x^2| - b_2\right)^2$$

$$+ \frac{e_3^2}{2}\left(\frac{|z^2|^2}{2} - |x^3| - b_3\right)^2. \qquad (15.121)$$

**Fig. 15.6** Toric diagrams for the $\mathbb{C}^2/(\mathbb{Z}_2 \times \mathbb{Z}_2)$ orbifold and its resolutions [38]. Besides these six triangulations, we have eight more triangulations. The existence of the curve can be determined by parameters

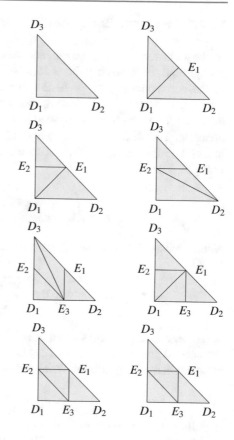

The first term can vanish only if $b_1 \geq 0$. The rest terms of the potential are minimized if

$$|x^2|^2 + |x^3|^2 = b_1 - b_2 - b_3 \geq 0, \tag{15.122}$$

in other words $b_1 \geq b_2 + b_3$. This is the condition for the existence of the curve $D_1 E_1$. Likewise, we may investigate the existence of points, which is related to the intersection numbers. The resulting geometry has various phases, shown in Fig. 15.6.

## 15.7 Non-perturbative Vacua

We may understand the effect of the shift vector associated with orbifold in terms of instanton background. In six dimensions.

### 15.7.1  Instanton Background

In the field theory, the phenomenon described by a shift vector is an instanton background. Each fixed point is locally $\mathbb{C}^d/\mathbb{Z}_N$ and is regarded as singular limit of the corresponding $A_{N+1}$ ALE space [5]. Then, there can be a nontrivial flat connection at the infinity, whose breaking is described by the shift vector. Collecting them, we have "the bulk," described by the common intersection of such symmetry breaking. The shift vector (13.6) has eigenvalues of the generator of the group at hand, SO(32) or $E_8 \times E_8$ [40]. At each order $N$ fixed point, we have $n_i$ eigenvalues $i/N$ [24, 40]

$$k_{\mathbb{Z}_N} = \frac{1}{8\pi^2} \int_{\mathbb{Z}_N \text{ALE}} \text{tr } F^2 = k_{\mathbb{Z}_N,\text{U}} + k_{\mathbb{Z}_N,\text{T}}, \tag{15.123}$$

where we have instanton number from the flat background

$$k_{\mathbb{Z}_N,\text{U}} = \sum_{i=0}^{N-1} \frac{i(N-i)}{2N} n_i, \tag{15.124}$$

where $n_i$ are the numbers of eigenstate component in the vector bundle and thus the second contribution in (15.123) is the usual second Chern class from the flat connection from the infinity. We also have an integer number

$$k_{\mathbb{Z}_N,\text{T}} = \text{integer}, \tag{15.125}$$

counting the "instanton number" in the twisted sectors. In field theory, we may put arbitrary fields on this fixed point; however, in string theory this number is fixed by the modular invariance condition.

Here and in what follows, the integration is done over the $A_{N-1}$ ALE space. An integer $K_N$ is determined by the field contents on this ALE space, which becomes twisted field in the orbifold picture. It is the shift vector that contains exactly this information, thus we have another equivalent expression to (15.123),

$$k_{\mathbb{Z}_N} = K_N + \frac{N}{2} \sum_{I=1}^{16} V_I (1 - V_I). \tag{15.126}$$

We may interpret that $K_N$ is the index of the localized fields at the tip of the ALE space.

Now, we wish to collect all the fixed point contributions to make that of the total $T^4/\mathbb{Z}_N$. We know the multiplicity $d_j$ of $\mathbb{Z}_{N/j}$ invariant fixed points. We can express the Euler number in terms of the shift vector in the $j$-th twisted sector $\phi_a^{(j)}$

$$24 = \frac{1}{8\pi^2} \int \text{tr } R^2 = \sum_{j \in N/\mathbb{Z}} d_j \frac{N}{j} \left(\frac{N}{j} + 1\right) \sum_{a=1}^{2} \phi_a^{(j)} \left(1 - \phi_a^{(j)}\right). \tag{15.127}$$

Therefore, we have the total instanton numbers

$$k_U = \frac{1}{8\pi^2} \int \text{Tr } F_U^2 = \sum_{j \in N/\mathbb{Z}} d_j \frac{N}{2j} \sum_{I=1}^{16} V_I^{(j)} \left(1 - V_I^{(j)}\right).$$
(15.128)

$$k_T = \frac{1}{8\pi^2} \int \text{Tr } F_T^2 = \sum_{j \in N/\mathbb{Z}} d_j K_{N/j}.$$
(15.129)

Here, again $V_I^{(j)}$ is $jV_I$ subtracted by an appropriate integer to lie it in the interval $[0, 1]$. This establishes the relationship between the modular invariance condition (7.50) and Bianchi identity (11.133). In the presence of Wilson lines, the degeneracy from $d_j$ disappears and we have independent contribution from local shift vectors at each fixed point. This establishes the relation between the Bianchi identity and modular invariance condition.

For a $\mathbb{Z}_3$ shift vector of a form $V = \frac{1}{3}(0^{n_0} \ 1^{n_1} \ 2^{n_2})$, we have contributions at each fixed point

$$k_{\mathbb{Z}_3} = K_3 + 3 \cdot \frac{1}{2} \cdot \frac{1}{3} \cdot \frac{2}{3} \cdot (n_1 + n_2).$$
(15.130)

Thus, in the bulk we have

$$k = 9K_3 + 3n_1 + 3n_2.$$
(15.131)

For a spinorial shift vector of the form $V = \frac{1}{6}(1^{n+} \ 3^{n_*})$, we have

$$k_{\mathbb{Z}_3} = K_3 + 3 \cdot \frac{1}{2} \left(n_+ \frac{1}{2} \cdot \frac{1}{2} + n_* \frac{3}{2} \cdot \frac{3}{2}\right).$$
(15.132)

For a $\mathbb{Z}_4$ shift vector of a form $V = \frac{1}{4}(0^{n_0} \ 1^{n_1} \ 2^{n_2} \ 3^{n_3})$, we have

$$k_{\mathbb{Z}_4} = K_4 + 4 \cdot \frac{1}{2} \left(\frac{2}{4} \cdot \frac{2}{4} \cdot n_2 + \frac{1}{4} \cdot \frac{3}{4} \cdot (n_1 + n_3)\right),$$
(15.133)

$$k_{\mathbb{Z}_2} = K_2 + 2 \cdot \frac{1}{2} \left(0 \cdot \frac{4}{4} \cdot n_2 + \frac{1}{2} \cdot \frac{1}{2} \cdot (n_1 + n_3)\right).$$
(15.134)

Here, the integral components in $2V$, counted by $n_2$, are replaced by zero. The total instanton number is

$$k = 4k_{\mathbb{Z}_4} + 6k_{\mathbb{Z}_2} = 4K_4 + 6K_2 + 2n_2 + 3n_1 + 3n_3,$$
(15.135)

which reproduces (15.126).

With these, we can connect the Bianchi identity with the modular invariance condition (15.53). Integrating this over K3, we have

$$0 = 24 - k_U - k_T. \tag{15.136}$$

This cancels six dimensional $SO(2n_0)$ gauge anomaly, with the chiral matter $(\mathbf{n_1} + \mathbf{n_{N-1}}, \mathbf{2n_0})$ under $SU(n_1 + n_{N-1}, 2n_0)$ (or $(\mathbf{2n_1}, \mathbf{2n_0})$ of $SO(2n_1, 2n_0)$ for $\mathbb{Z}_2$)

$$(2n_0 - 8) - n_1 - n_{N-1} - k_T = 0, \tag{15.137}$$

as long as the total instanton numbers should satisfy the condition (15.80)

$$k = k_U + k_T = 24.$$

It is because we can show the following using $\sum n_i = 16$ and (15.126)

$$
\begin{aligned}
k_T &= 24 - k_U \\
&= -8 + 32 - 2 \sum_{i=1}^{N-1} n_i - n_1 - n_{N-1} \\
&= -8 + 2n_0 - n_1 - n_{N-1}.
\end{aligned}
\tag{15.138}
$$

This proves the relation (15.137). The reason is that the total instanton number is related to the unbroken gauge group. Here, the only unbroken part is $SO(2n_0)$.

In understanding the consistent vacua, the $SO(2n_0)$ anomaly plays an important role. In six dimension, the chirality of gaugino in the vector multiplet is always opposite to the fermions in the hypermultiplets, so that the matter contents are constrained by anomalies. Six dimensional gauge anomalies can be cancelled, up to Green–Schwarz mechanism, if the anomaly polynomial has vanishing tr $F^4$ term [40]. From the contributions of gaugino

$$\text{Tr}\, F^4_{SO(2n)} = (2n - 8)\, \text{tr}_v\, F^4_{SO(2n)} + 6 \left( \text{tr}\, F^2_{SO(2n)} \right)^2, \tag{15.139}$$

$$\text{Tr}\, F^4_{SU(n)} = 2n\, \text{tr}\, F^4_{SU(n)} + 6 \left( \text{tr}\, F^2_{SU(n)} \right)^2, \tag{15.140}$$

we see that we need $(2n - 8)$ vectors and $2n$ fundamentals, respectively, to cancel $SO(n)$ and $SU(n)$ anomalies. Other representations may contribute, as we have summarized the decomposition in the Appendix A.

## 15.7.2 Non-perturbative String Vacua

Lastly, we study the small instanton transition and apply it to orbifold vacua [25,41]. Some gauge background can undergo transition and becomes instantons. This means by exchange of such instantons, vacua that has been regarded as different can in fact be connected. Also, we may obtain non-perturbative generalization of the modular invariance condition.

Note that the Bianchi identity can be understood as the magnetic equation of $B$ field

$$dH = 0 = d^2 B + \frac{1}{2} \, \mathrm{tr}_v \, R^2 - \frac{1}{2} \, \mathrm{tr}_v \, F^2, \qquad (15.141)$$

with the last two term being sources. We also note that the instanton number $k$ is integrally quantized. Remembering that instantons are parameterized by size, position, and global transformations. In particular, if some of instantons shrink to zero size, the field strength becomes delta functions

$$dH = 0 = d^2 B + \frac{1}{2} \, \mathrm{tr}_v \, R^2 - \frac{1}{2} \, \mathrm{tr}_v \, F'^2 - \sum \delta^{(4)}(y - y_i), \qquad (15.142)$$

where some of the component from $F$ are emitted so we have $F'$, and $y$ is the collective notation for the internal coordinates. This can be also regarded as the pointlike source in this space. Including the remaining space, this source describes 5-branes that are natural magnetic sources for the rank two tensor field $B_{MN}$ Indeed there is a phase transition between the instanton number of 5-branes [42]. If we have such phase transition in SO(32) heterotic string, it becomes 5-branes. SO-group is anomalous in six dimensions, as before, there should be a local source of anomaly cancellation. It is provided by gauge theory localized on 5-branes. If we have $n$ coincident 5-branes, there is Sp($2n$) localized gauge theory on the worldvolume, where in our convention Sp(1) = U(2). In $E_8 \times E_8$ gauge group; however, there is no anomaly and in fact there is no extra anomaly cancellation from the 5-branes. We have discussed in Sect. 6.4.4 that we have eleven dimensional theory in the strong coupling limit. These 5-branes can move into the bulk in the eleven dimensions and promoted to M5-branes. One interesting feature is that now the M2-brane can end on M5-branes.

If we go back to the original orbifold limit, emitting instanton reduces the shift vector. In other words, some components of the shift vector becomes the instantons. Decomposing

$$V = V_1 + V_2, \qquad (15.143)$$

the emitted instanton components are described by $V_2$, which describes also the recovered part.

The mass shell condition changes. Consider the twisted sector of a *perturbative, modular invariant* theory parameterized by $V$. Expanding the mass shell condition (7.37), we have

$$\frac{1}{4} M_L^2 = \frac{(P + V_1)^2}{2} + \tilde{N} + P \cdot V_2 + V_1 \cdot V_2 + \frac{1}{2} V_2^2 + E_0$$
$$= \frac{(P + V_1)^2}{2} + \tilde{N} + E_0 + \Delta E_0. \tag{15.144}$$

We may regard this formula as the mass shell condition for the twisted sector for a daughter vacuum with $V_1$, with instantons emitted. Then, the extra piece originating from $V_2$ becomes a non-perturbative correction in the zero-point energy

$$\Delta E_0 \equiv V_1 \cdot V_2 + \frac{1}{2} V_2^2. \tag{15.145}$$

A part of instantons described by the shift vector $V_2$ "condensates" in the CFT description. We assumed $P \cdot V_2 = 0$ because we want $\Delta E_0$ is a constant in the new, daughter vacuum, not depending on $P$.

The spectrum can still be calculated using the same CFT in the presence of 5-branes, using the same mass shell conditions and level matching condition $M_L^2 = M_R^2 = 0$ as those of the perturbative case [25, 41]. Since the inclusion of heterotic 5-branes does not affect the internal geometry of orbifold, we have no change in the spacetime part, including that of the right mover.

We can also have modified, generalized GSO projector (8.61) in the presence of heterotic 5-branes. Since the terms involving $V$ are modified as

$$(P + V) \cdot V - \frac{1}{2} V^2 = (P + V_1)^2 - \frac{1}{2} V_1^2 + \frac{1}{2} V_2^2 + V_1 \cdot V_2$$
$$= (P + V_1)^2 - \frac{1}{2} V_1^2 + \Delta E_0, \tag{15.146}$$

where again $P \cdot V_2 = 0$ is assumed. Therefore, we have modified, generalized GSO projector (8.61) for a non-perturbative shift vector $V$ taking into account heterotic 5-branes

$$e^{2\pi i \left( (P+V) \cdot V - (s + \phi + \rho_L - \rho_R) \cdot \phi - \frac{1}{2}(V^2 - \phi^2) + \Delta E_0 \right)}. \tag{15.147}$$

The extra phase is simply expressed by $\Delta E_0$. The 5-branes can only affect the CFT on the fixed point only quantitatively.

As a toy model, we consider transition from a perturbative vacuum of $G = U(8) \times SO(16)$ with the shift vector $V = \frac{1}{3}(1^8 \, 0^8)$ to a non-perturbative vacuum of

$G_1 = U(2) \times SO(28)$ with the shift vector $V_1 = \frac{1}{3}(1^2 \, 0^{14})$. The shift vector $V$ of $G$ is decomposed into $V_1$ of $G_1$ and an extra $V_2$ as

$$V = V_1 + V_2 = \frac{1}{3}\left(1^2 \, 0^{14}\right) + \frac{1}{3}\left(0^2 \, 1^6 \, 0^8\right). \tag{15.148}$$

Upon phase transition, the small instanton components described by $V_2$ is going to become heterotic 5-branes. Therefore, we are left with the remaining component $V_1$, describing a non-perturbative vacua with the gauge group $G_1$.

The shift vector $V_1$ gives rise to a vacua whose gauge group is $U(2) \times SO(28)$ with untwisted matter $(\mathbf{2}, \mathbf{28}) + 3(\mathbf{1}, \mathbf{1})$. When instantons shrink, they are embedded in the structure group $\mathbb{Z}_3 \subset U(1) \times SU(2)$ with the instanton number $k_U = 3n_1 = 6$. It can be emitted and become as many 5-branes.

In the twisted sector, on top of the usual zero-point energy

$$E_0 = -1 + 2 \cdot \frac{1}{2} \cdot \frac{1}{3} \cdot \left(1 - \frac{1}{3}\right) = -\frac{7}{9},$$

there is modification as in (15.145),

$$\Delta E_0 = 0 + \frac{1}{2}\left(\frac{1}{3}(1^6 \, 0^{10})\right)^2 = \frac{18}{54}. \tag{15.149}$$

Plugging these into the mass shell condition (15.144), we find the spectrum shown in Table 15.3. There is no charged representation. Since there is no $SO(28)$ vector or spinor, thus no instanton contribution comes from the twisted sector $k_T = 0$.

**Table 15.3** Some vacua of SO(32) string on $T^4/\mathbb{Z}_3$ orbifold

| Shift vector $V$ | Untwisted | $k_U$ |
|---|---|---|
| Group | Twisted | $k_T$ |
| | Heterotic 5 localized | $n$ |
| $\frac{1}{3}(1^2 \, 0^{14})$ | $(\mathbf{2}, \mathbf{28}; \mathbf{1}) + 2(\mathbf{1}, \mathbf{1}; \mathbf{1})$ | 6 |
| $U(2) \times SO(28) \times Sp(18)$ | $9(\mathbf{1}, \mathbf{1}; \mathbf{1}) + 18(\mathbf{1}, \mathbf{1}; \mathbf{1})$ | 0 |
| | $\frac{1}{2}(\mathbf{1}, \mathbf{28}; \mathbf{36}) + (\mathbf{2}, \mathbf{1}; \mathbf{36})$ | 18 |
| $\frac{1}{3}(1^2 \, 0^{14})$ | $(\mathbf{2}, \mathbf{28}) + 3(\mathbf{1}, \mathbf{1})$ | 6 |
| $U(2) \times SO(28)$ | $9(\mathbf{2}, \mathbf{28}) + 63(\mathbf{1}, \mathbf{1})$ | 18 |
| | No state | 0 |
| $\frac{1}{3}(1^8 \, 0^8)$ | $(\mathbf{8}, \mathbf{16}) + (\mathbf{28}, \mathbf{1}) + 2(\mathbf{1}, \mathbf{1})$ | 24 |
| $U(8) \times SO(16)$ | $9(\mathbf{28}, \mathbf{1}) + 18(\mathbf{1}, \mathbf{1})$ | 0 |
| | No state | 0 |

The parameters $k_U, k_T, n$ are, respectively, instanton numbers in the untwisted and twisted sectors, and the number of heterotic 5-branes

It is also interesting to compare this model with a perturbative model, having the same shift vector. In general, non-perturbative models cannot satisfy the modular invariance condition, but with the modification

$$\frac{(P+V)^2}{2} + \tilde{N} + \Delta E_0 - \frac{(s+\phi)^2}{2} - N + \frac{1}{2} = 0 \mod 1. \tag{15.150}$$

Since each term is proportional to $1/N$, we expect their cancellation gives an integer sum, expect the $V^2$ and $\phi^2$ terms. Thus, we require the modified modular invariance condition

$$\frac{V^2}{2} + \Delta E_0 - \frac{\phi^2}{2} \equiv 0 \mod \frac{1}{N}. \tag{15.151}$$

## Other Manifolds

According to Berger [43], a simply-connected, irreducible, and nonsymmetric manifold of dimension $d$, with Riemannian metric, is uniquely classified according to Table 15.4. In our heterotic string case, the Calabi–Yau threefold is our main interest in compactifying heterotic string, since it leaves $\mathcal{N} = 1$ SUSY in four dimensions. In view of string duality, unified theories require more dimensions, implying more compact dimensions and different special holonomies. For example, M- and F-theories are living in eleven and twelve dimensions, thus we introduce manifold of $G_2$ and SU(4) holonomy, respectively. The other side of the coin is that we can see string duality when we compactify merely one or two dimensions in these generalized theories.

## Flux Compactification

Recently, lots of studies are done under nontrivial $H \neq 0$ configurations, generally referred to as "flux compactification." In this case, Table 15.4 is not the complete list any more. This is so because if we turn on the fluxes then this special holonomy condition for unbroken supersymmetry does not hold. However, there is another condition compensating this. A more general analysis is reviewed in Ref. [44].

**Table 15.4** The Berger classification on holonomy group of a simply-connected, irreducible, and nonsymmetric manifold of dimension $d$ with Riemannian manifolds [6, 43]

| Real dimension | Holonomy group | Type |
|---|---|---|
| $d$ | SO($d$) | Orientable |
| $2n \geq 4$ | U($n$) | Kähler |
| $2n \geq 4$ | SU($n$) | Ricci-flat, Kähler |
| $4n \geq 8$ | Sp($n$)$\times$Sp(1) | Einstein |
| $4n \geq 8$ | Sp($n$) | Ricci-flat, Kähle |
| 7 | $G_2$ | Ricci-flat |
| 8 | Spin(7) | Ricci-flat |

## Exercise

▶ **Exercise 15.1** Show that the $A_{n-1}$ singularity (15.82) is smooth for $n = 1$.

## References

1. P. Candelas, G.T. Horowitz, A. Strominger, E. Witten, Vacuum configurations for superstrings. Nucl. Phys. **B258**, 46–74 (1985)
2. P. Candelas, Lectures on complex manifolds, in *Superstrings and grand unification* (1988)
3. M.B. Green, J.H. Schwarz, E. Witten, *Superstring Theory. Vol. 2: Loop Amplitudes, Anomalies and Phenomenology* (1988)
4. P.S. Aspinwall, K3 surfaces and string duality, in *Differential Geometry Inspired by String Theory* (1996), pp. 421–540. [1(1996)]
5. T. Eguchi, P.B. Gilkey, A.J. Hanson, Gravitation, gauge theories and differential geometry. Phys. Rep. **66**, 213 (1980)
6. D.D. Joyce, *Compact Manifolds with Special Holonomy* (Oxford University Press, Oxford, 2000)
7. S.S. Gubser, TASI lectures: special holonomy in string theory and M theory, in *Strings, branes and extra dimensions: TASI 2001: proceedings* (2002), pp. 197–233
8. M. Nakahara, *Geometry, Topology and Physics* (CRC Press, Boca Raton, 2003)
9. D.A. Cox, S. Katz, *Mirror Symmetry and Algebraic Geometry* (American Mathematical Society, Providence, 1999)
10. W. Fulton, *Introduction to Toric Varieties*, vol. 131 (Princeton University Press, Princeton, 1993)
11. D.A. Cox, J.B. Little, H.K. Schenck, *Toric Varieties* (American Mathematical Society, Providence, 2011)
12. R. Bott, L.W. Tu, *Differential Forms in Algebraic Topology* (Springer Science+Business Media, New York, 1982)
13. P. Griffiths, J. Harris, *Principles of Algebraic Geometry* (Wiley, Hoboken, 1978)
14. R. Hartshorne, *Algebraic Geometry*, vol. 52 (Springer Science & Business Media, New York, 2013)
15. T. Hubsch, *Calabi-Yau Manifolds: A Bestiary for Physicists* (World Scientific, Singapore, 1994)
16. G. Tian, S.-T. Yau, Three-dimensional algebraic manifolds with C(1) = 0 and Chi = -6. Conf. Proc. **C8607214**, 543–559 (1986)
17. V. Braun, Y.-H. He, B.A. Ovrut, T. Pantev, The exact MSSM spectrum from string theory. J. High Energy Phys. **5**, 043 (2006)
18. D. Mumford, Projective invariants of projective structures and applications, in *Proc. Internat. Congr. Mathematicians* (1963)
19. R. Friedman, *Algebraic Surfaces and Holomorphic Vector Bundles* (Springer, New York, 2012)
20. M.R. Douglas, B. Fiol, C. Romelsberger, Stability and BPS branes. J. High Energy Phys. **9**, 006 (2005)
21. H. Enger, C.A. Lutken, Nonlinear Yang-Mills instantons from strings are pi stable D-branes. Nucl. Phys. **B695**, 73–83 (2004)
22. S.K. Donaldson et al., An application of gauge theory to four-dimensional topology. J. Differ. Geom. **18**(2), 279–315 (1983)
23. K. Uhlenbeck, S.-T. Yau, On the existence of hermitian-yang-mills connections in stable vector bundles. Commun. Pure Appl. Math. **39**(S1), S257–S293 (1986)
24. M. Berkooz, R.G. Leigh, J. Polchinski, J.H. Schwarz, N. Seiberg, E. Witten, Anomalies, dualities, and topology of D = 6 N=1 superstring vacua. Nucl. Phys. **B475**, 115–148 (1996)
25. K.-S. Choi, T. Kobayashi, Transitions of orbifold vacua. J. High Energy Phys. **7**, 111 (2019)

26. R. Donagi, Y-H. He, B.A. Ovrut, R. Reinbacher, The Spectra of heterotic standard model vacua. J. High Energy Phys. **06**, 070 (2005)
27. M.F. Atiyah, I.M. Singer, The Index of elliptic operators. 1. Ann. Math. **87**, 484–530 (1968)
28. M.F. Atiyah, I.M. Singer, Dirac operators coupled to vector potentials. Proc. Nat. Acad. Sci. **81**, 2597–2600 (1984)
29. L. Alvarez-Gaume, P.H. Ginsparg, The structure of gauge and gravitational anomalies. Ann. Phys. **161**, 423 (1985). [Erratum: Annals Phys.171,233(1986); ,423(1984)]
30. O. Alvarez, I.M. Singer, B. Zumino, Gravitational anomalies and the family's index theorem. Commun. Math. Phys. **96**, 409 (1984)
31. T. Eguchi, A.J. Hanson, Asymptotically flat selfdual solutions to euclidean gravity. Phys. Lett. **74B**, 249–251 (1978)
32. T. Eguchi, A.J. Hanson, Selfdual solutions to euclidean gravity. Ann. Phys. **120**, 82 (1979)
33. S. Groot Nibbelink, T.-W. Ha, M. Trapletti, Toric resolutions of heterotic orbifolds. Phys. Rev. **D77**, 026002 (2008)
34. I. Shafarevich, *Basic Algebraic Geometry 2: Schemes and Complex Manifolds*, (3rd ed.), (Springer-Verlag, Berlin, 2013)
35. D. Lüst, S. Reffert, E. Scheidegger, S. Stieberger, Resolved toroidal orbifolds and their orientifolds. Adv. Theor. Math. Phys. **12**(1), 67–183 (2008)
36. P. Berglund, S.H. Katz, A. Klemm, Mirror symmetry and the moduli space for generic hypersurfaces in toric varieties. Nucl. Phys. B **456**, 153–204 (1995)
37. E. Witten, Phases of N=2 theories in two-dimensions. AMS/IP Stud. Adv. Math. **1**, 143–211 (1996)
38. S. Groot Nibbelink, Heterotic orbifold resolutions as (2,0) gauged linear sigma models. Fortsch. Phys. **59**, 454–493 (2011)
39. M. Blaszczyk, S. Groot Nibbelink, F. Ruehle, Gauged linear sigma models for toroidal orbifold resolutions. J. High Energy Phys. **5**, 053 (2012)
40. K.A. Intriligator, RG fixed points in six-dimensions via branes at orbifold singularities. Nucl. Phys. **B496**, 177–190 (1997)
41. G. Aldazabal, A. Font, L.E. Ibanez, A.M. Uranga, G. Violero, Nonperturbative heterotic D = 6, D = 4, N=1 orbifold vacua. Nucl. Phys. **B519**, 239–281 (1998)
42. E. Witten, Small instantons in string theory. Nucl. Phys. **B460**, 541–559 (1996)
43. M. Berger, Sur les groupes d'holonomie homogènes de variétés à connexion affine et des variétés riemanniennes. Bull. Soc. Math. France **83**, 279–330 (1955)
44. G. Dall'Agata, String compactifications with fluxes. Class. Quant. Grav. **21**, S1479–S1499 (2004)

# Flavor Physics

16

"Who ordered muon?", put forward after the discovery of muon (with the already known electron), was the culminating phrase on the flavor problem or the family problem. Muon behaved exactly in the same way in the QED and weak interactions except for their mass difference. This difference in masses is one of the flavor problems.

Not only in masses but also the way they interact is the key in the flavor puzzle. The first observation on this was noticed from the SU(3) symmetry of the currents before the advent of quark flavors, $u$, $d$, and $s$. The corresponding SU(3) charges, $F_i$ ($i = 1, 2, \ldots, 8$), are generators of the SU(3) transformation,

$$\Psi \to e^{i\alpha_i F_i} \Psi. \tag{16.1}$$

If we pick up a proper direction, say $2 \times 2$ matrix $X$, i.e. $F_X = \mathrm{diag}(X, 1)$ rearranged with the first two rows and columns for $X$, the unitary transformation can be written as $\mathrm{diag}(\cos \alpha_X, 1) + \mathrm{diag}(\sin \alpha_X \, i\sigma_y, 1)$. The rank-2 group SU(3) has two Cartan subalgebras generated by the isospin $T_3 (= F_3)$ and strangeness $S (= 2Y - B)$, with the octet charges shown in Fig. 16.1. Since there are two charge $+1$ operators, the transformation (16.1) can include $+$ charge operators as $e^{i(\alpha_\pi F_{1+i2} + \alpha_K F_{4+i5} + \cdots)}$, which therefore includes

$$U \ni i(\alpha_\pi F_{1+i2} + \alpha_K F_{4+i5}) + \cdots, \tag{16.2}$$

with $\alpha_\pi^2 + \alpha_K^2 = 1$, or $\tan \theta_C = \alpha_K / \alpha_\pi$ in terms of the Cabibbo angle $\theta_C$ [1]. Cabibbo compared the proton and $\Lambda$ decay rates with the muon decay rate. Muon and electron do not have isospin and strangeness quantum numbers and the unitary transformation on them must be just a phase, i.e. it transforms as $\mathbf{1}$ under SU(3) transformation. This observation was made before any of the quark models was suggested. But with the SU(3) transformation on hadrons, the neutral operators $F_{6\pm i7}$, together with $F_3$ and $F_8$, must be included also in Eq. (16.2), which is

© Springer Nature Switzerland AG 2020
K.-S. Choi, J. E. Kim, *Quarks and Leptons From Orbifolded Superstring*,
Lecture Notes in Physics 954, https://doi.org/10.1007/978-3-030-54005-0_16

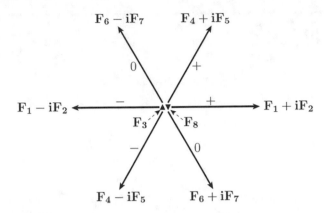

**Fig. 16.1** Eight generators of SU(3). The horizontal axis is the eigenvalue of $F_3$ and the vertical axis is the eigenvalue of $Y = \frac{S+B}{2} \sim$ (strangeness plus baryon number). Change of electromagnetic charges by the operations is shown with the red color

dangerous at that time because the strangeness changing neutral processes were not observed.

With the quark model, we introduced a L-handed weak doublet,

$$\begin{pmatrix} u \\ d \cos\theta_C + s \sin\theta_C \end{pmatrix}_L \tag{16.3}$$

which introduces a strangeness changing neutral current. This was resolved by the GIM mechanism, introducing the second doublet [2]. This was done before the discovery of the charm quark $c$. But in the unitary transformation in the form of Eq. (16.1), without any reference to quarks, one has to increase the symmetry group to SU(4). The third family quarks were discovered in 1978 and 1995 [3, 4].

In the leptonic sector, neutral processes were difficult to identify experimentally even though neutrino–antineutrino oscillation was suggested before as a possibility [5]. In 1962, the muon-type neutrino was discovered, completing the discovery of the second family of leptons [6], before any theoretical needs were proposed. Now flavor physics in the leptonic sector is parametrized by the MNS matrix.[1] Charged lepton $\tau$ was discovered in 1975 [7], and its partner $\nu_\tau$ was seen directly in 2000 [8].

In the standard model(SM), flavor physics is studied by the charged currents(CCs). Here, the CCs are left-handed(L-handed) but the currents, basically being of gauge interactions, are properly defined in the basis where all fermion masses are diagonalized. This introduces the Yukawa couplings by which all the

---

[1]The $3 \times 3$ neutrino mixing matrix is responsible for the oscillations between neutrinos and we use MNS. To write PMNS, we must use $6 \times 6$ matrix. But, we use the standard convention PMNS even for the $3 \times 3$ matrix.

SM fermions acquire masses by the Higgs mechanism. Therefore, we should study Yukawa couplings to pinpoint the origin of flavor physics.

For the L-handed fermions of the SM spectra, Eq. (2.7), we can unitarily transform by a $3 \times 3$ unitary matrices, $U$ and $V$, on the three members of $\Psi$. In this Chapter, we will use $U$ for acting on the L-handed fields and $V$ for acting on the R-handed fields. Let the Yukawa coupling matrices $Y^{(u,d,\nu,e)}$ be given for the mass matrices of up-type quarks, down-type quarks, neutrinos, and $Q_{em} = -1$ leptons, respectively. By the standard procedure, a number of phases can be absorbed by redefining fermion fields, for example,

$$Y^{(u,d,\nu,e)\,0} \to U^{(u,d,\nu,e)\dagger} Y^{(u,d,\nu,e)\,0} V^{(u,d,\nu,e)}, \quad \text{etc.} \tag{16.4}$$

on the Yukawa coupling matrices

$$\mathcal{L}_Y^{(u,d,\nu,e)} = \bar{u}_R^{(u,d,\nu,e)\,0} Y^{(u,d,\nu,e)0} q_L^0, \tag{16.5}$$

where superscript $^0$ refer to the original form and those without it is in the diagonalized mass bases,

$$\Psi^{(u,d,\nu,e)} = (U \text{ or } V)^{(u,d,\nu,e)} \Psi^{(u,d,\nu,e)\,0}. \tag{16.6}$$

In flavor physics from string compactification, our objective is to present the form $Y^{(u,d,\nu,e)0}$ of Eq. (16.5). Most easily, one may study just the renormalizable couplings. But, the possibility of nonrenormalizable interactions is always open, whose strength can still be sizable in string compactification. So, $Y^{(u,d,\nu,e)0}$ are the ones including sizable nonrenormalizable interactions also.

At field theory level, some continuous and discrete symmetries were proposed in the past but in string compactification there has not been any accepted scenario yet. We take the viewpoint that any suggestion is not enough if it fails in explaining even in just one component of the CKM and MNS matrices. So, let us begin by listing the experimentally determined CKM and PMNS matrix elements.

The CKM and PMNS matrices are given by[2]

$$V_{CKM} = (U^{(u)})(U^{(d)})^\dagger \tag{16.7}$$

$$V_{PMNS} = (U^{(\nu)})(U^{(e)})^\dagger, \tag{16.8}$$

where $U^{(u,d)}$ and $U^{(\nu,e)}$ are diagonalizing unitary matricies of L-handed quark and lepton fields, respectively.

---

[2]The usual definition in Ceccucci et al. on the CKM matrix [9] is the same as ours but the definition on the PMNS matrix in S. Petcov [9] is the opposite to ours.

## 16.1    Data on Flavor Physics

### 16.1.1  CKM Matrix

The CKM data in the PDG book is present by fitting to an approximate unitary matrix [10], which is not adequate in calculating the Jarlskog determinant because they did not use an exact CKM matrix. Since we try to determine the phase, we must use a parametrization where at least three elements do not contain the phase such that three real angles are determined. Therefore, we determine three real angles, using the exact Kim–Seo (KS) form [11] which will be discussed in Sect. 16.2.1. We will use the (11), (12), and (21) elements of $V$, which are[3]

$$(11) = 0.97420 \pm 0.00021, \ (12) = 0.2243 \pm 0.0005, \ (21) = -0.218 \pm 0.004.$$

Let us define the experimentally determined error bars of $\theta_{1,2,3}$ and $\alpha_{KS}$[4] as

$$\theta_1 = \bar{\theta}_1 + \delta_1, \ \theta_2 = \bar{\theta}_2 + \delta_2, \ \bar{\theta}_3 + \delta_3, \ \alpha_{KS} = \bar{\alpha}_{KS} + \delta_{KS}. \tag{16.9}$$

Then, we obtain

$$\bar{\theta}_1 = 13.0432^{\circ}, \ \bar{\theta}_2 = 14.9964^{\circ}, \ \bar{\theta}_3 = 6.3541^{\circ}, \tag{16.10}$$

$$\delta_1 = \pm 0.0533^{\circ}, \ \delta_2 = \pm 4.017^{\circ}, \ \delta_3 = \pm 2.364^{\circ}. \tag{16.11}$$

To determine the CP phase $\alpha_{KS}$, we use the formula for the Jarlskog angle. Since the shape of the Jarlskog triangle is the same in any parametrization, the $\mathscr{CP}$ violation error will be the same in any parametrization. So, for the error bars of $J$, i.e. $\delta_J$, we use the PDG value. The PDG gives $\delta_J/J = 0.15/3.18 = \frac{1}{21.2}$. If the formula gives $|\sin \bar{\alpha}_{KS}| > 1$, we use $\bar{\alpha}_{KS} = 90^{\circ}$. Since $J \simeq \frac{1}{8} \sin 2\bar{\theta}_1 \sin 2\bar{\theta}_2 \sin 2\bar{\theta}_3 \sin \bar{\theta}_1 (\sin \alpha_{KS})$, applying small errors in Eq. (16.11), we obtain $\delta(\sin \alpha_{KS}) = -\frac{1}{21.2} = -0.04717$ and $\delta(\cos \alpha_{KS}) = 0.28402$, leading to $\delta \alpha_{KS} = 16.5^{\circ}$. Therefore, we obtain the following numerical data on $V_{CKM}$ using

---

[3]T. Gerson and Y. Nir in [9].
[4]These parameters will be defined later in Eq. (16.24).

the error propagation method, which is useful for the KS form since all the first row elements are real [12],

$$
\begin{pmatrix}
0.97420 \pm 0.00021\,, & 0.22430 \pm 0.00137\,, & 0.02498 \pm 0.00926 \\[2ex]
-0.2180 \pm 0.00419\,, & \begin{aligned} &+0.93524 \pm 0.01653 \\ &-i(0.02864 \pm 0.00144)\sin\alpha_{KS}\,, \\ &+(0.02864 \pm 0.00144)\cos\alpha_{KS} \end{aligned} & \begin{aligned} &+0.10415 \pm 0.00321 \\ &+i(0.25717 \pm 0.05522)\sin\alpha_{KS} \\ &-(0.25717 \pm 0.05522)\cos\alpha_{KS} \end{aligned} \\[4ex]
\begin{aligned} &-i(0.05840 \pm 0.01529)\sin\alpha_{KS} \\ &-(0.05840 \pm 0.01529)\cos\alpha_{KS} \end{aligned}\,, & \begin{aligned} &-0.10690 \pm 0.00339 \\ &+i(0.25054 \pm 0.05234)\sin\alpha_{KS}\,, \\ &+(0.25054 \pm 0.05234)\cos\alpha_{KS} \end{aligned} & \begin{aligned} &+0.96001 \pm 0.01709 \\ &+i(0.02790 \pm 0.01037)\sin\alpha_{KS} \\ &+(0.02790 \pm 0.01037)\cos\alpha_{KS} \end{aligned}
\end{pmatrix}.
$$

$$(16.12)$$

In addition, we obtain $J = 1.364^{+0.599}_{-0.599} \times 10^{-3}$ which is considered as the experimentally determined value. The Jarlskog triangles of the quark sector are shown in Fig. 16.2.

### 16.1.2  Neutrino Oscillation and PMNS Matrix

Most data on flavor physics in the hadronic sector are from meson decays, thanks to the formation of K, D, and B mesons due to the confining QCD. On the other hand, leptons do not carry color by definition and in the leptonic sector there is no counterpart of "meson" of the hadronic sector. At best for the counterpart, we can consider processes involving two leptons, which is known as *neutrino oscillation*. Neutrino flavors (denoted by $\alpha$, $\beta$, etc.) of energy $E$ (in eV units) oscillate in

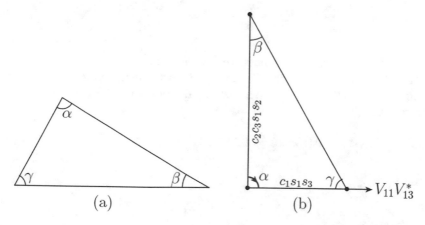

**Fig. 16.2** The Jarlskog triangle: (**a**) the usual layout in the PDG book, (**b**) the layout given in [13]. In (**b**), the invariant $J$ is given by $c_1 c_2 c_3 s_1^2 s_2 s_3 \sin\alpha$ in terms of the KS parametrization

vacuum as

$$P(\nu_\alpha \to \nu_\beta) = \delta_{\alpha\beta} - 4 \sum_{i>j} \mathrm{Re}(U^*_{\alpha i} U_{\beta i} U_{\alpha j} U^*_{\beta j}) \sin^2[1.27 \Delta m^2_{ij}(L/E)]$$

$$+ 2 \sum_{i>j} \mathrm{Im}(U^*_{\alpha i} U_{\beta i} U_{\alpha j} U^*_{\beta j}) \sin^2[2.54 \Delta m^2_{ij}(L/E)], \qquad (16.13)$$

where indices $i$, $j$ denote the mass eigenstates, and $L$ is the path length (in km units). Formulae for oscillation in matter involve more parameters [14,15]. The mostly used formulae for neutrino oscillation for the T2K experiments of the $\nu_\mu$ beam from the KEK are its survival probability

$$P(\nu_\mu \to \nu_\mu) \simeq 1 - \left( \cos^4 \theta_{13} \sin^2 2\theta_{23} + \sin^2 2\theta_{13} \sin^2 \theta_{23} \right) \sin^2 \left( \Delta m^2_{31} \frac{L}{4E} \right),$$
$$(16.14)$$

and its conversion probability to $\nu_e$

$$P(\nu_\mu \to \nu_e) \simeq \sin^2 2\theta_{13} \, \sin^2 \theta_{23} \, \frac{\sin^2[(1-x)\Delta]}{(1-x)^2}$$

$$+ \alpha \cos(\delta + \Delta) \sin 2\theta_{12} \, \sin 2\theta_{13} \, \sin 2\theta_{23} \, \frac{\sin[x\Delta]}{x} \, \frac{\sin[(1-x)\Delta]}{(1-x)} \qquad (16.15)$$

$$+ O(\alpha^2),$$

where the vacuum oscillation parameters $\alpha$ and $\Delta$ and the matter oscillation parameter $x$ are given by

$$\alpha = \left| \frac{\Delta m^2_{21}}{\Delta m^2_{31}} \right|, \quad \Delta = \frac{\Delta m^2_{31} L}{4E}, \quad x = \frac{2\sqrt{2} G_F N_e E}{\Delta m^2_{31}}, \qquad (16.16)$$

in terms of mass squared differences $\Delta m^2_{ij} = m^2_i - m^2_j$.

The 2019 fit [16, 17] using the PDG parametrization of the PMNS matrix obtained $\pm 1\sigma$ mass squared differences in the normal (NH) and inverted (IH) hierarchies as (Fig. 16.3)

$$\text{NH and IH}: \Delta m^2_{21} = 0.755^{+0.020}_{-0.016} \times 10^{-4}\,\mathrm{eV}^2, \qquad (16.17)$$

and

$$\text{NH}: \Delta m^2_{31} = 2.50^{+0.03}_{-0.03} \times 10^{-3}\,\mathrm{eV}^2,$$

$$\text{IH}: \Delta m^2_{31} = 2.42^{+0.03}_{-0.04} \times 10^{-3}\,\mathrm{eV}^2. \qquad (16.18)$$

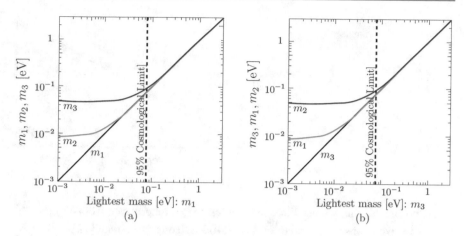

**Fig. 16.3** $m_1$, $m_2$, and $m_3$ vs. the lightest neutrino mass. (a) NH. (b) IH

The fit [17, 18] gives almost the same PMNS angles for the NH and IH mass squared differences, and hence these PMNS matrices are taken to be the same here. The leptonic data is not accurate enough in particular on the phase, i.e. $\delta = 241°^{+115°}_{-68°}$ for NH, and $\delta = 266°^{+61°}_{-58°}$ for IH. The Jarlskog determinant is about $3.5 \times 10^{-2} \sin \delta_{\text{PMNS}}$ in both cases, which ranges $[-3.5, +0.3] \times 10^{-2}$ for NH and $[-3.5, -1.6] \times 10^{-2}$ for IH. The neutrino data is not accurate enough to pinpoint the value $\delta_{\text{PMNS}}$.

We first determine three real angles, adopting the unitarity condition, for which we need just three data points. In fact, the familiar global analyses are actually overfitting. Because of the large uncertainty on the phase, we refrain from presenting error bars. Then, from the values of [17], we obtain

$$(11) = 0.6612, \quad (12) = 0.3154, \tag{16.19}$$

but in the PDG parametrization there is no other element without the phase, involving the first two family members. We try to use the absolute value of (21) element. Let us take the range $\delta_{\text{PDG}} = [225°, 315°]$ and $[45°, 135°]$. Now, from these experimental (11), (12), and (21) elements of $V_{\text{PMNS}}$, we can determine three real angles using the KS parametrization

$$\text{NH, IH}: \begin{cases} \bar{\Theta}_1 = 35.3948° \\ \bar{\Theta}_2 = 39.1733 - 55.9735° \text{ for } \delta_{\text{PDG}} = 225° - 315° \\ [\bar{\Theta}_2 = 39.1733 - 55.9735° \text{ for } \delta_{\text{PDG}} = 45° - 135°] \\ \bar{\Theta}_3 = 14.6965°, \end{cases} \tag{16.20}$$

where we used $\delta_{\text{PDG}}$ near $\frac{3\pi}{2}$ and $\frac{\pi}{2}$. For an illustration, we take $\delta_{\text{PDG}} = \frac{3\pi}{2}$, i.e. $\bar{\Theta}_2 = 30.9°$ and obtain the following.

**Fig. 16.4** A possible
leptonic Jarlskog triangle

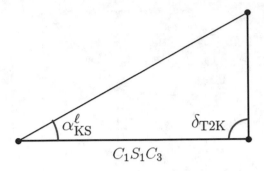

As an illustration, we use three real angles $\Theta_{1,2,3}$ determined from $\delta_{\text{PDG}} = \frac{3\pi}{2}$, i.e. for $\Theta_2$ we choose the median $\bar{\Theta}_2 = 47.4633°$. Thus, we obtain [12],

$$\begin{pmatrix} 0.81518^{+0.01174}_{-0.00978}, & 0.56026^{+0.01707}_{-0.01423}, & 0.14695^{+0.08925}_{-0.07438} \\[2ex] -0.39158^{+0.01371}_{-0.01475}, & \begin{matrix} +0.53308^{+0.02423}_{-0.02085} \\ -i(0.186939^{+0.11121}_{-0.09203})\sin\alpha^\ell_{\text{KS}} \\ +(0.186939^{+0.11121}_{-0.09203})\cos\alpha^\ell_{\text{KS}} \end{matrix}, & \begin{matrix} +0.13982^{+0.08188}_{-0.06726} \\ +i(0.71274^{+0.03291}_{-0.02853})\sin\alpha^\ell_{\text{KS}} \\ -(0.71274^{+0.03291}_{-0.02853})\cos\alpha^\ell_{\text{KS}} \end{matrix} \\[4ex] \begin{matrix} -i(0.42679^{+0.02029}_{-0.01895})\sin\alpha^\ell_{\text{KS}} \\ -(0.42679^{+0.02029}_{-0.01895})\cos\alpha^\ell_{\text{KS}} \end{matrix}, & \begin{matrix} -0.17152^{+0.10217}_{-0.08463} \\ +i(0.58101^{+0.02637}_{-0.02268})\sin\alpha^\ell_{\text{KS}} \\ +(0.58101^{+0.02637}_{-0.02268})\cos\alpha^\ell_{\text{KS}} \end{matrix}, & \begin{matrix} +0.653943^{+0.03034}_{-0.02636} \\ +i(0.15239^{+0.08883}_{-0.07281})\sin\alpha^\ell_{\text{KS}} \\ +(0.15239^{+0.08883}_{-0.07281})\cos\alpha^\ell_{\text{KS}} \end{matrix} \end{pmatrix}.$$

$$(16.21)$$

In this case, we obtain $J \simeq 3.34 \times 10^{-2}|\sin\alpha^\ell_{\text{KS}}|$ from any one out of the six possible products of the form given in (16.25). Comparing this KS value with the previous PDG value $3.5 \times 10^{-2}\sin\delta_{\text{PDG}}$, we have a relation $|\sin\alpha^\ell_{\text{KS}}| \simeq 0.56|\sin\delta_{\text{PMNS}}|$. If we take the initial central value of the T2K experimental range [19], i.e. $\delta_{\text{T2K}} = 270°$, we can draw the leptonic Jarlskog triangle as shown in Fig. 16.4. This is because both $|\alpha^\ell_{\text{KS}}| \simeq 30.9°$ and $|\bar{\delta}_{\text{T2K}}| \simeq 90°$ must belong to the corner angles of this triangle. The Jarlskog triangle of the lepton sector is shown in Fig. 16.4.

Note that for $\cos\alpha^\ell_{\text{KS}} \simeq 0$, the absolute values of $V^{\text{KS}}_{PMNS}$ is somewhat close to a tri-bimaximal form.

## 16.2 Theories on Flavor Physics in Field Theory

In the SM, we need at least three families of quarks and leptons,

$$\begin{pmatrix} u'^\alpha \\ d^\alpha \end{pmatrix}_L, \quad \begin{pmatrix} c'^\alpha \\ s^\alpha \end{pmatrix}_L, \quad \begin{pmatrix} t'^\alpha \\ b^\alpha \end{pmatrix}_L \qquad (16.22)$$

$$u_{\alpha L}^c, \ d_{\alpha L}^c, \quad c_{\alpha L}^c, \ s_{\alpha L}^c, \quad t_{\alpha L}^c, \ b_{\alpha L}^c$$

$$\begin{pmatrix} \nu_e' \\ e \end{pmatrix}_L, \quad \begin{pmatrix} \nu_\mu' \\ \mu \end{pmatrix}_L, \quad \begin{pmatrix} \nu_\tau' \\ \tau \end{pmatrix}_L \tag{16.23}$$

$$e_L^c, \qquad \mu_L^c, \qquad \tau_L^c$$

$$N_1^c, \qquad N_2^c, \qquad N_3^c,$$

where the observed heavy 3rd family members are colored red, and the primed fields are mixtures of mass eigenstate fields. There are 15 Weyl fields in one family. By adding $N_3^c$ to these, we have the representation **16** of SO(10). Because we will remove the flavor changing neutral current (FCNC) effects, flavor physics in the SM is described by CCs based on Eqs. (16.22) and (16.23). With the definition of Eq. (16.22) without the red colors, Eq. (16.3) corresponds to $u' = u \cos \theta_C - c \sin \theta_C$ and $c' = u \sin \theta_C + c \cos \theta_C$ in Eq. (16.22). Here, we choose the forms of Eqs. (16.22) and (16.23) because it is better to start with the bases of diagonalized $Q_{\rm em} = -1$ leptons. In this chapter, we use the Kim–Seo (KS) form for the CKM matrix of Eq. (16.22) [11],

$$V^{\rm KS} = \begin{pmatrix} c_1, & s_1 c_3, & s_1 s_3 \\ -c_2 s_1, & c_1 c_2 c_3 + s_2 s_3 e^{-i\alpha_{\rm KS}}, & c_1 c_2 s_3 - s_2 c_3 e^{-i\alpha_{\rm KS}} \\ -s_1 s_2 e^{+i\alpha_{\rm KS}}, & -c_2 s_3 + c_1 s_2 c_3 e^{+i\alpha_{\rm KS}}, & c_2 c_3 + c_1 s_2 s_3 e^{+i\alpha_{\rm KS}} \end{pmatrix},$$

$$\tag{16.24}$$

where $c_i$ and $s_i$ are cosines and sines of three real angles $\theta_i$ ($i = 1, 2, 3$) and $\alpha_{\rm KS}$ is the $\mathscr{CP}$ phase $\delta_{\rm CKM}$. The KS form is written such that the elements in the 1st row are all real, which makes it easy to draw the Jarlskog triangle with one side sitting on the horizontal axis. Furthermore, the (21) element is real and hence it is possible to determine three real angles from the experimental values of (11), (12), and (21) elements. For the PMNS matrix of Eq. (16.23), we use another four parameter set, $\Theta_i$ (giving corresponding $C_i$ and $S_i$) and $\alpha_{\rm KS}^\ell$.

If future data are not consistent with the representations given in Eqs. (16.22), (16.23), physics beyond the standard model (BSM) is needed. To find out the BSM contribution just above the electroweak scale, some discrepancy with the SM parametrization (16.22), (16.23) has to be observed.

## 16.2.1 Electroweak CP Violation

After the discovery of the electroweak $\mathscr{CP}$ violation in the neutral K meson decays [20] and from the recent B meson decays,[5] its final form in the SM has settled to a kind of Kobayashi–Maskawa form [21]. There can be numerous ways to parametrize the mixing angles with three families, which have led to a unique invariant for the weak $\mathscr{CP}$ violation. It is the Jarlskog determinant $J$ which is twice the area of the triangle shown in Fig. 16.2. The original form of $J$ is given by two elements of $V_{\text{CKM}}$ and two elements of $V_{\text{CKM}}^*$, which counts two sides of the triangle. One such side length is shown as the horizontal segment in Fig. 16.2(b) as $V_{11}V_{13}^*$. Referring these figures, one notices that one of the three angles, $\alpha$, $\beta$, or $\gamma$ can be used for the area of the triangle. For the parametrization in the PDG book $\gamma$ is chosen, and for the KS parametrization [11] $\alpha$ is used for the area (Fig. 16.3).

But a simple form, readable from the $3 \times 3$ CKM matrix itself [13] is given by

$$J = |\text{Im } V_{31} V_{22} V_{13}|, \text{ after making Det.}V \text{ real}, \tag{16.25}$$

which is the KS form of $J$. The CKM matrix given in the PDG book has a real determinant and the parametrization given in [11] also leads to a real number. The triangle in Fig. 16.2b uses the product of the 1st and 3rd columns, $V_{i1}^{\text{KS}} V_{i3}^{\text{KS}*}$ for which the three sides are given by $i = 1, 2, 3$. There are two more ways to make triangles, $V_{i2}^{\text{KS}} V_{i1}^{\text{KS}*}$ and $V_{i3}^{\text{KS}} V_{i2}^{\text{KS}*}$. The shapes of the other triangles are completely different from that of Fig. 16.2b but the invariant phase $\alpha$ also appear there. All these triangles give the identical value for $J$. Thus, $J$ is not a process dependent value but a theory dependent value. If we pick up the triangle $V_{i2}^{\text{KS}} V_{i1}^{\text{KS}*}$, two side lengths are $|V_{12}V_{11}^*| \simeq \lambda$ and $|V_{22}V_{21}^*| \simeq \lambda$ where $\sin\theta_C = O(\lambda)$. Therefore, the third length is $O(\lambda^{4-5})$ since $J \sim O(\lambda^{5-6})$. This implies that one angle in the CKM matrix must be close to $90°$, which is shown in Fig. 16.5. This angle is $\alpha$ in the fat triangle Fig. 16.2b and also in the thin triangle 16.5. $\alpha$ is the invariant angle in the quark sector, and it is proper to assign a unique name in the SM, $\delta_{\text{CKM}}$. The small side has a length at most $O(\lambda^4)$, which implies that two angles are close to $90°$. Namely, from trigonometry, if we have two long sides of length $\lambda$ and $\lambda + O(\lambda^4)$, the angle between them, $\epsilon$, is given for $\lambda \simeq 0.225$ by [12]

$$\cos\epsilon = 1 - O(\lambda^3) + \cdots \lesssim 0.9886 \rightarrow \epsilon \lesssim 8.65°. \tag{16.26}$$

One among $\alpha$, $\beta$, and $\gamma$ must be $\alpha_{\text{KS}}$, and $\epsilon$ cannot be $\alpha_{\text{KS}}$ since there is no angle close to 0 among them. The shape of this thin triangle is shown in Fig. 16.5, with an exaggrated $\epsilon$.

---

[5]See, T. Gerson and Y. Nir in [9].

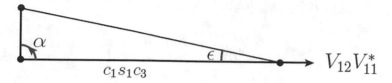

**Fig. 16.5** The Jarlskog triangle with the horizontal axis $V_{12}^{KS} V_{11}^{KS*}$

In terms of the KS angles $\theta_{1,2,3}$ and $\alpha_{KS}$, the Jarlskog determinant $J^{KS}$ is given by

$$J^{KS} = \frac{1}{8} |\sin 2\theta_1 \sin 2\theta_2 \sin 2\theta_3 \sin \theta_1 \sin \alpha_{KS}|. \tag{16.27}$$

The same formula also holds in the leptonic sector by replacing $\theta_{1,2,3} \rightarrow \Theta_{1,2,3}$ and $\alpha_{KS} \rightarrow \alpha_{KS}^\ell$.

In the leptonic sector, we define the KS form for the PMNS matrix with parameters $\Theta_i$ and the invariant phase $\alpha_{KS}^\ell$. The comments made for the quark sector also apply here, i.e. $J$ in the lepton sector are the same for all three Jarlskog triangles of leptons and $\alpha_{KS}^\ell \equiv \delta_{MNS}$ is the unique invariant lepton phase. Since we take the basis where $Q_{em} = -1$ leptons are in the mass eigenstate, the diagonalization procedure of lepton masses is on the neutrino masses. At the renormalizable Lagrangian level, there is no mass term of the neutrinos in the lepton doublets in Eq. (16.23).

The effective dimension 5 neutrino mass operators initially defined in terms of $^0$ superscripts of Eq. (16.6) are given by

$$\mathscr{L}_\nu = \ell^{0T} C^{-1} M_\nu^0 \ell^0, \tag{16.28}$$

where $C$ is the charge conjugation matrix and $M_\nu^0$ is the neutrino mass matrix. In the seesaw model, it is proportional to the square of the VEV of $H_u$ and suppressed by the singlet neutrino mass,

$$M_\nu^0 = \frac{\langle H_u \rangle^2}{M_N} Y^0, \tag{16.29}$$

where $Y^0$ is a dimensionless $3 \times 3$ matrix. Since $\Psi^{(\nu)0} = U^{(\nu)\dagger} \Psi^{(\nu)}$ from Eq. (16.6), in the mass eigenstate basis Eq. (16.28) becomes

$$\mathscr{L}_\nu = \nu^T U^{(\nu)*} C^{-1} M_\nu^0 U^{(\nu)\dagger} \nu. \tag{16.30}$$

Thus, $U^{(\nu)*}$ and $U^{(\nu)\dagger}$ diagonalize $M_\nu^0$,

$$U^{(\nu)*} M_\nu^0 U^{(\nu)\dagger} = \begin{pmatrix} m_1, & 0, & 0 \\ 0, & m_2, & 0 \\ 0, & 0, & m_3 \end{pmatrix} \tag{16.31}$$

The ansatz for the so-called tri-bimaximal mixing is [22]

$$\sim \begin{pmatrix} \frac{\sqrt{2}}{\sqrt{3}}, & \frac{1}{\sqrt{3}}, & 0 \\ -\frac{1}{\sqrt{6}}, & \frac{1}{\sqrt{3}}, & -\frac{1}{\sqrt{2}} \\ -\frac{1}{\sqrt{6}}, & \frac{1}{\sqrt{3}}, & \frac{1}{\sqrt{2}} \end{pmatrix} \tag{16.32}$$

but the recent determination of $V_{\text{PMNS}}$ in Eq. (16.21) is not close to the tri-bimaximal form. On the other hand, the approximate form of $V_{\text{CKM}}$ takes the form

$$\sim \begin{pmatrix} 1 & 0 & 0 \\ 0 & 1 & 0 \\ 0 & 0 & 1 \end{pmatrix} \tag{16.33}$$

which is consistent with the data given in Eq. (16.12).

The original $3 \times 3$ neutrino mass matrix contained 18 parameters. It is diagonalized to 3 parameters by 9 parameter matrix $U^{(\nu)\dagger}$. Note that all 9 parameters can be used since Eq. (16.31) does not cancel any phase unlike in the quark case. Out of 18 original parameters, there will remain 9 independent parameters which are counted as 3 masses and 4 PMNS angles and two more. These two more are named as two Majorana phases, say $\delta_{N2}$ and $\delta_{N3}$, which is usually written as a two parameter diagonal matrix $V_{\text{Maj}}$ which has the real determinant for Eq. (16.2) to be applicable. Thus, the full electroweak lepton currents are parametrized by

$$V^{\text{lept}} = V_{\text{PMNS}} V_{\text{Maj}}, \tag{16.34}$$

where

$$V_{\text{Maj}} = \begin{pmatrix} e^{-i(\delta_{N2}+\delta_{N3})} & 0 & 0 \\ 0 & e^{i\delta_{N2}} & 0 \\ 0 & 0 & e^{i\delta_{N3}} \end{pmatrix}, \tag{16.35}$$

which derives from the heavy neutrino masses of $N_1$, $N_2$, and $N_3$ implied by Eq. (16.29). These Majorana phases do not contribute to the Jarlskog triangle [See Exercise 16.2]. The leptonic Jarlskog triangle for $\delta_{\text{T2K}} = 270°$, viz. Eq. (16.21), is shown in Fig. 16.4. In this case, the invariant CP phase is $\alpha_{\text{KS}}^\ell$, i.e. any leptonic Jarlskog triangle contains the angle $\alpha_{\text{KS}}^\ell$.

**Table 16.1** Definition of lepton numbers in Type-II leptogeneses

|        | $\ell_L$ | $H_u$ | $H_d$ | $h_{u,d}$ | $N$ | $\mathcal{N}$ |
|--------|----------|-------|-------|-----------|-----|---------------|
| $L$    | +1       | −2    | +2    | 0         | −1  | +1            |
| VEV    | ×        | inert | inert | $v_{\text{ew}}\{s_\beta, c_\beta\}$ | × | × |

We introduced an inert Higgs $H_u$ carrying $L = -2$ with zero VEV and singlet leptons $\mathcal{N}$ carrying $L = +1$

### 16.2.2  $B$ and $L$ Generation

In the Universe, the global quantum numbers can be generated from the initially vanishing global quantum numbers if Sakharov's three conditions [23] are suitably applicable. One condition is that there should be $\mathcal{C}$ and $\mathcal{C}\mathcal{P}$ violation at the time when the generation process is active. Here, the baryon ($B$) and lepton ($L$) number generations are of our interest. Baryon number generation called *baryogenesis* was initially studied in GUT models [24, 25] at the GUT time scale in the Universe evolution, which was not favored because of the back reaction equilibrating $B$ and $\overline{B}$ at the GUT time scale. The nonequilibrium conditions in baryogenesis [26] triggered by heavy particle decays were considered under the names, *leptogenesis* [27] and *Q genesis* [28]. Especially, leptogenesis attracted a great deal of attention because heavy neutrinos can be easily added in GUT models as shown in Eq. (16.23). Here, the needed $\mathcal{C}\mathcal{P}$ violation parameter is not the one in the SM, $\delta_{\text{PMNS}}$, but those arising from the heavy neutrino sector $\delta_{N_1}$ and $\delta_{N_2}$. This idea of leptogenesis was discussed extensively.[6]

But, there is a mechanism that the $\mathcal{C}\mathcal{P}$ parameter of the electroweak scale can contribute in some leptogenesis models. Type-II leptogenesis [30] introduces more fields, especially scalar doublets which do not develop any VEV. Here, we briefly comment on the Type-II leptogenesis since it introduces $\delta_{\text{MNS}}$ and also the model incorporates all the needed conditions of the global number generation in the Universe. In the example of Table 16.1 [31], the lepton numbers for several fields are defined. We need an inert Higgs doublet $H_u$ carrying $L = -2$ and singlet leptons $\mathcal{N}$ carrying $L = +1$. Anyway, the fields $H_{u,d}$ and $\mathcal{N}$ introduced at high energy scale are not visible at low-energy scale. The lepton number is violated at the high energy scale by the heavy neutrino mass term. Let the lightest among the heavy neutrinos be $N_0$. For the lepton number generation, both $L$ violation and $\mathcal{C}\mathcal{P}$ violation are needed. So, interactions conserving $L$ are $N_0 \ell_L h_u$, $\mathcal{N}_0 \ell_L H_u$, $N_0 \mathcal{N}_0$, $H_u H_d$, $\cdots$, and interactions violating $L$ are $h_u^* H_u$, $N_0 N_0$, $\mathcal{N}_0 \mathcal{N}_0$, $\cdots$. In Type-II leptogenesis the needed $L$ violation is given by spin-0 bosons.

In the decay scenarios, two decay channels are required and both $L$ and $\mathcal{C}\mathcal{P}$ violation occurs on the left-hand side of the cut diagram,[7] which is known as the Nanopoulos–Weinberg theorem [32]. In Type-II leptogensis case, therefore, the

---

[6] See, for example, Fukugita and Yanagida [29].

[7] One can form a loop diagram by attaching (a)$^\dagger$ on the right-hand sides of (b), (c), and (d). The cut line is the line connecting two points attaching (a)$^\dagger$.

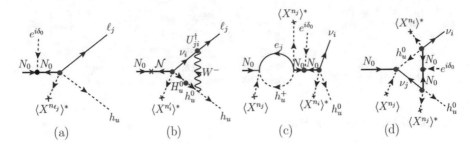

**Fig. 16.6** Feynman diagrams interfering in the $N_0$ decay: (**a**) the effective tree diagram with possible insertions of VEVs of the SM singlet fields $X^{n\ell_j}$, (**b**) diagrams containing $\delta_{MNS}$, (**c**) the wave function renormalizable diagrams, and (**d**) the vertex diagrams. $e^{i\delta_0}$ is a Majorana phase of $V_{\text{Maj}}$

same mother particle $N_0$ has two decaying channels with different lepton numbers, and the model satisfies the Nanopoulos–Weinberg theorem as shown by Fig. 16.6a and b. For reference, we show diagrams of the Type-I leptogenesis [27] in Fig. 16.6a, c, d [33].

How can we pinpoint that leptogenesis employs $\delta_{MNS}$? Since leptogenesis uses all the phases in $V^{\text{lept}}$, it is virtually impossible if we do not have any information on the Majorana phases $\delta_{N_2}$ and $\delta_{N_3}$. If any independent phenomena proves that there is at most one Majorana phase [34], then $\delta_{PMNS}$ must contribute to leptogenesis and predicts that there are additional particles as suggested in Table 16.1.

### 16.2.3 Discrete Symmetries

The first example of discrete symmetries applied to the flavor problem was permutation symmetry $S_3$ by Pakvasa and Sugawara [35]. In the calculable solutions of the strong $\mathscr{CP}$ problem, $\mathscr{CP}$ itself was used in most cases and $S_3$ was also used as discussed in Sect. 13.5.3 [36].

But, the main arena of applying discrete symmetries is in the flavor problem, which started with the parameter $\theta_C$ and became popular in recent years in explaining the PMNS matrix. The approximate form of the PMNS matrix as shown in Eq. (16.32) questions for some discrete symmetries. On the other hand, the approximate form of the CKM matrix as shown in Eq. (16.33) do not question such a symmetry but proposes an expansion in terms of a small number [37].

The discrete symmetry can be Abelian or non-Abelian. Both are required to be subgroups of continuous groups because we need a kind of discrete gauge symmetry to avoid cosmological problems [38].

Typical Abelian discrete groups are $\mathbb{Z}_n$ groups which are usually taken as subgroups of U(1). Because the Cabibbo angle is about $\frac{\pi}{12}$, the dodeca symmetry was considered for the quark mixing matrix [39, 40] and extended for the lepton mixing matrix [41].

Most discrete symmetries studied for the flavor problem are non-Abelian, which is extensively reviewed by a Japanese group [42, 43]. The mother gauge group must be non-Abelian, as SU(3) is the mother gauge group of $\Delta(3n^2)^8$ and $\Delta(6n^2)$ [45]. $\Delta(27)$ was used for the PMNS matrix [46, 47]. Most non-Abelian discrete groups applied for neutrino mixing are subgroups of SU(3). The subgroup $\Delta(6n^2)$ of SU(3) contains the familiar $S_3$ for $n = 1$ and $S_4$ for $n = 2$. Before considering $A_4 \subset S_4$, let us briefly list some references on dihedral groups $D_n$.

All $D_n$ groups are non-Abelian apart from $D_1(\mathbb{Z}_2)$ and $D_2(\mathbb{Z}_2 \times \mathbb{Z}_2)$. Group theory of dihedral groups is discussed in [48] and $D_7$ and $D_{14}$ are applied to the quark mixing in [49]. The number 14 can be roughly guessed from the magnitude of $\pi/\theta_C \simeq 13.8$. But to fit to the PMNS matrix, much effort is needed. From a $\mathbb{Z}_{6-\mathrm{II}}$ orbifold compactification, the discrete symmetry $D_4$ is shown to be possible [50]. But such $D_4$ cannot be obtained from $\mathbb{Z}_{12-\mathrm{I}}$ orbifold since it does not have a 180° or 90° rotation in a two-torus as one can see the number of fixed points discussed in Chap. 5.

The use of permutation symmetry $S_3$ [35] can be said "obvious" colloquially in a sense because three families have the same gauge interactions. Nevertheless, the representation of $S_3$ does not contain **3**, but **2** and **1**. This fact is the basis of $S_3$ toward the bimaximal mixing of $\nu_\mu$ and $\nu_\tau$ [51, 52]. Except in the bimaximal mixing in one column of the PMNS matrix, the other angles are not fixed with $S_3$. Somehow, representation **3** is needed to have a permutation symmetry of three objects. $S_4$ contains representation **3** and indeed it was used for neutrino masses [53, 54]. To have a bimaximal part one has to massage the model. Permutation of 4 objects has 4! (=24) elements. If we use only half of these, namely by choosing only cyclic permutations and discarding anti-cyclic permutations, we have 12 elements. Recently, it was shown that $A_4$ is predicted with three identical elements in the PMNS matrix [55]. One may guess that this choice have one column to contain **2** to realize the bimaximal mixing. Then, one column (say the 2nd) is exactly permutable giving the identical numbers, and another column (say the 3rd) containing **2** is exactly permutable with two identical numbers and one zero. Indeed, this half elements of $S_4$ was used in the Yukawa couplings and the potential to obtain tri-bimaximal mixing [56, 57]. It can be identified with the group $A_4$ which acts as the discrete rotation of a regular tetrahedron in three space dimensions. A regular tetrahedron $ABCD$ is shown in Fig. 16.7. Changing vertices gives 4! ways. The reflection over the facing plane gives another 24 which are identical to the original set. So, these 24 can be distinguished by the order, i.e. cyclic ones and anti-cyclic ones. One such anti-cyclic permutation is shown by the red color. Namely, half of them has one definite cyclic property. These 12 elements with a definite cyclic quantum number are the permutations defined by the double covering group $A_4$. Usefulness of $A_4$ rather than $S_4$ can be guessed from the maximally parity violating interactions of the L-handed light neutrinos. Somehow the geometrical meaning of the tetrahedron incodes the L-handed chirality of three light neutrinos. Of course,

---

[8]Mathematical techniques are discussed in Luhn et al. [44].

**Fig. 16.7** A regular
tetrahedron $ABCD$.
Reflecting only point $A$ over
the plane $BCD$ gives an
anti-cyclic shape $A'CBD$

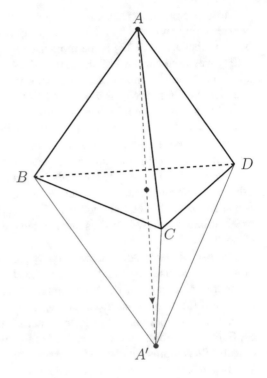

we should include the effects of Higgs fields for the full interactions of neutrinos, but
the SM also obtains the maximally parity violating interactions from the effects of
Higgs fields. So, Higgs fields must be given with certain cyclic quantum numbers. In
this spirit, phenomenology on neutrino masses has been studied in nunerous papers
based on $A_4$ symmetry [56, 58–66].

Related to the field theory orbifold of Chap. 5, Refs. [67,68] showed a possibility
that $A_4$ appears as the remnant of the reduction from 6D to 4D spacetime symmetry.

To connect the PMNS matrix to the CKM matrix, there must be relations between
quarks and leptons. One obvious connection is the quark and lepton unification, i.e.
the $A_4$ applications as performed in GUTs [59, 69, 70]. For this purpose, it is better
to take bases such that both the lower components of Eqs. (16.22) and (16.23) are
diagonalized, or vice versa, such that the currents coupled to $W_\mu^+$ are given by

$$+\frac{g_2}{2\sqrt{2}}\left(\bar{u}^0\ \bar{c}^0\ \bar{t}^0\right)_L \gamma^\mu(1+\gamma_5)\begin{pmatrix} d \\ s \\ b \end{pmatrix}_L W_\mu^+$$

$$+\frac{g_2}{2\sqrt{2}}\left(\bar{\nu}_e\ \bar{\nu}_\mu\ \bar{\nu}_\tau\right)_L \gamma^\mu(1+\gamma_5)\begin{pmatrix} e \\ \mu \\ \tau \end{pmatrix}_L W_\mu^+, \qquad (16.36)$$

which is the reason that we mixed the upper components of Eqs. (16.22) and (16.23). In this bases, it was simple to count the number of independent variables as done in the end of Sect. 16.2.1. Both in the anti-SU(5) and SU(5) GUTs, $(e, \mu, \tau)^T$ appears in $(\mathbf{5}_1, \mathbf{5}_2, \mathbf{5}_3)^T$ and $(\bar{\nu}_e, \bar{\nu}_\mu, \bar{\nu}_\tau)$ appears in $(\bar{\mathbf{5}}_1, \bar{\mathbf{5}}_2, \bar{\mathbf{5}}_3)$. Similarly, $(d, s, b)^T$ appears in $(\overline{\mathbf{10}}_1, \overline{\mathbf{10}}_2, \overline{\mathbf{10}}_3)^T$. The anti-SU(5) Yukawa couplings for the $Q_{em} = -\frac{1}{3}$ quarks are of the form $\overline{\mathbf{10}}_i C^{-1} Y_{ij}^{(d)} \overline{\mathbf{10}}_j \cdot \mathbf{5}_{H_d}$. For the $Q_{em} = -1$ leptons, they are of the form $\bar{\mathbf{1}}_i C^{-1} Y_{ij}^{(d)} \mathbf{5}_j \cdot \mathbf{5}_{H_d}$ where $\bar{\mathbf{1}}$ contains $e$, $\mu$, and $\tau$. $\mathbf{5}_{H_d}$ containing the electroweak doublet Higgs $H_d$ gives mass to $Q_{em} = -\frac{1}{3}$ quarks and $Q_{em} = -1$ leptons. We chose that the mass matrices of these lower component fermions are diagonalized. The $Q_{em} = +\frac{2}{3}$ quarks obtain mass by $\mathbf{5}_i C^{-1} Y_{ij}^{(u)} \overline{\mathbf{10}}_j \cdot \mathbf{5}_{H_u}$, and neutrinos obtain mass by $\overline{\mathbf{10}}_i C^{-1} Y_{ij}^{(\nu)} \overline{\mathbf{10}}_j \cdot \frac{\mathbf{5}_{H_u} \mathbf{5}_{H_u}}{M_{\rm GUT}^2} \mathbf{10}_{H_{\rm GUT}}$ where $\mathbf{10}_{H_{\rm GUT}}$ breaks the anti-SU(5) down to the SM gauge group. Since both neutrinos and $u$ quarks are in $\overline{\mathbf{10}}_i$, basically $Y_{ij}^{(\nu)}$ and $Y_{ij}^{(u)}$ relate the PMNS and CKM matrices. Depending on the discrete symmetries restricting $Y_{ij}^{(\nu)}$ and $Y_{ij}^{(u)}$, the way they are related is given.

In the literature, the so-called quark–lepton complementarity was suggested to achieve the phenomenological relation [71, 72],

$$\theta_{\rm sol} + \theta_C \simeq 33^\circ + 13^\circ \simeq \frac{\pi}{4}, \qquad (16.37)$$

where $\theta_{\rm sol}$ is the solar neutrino mixing angle between $\nu_e$ and $\nu_\mu$. As given in Eq. (16.39), (16.40), $V_{\rm CKM}$ is given by $U^{(u)} U^{(d)\dagger}$, and $V_{\rm PMNS}$ is given by $U^{(\nu)} U^{(e)\dagger}$, leading to $U^{(e)} = U^{(\nu)} V_{\rm PMNS}^\dagger$ and $U^{(d)} = U^{(u)} V_{\rm CKM}$. Our GUT realation is $U^{(d)} = U^{(e)\dagger}$ both of which are diagonalized. The quark–lepton complementarity put an ansatz $V_{\rm CKM} V_{\rm PMNS} = U^{(u)\dagger} U^{(\nu)}$, and it is possible to obtain the relation (16.37). In fact, our scheme of diagonalized $U^{(d)}$ and $U^{(e)}$ gives the same relation. Flavor symmetries can be introduced in extended GUTs also. Recently, flavor matrices with an $A_4$ symmetry in an anti-SU(5) model have been studied [73] [More references on Yukawa couplings in string theory can be found here].

### 16.2.4 Continuous Symmetries

Continuous horizontal (family) symmetries were also considered. For three families, SU(3) and SO(3) can be considered. For SU(3) if gauged, one has to consider anomalies in addition, which makes it tedious to work out and just we present a relatively recent reference here [74] where the relation (16.37) was also discussed.

## 16.3    Flavors from String Compactification

At the current stage of the SM, the most appealing theoretical issue is understanding the flavor puzzle. Therefore, in this last chapter, we present several issues in the flavor problem. In Sect. 16.1, we summarized the flavor data at some length and presented the results in particular in the matrix forms so that a possible string compactification can be compared to the quark and lepton flavor matrices presented in Eqs. (16.12) and (16.21).

One flavor puzzle has been, "Why are there three quark and lepton families?" This problem on the number of families is usually framed in GUTs in field theory because a GUT introduces the same number of families in the quark and lepton sectors. This issue has been discussed in Chap. 14. The condition for the number of families is the absence of anomalies.

From string theory, the most advanced search obtaining three families is in the orbifold compactification of the heterotic string $E_8 \times E_8'$. Chiefly because of the lack of an adjoint representation of a GUT group at the level-1 construction, standard-like models with three SM families have been looked for as discussed in detail in Chap. 13.

For GUT groups from orbifold compactification, it has been noted that anti-SU(5) GUT is promising because a kind of fundamental representations, the antisymmetric tensors $\mathbf{10}$ and $\overline{\mathbf{10}}$, can achieve spontaneous symmetry breaking of SU(5)×U(1) down to the SM gauge group SU(3)×SU(2)×U(1). Distinguishing a string SM from string GUT from experimental data will be very difficult. Even if proton decay is observed in the future, one cannot pinpoint easily it to "O, a GUT theory has shown up," because the dimension-5 $B$ violating operators in SUSY SMs can trigger a similar proton decay rate also. If it is so difficult to distinguish phenomenologically a GUT from the SM, how can we convince string compactification has been realized?

As is the case in every issue for models with Planck scale string tension, string compactification aims at best for a consistent framework toward the electroweak SM. Maybe, one can go a step further than just obtaining three families. It is at the place of reasonably fitting to the data presented in Eqs. (16.12) and (16.21). For this purpose, three family GUT models are preferred because of smaller number of Yukawa couplings in SUSY GUTs than in SUSY SMs.

The numbers which are useful for data fitting in string compactification are the fundamental Yukawa coupling matrices, $Y^{(u,d,v,e)\,0}$, shown in Eq. (16.5),

$$Y^{(u,d,v,e)\,0} = U^{(u,d,v,e)}\,Y^{(u,d,v,e)}\,V^{(u,d,v,e)\,\dagger}, \tag{16.38}$$

from which, using the L-hand matrices $U^{(u,d,v,e)}$ only, the CKM and PMNS matrices are obtained as shown in Eqs. (16.39) and (16.40). In our set-up, $Q_{em} = -1$

leptons and $Q_{em} = -\frac{1}{3}$ quarks are taken as mass eigenstates. So, we take $U^{(d,e)} = \mathbf{1}$ and $V^{(d,e)} = \mathbf{1}$. In the same set-up, therefore, we have

$$U^{(u)} = V_{CKM} \tag{16.39}$$

$$U^{(v)} = V_{PMNS}, \tag{16.40}$$

and

$$Y^{(u)\,0} = V_{CKM} Y^{(u)} V^{(u)\,\dagger}, \tag{16.41}$$

$$Y^{(v)\,0} = V_{PMNS} Y^{(v)} V^{(v)\,\dagger}, \tag{16.42}$$

where $Y^{(u)}$ and $Y^{(v)}$ are proportional to the mass eigenvalues. In the NH for neutrinos, therefore, we take $Y^{(u)}$, $Y^{(v)}_{NH}$, and $Y^{(v)}_{IH}$ as

$$M_{\frac{2}{3}} \propto \begin{pmatrix} \frac{m_u}{m_t}, & 0, & 0 \\ 0, & \frac{m_c}{m_t}, & 0 \\ 0, & 0, & 1 \end{pmatrix}, NH \propto \begin{pmatrix} \frac{m_{v_1}}{m_{v_3}}, & 0, & 0 \\ 0, & \frac{m_{v_2}}{m_{v_3}}, & 0 \\ 0, & 0, & 1 \end{pmatrix}, IH \propto \begin{pmatrix} 1, & 0, & 0 \\ 0, & \frac{m_{v_2}}{m_{v_1}}, & 0 \\ 0, & 0, & \frac{m_{v_3}}{m_{v_1}} \end{pmatrix}. \tag{16.43}$$

Now, there is a freedom to write R-hand unitary matrices $V^{(u)}$ and $V^{(v)}$. It is the problem of matching which R-handed fields are matched to which L-handed fields. In the standard-like models, there are too many possibilities. Even in the anti-SU(5) GUT, there are several possibilities. The common scenario is

$$u_L^c, \quad c_L^c, \quad t_L^c$$
$$\downarrow \quad \downarrow \quad \downarrow \tag{16.44}$$
$$d_L, \quad s_L, \quad b_L$$

in which case we can write $V^{(u)} = \mathbf{1}$. A similar consideration applies to $V^{(v)}$ also.

For $Y^{(u)}$ given in Eq. (16.43), we have the following $Y^{(u)\,0}$ to be determined at the compactification scale [12],

$$\begin{pmatrix} 1.216^{+0.276}_{-0.146} \cdot 10^{-5}, & (1.647 \pm 0.028) \cdot 10^{-3}, & (2.500 \pm 0.925) \cdot 10^{-2} \\ \\ -2.722^{+0.620}_{-0.332} \cdot 10^{-6}, & \begin{matrix} +(6.866 \pm 0.154) \cdot 10^{-3} \\ +(2.102 \pm 0.596) \cdot 10^{-4} e^{-i\alpha}, \end{matrix} & \begin{matrix} +(0.104 \pm 0.003) \\ -i(0.257 \pm 0.0552) e^{-i\alpha} \end{matrix} \\ \\ -(7.291^{+2.526}_{-0.425}) \cdot 10^{-7} e^{i\alpha}, & \begin{matrix} -(7.848 \pm 0.239) \cdot 10^{-4} \\ +(1.839 \pm 0.385) \cdot 10^{-3} e^{i\alpha}, \end{matrix} & \begin{matrix} +(0.9600 \pm 0.0171) \\ +(0.0028 \pm 0.0010) e^{i\alpha} \end{matrix} \end{pmatrix} \tag{16.45}$$

where $\alpha = \alpha_{KS}$.

## 16.4   CP Violation from String

All couplings of 4D effective fields after compactification are determined by giving VEVs to all the moduli fields. Therefore, 4D couplings are generated spontaneously. In this sense, symmetry breaking can be of spontaneous in origin. However, it is proper to define the coupling constants first and talk about spontaneous symmetry breaking. This is achieved only after inserting the VEVs of moduli fields.

In Ref. [75], it has been shown that four dimensional $\mathscr{CP}$ can be a discrete gauge symmetry in theories with dimensional compactification, if the original number of Minkowski dimensions equals $8k+1$, $8k+2$, or $8k+3$, with the condition of inner automorphism. Among Lie groups, this condition is satisfied only for $E_8$, $E_7$, $SO(2n + 1)$, $SO(4n)$, $Sp(2n)$, $G_2$, and $F_4$.

The superstring dimension 10D belongs here. Anyway, if $\mathscr{CP}$ turns out to be a discrete symmetry from string compactification, it must be a consistent discrete symmetry, i.e. the discrete gauge symmetry included, without a need to resort to Ref. [75].

▶ **Exercise 16.1** Without using the quarks, such as Eq. (16.3), show that there is no strangeness changing neutral current effects with Cabibbo's original idea but with SU(4) currents.

▶ **Exercise 16.2** Using the KS form for the CKM matrix, Eq. (16.24), prove that any Jarlskog triangle in the quark sector has one angle as $\delta_{CKM}$.

▶ **Exercise 16.3** Show that the MNS matrix $V^{lept}$ of Eq. (16.35) gives the same Jarlskog determinant as in the case with no Majorana phases,

$$J = \frac{1}{8} \sin 2\Theta_1 \sin 2\Theta_2 \sin 2\Theta_3 \sin \Theta_1 \sin \delta_{MNS}. \tag{16.46}$$

▶ **Exercise 16.4** Obtain the matrix given in Eq. (16.20), using the angles given in Eq. (16.21) for $\bar{\Theta}_2 = 30.9°$.

▶ **Exercise 16.5** List the other possibilites, beyond (16.44), of matching R-handed quarks to L-handed quarks.

## References

1. N. Cabibbo, Unitary symmetry and leptonic decays. Phys. Rev. Lett. **10**, 531–533 (1963)
2. S.L. Glashow, J. Iliopoulos, L. Maiani, Weak interactions with lepton-hadron symmetry. Phys. Rev. D2, 1285–1292 (1970)
3. F. Abe et al., Observation of top quark production in $\bar{p}p$ collisions. Phys. Rev. Lett. **74**, 2626–2631 (1995)
4. S. Abachi et al., Observation of the top quark. Phys. Rev. Lett. **74**, 2632–2637 (1995)

5. B. Pontecorvo, Inverse beta processes and nonconservation of lepton charge. Sov. Phys. JETP **7**, 172–173 (1958) [Zh. Eksp. Teor. Fiz. **34**, 247 (1957)]
6. G. Danby, J.M. Gaillard, K.A. Goulianos, L.M. Lederman, N.B. Mistry, M. Schwartz, J. Steinberger, Observation of high-energy neutrino reactions and the existence of two kinds of neutrinos. Phys. Rev. Lett. **9**, 36–44 (1962)
7. M.L. Perl et al., Evidence for anomalous lepton production in e+ - e− annihilation. Phys. Rev. Lett. **35**, 1489–1492 (1975)
8. K. Kodama et al., Observation of tau neutrino interactions. Phys. Lett. **B504**, 218–224 (2001)
9. M. Tanabashi et al., Review of particle physics. Phys. Rev. **D98**(3), 030001 (2018)
10. L. Wolfenstein, Parametrization of the Kobayashi-Maskawa matrix. Phys. Rev. Lett. **51**, 1945 (1983)
11. J.E. Kim, M.-S. Seo, Parametrization of the CKM matrix. Phys. Rev. **D84**, 037303 (2011)
12. J.E. Kim, S-J. Kim, S. Nam, M. Shim, Jarlskog Determinant and Data on Flavor Matrices. Mod. Phys. Lett. **A35**, (2020)
13. J.E. Kim, D.Y. Mo, S. Nam, Final state interaction phases obtained by data from CP asymmetries. J. Korean Phys. Soc. **66**(6), 894–899 (2015)
14. L. Wolfenstein, Neutrino oscillations in matter. Phys. Rev. **D17**, 2369–2374 (1978)
15. S.P. Mikheyev, A.Yu. Smirnov, Resonance amplification of oscillations in matter and spectroscopy of solar neutrinos. Sov. J. Nucl. Phys. **42**, 913–917 (1985) [305 (1986)]
16. G.-J. Ding, N. Nath, R. Srivastava, J.W.F. Valle, Status and prospects of 'bi-large' leptonic mixing. Phys. Lett. B **796**, 162–167 (2019)
17. D.V. Forero, M. Tortola, J.W.F. Valle, Neutrino oscillations refitted. Phys. Rev. **D90**(9), 093006 (2014)
18. P.F. de Salas, D.V. Forero, C.A. Ternes, M. Tortola, J.W.F. Valle, Status of neutrino oscillations 2018: 3σ hint for normal mass ordering and improved CP sensitivity. Phys. Lett. B **782**, 633–640 (2018)
19. K. Abe et al., Combined analysis of neutrino and antineutrino oscillations at T2K. Phys. Rev. Lett. **118**(15), 151801 (2017)
20. J.H. Christenson, J.W. Cronin, V.L. Fitch, R. Turlay, Evidence for the $2\pi$ Decay of the $K_2^0$ Meson. Phys. Rev. Lett. **13**, 138–140 (1964)
21. M. Kobayashi, T. Maskawa, CP violation in the renormalizable theory of weak interaction. Prog. Theor. Phys. **49**, 652–657 (1973)
22. P.F. Harrison, W.G. Scott, Symmetries and generalizations of tri - bimaximal neutrino mixing. Phys. Lett. **B535**, 163–169 (2002)
23. A.D. Sakharov, Violation of CP invariance, C asymmetry, and baryon asymmetry of the universe. Pisma Zh. Eksp. Teor. Fiz. **5**, 32–35 (1967) [Usp. Fiz. Nauk **161**(5), 61 (1991)]
24. M. Yoshimura, Unified gauge theories and the Baryon number of the universe. Phys. Rev. Lett. **41**, 281–284 (1978) [Erratum: Phys. Rev. Lett. **42**, 746 (1979)]
25. E.W. Kolb, M.S. Turner, Grand unified theories and the origin of the Baryon asymmetry. Ann. Rev. Nucl. Part. Sci. **33**, 645–696 (1983)
26. D. Toussaint, S.B. Treiman, F. Wilczek, A. Zee, Matter - antimatter accounting, thermodynamics, and black hole radiation. Phys. Rev. **D19**, 1036–1045 (1979)
27. M. Fukugita, T. Yanagida, Baryogenesis without grand unification. Phys. Lett. **B174**, 45–47 (1986)
28. H.D. Kim, J.E. Kim, T. Morozumi, A New mechanism for baryogenesis living through electroweak era. Phys. Lett. **B616**, 108–113 (2005)
29. M. Fukugita, T. Yanagida, *Physics of Neutrinos and Application to Astrophysics* (Springer, Berlin, 2003)
30. L. Covi, J.E. Kim, B. Kyae, S. Nam, Leptogenesis with high-scale electroweak symmetry breaking and an extended Higgs sector. Phys. Rev. **D94**(6), 065004 (2016)
31. J.E. Kim, Type-II leptogenesis, in *PoS, ICHEP2016*, vol. 107 (2016)
32. D.V. Nanopoulos, S. Weinberg, Mechanisms for cosmological Baryon production. Phys. Rev. **D20**, 2484 (1979)

33. A. Abada, S. Davidson, F.-X. Josse-Michaux, M. Losada, A. Riotto,   Flavor issues in leptogenesis. J. Cosmol. Astropart. Phys. **0604**, 004 (2006)
34. P.H. Frampton, S.L. Glashow, T. Yanagida, Cosmological sign of neutrino CP violation. Phys. Lett. **B548**, 119–121 (2002)
35. S. Pakvasa, H. Sugawara,  Discrete symmetry and Cabibbo angle.  Phys. Lett. **B73**, 61–64 (1978)
36. G. Segre, H.A. Weldon,  Natural suppression of strong $P$ and $T$ violations and calculable mixing angles in SU(2) X U(1). Phys. Rev. Lett. **42**, 1191 (1979)
37. C.D. Froggatt, H.B. Nielsen,  Hierarchy of quark masses, Cabibbo angles and CP violation. Nucl. Phys. **B147**, 277–298 (1979)
38. L.M. Krauss, F. Wilczek, Discrete gauge symmetry in continuum theories. Phys. Rev. Lett. **62**, 1221 (1989)
39. J.E. Kim, M.-S. Seo, Quark and lepton mixing angles with a dodeca-symmetry. J. High Energy Phys. **02**, 097 (2011)
40. J.E. Kim, The CKM matrix with maximal CP violation from Z(12) symmetry. Phys. Lett. **B704**, 360–366 (2011)
41. J.E. Kim, M.-S. Seo,  Parametrization of PMNS matrix based on dodeca-symmetry.  Int. J. Mod. Phys. **A27**, 1250017 (2012)
42. H. Ishimori, T. Kobayashi, H. Ohki, Y. Shimizu, H. Okada, M. Tanimoto, Non-Abelian discrete symmetries in particle physics. Prog. Theor. Phys. Suppl. **183**, 1–163 (2010)
43. H. Ishimori, T. Kobayashi, H. Ohki, H. Okada, Y. Shimizu, M. Tanimoto,  An introduction to non-Abelian discrete symmetries for particle physicists. Lect. Notes Phys. **858**, 1–227 (2012)
44. C. Luhn, S. Narsi, P. Ramond, The flavor group $\Delta(3n^2)$. J. Math. Phys. **48**, 073501 (2007)
45. J.A. Escobar, C. Luhn, The flavor group delta(6n**2). J. Math. Phys. **50**, 013524 (2009)
46. E. Ma,  Neutrino mass matrix from delta(27) symmetry.  Mod. Phys. Lett. **A21**, 1917–1921 (2006)
47. E. Ma,  Near tribimaximal neutrino mixing with Delta(27) symmetry. Phys. Lett. **B660**, 505–507 (2008)
48. A. Blum, C. Hagedorn, M. Lindner,   Fermion masses and mixings from dihedral flavor symmetries with preserved subgroups. Phys. Rev. **D77**, 076004 (2008)
49. A. Blum, C. Hagedorn, A. Hohenegger, theta(C) from the dihedral flavor symmetries D(7) and D(14). J. High Energy Phys. **03**, 070 (2008)
50. T. Kobayashi, H.P. Nilles, F. Ploger, S. Raby, M. Ratz, Stringy origin of non-Abelian discrete flavor symmetries. Nucl. Phys. **B768**, 135–156 (2007)
51. P.F. Harrison, D.H. Perkins, W.G. Scott, Threefold maximal lepton mixing and the solar and atmospheric neutrino deficits. Phys. Lett. **B349**, 137–144 (1995)
52. K. Kang, J.E. Kim, P. Ko,  A simple modification of the maximal mixing scenario for three light neutrinos. Z. Phys. **C72**, 671–675 (1996)
53. W. Grimus, L. Lavoura, P.O. Ludl,  Is S(4) the horizontal symmetry of tri-bimaximal lepton mixing? J. Phys. **G36**, 115007 (2009)
54. F. Bazzocchi, L. Merlo, S. Morisi, Fermion masses and mixings in a S(4)-based model. Nucl. Phys. **B816**, 204–226 (2009)
55. P.H. Frampton, J.E. Kim, Anticorrelation of Mass and Mixing Angle Hierarchies. J. Korean Phys. Soc. **76**(8), 695–700 (2020)
56. E. Ma, G. Rajasekaran, Softly broken A(4) symmetry for nearly degenerate neutrino masses. Phys. Rev. **D64**, 113012 (2001)
57. K.S. Babu, E. Ma, J.W.F. Valle,  Underlying A(4) symmetry for the neutrino mass matrix and the quark mixing matrix. Phys. Lett. **B552**, 207–213 (2003)
58. A. Zee,  Obtaining the neutrino mixing matrix with the tetrahedral group. Phys. Lett. **B630**, 58–67 (2005)
59. G. Altarelli, F. Feruglio,  Tri-bimaximal neutrino mixing, A(4) and the modular symmetry. Nucl. Phys. **B741**, 215–235 (2006)
60. E. Ma, Tribimaximal neutrino mixing from a supersymmetric model with A4 family symmetry. Phys. Rev. **D73**, 057304 (2006)

61. K.S. Babu, X.-G. He, Model of geometric neutrino mixing (2005)
62. X.-G. He, Y.-Y. Keum, R.R. Volkas, A(4) flavor symmetry breaking scheme for understanding quark and neutrino mixing angles. J. High Energy Phys. **04**, 039 (2006)
63. G. Altarelli, F. Feruglio, Y. Lin, Tri-bimaximal neutrino mixing from orbifolding. Nucl. Phys. **B775**, 31–44 (2007)
64. I. de Medeiros Varzielas, S.F. King, G.G. Ross, Neutrino tri-bi-maximal mixing from a non-Abelian discrete family symmetry. Phys. Lett. **B648**, 201–206 (2007)
65. F. Bazzocchi, S. Kaneko, S. Morisi, A SUSY A(4) model for fermion masses and mixings. J. High Energy Phys. **03**, 063 (2008)
66. M. Hirsch, S. Morisi, J.W.F. Valle, A4-based tri-bimaximal mixing within inverse and linear seesaw schemes. Phys. Lett. **B679**, 454–459 (2009)
67. G. Altarelli, F. Feruglio, Tri-bimaximal neutrino mixing from discrete symmetry in extra dimensions. Nucl. Phys. **B720**, 64–88 (2005)
68. F. Feruglio, C. Hagedorn, Y. Lin, L. Merlo, Tri-bimaximal neutrino mixing and quark masses from a discrete flavour symmetry. Nucl. Phys. **B775**, 120–142 (2007) [Erratum: Nucl. Phys. **B836**,127 (2010)]
69. M. Raidal, Relation between the neutrino and quark mixing angles and grand unification. Phys. Rev. Lett. **93**, 161801 (2004)
70. S.F. King, M. Malinsky, A(4) family symmetry and quark-lepton unification. Phys. Lett. **B645**, 351–357 (2007)
71. H. Minakata, A.Yu. Smirnov, Neutrino mixing and quark-lepton complementarity. Phys. Rev. **D70**, 073009 (2004)
72. J.E. Kim, J.-C. Park, Quantum numbers of heavy neutrinos, tri-bi-maximal mixing through double seesaw with permutation symmetry, and comment on theta (sol) + theta(c) = pi/4. J. High Energy Phys. **05**, 017 (2006)
73. P.H. Frampton, J.E. Kim, S-J. Kim, S. Nam, Tetrahedral $A_4$ Symmetry in Anti-SU(5) GUT. Phys. Rev. **D101**, 055022 (2020)
74. Ivo de Medeiros Varzielas, Graham G. Ross, SU(3) family symmetry and neutrino bi-tri-maximal mixing. Nucl. Phys. B. **733**, 31–47 (2006)
75. K.-W. Choi, D.B. Kaplan, A.E. Nelson, Is CP a gauge symmetry? Nucl. Phys. **B391**, 515–530 (1993)

# Other Constructions

**17**

In this book, we emphasized the chiral nature of the SM. The orbifold construction has been discussed toward obtaining three family models from the $E_8 \times E_8$ heterotic string. But there are other methods also for obtaining 4D string models. Among these, we review very briefly on the fermionic construction and the intersecting brane setup. Since obtaining three families is one of the most important objectives going beyond 4D, we discuss three other constructions, fermionic construction, intersecting brane models, and F-theory, up to the point of introducing the possibility for three families as the first step to realistic model buildings.

## 17.1 Fermionic Construction

We focused on the bosonic description so far. On the other hand, there exists an equivalent fermionic description: in two dimensions, two (anti)holomorphic Majorana–Weyl (MW) fermions, $\lambda_1(z)$ and $\lambda_2(z)$, are equivalent to one periodic, (anti)holomorphic boson $\partial X(z)$. Since the observed spacetime is four, we may compact all the other dimensions on torus and treat all the field democratically. The resulting fermions are $44(= 2 \cdot (26 - 4))$ left movers. The right mover is superstring, so for each dimension, we have one holomorphic boson and one MW fermion, or equivalent, three MW fermions, so that there are $18(= 3 \cdot (10 - 4))$ right movers.

With the critical radius $\sqrt{\alpha'}$, the symmetry is enhanced to

$$SO(44) \times U(1)^6.$$

In the right mover, we cannot make the self-dual lattice or modular invariant partition function, so at best we may have $U(1)^6$. Note that fermions in the noncompact dimensions should be considered as well.

We may regard the orbifolding as simply assigning nontrivial boundary conditions. In this context, the $E_8 \times E_8$ heterotic string in 10D is the first example of

© Springer Nature Switzerland AG 2020
K.-S. Choi, J. E. Kim, *Quarks and Leptons From Orbifolded Superstring*,
Lecture Notes in Physics 954, https://doi.org/10.1007/978-3-030-54005-0_17

fermionic construction: we partitioned 32 fermions into two sets of 16 fermions each and assigned different boundary conditions. That was the only way of modular invariant sets with 32 fermions. With more and more fermions, the number of possibility grows in a geometrical progression.

### 17.1.1 Rules

Conventionally we define the twist vector $W = (w_i)$ as $w_i = 1 - V_i$, where $V_i$ is the shift vector as before

$$\lambda_i(2\pi) = e^{\pi i(1-w_i)}\lambda_i(0). \tag{17.1}$$

The fermions are Majorana–Weyl thus assumes $w_i = 1$ or $0$ only, corresponding to Ramond (R) or Neveu–Schwarz (NS) sector, respectively. We will later complexify them and generalize the boundary condition. Now we have vectors with $20 + 44$ entries and name the twisted sector with them. We divide the vector using a bar; The standard convention is to place the right mover boundary conditions on the left side of the bar,

$$(w_1 \ w_2 \ \cdots \ w_{20}|\tilde{w}_1 \ \tilde{w}_2 \ \cdots \ \tilde{w}_{44}). \tag{17.2}$$

The nontrivial restrictions are the GSO projection and the modular invariance. These conditions are extracted in a most straightforward way from the partition function [1–3]. We just present the rules in their simplest forms [4] and explain their origins:

1. **Modular invariance:** For two sectors specified by the twist vectors $W_1$ and $W_2$, then there must exist a sector specified by

$$W_1 + W_2. \tag{17.3}$$

We define the sum modulo 2 by the periodicity of $w_i$ as in (17.1), thus $1 + 1 = 0$. This seems contradictory, since two successive peroidic (R) boundary condition give antiperiodic (NS) one. However, this take into account of the spin structure considered in Sect. 6.2.3: the extra phase $e^{\pi i}$ is needed for the periodic (R) boundary condition in the $\tau$ direction.

   In addition, like the untwisted sector discussed in Chap. 5, we require that there exist a trivial sector, or an untwisted sector, where all the fermions are subject to the NS or the R boundary condition,

$$(0^{20}|0^{44}), \quad (1^{20}|1^{44}). \tag{17.4}$$

The former is redundant because the vector

$$W_0 = (1^{20}|1^{44})$$ (17.5)

gives rise to it by the condition (17.3).

2. **GSO projection:** The Gliozzi–Scherk–Olive (GSO) projection projects out half of the spinors of definite chirality. Before orbifolding it is defined by (6.146), (6.154). Now in the presence of many shift vectors, we have an additional rule: if there are the coinciding Ramond boundary conditions "1" between two sectors, in each sector we just project out half of the corresponding spinors. Which half we project out is our freedom of choice. In the literature, it is called the choice of correlation coefficient of the GSO projection. We will show this in the following examples.

3. **Triplet constraint:** For the right movers, the fermionic worldsheet energy-momentum tensor is

$$T_F = 2i \sum_{\mu=1}^{2} \psi^\mu \partial_{\bar{z}} X^i + i \sum_{i=1}^{6} \lambda_1^i \lambda_2^i \lambda_3^i.$$ (17.6)

The first term comes from the noncompact two dimensions in the light-cone gauge and the second term comes from the compact six dimensions. Thus we partition 20 right moving fermions as

$$\{(\psi^1 \psi^2)\,(\lambda_1^1 \lambda_2^1 \lambda_3^1) \cdots (\lambda_1^6 \lambda_2^6 \lambda_3^6)\}.$$ (17.7)

It is consistent only when every term in (17.6) has the same boundary condition. So,

$$w^\mu = w_1^i + w_2^i + w_3^i, \quad \text{mod } 2$$ (17.8)

for each $\mu = 1, 2$ and $i = 1, \ldots, 6$. Here, $w^\mu$ is the boundary conditions for the $\psi^\mu$, and $w_j^i$'s are those of $\lambda_j^i$.

In the symmetric orbifold, we assign the same boundary condition on the corresponding left movers $\tilde{\lambda}_j^i$.

The spectrum is obtained from the mass shell condition

$$M_L^2 = \tilde{N} - \tilde{c} = 0,$$ (17.9)

$$M_R^2 = N - c = 0,$$ (17.10)

with the zero point energies $\tilde{c}$ and $c$ are calculated from (7.14). Note that the sign is the opposite of that of the boson given in Eq. (7.14). We have contribution $-\frac{1}{48}$ and $\frac{1}{24}$ from the entries of 0 and 1, respectively.

## 17.1.2 Models

### Original $\mathrm{E}_8 \times \mathrm{E}_8$ Heterotic String

As an example, we first treat the original ten dimensional $\mathrm{E}_8 \times \mathrm{E}_8$ heterotic string. $W_0$ is always present. We specify two vectors, $W_1$ and $W_2$,

$$W_0 = (1^8 | 1^{32})$$
$$W_1 = (1^8 | 0^{32}) \tag{17.11}$$
$$W_2 = (0^8 | 1^{16} 0^{16}).$$

We have $2^3 = 8$ twisted sectors in total. The $W_1$ and $2W_i \equiv W_i + W_i = (0^8 | 0^{32})$ sectors have the same left movers, which is our primary interest. Its zero point energy is $-1 = -\frac{1}{24} \cdot 8 + (-\frac{1}{48}) \cdot 32$ from the left sector only. When combined with the right movers, we have the adjoint $(\mathbf{120}, \mathbf{1}) \oplus (\mathbf{1}, \mathbf{120})$ of $\mathrm{SO}(16) \times \mathrm{SO}(16)$,

$$\tilde{\lambda}^i_{-1/2} \tilde{\lambda}^j_{-1/2} |0\rangle, \quad i, j = 1, \ldots, 16, \quad \text{or } i, j = 17, \ldots, 32,$$

and in addition the conventional graviton, dilation, and antisymmetric tensor fields. We have no crossed sectors such as the set one from $i = 1, \ldots, 16$, and the other from $j = 17, \ldots, 32$, because the original GSO projection (6.146) before orbifolding. Adding $W_1$ to each sector, we have superpartners because of the change in the right mover boundary conditions.

In the $W_2$ sector, we have the zero point energy $0 = -\frac{1}{24} \cdot 8 + \frac{1}{48} \cdot 16$. The states are constructed by successive applications of creation operators of the form

$$|s_i\rangle, \quad i = 1, \ldots, 8,$$

where the spinorial $s$ is generated by $\tilde{\lambda}^i_0$. They seem to be spinorial of dimension $2^8$. However, this $W_2$ has a Ramond overlap with $W_0$ (some entries have the common 1s), thus half the states are projected out. We have $(\mathbf{128}, \mathbf{1})$. For the $W_0 + W_1 + W_2$ sector, we obtain $(\mathbf{1}, \mathbf{128})$ in a similar manner. As we know, these representations $(\mathbf{120}, \mathbf{1})$, $(\mathbf{128}, \mathbf{1})$, $(\mathbf{1}, \mathbf{120})$, and $(\mathbf{1}, \mathbf{128})$ are not separable and we have an enhanced gauge symmetry with the adjoint $\mathbf{248}$s. Namely, the gauge group is $\mathrm{E}_8 \times \mathrm{E}_8$.

## Flipped SU(5) Model

As an example, let us consider the flipped SU(5) model of [3], the so-called revamped version. They are the nine vectors $1, S, \zeta, b_1, b_2, b_3, b_4, b_5$, and $\alpha$ with one constraint.[1] Let us start the discussion with the following three vectors:

$$1 = \left( (1^2)(1^6)(1^{12})|(1^{16})(1^{12})(1^{16}) \right)$$

$$S = \left( (1^2)(1^6)(0^{12})|(0^{16})(0^{12})(0^{16}) \right)$$

$$\zeta = \left( (0^2)(0^6)(0^{12})|(0^{16})(0^{12})(1^{16}) \right).$$

This is essentially the same as (17.11). Focusing on the *left movers*, the only difference from the 10D example is that in 4D we have twelve more entries. We interpret that the first and the last 16 entries are going to describe $E_8 \times E_8$. The 12 entries in the middle are interpreted as compactification of six dimensions, equivalent to 12 fermionic degrees of freedom. Thus we may interpret this piece as *orbifolding*. The $S$ sector[2] just provide the superpartners, as $W_1$ did in the 10D example. The $\zeta$ sector has zero point energy 0, as $W_2$ did in 10D, providing $(\mathbf{1}, \mathbf{128})$ of $SO(28) \times SO(16)$. The untwisted sector has the zero point energy $-1$ providing $(\mathbf{378}, \mathbf{1})$ and $(\mathbf{1}, \mathbf{120})$. These together make up the adjoint representations of $SO(28) \times E_8$. Note that we have an enhanced gauge group

$$SO(28) \times E_8 \times U(1)^6$$

by taking the critical radii $R = \sqrt{\alpha'}$.

Now introduce more vectors to break the group

$$b_1 = \left( (1^2)(1^20^20^2)(1^40^40^4)|(1^{10})(1^20^20^2)(\mathbf{1^40^40^4})(0^{16}) \right)$$

$$b_2 = \left( (1^2)(0^21^20^2)(0^41^40^4)|(1^{10})(0^21^20^2)(\mathbf{0^41^40^4})(0^{16}) \right)$$

$$b_3 = \left( (1^2)(0^20^21^2)(0^40^41^4)|(1^{10})(0^20^21^2)(\mathbf{0^40^41^4})(0^{16}) \right),$$

---

[1] For comparison, we follow the names of fields used in [3]

$$\left( (\chi^\mu)(\chi^1 y^1 \omega^1 \cdots \chi^6)(y^1 \cdots y^6 \omega^1 \cdots \omega^6)|(\bar\psi^1 \cdots \bar\psi^5 \bar\eta^1 \cdots \bar\eta^3)(\bar y^1 \cdots \bar y^6 \bar\omega^1 \cdots \bar\omega^6)(\bar\phi^1 \cdots \bar\phi^8) \right).$$

The fields $\chi, \bar\chi, y, \bar y, \omega$, and $\bar\omega$ are real fields and the $\bar\eta, \bar\psi$, and $\bar\psi$ are complex fields counted up to 20 for the right movers and up to 44 for the left movers.

[2] It should be the $S$ boundary condition, but "sector" is used here to follow the terminology in the fermionic construction.

where we used bold-faced numbers for the middle 12 entries corresponding to the left mover boundary conditions. Conditions $b_1$, $b_2$, and $b_3$ have a cyclic symmetry and note that every right mover has the same 12 components with the corresponding left mover entries. This corresponds to a symmetric orbifold. Note that there exists a relation $\zeta = b_1+b_2+b_3+1$, thus there will be eight sectors instead of nine. We may not need $\zeta$ any more, but its explicit form simplifies the discussion. Only $b_i + b_j$ and $1 + b_i + b_j$ sectors, with ($i \neq j$, thus six in total), have non-positive zero point energy $\tilde{c} = -1$ and 0, respectively, and hence a possibility for massless spectra. In fact, we can check by considering $\lambda^i_{-1/2}$ and $\lambda^j_{-1/2}$ that the group is broken to

$$SO(10) \times SO(6)^3 \times E_8 \times U(1)^6.$$

Considering the matter spectrum, we have three equivalent sectors $b_1$, $b_2$, and $b_3$. In each sector, there are $(2^5, 2^3) = (32, 8)$ states, $(\mathbf{16}, \mathbf{4})+(\mathbf{16}, \bar{\mathbf{4}})+(\overline{\mathbf{16}}, \mathbf{4})+(\overline{\mathbf{16}}, \bar{\mathbf{4}})$. Combined with right movers, half of them are the $\mathscr{CPT}$ conjugate the other half. Thus we have the spectrum

$$(\mathbf{16}, \mathbf{4}) + (\mathbf{16}, \bar{\mathbf{4}}).$$

Similarly the right movers have $2^4$ multiplets, yet we count only half of them because a complete Weyl fermion is made of two helicity states. We see that each $b_i$, due to the nonvanishing $1^2$ in the middle bracket, reduces the number the half and, at the same time, breaks half of the supersymmetries. However, only two of them among three $b_i$ are independent directions, leading to $\mathcal{N} = 4 \cdot \frac{1}{2} \cdot \frac{1}{2} = 1$ supersymmetry and 2 multiplet. We have two $(\mathbf{16}, \mathbf{4}) + (\mathbf{16}, \bar{\mathbf{4}})$s in each $b_1$, $b_2$, and $b_3$ sector, thus we have total 48 generations ($\mathbf{16}$s) of $SO(10)$.

We have geometric interpretation: we can check that in the 12 component orbifold section (=the bold-faced ones) in the left movers, each $b_i$ corresponds to a $\mathbb{Z}_2$ orbifold action. The $b_3$ and $b_1$ action correspond to

$$\theta = \tfrac{1}{2}(1\ 1\ 0), \quad \omega = \tfrac{1}{2}(0\ 1\ 1), \tag{17.12}$$

respectively.[3] Here, $b_2$ is generated by the other two, $\theta\omega = \tfrac{1}{2}(1\ 0\ 1)$: we have a $\mathbb{Z}_2 \times \mathbb{Z}_2$ symmetry. The important thing to note is that we have the family number as a multiple of three because of the triple symmetries in exchanging $b_i$'s. Basically, this can be interpreted as three fixed points of $\omega, \theta, \omega\theta$. In view of the forms $b_1, b_2$, and $b_3$, 6 real dimensions are split in such a way that they are grouped into three by complexifying them and furthermore they should give only three fixed points. Then, the number of families is a multiple of three. Here, the number of complex compact dimensions appears as $\tfrac{1}{2} \cdot 6 = 3$. Still we have not obtained three families

---

[3]The $(1, 1, 0)$ corresponds to a $\mathbb{Z}_2$ orbifold in the 4–5 and 6–7 tori and $(0, 1, 1)$ corresponds to another $\mathbb{Z}_2$ orbifold in the 6–7 and 8–9 tori, as discussed in Chap. 5.

yet. This set $b_1$, $b_2$, and $b_3$ is called the NAHE set [3]. So far the NAHE set respects the $\mathbb{Z}_2 \times \mathbb{Z}_2$ symmetry.

We proceed to introduce more vectors to have a realistic model,

$$b_4 = \left((1^2)(1^20^20^2)(110010001000)|(1^{10})(1^20^20^2)(110010001000)(0^{16})\right)$$

$$b_5 = \left((1^2)(0^21^20^2)(100011000100)|(1^{10})(0^21^20^2)(100011000100)(0^{16})\right)$$

$$\alpha = \left((1^2)(0^20^20^2)(100000100000)|(\tfrac{1}{2}^{10})(\tfrac{1}{2}^2\tfrac{1}{2}^2\tfrac{1}{2}^2)(100000100000)(\tfrac{1}{2}^8 1^4 0^4)\right).$$

Here, we consider more general boundary conditions. We can complexify a pair of fermions as $\tilde{\lambda}^{I+} = \tilde{\lambda}^{2I-1} + i\tilde{\lambda}^{2I}$ and consider more general boundary conditions,

$$\tilde{\lambda}^{I+} \to e^{2\pi i V_I} \tilde{\lambda}^{I+} \tag{17.13}$$

with the 16 component vector $V$. Conditions $b_4$ and $b_5$ give rise to Higgs bosons and the last shift $\alpha$ breaks SO(10) down to the flipped SU(5). But, we can see that the $\mathbb{Z}_2 \times \mathbb{Z}_2$ symmetry is not present anymore. Note that the right movers have the same form as before, and hence do not touch upon the $\mathcal{N} = 1$ supersymmetry. But the other right moving entries are different, reducing the number of generations down from 48 to 6. However, there is an independent GSO projection provided by $2\alpha$ which also reduce the number of generations by two, thus we have three generations.

The final gauge group contains flipped SU(5):

$$[SU(5) \times U(1)] \times U(1)^4 \times SO(10) \times SO(6).$$

The rank is reduced by 2, by the shift $\alpha$. Now the spectrum contains complete families $\mathbf{1} + \mathbf{5} + \mathbf{10}$, and $\bar{\mathbf{1}} + \bar{\mathbf{5}} + \overline{\mathbf{10}}$. By a GSO projection, we keep $\mathbf{1} + \bar{\mathbf{5}} + \mathbf{10}$ so that they form a flipped SU(5) spectrum. It has been shown that the spectrum $\mathbf{1} + \bar{\mathbf{5}} + \mathbf{10}$ has the U(1) charge ratio $5 : -3 : 1$ of the flipped SU(5) [3]. A flipped SU(5) model can be broken down to the SM by VEVs of $\mathbf{10}_H$ and $\overline{\mathbf{10}}_{\mathbf{H}}$. The Higgs fields $\mathbf{10}_H + \overline{\mathbf{10}}_H$ needed for the GUT breaking and $\mathbf{5}_H + \bar{\mathbf{5}}_H$ needed for the electroweak symmetry breaking are also present. This model, however, has a U(1) and hence there appear exotic multiplets (singlets and $\mathbf{10}$ of different SO(10) and $\mathbf{4}$, $\mathbf{6}$, etc.) with unfamiliar U(1) charges. Of course, the extra U(1)'s beyond the U(1) in the flipped SU(5) may be required to be broken. One such example, the Fayet–Illiopoulos mechanism with the anomalous U(1) was discussed in Chap. 9.

In this kind of compactification, 44 fermions are equal, and we are not obliged to use the same twisting to both left and right movers. Then, we have the fermionic version of the asymmetric orbifold.

## 17.2  Intersecting Branes

In Sect. 6.4 we studied that open strings can describe non-Abelian gauge theory. We can assign non-Abelian charge at both ends of the open string, or in $T$-dual space they become the locations of the D-branes.

If we make a stack of D-branes with $n$ slices, on the worldvolume where the brane stacks are located, we have $U(n)$ gauge theory. The resulting low-energy fields belong to an adjoint representation, transforming like $\mathbf{n} \times \bar{\mathbf{n}} = (\mathbf{n^2 - 1}) + \mathbf{1}$ of $U(n)$.

We can consider two parallel stacks of branes. For instance, if the two stacks contain $n_1$ and $n_2$ slices of branes, they, respectively, support $U(n_1)$ and $U(n_2)$ gauge group. An open string stretched between the two stacks have now bifundamental representation $(\mathbf{n_1}, \mathbf{n_2})$. The mass of the lightest mode is proportional to the interval length. If the two stacks come close and become coincident, we have gauge symmetry enhancement

$$U(n_1 + n_2) \leftrightarrow U(n_1) \times U(n_2). \tag{17.14}$$

The opposite process is also possible and it is nothing but the Higgs mechanism.

It is interesting to note that, if we only consider the non-Abelian group $SU(3)_c \times SU(2)_L$, all the Standard Model fields (2.7) are bifundamental. For quarks and leptons, however, we need *chiral* representations. If two stacks are intersecting at angles, we can obtain chiral fermion [5]. The string stretched between the two are localized at the intersecting point to minimize the energy. Quantization gives the massless chiral fermions of bifundamental $(\mathbf{n_1}, \mathbf{n_2})$.

In Fig. 17.1, we can construct a stacks of branes giving rise to the fields with the SM quantum numbers. There we have four $U(n)$ gauge groups. Since the $U(1)$ factors decouple as $U(n) = U(1) \times SU(n)$, the $U(1)$ charges for matter fields are provided by some linear combinations these. If the SM fields were charged

**Fig. 17.1** Example for a standard-like model configuration

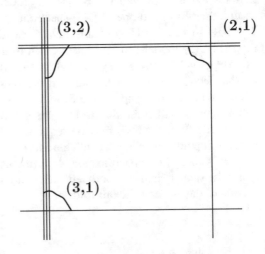

under more than two non-Abelian groups, we cannot make an intersecting brane model, contrary to the case in heterotic string models. It follows from a simple observation that a non-Abelian horizontal (family) symmetry cannot be introduced from intersecting brane models since one cannot introduce an additional set of branes intersecting at the same intersection.

Now, to have our 4D we need to compactify six dimensions. In a typical realistic setup, we compactify six dimension as before and consider 3-cycle $[\pi_A]$ on which D6-brane stacks wrap. With some nontrivial cycle, we can make more than one intersection points. Decomposing six torus as $T^2 \times T^2 \times T^2$ and denoting the wrapping number

$$[\pi_A] = \prod_{a=1}^{3} (l_A^a, m_A^a) \tag{17.15}$$

along each unit direction $a, b$ of the $i$th 2-torus, we have the intersection number

$$I_{AB} = \prod_{a=1}^{3} (l_A^a m_B^a - m_A^a l_B^a). \tag{17.16}$$

For instance, three intersection points can explain *three families*, because the same fermions appear as many times. An example is shown in Fig. 17.2. Like in the heterotic string on orbifold, this also has the same merit: the distance explains the hierarchy between Yukawa interactions. The more distant the intersection points are separated, the more (exponentially) suppressed the interaction between them.

**Ramond–Ramond Tadpole Cancellation**
We can do bottom-up construction of the model as long as the global consistency condition is satisfied. Recall that in the heterotic string the consistency condition came from the modular invariance of one-loop amplitude, which gives no divergence in the triangle diagram in four dimensions. In the open string, a potential divergence lies in the RR tadpole diagram. For vanishing divergence, the sum of the RR charge should be cancelled in the transverse compact dimension to the cycle. The condition is [6]

$$\sum_A n_A [\pi_A] = 0. \tag{17.17}$$

It is obtained from the consistency condition of equations of motion. Physically, it is understood in terms of the (RR) charge conservation in the compact space. In the open string, the corresponding diagram is sum of cylinder and its modifications by unorientation [7].

With D-branes only, it is hard to cancel the RR charges. We can use anti-D-branes which are D-branes wrapped on the same cycle with the opposite orientations

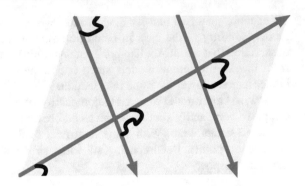

**Fig. 17.2** Two one-cycles with wrapping numbers $(1, 1)$ and $(1, -2)$. The intersection number $1 \cdot 1 - 1 \cdot (-2) = 3$ accounts for the number of families

and have the opposite RR charge; however, the brane then attract each other, destined to be annihilated. They also break the desired supersymmetry, as seen from (6.262).The cure can come from the introduction of *orientifold*. Consider the worldsheet orientation reversal in (6.266). Indeed this operation interchanges the left and right mover. In the $T$-dual coordinate $X' = X'_L(\tau + \sigma) - X'_R(\tau - \sigma)$, this operation accompanies a spacetime $\mathbb{Z}_2$ reflection

$$R : X' \to -X', \tag{17.18}$$

(and technically also adds a $G$-parity action $(-1)^F$ in (6.142) with right movers). As a result, the dual direction is now a $S/\mathbb{Z}_2$ orbifold. We count only D-branes in the fundamental region, or interval. For instance, to have a bifundamental $(\mathbf{n_1}, \overline{\mathbf{n_2}})$, we need two stacks of branes $n_1$ and $n_2$ in the fundamental region. Also due to the $\Omega R(-1)^F$ identification "reflect" the situation in the fundamental domain with respect to the fixed point. So, if we have a stack of brain in the interval, we also have a mirror stack at the reflected point. This gives rise to a new open string ending between the original stack and the mirror stack. Since the two stacks have the opposite orientation, the low-energy field carries the bifundamental charge $(\mathbf{n_1}, \mathbf{n_2})$ not anti-fundamental. We say that there is an orientifold plane, or O$p$-plane at the fixed point, as shown in Fig. 17.3. If an open string connects between one stack and its mirror stack, the resulting field has both symmetric and antisymmetric representation. Thus, if an $n$ stack of branes are placed at the orientifold plane, the resulting gauge theory becomes SO($2n$).

Like D-branes, the orientifold is charged under NSNS and RR. We can calculate the brane tension by the amplitude of closed string exchange between the orientifold and the D-brane or between D-brane and mirror brane. The resulting tension is

$$T_p^O = \pm 2^{p-5} T_p^D. \tag{17.19}$$

When we compactify one dimension, the worldsheet parity $\mathbb{Z}_2$ induces two fixed points as discussed in Chap. 3, thus we have a double O($p-1$)s. So, for every $p$, we always need 16 D-branes to cancel the RR charges of orientifolds. The nonvanishing sum of NSNS charges simply implies the time dependent dynamics.

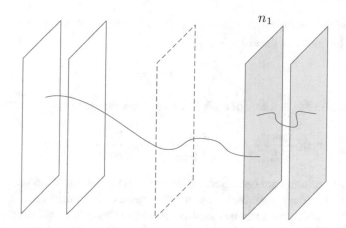

**Fig. 17.3** Orientifold plane (center, dotted) and mirror branes

The RR charge cancellation leads to the anomaly cancellation in the low-energy theory. For example, $SU(n_A)^3$ anomaly is proportional to the number of $\mathbf{n}_A$ minus number of $\overline{\mathbf{n}}_A$, hence proportional to

$$\sum_B I_{AB} n_B = [\pi_A] \cdot \sum_B n_B [\pi_B] = 0$$

coming from the above condition (17.17). Generically, the rest of anomalies such as $U(1)^3$ or $SU(n_a)^2 \times U(1)$ are not cancelled in the same way because like in the heterotic string case there are potential anomalous $U(1)$s. They are cancelled by a generalized Green–Schwarz mechanism, here the antisymmetric tensors responsible for axions are provided by the RR tensor fields in the compact space. There are more than one such anomalous $U(1)$s.

**Supersymmetry**

We comment that the supersymmetry condition [5] is the same with that of the heterotic string on orbifold (3.73). Here the condition is parametrized by three angles $\phi_i$, tilted angle from the unit direction in each $i$th two-torus. This tilts the brane and the supersymmetry generator as

$$\tilde{Q} + RPR^{-1}Q \tag{17.20}$$

with the eigenvalues of $R$ being $e^{2\pi i \phi_i}$. The unbroken symmetry is the common intersection $P = RPR^{-1}$. The condition reduces to the invariant part of $Q$ under $R^2$, in the spinorial representation $s$

$$s \cdot 2\phi = 0 \tag{17.21}$$

for at least one combination of $s = ([\frac{1}{2}^4])$ with even numbers of minus sign. For example, taking a basis $s = (\frac{1}{2} \frac{1}{2} \frac{1}{2} \frac{1}{2})$,

$$\phi_1 + \phi_2 + \phi_3 = 0$$

We can rewrite this in terms of radii $R_i$ of six-torus and three cycle numbers

$$\arctan \frac{R_2 m_1}{R_1 l_1} + \arctan \frac{R_4 m_2}{R_3 l_2} + \arctan \frac{R_6 m_3}{R_5 l_3} = 0. \tag{17.22}$$

The merit of non-supersymmetric model is that we can interpret the Higgs field as tachyon. It signals instability and triggers geometric transition to the setup of lower energy which typically has smaller gauge symmetry. This is interpreted as the tachyon condensation or the brane recombination. There is no hierarchy problem since the Yang–Mills field coupling can be arbitrarily lowered by adjusting the compactification size

$$M_{\text{Pl}}^2 g_{\text{YM}}^2 = \frac{M_s^{11-p} V_X}{g_s}$$

with $V_X$ being the volume of the space spanned by 3-cycle. However, it suffers another hierarchy problem because of the size $V_X$.

Instead of torus, we can use the Calabi–Yau manifold. Locally the SUSY condition is essentially the same and we have SUSY preserving cycle called special Lagrangian cycle.

### 17.2.1 Models

First, we discuss the simple non-supersymmetric model given in Ref. [6]. It is bases on Type IIA string theory compactified on $T^6$. We have four stacks of D6 branes as given in Table 17.1. With Relation (17.16), we also have the intersection number $I$ giving the number of families. The resulting spectrum is shown in Table 17.2. With a number of intersecting branes, we have the gauge group,

$$U(3)_a \times U(2)_b \times U(1)_c \times U(1)_d.$$

**Table 17.1** An example of D6-brane, Ref. [6], with wrapping numbers giving a SM spectrum

| Stack $A$ | # slices | $(l_A^1, m_A^1)$ | $(l_A^2, m_A^2)$ | $(l_A^3, m_A^3)$ |
|---|---|---|---|---|
| a | 3 | $(1, 0)$ | $(2, 1)$ | $(1, 1/2)$ |
| b | 2 | $(0, -1)$ | $(1, 0)$ | $(1, 3/2)$ |
| c | 1 | $(1, 3)$ | $(1, 0)$ | $(0, 1)$ |
| d | 1 | $(1, 0)$ | $(0, -1)$ | $(1, 3/2)$ |

**Table 17.2**  The standard model spectrum of Table 17.1 and their U(1) charges

| Intersection | $I$ | Matter | | $Q_a$ | $Q_b$ | $Q_c$ | $Q_d$ | Y |
|---|---|---|---|---|---|---|---|---|
| (ab) | 1 | $q_L$ | $(\mathbf{3}, \mathbf{2})$ | 1 | $-1$ | 0 | 0 | 1/6 |
| (ab*) | 2 | $q_L$ | $2(\mathbf{3}, \mathbf{2})$ | 1 | 1 | 0 | 0 | 1/6 |
| (ac) | $-3$ | $u_R$ | $3(\overline{\mathbf{3}}, \mathbf{1})$ | $-1$ | 0 | 1 | 0 | $-2/3$ |
| (ac*) | $-3$ | $d_R$ | $3(\overline{\mathbf{3}}, \mathbf{1})$ | $-1$ | 0 | $-1$ | 0 | 1/3 |
| (bd*) | $-3$ | $l_L$ | $3(\mathbf{1}, \mathbf{2})$ | 0 | $-1$ | 0 | $-1$ | $-1/2$ |
| (cd) | $-3$ | $e_R$ | $3(\mathbf{1}, \mathbf{1})$ | 0 | 0 | $-1$ | 1 | 1 |
| (cd*) | 3 | $n_R$ | $3(\mathbf{1}, \mathbf{1})$ | 0 | 0 | 1 | 1 | 0 |

The intersection bd has vanishing intersection number

To cancel the total RR charges, O6 orientifold planes are introduced. Because of this O6, we have mirror branes for which we use a star(*). They sit at the mirror image points of the original branes, and are complex conjugate representations. Compare (ab) intersection with (ab*).

With the U(1) charge assignments of Table 17.2, we can interpret $Q_a$, $Q_b$, $Q_c$, $Q_d$ as the baryon number, the lepton number, the right-handed isospin, and a Pecci–Quinn type charge, respectively. This is a generic feature because, for example, the U(1) from the strong interaction group U(3) = U(1) × SU(3) must be a significant contributor to the baryon number symmetry U(1)$_B$, because only strongly interacting quarks are transforming nontrivially under the strong gauge group. The lepton number in the above model is conserved, so Majorana type neutrino masses are not possible, and hence a seesaw mechanism is not allowed. Such U(1) symmetries are local, and a potential anomaly is cancelled by the GS mechanism. The corresponding gauge bosons become massive but the symmetries still persists as global symmetries, which suppress rapid proton decay. The non-anomalous hypercharge generator is

$$Y = \frac{1}{6}Q_a - \frac{1}{2}Q_c + \frac{1}{2}Q_d.$$

Instead of torus, we may use orbifold to obtain more realistic models. For instance, we may introduce $T^6/\mathbb{Z}_N$ or $T^6/(\mathbb{Z}_N \times \mathbb{Z}_M)$ orbifold. As in the heterotic string case, the supersymmetry is broken by holonomy. D-branes and orientifolds can break further supersymmety, but a judicious choice can break only half of the supersymmetry yielding $\mathcal{N} = 1$ in four dimensions. There are three-generation models in Refs. [8–11].

## 17.3  Magnetized Brane

So far we have obtained chirality by projecting out chiral component of a fermion. There is another way to obtain chirality on flat geometry. Under constant magnetic flux background the solution to Dirac equation selects one chirality [12, 13].

Consider a gauge U($n$) gauge field $A_M$ on torus $T^2$ having the coordinates $x^5, x^6$. We may turn on a constant magnetic flux

$$F_{45} = f = \text{const.}$$

We have freedom to fix the gauge as

$$A_5 = 0, \qquad A_6 = f\mathbf{1}_n x^4. \tag{17.23}$$

This background does not break Lorentz symmetry in four dimensions. The expectation value for $A_6$ is non-constant but linearly grow with $x^5$. At first sight it seems not preserve the periodicity of the torus $x^5 \sim x^5 + e^5$. However, we only need to have periodicity up to the global symmetry. In fact the consistency condition requires (5.106) [12, 13]

$$\frac{1}{2\pi} \oint dx \,\text{tr}\, A = \frac{1}{2\pi} \int_{T^2} d^2x \,\text{tr}\, f\mathbf{1}_n = \text{integer}. \tag{17.24}$$

Thus the magnetic flux is quantized. So we may take $f = m/n$ with an integer $m$.

The background

$$\frac{F_{45}}{2\pi} = \begin{pmatrix} \frac{m_1}{n_1}\mathbf{1}_{n_1} & 0 \\ 0 & \frac{m_2}{n_2}\mathbf{1}_{n_2} \end{pmatrix}, \tag{17.25}$$

breaks U($n$) down to U($n_1$) $\times$ U($n_2$). It is convenient to define an "intersection number"

$$I = m_1 n_2 - m_2 n_1, \qquad \tilde{I} = I/(n_1 n_2). \tag{17.26}$$

We reduce the six dimensional gauginos as

$$\lambda(x^\mu, y) = \sum \lambda^{(n)}(x) \otimes \chi^{(n)}(y). \tag{17.27}$$

Here $\lambda^{(n)}$ are four component spinors and $\chi^{(n)}(y)$ are two component spinors

$$\chi(y) = \begin{pmatrix} \chi_+(y) \\ \chi_-(y) \end{pmatrix}. \tag{17.28}$$

From now on we focus on zero modes $n = 0$ and suppress the index.

The gamma matrix along these directions are

$$\Gamma^4 = \mathbf{1}_4 \otimes \sigma^1, \qquad \Gamma^5 = \mathbf{1}_4 \otimes \sigma^2. \tag{17.29}$$

They are further decomposed into the blocks in the group space

$$\chi_{\pm}(y) = \begin{pmatrix} \chi_{\pm}^{11}(y) & \chi_{\pm}^{12}(y) \\ \chi_{\pm}^{21}(y) & \chi_{\pm}^{22}(y) \end{pmatrix} \tag{17.30}$$

in the sense of (17.25). The fields $\chi_{\pm}^{11}$ and $\chi_{\pm}^{22}$ transform as adjoint representations of the unbroken gauge group $U(n_1) \times U(n_2)$. Then $\chi_{\pm}^{12}$ and $\chi_{\pm}^{21}$ correspond to bi-fundamental matter fields, $(\mathbf{n_1}, \mathbf{\bar{n}_2})$ and $(\mathbf{\bar{n}_1}, \mathbf{n_2})$.

The Dirac equations for the gaugino zero modes are obtained as

$$\begin{pmatrix} \partial\chi_+^{11} & [\partial + 2\pi \tilde{I} x^5]\chi_+^{12} \\ [\partial + 2\pi \tilde{I} x^4]\chi_+^{21} & \partial\chi_+^{22} \end{pmatrix} = 0, \tag{17.31}$$

$$\begin{pmatrix} \bar{\partial}\chi_-^{11} & [\bar{\partial} - 2\pi \tilde{I} x^5]\chi_-^{12} \\ [\bar{\partial} - 2\pi \tilde{I} x^4]\chi_-^{21} & \bar{\partial}\chi_-^{22} \end{pmatrix} = 0, \tag{17.32}$$

where $\partial = \partial_5 + i\partial_6$ and $\bar{\partial} = \partial_5 - i\partial_6$.

For $I > 0$, the solutions $\chi_+^{12}$ and $\chi_-^{21}$ have normalizable zero modes, while $\chi_-^{12}$ and $\chi_+^{21}$ do not. The normalizable wave function is obtained as [14]

$$\Theta^j(x^5, x^6) = N_j e^{-M\pi y_5^2} \vartheta \begin{bmatrix} j/I \\ 0 \end{bmatrix} (iI| I(x^5 + ix^6)), \quad j = 0, 1, \cdots, I-1, \tag{17.33}$$

where $N_j$ is a normalization constant. It is convenient to extended Jacobi theta function by replacing $\beta$ with $\nu + \beta$ as

$$\vartheta \begin{bmatrix} \alpha \\ \beta \end{bmatrix} (\nu|\tau) \equiv \vartheta \begin{bmatrix} \alpha \\ \nu + \beta \end{bmatrix} (\tau). \tag{17.34}$$

It has similar transformation property

$$\mathcal{T} : \vartheta \begin{bmatrix} \alpha \\ \beta \end{bmatrix} (\nu|\tau + 1) = e^{i\pi(\alpha^2 - \alpha)} \vartheta \begin{bmatrix} \alpha \\ \alpha + \beta - 1/2 \end{bmatrix} (\nu|\tau), \tag{17.35}$$

$$\mathcal{S} : \vartheta \begin{bmatrix} \alpha \\ \beta \end{bmatrix} \left(\frac{\nu}{\tau} \Big| \frac{-1}{\tau}\right) = \sqrt{-i\tau} e^{2\pi i(\nu^2/2\tau + \alpha\beta)} \vartheta \begin{bmatrix} \beta \\ -\alpha \end{bmatrix} (\nu|\tau). \tag{17.36}$$

Under transformation of argument $\nu$

$$\vartheta \begin{bmatrix} \alpha \\ \beta \end{bmatrix} (\nu + 1|\tau) = e^{2\pi i\alpha} \vartheta \begin{bmatrix} \alpha \\ \beta \end{bmatrix} (\nu|\tau) \tag{17.37}$$

$$\vartheta \begin{bmatrix} \alpha \\ \beta \end{bmatrix} (\nu + \tau|\tau) = e^{-2\pi i(\beta + \nu + \tau/2)} \vartheta \begin{bmatrix} \alpha \\ \beta \end{bmatrix} (\nu|\tau). \tag{17.38}$$

In dealing with shift vectors, it is convenient to use

$$\vartheta \begin{bmatrix} \alpha+1 \\ \beta \end{bmatrix} (\nu|\tau) = \vartheta \begin{bmatrix} \alpha \\ \beta \end{bmatrix} (\nu|\tau) \tag{17.39}$$

$$\vartheta \begin{bmatrix} \alpha \\ \beta+1 \end{bmatrix} (\nu|\tau) = e^{2\pi i \alpha} \vartheta \begin{bmatrix} \alpha \\ \beta \end{bmatrix} (\nu|\tau). \tag{17.40}$$

They are easily read off from the sum and the product definitions, respectively.

Conversely, when $I < 0$, $\chi_-^{12}$ and $\chi_+^{21}$ become exclusively normalizable zero modes. In both cases the number of normalizable zero modes is $|I|$. We obtain chiral theory. In this way can obtain three generalization of bifundamentals if we turn on magnetic flux such that $I = 3$. We have definite chirality depending on the sign of $I$. Later we will see that this is dual to intersecting brane models, which we discuss in Sect. 17.2, and it shares many interesting feature like the chirality and the number of zero modes.

We may generalize this in two ways. First, we may consider breaking $U(n)$ into more than two gauge groups. The discussion applies the same if we consider a $2 \times 2$ block diagonal matrices. Interestingly all the SM fermions are bifundamental under the non-Abelian gauge group. They may have nontrivial U(1) charges and we can obtain desired hypercharge. Also we can generalize it to higher dimensions. The discussion also applies if we consider each two-torus separately. The number of generations $I$ is now multiplied as $I_{45}I_{67}I_{89}$ where the subscripts denote the two-torus directions. Combining this, we may turn on a nontrivial flux that is not magnetic flux. We may start from type I string with SO(32) and turn on magnetic fluxes to obtain this.

We may have $F \wedge F$ which is not decomposed into $F$. This is the instanton solution and its generalization that we discuss in Chap. 15.

## 17.4  F-theory

Finally, we introduce the most unified framework describing the open and closed strings at the same time. Recall that the open string description nicely describes the bifundamental quantum numbers. However, closed string theory nicely described the unification structure using the $E_n$ series of GUT groups.

We have seen a strange coincidence that the Narain lattice in eight dimensions

$$\Gamma_{16} \times \Gamma_{16} \times \Gamma_{1,1} \times \Gamma_{1,1} \times \Gamma_{1,1}$$

and the intersection matrix of the harmonic two forms in (15.31) are the same. This can be nicely explained by M-theory and F-theory we discuss here.

While in the heterotic side we have extra left-moving worldsheet fields for the current algebra taking into account the non-Abelian gauge theory, we have no extra degrees of freedom on the other side. The secret is that the compact K3 in each has singular geometry. Wrapped string/M2/D3 on the $A$, $D$, $E$ singularity gives rise to massless gauge fields whose quantum number has same connected structure of the singularity.

Recall the field contents of IIB supergravity: there are RR tensor fields $C_0, C_2, C_4$ of ranks 0, 2, 4, respectively, as well as dilaton $\phi$ and NSNS tensor field $B$. In particular the rank 4 field is self-Hodge-dual in ten dimensions. Recall that the $C_0$ is a pseudoscalar and the expectation value of $e^\phi$ gives the open string coupling. We may complexify them as

$$\tau \equiv C_0 + i e^{-\phi}. \tag{17.41}$$

The action has $SL(2, \mathbb{R})$ symmetry

$$\tau \rightarrow \frac{a\tau + b}{c\tau + d}, \tag{17.42}$$

$$\begin{pmatrix} H \\ F \end{pmatrix} \rightarrow \begin{pmatrix} a & b \\ c & d \end{pmatrix} \begin{pmatrix} H \\ F \end{pmatrix}, \tag{17.43}$$

with $a, b, c, d \in \mathbb{R}$ and $ad - bc = 1$. Other fields are invariant. Here $H = dB$, $F = dC_2$ are the field strengths of the NSNS and RR two forms, respectively. A quantum version of this, $SL(2, \mathbb{Z})$, for which $a, b, c, d$ are integers, is also the symmetry of IIB *string* theory.

Note that this is also the same symmetry of the torus: the complex structure transforms in the same ways as (17.42). The SUGRA action obeying this symmetry is identical to dimensionally reduced theory if we identify the axio-dilaton field (17.41) as the complex structure of a two-torus. Thus type IIB string may be understood as a dimensionally reduced theory of a certain twelve dimensional theory. We call this F-theory [15]. This torus is a virtual whose one complex structure $\tau$ is identified as axio-dilaton and has no area. Note that we have no full twelve dimensional spacetime with the Lorentz signature $(1, 11)$ because then the degrees of freedom of the gravitino is too large. Rather, two of them are $T$-dual to each other, so we are not able to see both of them at the same time [16–18].

We may understand the geometry and its relation to gauge theory using algebraic description, discussed in Sect. 15.5. The torus is described by Weierstrass equation

$$y^2 = x^3 + fx + g. \tag{17.44}$$

That is, the torus is the complex curve satisfying this equation in $\mathbb{C}^2$ with coordinates $x, y$, including the point at the infinity. For this reason, the two-torus is also called the elliptic curve. One way to see this is to express it as

$$y = \pm[(x - x_1)(x - x_2)(x - x_3)]^{1/2},$$

where $x_1, x_2, x_3$ are three roots of Eq. (17.44) with $y = 0$. Here $y$ is not a single-valued function: we have order two branch point at each $x_i$. To have global understanding, we may take two branch cuts connecting $x_1$-$x_2$ and $x_3$-$Infty$. Once we cross the branch cut

$$(x - x_i) \rightarrow e^{2\pi i}(x - x_i), \quad i = 1, 2, 3,$$

$y$ acquires a phase $e^{\pi i} = -1$. We understand it as that we have two Riemann sheets, each of which is a sphere if we include the point at infinity. Once we cross the branch cut we go to the other sheet. In effect two spheres are connected by two tubes, so the total topology is the torus.[4]

We may also define F-theory as the M-theory compactified on a torus $T^2$ in the vanishing volume limit. Due to $T$-duality, we have one extra dimension. So it makes sense to have a torus without volume. We obtain four dimensional theory if we compactify F-theory on a Calabi–Yau fourfold. Since we need the $T^2$ structure the Calabi–Yau manifold should have elliptically fibered. We define fiber bundle with the base and fiber. Locally the fiber is vector space but vary so the geometry is not direct product.

The complex structure $\tau$ is related to the parameters $f$ and $g$ through Klein's $j$-function

$$j(\tau) = \frac{(12f)^3}{4f^3 + 27g^2}. \tag{17.45}$$

The denominator is the discriminant

$$\Delta = 4f^3 + 27g^2 \tag{17.46}$$

of the cubic function, that is the RHS of Eq. (17.44).

F-theory compactified on a K3 is dual to heterotic string compactified on torus. We may construct K3 as elliptic fibration (because torus is described by elliptic curve) over a $\mathbb{CP}^1$ (two sphere). Due to fibration, $f$ and $g$ are not functions anymore, because they vary once we transport around the base. However, the topology of the base restricts the consistent condition like Dirac monopole quantization. Thus $f$ and $g$ are sections of $K^{-4}$ and $K^{-6}$, respectively. Here $K$ is the canonical line bundle of the base, which is generalization of the volume form. It is because $x$ and $y$ are sections of the line bundles $\mathscr{L}^3$ and $\mathscr{L}^2$ to admit good sections. The Chern class of the total manifold is induced

$$c_1(X) = \pi^*(\mathscr{L} \otimes K), \tag{17.47}$$

---

[4] Another way to see this is that a homogeneous degree $n + 2$ polynomial in $\mathbb{P}^{n+1}$ describes the Calabi–Yau $n$-fold. Using homogeneous coordinate we show that any cubic equation can be completed to Weierstrass form.

**Table 17.3** Kodaira
classification of singularities

| ord $f$ | ord $g$ | ord $\Delta$ | Name | Algebra |
|---|---|---|---|---|
| $\geq 0$ | $\geq 0$ | 0 | $I_0$ | – |
| 0 | 0 | $k \geq 1$ | $I_k$ | $A_{k-1}, C_k$ |
| $\geq 1$ | 1 | 2 | II | – |
| 1 | $\geq 2$ | 3 | III | $A_1$ |
| $\geq 2$ | 2 | 4 | IV | $A_1, A_2$ |
| $\geq 2$ | $\geq 3$ | 6 | $I_0^*$ | $D_4, B_3, G_2$ |
| 2 | 3 | $k \geq 7$ | $I_{k-6}^*$ | $D_{k-2}, B_{k-3}$ |
| $\geq 3$ | 4 | 8 | IV* | $E_6, F_4$ |
| 3 | $\geq 5$ | 9 | III* | $E_7$ |
| $\geq 4$ | 5 | 10 | II* | $E_8$ |

The corresponding 7-branes supports the algebra, depending on further splitting or monodromy conditions [32]

up to some irrelevant contributions [19]. We thus have vanishing first Chern class if

$$\mathcal{L} = K^{-1}. \tag{17.48}$$

If we have vanishing discriminant, meaning that the elliptic fiber becomes singular, we have symmetry enhancement. We interpret that we have 7-brane at the locus. Like that in the quadratic function, it tells us how many zeros. For $\Delta > 0$, we have three real solutions. $\Delta = 0$ gives double points, which means two of the above three solutions become coincident. In view of the torus, one of the cycle is pinched off and the resulting torus becomes singular. If we fiber this over $\mathbb{CP}^1$. Expanding the $j$ function we have

$$j(\tau) = e^{-2\pi i \tau} + 744 + 196884 e^{2\pi i \tau} + \dots . \tag{17.49}$$

If we blow-up this singularity, we may obtain the same geometry. This shape of singularities can be classified by the local behavior. We have displayed this in Table 17.3. For $E_8$, as long as the order of $g$ is 5, the leading order is dominated by $\Delta \simeq g^2 \simeq z^{10}$ as long as the order of $f$ is equal or greater than 4. However, for other groups like $A_{n-1}$ and $D_n$, there is cancellation between leading order terms in $4f^3$ and $27g^2$ making the order of $\Delta$ nontrivial.

Roughly the vanishing order of $\Delta$, ord$\Delta$, is the number of 7-branes. For $E_8$ we have ten 7-branes. For $A_{n-1}$ we have $n$ branes.[5]

---

[5]The $j$ function may not vanish in the vanishing limit of the discriminant locus. For $A_{n-1}$ and $D_{n+4}$ we have $j \to \infty$. For III and III* we have $j \to 1728$. For others $j \to 0$.

## 17.4.1 Models

If we have 7-branes describing SU(5) we have SU(5) GUT. To track monodromy, it is convenient to use the expanded form, or Tate form

$$y^2 + a_1 xy + a_3 y = a_2 x^2 + a_4 x + a_6,\tag{17.50}$$

where $a_i$ are sections of $K_B^{-i}$. The classification is done and the result is shown in Table 17.4.

For concrete discussion, we construct an $SU(5)$ singularity

$$y^2 + b_5 xy + b_3 z^2 y = b_4 z x^2 + b_2 x + b_0,\tag{17.51}$$

**Table 17.4** Singularities identified by the orders of the coefficients of Weierstrass equation in the Tate form (17.44) and (17.50) [32]

| Type | Group | $a_1$ | $a_2$ | $a_3$ | $a_4$ | $a_6$ | $\Delta$ | $f$ | $g$ |
|------|-------|-------|-------|-------|-------|-------|----------|-----|-----|
| $I_0$ | Smooth | 0 | 0 | 0 | 0 | 0 | 0 | 0 | 0 |
| $I_1$ | U(1) | 0 | 0 | 1 | 1 | 1 | 1 | 0 | 0 |
| $I_2$ | SU(2) | 0 | 0 | 1 | 1 | 2 | 2 | 0 | 0 |
| $I_{2k-1}^{ns}$ | Unconven. | 0 | 0 | $k$ | $k$ | $2k-1$ | $2k-1$ | 0 | 0 |
| $I_{2k-1}^{s}$ | SU(2k−1) | 0 | 1 | $k-1$ | $k$ | $2k-1$ | $2k-1$ | 0 | 0 |
| $I_{2k}^{ns}$ | Sp(k) | 0 | 0 | $k$ | $k$ | $2k$ | $2k$ | 0 | 0 |
| $I_{2k}^{s}$ | SU(2k) | 0 | 1 | $k$ | $k$ | $2k$ | $2k$ | 0 | 0 |
| II | — | 1 | 1 | 1 | 1 | 1 | 2 | 1 | 1 |
| III | SU(2) | 1 | 1 | 1 | 1 | 2 | 3 | 1 | 1 |
| $IV^{ns}$ | Unconven. | 1 | 1 | 1 | 2 | 2 | 4 | 1 | 1 |
| $IV^{s}$ | SU(3) | 1 | 1 | 1 | 2 | 3 | 4 | 1 | 1 |
| $I_0^{*\,ns}$ | $G_2$ | 1 | 1 | 2 | 2 | 3 | 6 | 2 | 3 |
| $I_0^{*\,ss}$ | SO(7) | 1 | 1 | 2 | 2 | 4 | 6 | 2 | 3 |
| $I_0^{*\,s}$ | SO(8)* | 1 | 1 | 2 | 2 | 4 | 6 | 2 | 3 |
| $I_{2k-3}^{*\,ns}$ | SO(4k+1) | 1 | 1 | $k$ | $k+1$ | $2k$ | $2k+3$ | 2 | 3 |
| $I_{2k-3}^{*\,s}$ | SO(4k+2) | 1 | 1 | $k$ | $k+1$ | $2k+1$ | $2k+3$ | 2 | 3 |
| $I_{2k-2}^{*\,ns}$ | SO(4k+3) | 1 | 1 | $k+1$ | $k+1$ | $2k+1$ | $2k+4$ | 2 | 3 |
| $I_{2k-2}^{*\,s}$ | SO(4k+4)* | 1 | 1 | $k+1$ | $k+1$ | $2k+1$ | $2k+4$ | 2 | 3 |
| $IV^{*\,ns}$ | $F_4$ | 1 | 2 | 2 | 3 | 4 | 8 | 3 | 4 |
| $IV^{*\,s}$ | $E_6$ | 1 | 2 | 2 | 3 | 5 | 8 | 3 | 4 |
| $III^*$ | $E_7$ | 1 | 2 | 3 | 3 | 5 | 9 | 3 | 5 |
| $II^*$ | $E_8$ | 1 | 2 | 3 | 4 | 5 | 10 | 3 | 5 |
| Non-min | – | 1 | 2 | 3 | 4 | 6 | 12 | 4 | 6 |

Here $k \geq 2$, and the starred ones have a further condition

where we have the discriminant locus $S = \{z = 0\}$ supporting the SU(5) singularity and is the section of the normal bundle of $S$ in $B$. A normal bundle is the tangent bundle of $B$ modded out by the tangent bundle of $S$ thus measures the departure from the $S$. This means that we have chosen $a_i$s as above and we have only displayed the leading order terms thus it is local description. We have higher order terms in $z$ in each coefficients. For instance

$$a_1 = b_5 + b_{5,1}z + b_{5,2}z^2 + \dots \tag{17.52}$$

Completing the $y$ term and absorbing the $x^2$ term, we have the original form (17.44)[6] We find the discriminant

$$\Delta = b_5^4(b_0 b_5^2 - b_2 b_3 b_5 + b_3^2 b_4)z^5 + O(z^6). \tag{17.53}$$

The leading order $z^5$ and the coefficient vanishes at the zeros of the following:

$$R_{10} \equiv b_5, \tag{17.54}$$

$$R_{\bar{5}} \equiv b_0 b_5^2 - b_2 b_3 b_5 + b_3^2 b_4. \tag{17.55}$$

Each equation defines a hypersurface describing curve, so we define matter curves as vanishing loci of these. This reminds us the intersection between two brane stacks, discussed in Sect. 17.2. If we focus on one brane stack describing U($n_1$), the intersection from the other describing U($n_2$) looks like a (complex) codimension one hypersurface, which is curve in our case. We have bifundamental representation $(\mathbf{n_1}, \mathbf{\bar{n}_2})$ of the two groups, which can be understood as off-diagonal components of the adjoint of the unified group U($n_1 + n_2$) $\to$ U($n_1$) $\times$ U($n_2$). Here the same happens and we may have other groups than U($n$).

---

[6]Conventional parametrizations are as follows;

$$b_2 = a_1^2 + 4a_2,$$

$$b_4 = a_1 a_3 + 2a_4,$$

$$b_6 = a_3^2 + 4a_6,$$

$$b_8 = a_1^2 a_6 - a_1 a_3 a_4 + 4a_2 a_6 + a_2 a_3^2 - a_4^2,$$

$$f = b_2^2 - 24b_4,$$

$$g = -b_2^3 + 36b_2 b_4 - 216b_6,$$

$$\Delta = -b_2^2 b_8 - 8b_4^3 - 27b_6^2 + 9b_2 b_4 b_6.$$

We have gauge symmetry enhancement to SO(10) at the locus $R_{10} = 0$, under which the adjoint branches

$$45 \rightarrow 24 + 10 + \overline{10} + 1,$$

where **10** is the rank-two tensor representation of the SU(5) We have chiral zero mode. In view of Eq. (17.52), setting $b_5 = 0$ makes the leading order term of $a_1$ to be $b_{5,1}z$. Checking Table 17.4, we confirm that indeed the resulting singularity is $I_5^s$ describing the $SO(10)$. Also we have enhancement to SU(6) at $R_{\overline{5}} = 0$, under which the adjoint branches

$$35 \rightarrow 24 + 5 + \overline{5} + 1.$$

At the common intersections $R_{10} = R_{\overline{5}}$. They are codimension three points, along which we have gauge symmetry enhancement to $E_6$ which embraces all the $SU(6)$ and $SO(10)$. We have localized Yukawa coupling.

$$Y_u \sim \mathbf{10} \cdot \overline{\mathbf{5}} \cdot \overline{\mathbf{5}}, \quad Y_d \sim \mathbf{10} \cdot \mathbf{10} \cdot \mathbf{5}. \tag{17.56}$$

In the perturbative intersecting brane construction we cannot have the latter coupling because we should totally antisymmetrize the tensor indices. In the F-theory construction, this is inherited from $E_6$ coupling. F-theory naturally describes exceptional gauge group. Also we may have this term in nonperturbative way.

Wave functions are fuzzily localized at the matter curves and they have overlap at the Yukawa points.

Finally, we need to obtain the chiral spectrum in four dimensions. The matter curve is six dimensional so it becomes vectorlike. To have chirality, we turn on magnetic flux discussed in Sect. 17.3. There we have seen that a magnet flux on the flux also gives rise to a zero mode of one chirality exclusively. Here in F-theory we have a universal source of the magnetic field, which is three-form tensor in M-theory. We have Cartan subalgebra from the expansion in terms of harmonic two forms

$$C_{MNP} = \sum A_M \wedge \omega_{NP}, \tag{17.57}$$

where the sum is over the rank. We may turn on magnetic flux $\partial_{[M} A_{N]}$ from the field strength $G = dC$. This induces magnetic fluxes to all the branes. Using the index theorem, the number of zero modes is easily counted as

$$n_{\mathbf{R}} - n_{\overline{\mathbf{R}}} = \int_{\mathscr{S}_{\mathbf{R}}} G, \tag{17.58}$$

where $\mathscr{S}_{\mathbf{R}}$ is the matter surface, a $\mathbb{CP}^1$ fibration over the corresponding matter curve.

Instead of having the GUT group, we may have 7-brane describing $SU(3) \times SU(2) \times U(1)$ gauge group [20–24]. To construct $U(1)$, we need special setup. The fiber should admit more section than zero section. By theorem of Mordell and Weil we may have $\mathbb{Z}$ and $\mathbb{Z}_n$. We have as many $U(1)$ gauge groups as the rank of the Mordell–Weil group $\mathbb{Z}$s [25–27]. The discrete group is related to discrete symmetry.

## Duality with Heterotic String

One way to construct the stable vector bundle (15.52) is to construct spectral cover [28]. We follow the discussion in Refs. [29–31]. On a torus, we may turn on constant gauge field (flat bundle) or Wilson line. If we fiber this torus over a base, the value of the bundle may vary over the base. If this process is adiabatic, we can construct a stable vector bundle on the total manifold.

We may have one to one correspondence between line bundle and the position of real codimension two brane. This is essentially $T$-duality, although it is not well-defined in the heterotic string. Taking $T$-dual along one of the directions of $T^2$, the constant gauge field becomes a pointlike brane in the dual torus $\check{T}^2$. Fibering over the common base we hay have spectral cover. In the heterotic side the background breaks the gauge group $E_8 \times E_8$. On the other hand, in the F-theory side, the spectral cover controls the broken gauge group like an extra stacks of 7-branes.

This unifies the open and the closed string description. In Sect. 6.4.4, we have seen that shrinking a cylindrical M2-brane can give both closed and open strings depending on the direction. This M2-brane is lifted to the above D3-branes. and open. Shrinking along the axis we obtain an open string, which ends at the two $E_8$s at the end of the interval, which is drawn in Fig. 6.7.

We have seen various string constructions. Unification of string theory not only provides us various tools for constructing the Standard Model but also enables us to understand it in totally different ways.

# References

1. H. Kawai, D.C. Lewellen, S.H. Henry Tye, Construction of fermionic string models in four-dimensions. Nucl. Phys. B **288**, 1 (1987)
2. I. Antoniadis, J.R. Ellis, J.S. Hagelin, D.V. Nanopoulos, GUT model building with fermionic four-dimensional strings. Phys. Lett. **B205**, 459–465 (1988)
3. I. Antoniadis, J.R. Ellis, J.S. Hagelin, D.V. Nanopoulos, An improved SU(5) x U(1) model from four-dimensional string. Phys. Lett. **B208**, 209–215 (1988) [Addendum: Phys. Lett. **B213**, 562 (1988)]
4. J.D. Lykken, Four-dimensional superstring models, in *Proceedings, Summer School in High-Energy Physics and Cosmology*, Trieste, June 12–July 28 (1995), pp. 673–715
5. M. Berkooz, M.R. Douglas, R.G. Leigh, Branes intersecting at angles. Nucl. Phys. B **480**, 265–278 (1996)
6. L.E. Ibanez, F. Marchesano, R. Rabadan, Getting just the standard model at intersecting branes. J. High Energy Phys. **11**, 002 (2001)
7. C. Angelantonj, A. Sagnotti, Open strings. Phys. Rep. **371**, 1–150 (2002) [Erratum: Phys. Rep. **376**(6), 407 (2003)]

8. M. Cvetic, G. Shiu, A.M. Uranga, Three family supersymmetric standard-like models from intersecting brane worlds. Phys. Rev. Lett. **87**, 201801 (2001)
9. M. Cvetic, G. Shiu, A.M. Uranga, Chiral four-dimensional N=1 supersymmetric type 2A orientifolds from intersecting D6 branes. Nucl. Phys. B **615**, 3–32 (2001)
10. D. Cremades, L.E. Ibanez, F. Marchesano, Yukawa couplings in intersecting D-brane models. J. High Energy Phys. **07**,038 (2003)
11. F. Marchesano, G. Shiu, MSSM vacua from flux compactifications. Phys. Rev. D **71**, 011701 (2005)
12. G. 't Hooft, Some twisted selfdual solutions for the Yang-Mills equations on a hypertorus. Commun. Math. Phys. **81**, 267–275 (1981)
13. Z. Guralnik, S. Ramgoolam, From 0-branes to torons. Nucl. Phys. **B521**, 129–138 (1998)
14. D. Cremades, L.E. Ibanez, F. Marchesano, Computing Yukawa couplings from magnetized extra dimensions. J. High Energy Phys. **05**, 079 (2004)
15. C. Vafa, Evidence for F theory. Nucl. Phys. B **469**, 403–418 (1996)
16. K.-S. Choi, Supergravity in twelve dimension. J. High Energy Phys. **09**, 101 (2015)
17. D.S. Berman, C.D.A Blair, E. Malek, F.J. Rudolph, An action for F-theory: SL(2)$\mathbb{R}^+$ exceptional field theory. Class. Quant. Grav. **33**(19), 195009 (2016)
18. K.-S. Choi, Twelve-dimensional Effective Action and T-duality. Eur. Phys. J. C **75**(5), 202 (2015)
19. P.S. Aspinwall, K3 surfaces and string duality, in *Differential Geometry Inspired by String Theory* (1996), pp. 421–540. [1 (1996)]
20. K.-S. Choi, T. Kobayashi, Towards the MSSM from F-theory. Phys. Lett. B **693**, 330–333 (2010)
21. K.-S. Choi, SU(3) x SU(2) x U(1) vacua in F-theory. Nucl. Phys. B **842**, 1–32 (2011)
22. K.-S. Choi, On the standard model group in F-theory. Eur. Phys. J. C **74**, 2939 (2014)
23. L. Lin, T. Weigand, Towards the standard model in F-theory. Fortsch. Phys. **63**(2), 55–104 (2015)
24. M. Cvetič, L. Lin, M. Liu, P.-K. Oehlmann, An F-theory realization of the chiral MSSM with $\mathbb{Z}_2$-parity. J. High Energy Phys. **09**, 089 (2018)
25. K.-S. Choi, H. Hayashi, U(n) spectral covers from decomposition. J. High Energy Phys. **06**, 009 (2012)
26. D.R. Morrison, D.S. Park, F-theory and the Mordell-Weil group of elliptically-fibered Calabi-Yau threefolds. J. High Energy Phys. **10**, 128 (2012)
27. V. Braun, D.R. Morrison, F-theory on genus-one fibrations. J. High Energy Phys. **08**, 132 (2014)
28. R. Friedman, J. Morgan, E. Witten, Vector bundles and F theory. Commun. Math. Phys. **187**, 679–743 (1997)
29. M. Bershadsky, A. Johansen, T. Pantev, V. Sadov, On four-dimensional compactifications of F theory. Nucl. Phys. B **505**, 165–201 (1997)
30. R. Donagi, M. Wijnholt, Model building with F-theory. Adv. Theor. Math. Phys. **15**(5), 1237–1317 (2011)
31. R. Donagi, M. Wijnholt, Breaking GUT groups in F-theory. Adv. Theor. Math. Phys. **15**(6), 1523–1603 (2011)
32. M. Bershadsky, K.A. Intriligator, S. Kachru, D.R. Morrison, V. Sadov, C. Vafa, Geometric singularities and enhanced gauge symmetries. Nucl. Phys. **481**, 215–252 (1996)

# Useful Tables for Model Building

**A**

In this appendix, we collect useful tables for model building. Firstly, we list Tables 3.4 and 3.3 discussed in Chap. 3. Then, we list the degeneracy factor $\tilde{\chi}(\theta^i, \theta^j)$ which counts the GSO allowed multiplicity. Compared to the geometrical study for each case, using it is rather straightforward. Finally, we repeat Table 9.2, constraining Wilson line conditions.

The shift vectors $\phi$ and the number of fixed points $\chi$ for possible six dimensional orbifolds allowing $\mathcal{N} = 1$ supersymmetry are given in Table A.1.

The $\mathcal{N} = 1$ supersymmetry can be obtained also by a product orbifold $\mathbb{Z}_N \times \mathbb{Z}_M$, where $\mathbb{Z}_N$ and $\mathbb{Z}_M$ are chosen from the following table:

| P | Grid unit | $\phi$ | $\chi$ |
|---|---|---|---|
| $\mathbb{Z}_2$ | Half plane | $\frac{1}{2}(1\ 1)$ | 16 |
| $\mathbb{Z}_3$ | Hexagon | $\frac{1}{3}(1\ 1)$ | 9 |
| $\mathbb{Z}_4$ | Square | $\frac{1}{4}(1\ 1)$ | 1 |
| $\mathbb{Z}_6$ | Triangle | $\frac{1}{6}(1\ 1)$ | 1 |

For prime orbifolds, the degeneracy factor for the allowed massless fields is simply given by the number of fixed points. Massless fields are located either in the bulk or at the fixed points. So, it is clear in $\mathbb{Z}_3$ and $\mathbb{Z}_7$ orbifolds to locate geometrically the massless fields. For nonprime orbifolds, however, the allowed massless states are some combinations of functions located at fixed points. Thus, the calculation of the degeneracy factor is more involved, as given in the $\mathbb{Z}_{12}$ example of Chap. 14.

The $i^{\text{th}}$ twisted sector has the multiplicity

$$P_{\theta^k} = \frac{1}{N} \sum_{l=0}^{N-1} \tilde{\chi}(\theta^k, \theta^l) \left(\Delta_{\theta^k}\right)^l \tag{A.1}$$

© Springer Nature Switzerland AG 2020
K.-S. Choi, J. E. Kim, *Quarks and Leptons From Orbifolded Superstring*,
Lecture Notes in Physics 954, https://doi.org/10.1007/978-3-030-54005-0

**Table A.1**  Possible six dimensional orbifolds allowing for $\mathcal{N} = 1$ supersymmetry

| P | Coxeter lattice | Twist(s) | $\chi$ |
|---|---|---|---|
| $\mathbb{Z}_3$ | SU(3)$^3$ | $\frac{1}{3}(2\,1\,1)$ | 27 |
| $\mathbb{Z}_4$ | SU(2)$^2 \times$ SO(5)$^2$ | $\frac{1}{4}(2\,1\,1)$ | 16 |
| | SU(2) $\times$ SU(4) $\times$ SO(5) | | |
| | SU(4)$^2$ | | |
| $\mathbb{Z}_6$-I | SU(3) $\times$ G$_2^2$ | $\frac{1}{6}(2\,1\,1)$ | 3 |
| | [SU(3)$^2$] $\times$ G$_2$ | | |
| $\mathbb{Z}_6$-II | SU(2) $\times$ SU(6) | $\frac{1}{6}(3\,2\,1)$ | 12 |
| | SU(3) $\times$ SO(8) | | |
| | SU(2) $\times$ SU(3) $\times$ SO(7) | | |
| | SU(2)$^2 \times$ SU(3) $\times$ G$_2$ | | |
| | SU(2)$^2 \times$ [SU(3)$^2$] | | |
| $\mathbb{Z}_7$ | SU(7) | $\frac{1}{7}(3\,2\,1)$ | 7 |
| $\mathbb{Z}_8$-I | SO(9) $\times$ SO(5)* | $\frac{1}{8}(3\,2\,1)$ | 4 |
| | [SU(4)$^2$] | | |
| $\mathbb{Z}_8$-II | SU(2)$^2 \times$ SO(9) | $\frac{1}{8}(4\,3\,1)$ | 8 |
| | SU(2) $\times$ SO(10)* | | |
| $\mathbb{Z}_{12}$-I | E$_6$ | $\frac{1}{12}(5\,4\,1)$ | 3 |
| | SU(3) $\times$ F$_4$ | | |
| $\mathbb{Z}_{12}$-II | SU(2)$^2 \times$ F$_4$ | $\frac{1}{12}(6\,5\,1)$ | 4 |
| $\mathbb{Z}_2 \times \mathbb{Z}_2$ | SU(2)$^6$ | $\frac{1}{2}(1\,1\,0), \frac{1}{2}(1\,0\,1)$ | 32 |
| $\mathbb{Z}_2 \times \mathbb{Z}_4$ | SO(5) $\times$ SU(2)$^2 \times$ SO(5) | $\frac{1}{2}(1\,1\,0), \frac{1}{4}(1\,0\,1)$ | 16 |
| $\mathbb{Z}_3 \times \mathbb{Z}_3$ | SU(3)$^3$ | $\frac{1}{3}(1\,1\,0), \frac{1}{3}(1\,0\,1)$ | 27 |
| $\mathbb{Z}_2 \times \mathbb{Z}_6$-I | G$_2 \times$ SU(2)$^2 \times$ G$_2$ | $\frac{1}{2}(1\,1\,0), \frac{1}{6}(1\,0\,1)$ | 4 |
| $\mathbb{Z}_2 \times \mathbb{Z}_6$-II | G$_2 \times$ SU(3) $\times$ G$_2$ | $\frac{1}{2}(1\,1\,0), \frac{1}{6}(2\,1\,1)$ | 3 |
| $\mathbb{Z}_4 \times \mathbb{Z}_4$ | SO(5)$^3$ | $\frac{1}{4}(1\,1\,0), \frac{1}{4}(1\,0\,1)$ | 8 |
| $\mathbb{Z}_3 \times \mathbb{Z}_6$ | G$_2 \times$ SU(3) $\times$ G$_2$ | $\frac{1}{3}(1\,1\,0), \frac{1}{6}(1\,0\,1)$ | 3 |
| $\mathbb{Z}_6 \times \mathbb{Z}_6$ | G$_2^3$ | $\frac{1}{6}(1\,1\,0), \frac{1}{6}(1\,0\,1)$ | 1 |

The lattice within the [ ] bracket involves further modding by outer automorphisms and thus irreducible. On each line, the lattice has the same order of the entries of the twist vectors, except ones with asterisk (*). The point group is P and the number of fixed points is $\chi$. $\mathbb{Z}_2 \times \mathbb{Z}_3$ is missing because it is identical to $\mathbb{Z}_6$-II

where $\Delta_{\theta k}$ is the phase of the state, $|L - \text{movers}\rangle \otimes |R - \text{movers}\rangle$. In the standard case of taking $k = 0$ as the untwisted sector, the $k^{\text{th}}$ twisted sector has

$$\Delta_{\theta k} = e^{2\pi i[(P+kV)\cdot V - (\tilde{s}+k\phi+\rho_R-\rho_L)\cdot\phi+\frac{k}{2}(V^2-\phi^2)]}$$

where $\tilde{s}$ is the last three entries of $s = \{s_0; \tilde{s}\}$. The helicity $\oplus$ or $\ominus$ is determined from $|R - \text{movers}\rangle$, by the first component of $s = \{s_0; \tilde{s}\}$. $s$ has half integers such as $= [+ + ++]$ with [    ] indicating even number of sign flips, i.e. in this example even number of minus signs and all possible permutations.

As noted above, for prime orbifolds $\mathbb{Z}_N (N = 3, 7)$ we have the $(k, l)$ independent $\tilde{\chi}(\theta^k, \theta^l) = \chi$, and hence the multiplicity is $\chi(\Delta^0 + \Delta^1 + \cdots \Delta^{N-1})/N$ which is 0 for $\Delta = e^{2\pi i k/N}(k = 1, \cdots, N - 1)$ and equivalent to the Euler number $\chi$ for the case of $\Delta = 1$.

For nonprime orbifolds, one has to perform a nontrivial calculation based on (A.1). For this purpose, the degeneracy factors $\tilde{\chi}(\theta^k, \theta^j)$ calculated by Eq. (8.62) are given below:

$$\tilde{\chi}(\theta^k, \theta^l)$$

| $\mathbb{Z}_3$ | $k$ | $l =$ 0 | 1 | 2 |
|---|---|---|---|---|
| | 1 | 27 | 27 | 27 |

| $\mathbb{Z}_4$ | $k$ | $l =$ 0 | 1 | 2 | 3 |
|---|---|---|---|---|---|
| | 1 | 16 | 16 | 16 | 16 |
| | 2 | 16 | 4 | 16 | 4 |

| | $k$ | $l =$ 0 | 1 | 2 | 3 | 4 | 5 |
|---|---|---|---|---|---|---|---|
| $\mathbb{Z}_6$-I | 1 | 3 | 3 | 3 | 3 | 3 | 3 |
| $\frac{1}{6}(2\,1\,1)$ | 2 | 27 | 3 | 27 | 3 | 27 | 3 |
| | 3 | 16 | 1 | 1 | 16 | 1 | 1 |
| $\mathbb{Z}_6$-II | 1 | 12 | 12 | 12 | 12 | 12 | 12 |
| | 2 | 9 | 3 | 9 | 3 | 9 | 3 |
| | 3 | 16 | 4 | 4 | 16 | 4 | 4 |

| $\mathbb{Z}_7$ | $k$ | $l =$ 0 | 1 | 2 | 3 | 4 | 5 | 6 |
|---|---|---|---|---|---|---|---|---|
| | 1 | 7 | 7 | 7 | 7 | 7 | 7 | 7 |
| | 2 | 7 | 7 | 7 | 7 | 7 | 7 | 7 |
| | 3 | 7 | 7 | 7 | 7 | 7 | 7 | 7 |

| | $k$ | $l =$ 0 | 1 | 2 | 3 | 4 | 5 | 6 | 7 |
|---|---|---|---|---|---|---|---|---|---|
| $\mathbb{Z}_8$-I | 1 | 4 | 4 | 4 | 4 | 4 | 4 | 4 | 4 |
| | 2 | 16 | 4 | 16 | 4 | 16 | 4 | 16 | 4 |
| | 3 | 4 | 4 | 4 | 4 | 4 | 4 | 4 | 4 |
| | 4 | 16 | 2 | 4 | 2 | 16 | 2 | 4 | 2 |
| $\mathbb{Z}_8$-II | 1 | 8 | 8 | 8 | 8 | 8 | 8 | 8 | 8 |
| | 2 | 4 | 2 | 4 | 2 | 4 | 2 | 4 | 2 |
| | 3 | 8 | 8 | 8 | 8 | 8 | 8 | 8 | 8 |
| | 4 | 16 | 2 | 4 | 2 | 16 | 2 | 4 | 2 |

| | $k$ | $l =$ 0 | 1 | 2 | 3 | 4 | 5 | 6 | 7 | 8 | 9 | 10 | 11 |
|---|---|---|---|---|---|---|---|---|---|---|---|---|---|
| $\mathbb{Z}_{12}$-I | 1 | 3 | 3 | 3 | 3 | 3 | 3 | 3 | 3 | 3 | 3 | 3 | 3 |
| | 2 | 3 | 3 | 3 | 3 | 3 | 3 | 3 | 3 | 3 | 3 | 3 | 3 |
| | 3 | 4 | 1 | 1 | 4 | 1 | 1 | 4 | 1 | 1 | 4 | 1 | 1 |
| | 4 | 27 | 3 | 3 | 3 | 27 | 3 | 3 | 3 | 27 | 3 | 3 | 3 |
| | 5 | 3 | 3 | 3 | 3 | 3 | 3 | 3 | 3 | 3 | 3 | 3 | 3 |
| | 6 | 16 | 1 | 1 | 4 | 1 | 1 | 16 | 1 | 1 | 4 | 1 | 1 |
| $\mathbb{Z}_{12}$-II | 1 | 4 | 4 | 4 | 4 | 4 | 4 | 4 | 4 | 4 | 4 | 4 | 4 |
| | 2 | 1 | 1 | 1 | 1 | 1 | 1 | 1 | 1 | 1 | 1 | 1 | 1 |
| | 3 | 16 | 4 | 4 | 16 | 4 | 4 | 16 | 4 | 4 | 16 | 4 | 4 |
| | 4 | 9 | 1 | 1 | 1 | 9 | 1 | 1 | 1 | 9 | 1 | 1 | 1 |
| | 5 | 4 | 4 | 4 | 4 | 4 | 4 | 4 | 4 | 4 | 4 | 4 | 4 |
| | 6 | 16 | 1 | 1 | 4 | 1 | 1 | 16 | 1 | 1 | 4 | 1 | 1 |

They are the number of simultaneous fixed points under $\theta^k$ and $\theta^l$, *neglecting fixed tori*. For instance, in the $\mathbb{Z}_4$ orbifold, we have $\tilde{\chi}(\theta, \theta^l) = 16$ since the fixed points under $\theta$ are always those of $\theta^l$, $l = 0, 1, 2, 3$. However the converse is not true. For $\theta^2$ we have a fixed torus when $2\phi_a$ has integral element. Not counting this, we have the 16 fixed points under $\theta^2$, giving $\tilde{\chi}(\theta, 1) = 16$. Among them, we have $\tilde{\chi}(\theta, \theta^2) = 4$ fixed points that are also invariant under $\theta$.

For calculating masses in the $k^{\text{th}}$ twisted sector, we need the vacuum energy contribution. The weights for massless states satisfy

$$(P + \tilde{V})^2 = 2(1 - \zeta_{N,k}) - 2\tilde{N}_L. \tag{A.2}$$

where $\zeta_{N,k} = \frac{1}{4} \sum_i \phi_i (1 - \phi_i)$ with $\phi_i$ lattice-shifted such that $0 \leq \phi_i \leq 1$. The present $(1 - \zeta)$ corresponds to $\tilde{c}$ of Chaps. 6 and 7. Below we list the vacuum energy $2(1 - \zeta_{N,k})$ for the left movers,

| $-2\tilde{c} = -2(1 - \zeta_{N,k})$ | | | | | | |
|---|---|---|---|---|---|---|
| $k =$ | 1 | 2 | 3 | 4 | 5 | 6 |
| $\mathbb{Z}_{12} - \text{I}$ | $\frac{210}{144}$ | $\frac{216}{144}$ | $\frac{234}{144}$ | $\frac{192}{144}$ | $\frac{210}{144}$ | $\frac{216}{144}$ |
| $\mathbb{Z}_{12} - \text{II}$ | $\frac{206}{144}$ | $\frac{248}{144}$ | $\frac{198}{144}$ | $\frac{224}{144}$ | $\frac{206}{144}$ | $\frac{216}{144}$ |
| $\mathbb{Z}_8 - \text{I}$ | $\frac{94}{64}$ | $\frac{88}{64}$ | $\frac{94}{64}$ | $\frac{96}{64}$ | | |
| $\mathbb{Z}_8 - \text{II}$ | $\frac{90}{64}$ | $\frac{104}{64}$ | $\frac{90}{64}$ | $\frac{96}{64}$ | | |
| $\mathbb{Z}_7$ | $\frac{10}{7}$ | $\frac{10}{7}$ | $\frac{10}{7}$ | | | |
| $\mathbb{Z}_6 - \text{I}$ | $\frac{54}{36}$ | $\frac{48}{36}$ | $\frac{54}{36}$ | | | |
| $\mathbb{Z}_6 - \text{II}$ | $\frac{7}{18}$ | $\frac{5}{9}$ | $\frac{1}{2}$ | | | |
| $\mathbb{Z}_4$ | $\frac{22}{16}$ | $\frac{24}{16}$ | | | | |
| $\mathbb{Z}_3$ | $\frac{4}{3}$ | | | | | |
| $\mathbb{Z}_2(6D)$ | $\frac{3}{2}$ | | | | | |

For the vacuum energy of right movers, we need

| $-2c = -2\tilde{c} + 1$ | | | | | | |
|---|---|---|---|---|---|---|
| $k =$ | 1 | 2 | 3 | 4 | 5 | 6 |
| $\mathbb{Z}_{12} - \mathrm{I}$ | $\frac{11}{24}$ | $\frac{1}{2}$ | $\frac{5}{8}$ | $\frac{1}{3}$ | $\frac{11}{24}$ | $\frac{1}{2}$ |
| $\mathbb{Z}_{12} - \mathrm{II}$ | $\frac{31}{72}$ | $\frac{13}{18}$ | $\frac{3}{8}$ | $\frac{5}{9}$ | $\frac{31}{72}$ | $\frac{1}{2}$ |
| $\mathbb{Z}_8 - \mathrm{I}$ | $\frac{15}{32}$ | $\frac{3}{8}$ | $\frac{15}{32}$ | $\frac{1}{2}$ | | |
| $\mathbb{Z}_8 - \mathrm{II}$ | $\frac{13}{32}$ | $\frac{5}{8}$ | $\frac{13}{32}$ | $\frac{1}{2}$ | | |
| $\mathbb{Z}_7$ | $\frac{3}{7}$ | $\frac{3}{7}$ | $\frac{3}{7}$ | | | |
| $\mathbb{Z}_6 - \mathrm{I}$ | $\frac{1}{2}$ | $\frac{1}{3}$ | $\frac{1}{2}$ | | | |
| $\mathbb{Z}_6 - \mathrm{II}$ | $\frac{7}{18}$ | $\frac{5}{9}$ | $\frac{1}{2}$ | | | |
| $\mathbb{Z}_4$ | $\frac{3}{8}$ | $\frac{1}{2}$ | | | | |
| $\mathbb{Z}_3$ | $\frac{1}{3}$ | | | | | |
| $\mathbb{Z}_2(6D)$ | $\frac{1}{2}$ | | | | | |

where the vacuum energy contributions from the R and NS sectors add up. For example, the $k = 1$ sector of $\mathbb{Z}_{12}$-I leads to

$$c = 4f(0) + 2f(0) + 2f\left(\frac{5}{12}\right) + 2f\left(\frac{4}{12}\right) + 2f\left(\frac{1}{12}\right) = -\frac{11}{48}.$$

Finally, the Wilson line conditions are given in Tables A.2 and A.3.

**Table A.2** Constraints on Wilson lines for $\mathbb{Z}_N$ orbifold

| P | Lattice | Order | Condition |
|---|---|---|---|
| $\mathbb{Z}_3$ | $SU(3)^3$ | $3a_1 \approx 0,\ 3a_3 \approx 0,\ 3a_5 \approx 0$ | $a_1 \approx a_2,\ a_3 \approx a_4,\ a_5 \approx a_6$ |
| $\mathbb{Z}_4$ | $SU(2)^2 \times SO(5)^2$ | $2a_1 \approx 0,\ 2a_2 \approx 0,$ $2a_3 \approx 0,\ 2a_5 \approx 0$ | $a_3 \approx a_4,\ a_5 \approx a_6$ |
| | $SU(2) \times SU(4) \times SO(5)$ | $2a_1 \approx 0,\ 4a_2 \approx 0,\ 2a_5 \approx 0$ | $a_2 \approx a_3 \approx a_4,\ a_5 \approx a_6$ |
| | $SU(4)^2$ | $4a_1 \approx 0,\ 4a_4 \approx 0$ | $a_1 \approx a_2 \approx a_3,\ a_4 \approx a_5 \approx a_6$ |
| $\mathbb{Z}_6$-I | $SU(3) \times G_2^2$ | $3a_1 \approx 0$ | $a_1 \approx a_2$ |
| | $[SU(3)]^2 \times G_2$ | $3a_1 \approx 0$ | $a_1 \approx a_2 \approx a_3 \approx a_4$ |
| $\mathbb{Z}_6$-II | $SU(2) \times SU(6)$ | $2a_1 \approx 0,\ 6a_2 \approx 0$ | $a_2 \approx a_3 \approx a_4 \approx a_5 \approx a_6$ |
| | $SU(3) \times SO(8)$ | $3a_1 \approx 0,\ 2a_5 \approx 0$ | $a_1 \approx a_2,\ a_3 \approx a_4 \approx 0,\ a_5 \approx a_6$ |
| | $SU(2) \times SU(3) \times SO(7)$ | $2a_1 \approx 0,\ 3a_2 \approx 0$ | $a_2 \approx a_3,\ a_4 \approx a_5 \approx a_6 \approx 0$ |
| | $SU(2)^2 \times SU(3) \times G_2$ | $3a_1 \approx 0,\ 2a_3 \approx 0,\ 2a_4 \approx 0$ | $a_1 \approx a_2,\ a_5 \approx a_6 \approx 0$ |
| | $SU(2)^2 \times [SU(3)^2]$ | $3a_1 \approx 0,\ 2a_3 \approx 0,\ 2a_4 \approx 0$ | $a_1 \approx a_2,\ a_5 \approx a_6 \approx 0$ |
| $\mathbb{Z}_7$ | $SU(7)$ | $7a_1 \approx 0$ | $a_1 \approx a_2 \approx a_3 \approx a_4 \approx a_5 \approx a_6$ |
| $\mathbb{Z}_8$-I | $SO(9) \times SO(5)^*$ | $2a_1 \approx 0,\ 2a_6 \approx 0$ | $a_1 \approx a_2 \approx a_3 \approx a_4,\ a_5 \approx a_6$ |
| | $[SU(4)^2]$ | $4a_1 \approx 0$ | $a_1 \approx a_2 \approx a_3 \approx a_4 \approx a_5 \approx a_6$ |
| $\mathbb{Z}_8$-II | $SU(2)^2 \times SO(9)$ | $2a_1 \approx 2a_2 \approx 2a_3 \approx 0$ | $a_3 \approx a_4 \approx a_5 \approx a_6$ |
| | $SU(2) \times SO(10)^*$ | $2a_1 \approx 0,\ 2a_2 \approx 0$ | $a_2 \approx a_3 \approx a_4 \approx a_5 \approx a_6$ |
| $\mathbb{Z}_{12}$-I | $E_6$ | $3a_1 \approx 0$ | $a_1 \approx a_2 \approx a_3 \approx a_4 \approx a_5 \approx a_6$ |
| | $SU(3) \times F_4$ | $3a_1 \approx 0$ | $a_1 \approx a_2$ |
| $\mathbb{Z}_{12}$-II | $SU(2)^2 \times F_4$ | $2a_1 \approx 2a_2 \approx 0$ | |

They depend on the choice of the lattices, classified in Table 3.4. Conventions of shift vectors are given in (3.4) and the lattices follow the same orders. The sign "$\approx$" means equivalence up to lattice translation. The order of $F_4$ and $G_2$ Coxeter group is 1, so that we cannot turn on nontrivial Wilson line. On each line, the lattice has the same order of the entries of the twist vectors, except ones with asterisk (*)

**Table A.3** Constraints on Wilson lines for $\mathbb{Z}_N \times \mathbb{Z}_M$ orbifold

| P | Lattice | Order | Condition |
|---|---|---|---|
| $\mathbb{Z}_2 \times \mathbb{Z}_2$ | $SU(2)^6$ | $2a_1 \approx 0,\ 2a_2 \approx 0,\ 2a_3 \approx 0,$ $2a_4 \approx 0,\ 2a_5 \approx 0,\ 2a_6 \approx 0$ | |
| $\mathbb{Z}_2 \times \mathbb{Z}_4$ | $SO(5) \times SU(2)^2 \times SO(5)$ | $2a_1 \approx 0,\ 2a_3 \approx 0,$ $2a_4 \approx 0,\ 2a_5 \approx 0$ | $a_1 \approx a_2,\ a_5 \approx a_6$ |
| $\mathbb{Z}_3 \times \mathbb{Z}_3$ | $SU(3)^3$ | $3a_1 \approx 0,\ 3a_3 \approx 0,\ 3a_5 \approx 0$ | $a_1 \approx a_2,\ a_3 \approx a_4,\ a_5 \approx a_6$ |
| $\mathbb{Z}_2 \times \mathbb{Z}_6$-I | $G_2 \times SU(2)^2 \times G_2$ | $2a_3 \approx 0,\ 2a_4 \approx 0$ | |
| $\mathbb{Z}_2 \times \mathbb{Z}_6$-II | $G_2 \times SU(3) \times G_2$ | $3a_3 \approx 0$ | $a_3 \approx a_4$ |
| $\mathbb{Z}_4 \times \mathbb{Z}_4$ | $SO(5)^3$ | $2a_1 \approx 0,\ 2a_3 \approx 0,\ 2a_5 \approx 0$ | $a_1 \approx a_2,\ a_3 \approx a_4,\ a_5 \approx a_6$ |
| $\mathbb{Z}_3 \times \mathbb{Z}_6$ | $G_2 \times SU(3) \times G_2$ | $3a_3 \approx 0$ | $a_3 \approx a_4$ |
| $\mathbb{Z}_6 \times \mathbb{Z}_6$ | $G_2^3$ | | |

For conventions, see Table A.2

# Index

© Springer Nature Switzerland AG 2020
K.-S. Choi, J. E. Kim, *Quarks and Leptons From Orbifolded Superstring*,
Lecture Notes in Physics 954, https://doi.org/10.1007/978-3-030-54005-0